Selected Titles in Th

210 V. Kumar Murty and Michel Waldschmidt, Editors, Number theory, 1998

209 Steven Cox and Irena Lasiecka, Editors, Optimization methods in partial differential equations, 1997

208 Michel L. Lapidus, Lawrence H. Harper, and Adolfo J. Rumbos, Editors, Harmonic analysis and nonlinear differential equations: A volume in honor of Victor L. Shapiro, 1997

207 Yujiro Kawamata and Vyacheslav V. Shokurov, Editors, Birational algebraic geometry: A conference on algebraic geometry in memory of Wei-Liang Chow (1911–1995), 1997

206 Adam Korányi, Editor, Harmonic functions on trees and buildings, 1997

205 Paulo D. Cordaro and Howard Jacobowitz, Editors, Multidimensional complex analysis and partial differential equations: A collection of papers in honor of François Treves, 1997

204 Yair Censor and Simeon Reich, Editors, Recent developments in optimization theory and nonlinear analysis, 1997

203 Hanna Nencka and Jean-Pierre Bourguignon, Editors, Geometry and nature: In memory of W. K. Clifford, 1997

202 Jean-Louis Loday, James D. Stasheff, and Alexander A. Voronov, Editors, Operads: Proceedings of Renaissance Conferences, 1997

201 J. R. Quine and Peter Sarnak, Editors, Extremal Riemann surfaces, 1997

200 F. Dias, J.-M. Ghidaglia, and J.-C. Saut, Editors, Mathematical problems in the theory of water waves, 1996

199 G. Banaszak, W. Gajda, and P. Krasoń, Editors, Algebraic K-theory, 1996

198 Donald G. Saari and Zhihong Xia, Editors, Hamiltonian dynamics and celestial mechanics, 1996

197 J. E. Bonin, J. G. Oxley, and B. Servatius, Editors, Matroid theory, 1996

196 David Bao, Shiing-shen Chern, and Zhongmin Shen, Editors, Finsler geometry, 1996

195 Warren Dicks and Enric Ventura, The group fixed by a family of injective endomorphisms of a free group, 1996

194 Seok-Jin Kang, Myung-Hwan Kim, and Insok Lee, Editors, Lie algebras and their representations, 1996

193 Chongying Dong and Geoffrey Mason, Editors, Moonshine, the Monster, and related topics, 1996

192 Tomek Bartoszyński and Marion Scheepers, Editors, Set theory, 1995

191 Tuong Ton-That, Kenneth I. Gross, Donald St. P. Richards, and Paul J. Sally, Jr., Editors, Representation theory and harmonic analysis, 1995

190 Mourad E. H. Ismail, M. Zuhair Nashed, Ahmed I. Zayed, and Ahmed F. Ghaleb, Editors, Mathematical analysis, wavelets, and signal processing, 1995

189 S. A. M. Marcantognini, G. A. Mendoza, M. D. Morán, A. Octavio, and W. O. Urbina, Editors, Harmonic analysis and operator theory, 1995

188 Alejandro Adem, R. James Milgram, and Douglas C. Ravenel, Editors, Homotopy theory and its applications, 1995

187 G. W. Brumfiel and H. M. Hilden, $SL(2)$ representations of finitely presented groups, 1995

186 Shreeram S. Abhyankar, Walter Feit, Michael D. Fried, Yasutaka Ihara, and Helmut Voelklein, Editors, Recent developments in the inverse Galois problem, 1995

185 Raúl E. Curto, Ronald G. Douglas, Joel D. Pincus, and Norberto Salinas, Editors, Multivariable operator theory, 1995

184 L. A. Bokut′, A. I. Kostrikin, and S. S. Kutateladze, Editors, Second International Conference on Algebra, 1995

(Continued in the back of this publication)

CONTEMPORARY MATHEMATICS

210

Number Theory

Ramanujan Mathematical Society
January 3–6, 1996
Tiruchirapalli, India

V. Kumar Murty
Michel Waldschmidt
Editors

American Mathematical Society
Providence, Rhode Island

Editorial Board

Dennis DeTurck, managing editor

Andy Magid Michael Vogelius
Clark Robinson Peter M. Winkler

The international conference on Discrete Mathematics and Number Theory was held January 3–6, 1996, in Tiruchirapalli, India, to observe the tenth anniversary of the Ramanujan Mathematical Society. Support was provided by the Government of India through the Department of Science and Technology, the Council of Scientific and Industrial Research, and the National Board for Higher Mathematics, as well as from the International Mathematical Union, the International Centre for Theoretical Physics, the Centre National de la Recherche Scientifique, and James Vaughn, Jr.

1991 *Mathematics Subject Classification.* Primary 11Gxx, 11Fxx, 11Axx, 11Jxx, 11Nxx; Secondary 11Pxx, 11Mxx, 11Kxx, 11Dxx, 14Kxx, 14Gxx.

Library of Congress Cataloging-in-Publication Data
Number theory : discrete mathematics and number theory, January 3–6, 1996, Tiruchirapalli, India / V. Kumar Murty, Michel Waldschmidt, editors.
  p. cm. — (Contemporary mathematics ; 210)
  Includes bibliographical references.
  ISBN 0-8218-0606-8 (alk. paper)
  1. Number theory—Congresses. I. Murty, Vijaya Kumar, 1956– . II. Waldschmidt, Michel, 1946– . III. Series: Contemporary mathematics (American Mathematical Society) ; v. 210.
QA241.N8678 1997
512′.7—dc21   97-30207
                                                                                                                  CIP

Copying and reprinting. Material in this book may be reproduced by any means for educational and scientific purposes without fee or permission with the exception of reproduction by services that collect fees for delivery of documents and provided that the customary acknowledgment of the source is given. This consent does not extend to other kinds of copying for general distribution, for advertising or promotional purposes, or for resale. Requests for permission for commercial use of material should be addressed to the Assistant to the Publisher, American Mathematical Society, P. O. Box 6248, Providence, Rhode Island 02940-6248. Requests can also be made by e-mail to reprint-permission@ams.org.

Excluded from these provisions is material in articles for which the author holds copyright. In such cases, requests for permission to use or reprint should be addressed directly to the author(s). (Copyright ownership is indicated in the notice in the lower right-hand corner of the first page of each article.)

© 1998 by the American Mathematical Society. All rights reserved.
The American Mathematical Society retains all rights
except those granted to the United States Government.
Printed in the United States of America.

∞ The paper used in this book is acid-free and falls within the guidelines
established to ensure permanence and durability.
Visit the AMS home page at URL: http://www.ams.org/

10 9 8 7 6 5 4 3 2 1   03 02 01 00 99 98

Contents

Part I. Arithmetic Algebraic Geometry — 1

Relative splitting of one-motives
  DANIEL BERTRAND — 3

The Hasse principle in a pencil of algebraic varieties
  J.-L. COLLIOT-THÉLÈNE — 19

Stark-Heegner points over real quadratic fields
  H. DARMON — 41

On the Hecke action on the cohomology of Hilbert-Blumenthal surfaces
  FRED DIAMOND — 71

The ABC conjecture and exponents of class groups of quadratic fields
  M. RAM MURTY — 85

Heights of points on subvarieties of $\mathbb{G}_m^n$
  WOLFGANG SCHMIDT — 97

A note on a geometric analogue of Ankeny-Artin-Chowla's conjecture
  JING YU AND JIU-KANG YU — 101

Part II. Automorphic Forms — 107

Determinant of representations of division algebras of prime degree over local fields
  S. JAYASREE AND R. TANDON — 109

Mod p modular forms
  CHANDRASHEKHAR KHARE — 135

On exceptions of integral quadratic forms
  A. K. LAL AND B. RAMAKRISHNAN — 151

A brief survey on the theta correspondence
  DIPENDRA PRASAD — 171

Deformations of Siegel-Hilbert Hecke eigensystems and their Galois representations
  J. TILOUINE — 195

Part III. Elementary and Analytic Number Theory   **227**

A new key identity for Göllnitz' (Big) partition theorem
    KRISHNASWAMI ALLADI AND GEORGE E. ANDREWS   229

Some local-convexity theorems for the zeta-function-like analytic functions - III
    R. BALASUBRAMANIAN AND K. RAMACHANDRA   243

Perfect powers in products of terms in an arithmetical progression IV
    R. BALASUBRAMANIAN AND T. N. SHOREY   257

On some connections between probability and number theory
    JEAN-MARC DESHOUILLERS   265

Inhomogeneous minima of a class of quaternary quadratic forms of type (2,2)
    MADHU RAKA AND URMILA RANI   275

On an arithmetical inequality, II
    S. SRINIVASAN   299

Part IV. Transcendental Number Theory   **303**

Algebraic numbers close to 1: results and methods
    FRANCESCO AMOROSO   305

Transcendence in positive characteristic
    W. DALE BROWNAWELL   317

Minorations des hauteurs normalisées des sous-variétés de variétés abéliennes
    SINNOU DAVID AND PATRICE PHILIPPON   333

Klein polygons and geometric diagrams
    GILLES LACHAUD   365

Sails and Klein polyhedra
    GILLES LACHAUD   373

Automata and transcendence
    DINESH S. THAKUR   387

Preface

The Ramanujan Mathematical Society was founded in 1985 in Tiruchirapalli, India. To observe its tenth anniversary, an international conference on Discrete Mathematics and Number Theory was held during January 3–6, 1996. Approximately 200 mathematicians from around the world converged on Tiruchirapalli to participate in this conference.

This volume contains the proceedings of the Number Theory component of the conference. (The proceedings of the Discrete Mathematics component will be published in a special issue of the journal *Discrete Mathematics*.) The Number Theory component was divided into four sessions: Automorphic forms, Arithmetic algebraic geometry, Transcendental number theory, and Elementary and Analytic number theory. We are pleased that almost all of the speakers contributed papers for this volume, including the four plenary speakers: Dipendra Prasad, M. Ram Murty, W. Dale Brownawell, and J.-M. Deshouillers.

We would like to thank the Ramanujan Mathematical Society, and especially R. Balakrishnan, for inviting us to organize the Number Theory component of the Scientific Programme. Prof. Balakrishnan's constant enthusiasm for our efforts, together with the administrative support which was provided through the Society, were greatly appreciated. We are also grateful to Prof. C. S. Seshadri of the SPIC Science Foundation for help in communicating with local organizers.

Funding for the conference came from the Government of India through the Department of Science and Technology, the Council of Scientific and Industrial Research, and the National Board for Higher Mathematics, as well as from the International Mathematical Union, the International Centre for Theoretical Physics, the Centre National de la Recherche Scientifique, and James Vaughn, Jr. We heartily thank all of them for their support.

We would also like to thank Dennis DeTurck, Editor of Contemporary Mathematics, for encouraging us to publish in this series, and Christine Thivierge and Donna Harmon of the AMS Editorial Staff for their help in the production process. And of course we would like to thank all of the authors who contributed to this volume.

It is hoped that in some small measure, this conference has set a precedent, both in style and quality, of international meetings organized by the Ramanujan Mathematical Society in various locations in India. We were heartened to see the number of participants who had come from many different parts of India and the world. To us, it was a confirmation that good science can cut across all geographical, political, and cultural boundaries and bring people together.

Toronto and Paris   V. Kumar Murty and Michel Waldschmidt
May 1997

Part I

ARITHMETIC ALGEBRAIC GEOMETRY

Relative splitting of one-motives

by Daniel Bertrand

In a previous paper [3], we introduced two families of one-motives M over a number field k: those whose associated Galois representations are "relatively split", and those with "relative good reduction at all places of k". We showed that the first family is included in the second one, and conjectured that they coincide when k is totally real. In the first two parts of the present paper, we (hope to) shed new light on these notions by:

i) defining the elements M of the first family in terms of the weight filtration of their Mumford-Tate group $P(M)$, viz.: the subgroup $W_{-2}(P(M))$ of $\mathrm{Hom}(X \otimes Y, \mathbb{Q}(1))$ induced by M vanishes.

ii) characterizing the second family through the category $MM(S)$ (introduced in [10]) of mixed motives over S, viz.: the class $c(M)$ provided by M in $\mathrm{Ext}^2_{MM(S)}(X \otimes Y(0), \mathbb{Q}(1))$ vanishes.

Here, X and Y are constant groups attached to the one-motive M, and S is the 'compactification' of the ring of integers of k.

The conjectural implication (ii) $\to$ (i) for totally real k's is related to a natural problem in transcendence theory. While we make no progress on the case relevant to the conjecture, we do prove in the last section of the paper an analogous result for vector group extensions of abelian varieties with real multiplications.

§1. Relative global splitting of the Betti and ℓ-adic realizations.

We first recall various 'visions' of one-motives ([9]; see Figures 1 and 2 at the end of the paper), and fix notations in the process. Thus, k is a number field, A is an abelian variety over k, with dual A^V, a rigidification above 0 is given on the Poincaré biextension $\mathcal{P}$ of $A \times A^V$ by $\mathbb{G}_m$, Θ' and Θ are split tori over k, whose groups of characters are denoted by X, Y, and u, v are homomorphisms from X, Y into $A(k), A^V(k)$. To any trivialization ψ: $X \times Y \to (u \times v)^*\mathcal{P}$ of the pull-back of $\mathcal{P}$ by $(u \times v)$, one naturally associates a one-motive $M/k = M_\psi$, concretely given by a complex $[U : X \to G]$, where G is the unique extension

---

1991 *Mathematics Subject Classification.* Primary: 14G10; Secondary: 11G40.

of A by Θ such that for any (x,y) in $X \times Y$, y_*G is the extension of A by $\mathbb{G}_m$ parametrized by $v(y)$ and $y_*U(x)$ is the point on the fiber of y_*G above $u(x)$ corresponding to $\psi(x,y)$ (cf. Fig. 1 for a special case). We denote by $N := [u : X \to A]$ the quotient of M by $[0 \to \Theta]$, and by $M^\vee$ the Cartier dual of M. Throughout the paper, we view one-motives as objects in the category of one-motives *up to isogenies*. (An isogeny between two one-motives $M = [U : X \to G]$ and $M' = [U : X' \to G']$ is a morphism (f,g) of complexes such that $f : X \to X'$ is an injection with finite cokernel, and $g : G \to G'$ is a surjection with finite kernel.)

Fix a complex embedding σ of k. By base change, M/k defines a one-motive $M = M_\sigma$ over $\mathbb{C}$, whose Betti realization (with coefficients in $\mathbb{Q}$) carries a natural mixed $\mathbb{Q}$-Hodge structure TM. Its weight filtration is given by:

$$W_{-3}(TM) = 0, \quad W_{-2}(TM) = H_1(\Theta, \mathbb{Q}), \quad W_{-1}(TM) = H_1(G, \mathbb{Q}), \quad W_0(TM) = TM,$$

with associated graded pieces:

$$Gr_{-1}(TM) = H_1(A, \mathbb{Q}), Gr_0(TM) = X \otimes \mathbb{Q};$$

$$TM/W_{-2}(TM) = TN$$

(cf. Fig. 2). The natural map α from TM to $\mathrm{Lie}(G)$, whose image can be described as the space of logarithms of $U(X \otimes \mathbb{Q})$, extends to a $\mathbb{C}$-linear map on $TM \otimes \mathbb{C}$, whose kernel F^0 defines the Hodge filtration of $TM \otimes \mathbb{C}$. Denoting the Tate structure by $\mathbb{Q}(1)$, we identify $T(\Theta)$ with $\mathrm{Hom}(Y, \mathbb{Q}) \otimes \mathbb{Q}(1) = Y^\vee \otimes \mathbb{Q}$.

The category of mixed $\mathbb{Q}$-Hodge structures is tannakian, with neutral object $\mathbb{Q}(0)$. The Mumford-Tate group $P_\sigma(M)$ of TM_σ is the connected algebraic group attached to the tannakian subcategory generated by TM_σ. Assuming $W_{-1}(TM_\sigma) \neq 0$, we may alternatively describe $P_\sigma(M)$ as the stabilizer in $GL(TM_\sigma)$ of all Hodge tensors, i.e. of all morphisms from $\mathbb{Q}(0)$ to the tensor algebra generated by $TM_\sigma, \mathbb{Q}(1)$ and their duals (cf. [15] IV.1.2; [1] Lemma 2.a). In view of [6], $P_\sigma(M)$ depends on σ only up to an inner twist, and we shall drop σ in this notation. The weight filtration on TM induces a natural filtration on $P(M)$, whose first step is the unipotent radical of $P(M)$:

$$W_{-1}(P(M)) = \{g \in P(M), (id - g)(H_1(\Theta)) = 0,$$

$$(id - g)(H_1(G)) \subset H_1(\Theta), (id - g)(TM) \subset H_1(G)\}.$$

But we shall here be mainly concerned with its *second step* :

$$W_{-2}(P(M)) = \{g \in P(M), (id-g)(H_1(G)) = 0, (id-g)(TM) \subset \text{Hom}(Y, \mathbb{Q}(1))\},$$

which may be viewed as a subgroup of the group $\text{Hom}(X, \text{Hom}(Y, \mathbb{Q}(1))) = \text{Hom}(X \otimes Y, \mathbb{Q}(1))$ of bilinear maps on $X \times Y$ with values in $\mathbb{Q}(1)$.

Finally, fix an algebraic closure $\tau : k \to k^{\text{alg}}$ of k and a prime number ℓ . and let $T_\ell(M)$ be the corresponding ℓ-adic realization of M, tensored with $\mathbb{Q}_\ell$. The image in $GL(T_\ell(M))$ of the group $Gal(k^{\text{alg}}/k)$ in its natural action on $T_\ell(M)$ is a Lie group, which we denote by $\Pi_\ell(M)$ (up to a twist, it does not depend on τ). Again, $\Pi_\ell(M)$ comes equipped with a filtration, whose second step

$$W_{-2}(\Pi_\ell(M)) = \{g \in \Pi_\ell, (id-g)(T_\ell(G)) = 0, (id-g)(T_\ell(M)) \subset T_\ell(\Theta) = \text{Hom}(Y, \mathbb{Q}_\ell(1))\}$$

may be described (cf. [13]) as the Galois group of the field of definition of $U(X \otimes \mathbb{Q}_\ell/Z_\ell)$ over the field

$$k(W_{-1}T_\ell(M \times M^V)) := k(A_{tor} \otimes \mathbb{Q}_\ell/Z_\ell), u(X \otimes \mathbb{Q}_\ell/Z_\ell), v(Y \otimes \mathbb{Q}_\ell/Z_\ell));$$

The latter one can also be viewed as the field $K(G_{tor} \otimes \mathbb{Q}_\ell/Z_\ell, u(X \otimes \mathbb{Q}_\ell/Z_\ell))$.

To state the result of this section, we need a slight extension of the notion of isotropic abelian variety given in [3]. We shall say that the (isogeny class of the one-motive) $M/k = M_\psi$ is *isotropic* if for all (x,y) in $X \times Y$ the following *two* conditions hold (cf. Fig. 1):

i) the restriction of the Poincaré bundle on $A \times A^V$ to the connected component $B_{x,y}$ of the algebraic group generated by $(u(x), v(y))$ in $A \times A^V$ is algebraically equivalent to 0. Since $\mathcal{P}$ is symmetric (and rigidified), its restriction to $B_{x,y}$ then acquires (up to a 2-isogeny) a canonical splitting ξ ;

ii) if β denotes the restriction of $(u \times v)$ to the group Z generated by a sufficiently large multiple of (x, y) in $X \times Y$, the trivializations $\psi|_Z$ and $\beta^*\xi$ of $\beta^*\mathcal{P}$ coincide. (In view of [3], Prop. 1, these conditions are indeed invariant under isogenies.)

We then have

Theorem 1. *Let M/k be a one-motive, with Mumford-Tate group $P(M)$, and let $\Pi_\ell(M)$ be the Galois group cut out by $T_\ell(M)$, for some prime number ℓ. The following conditions are equivalent:*

(i) $W_{-2}(P(M)) = 0$;

(ii) $W_{-2}(\Pi_\ell(M)) = 0$;

(iii) M *is isotropic.*

Condition (ii) expresses the fact that the point $U(x)$ of $G(k)$ is infinitely ℓ-divisible over the field $k(W_{-1}T_\ell(M \times M^V))$. This was our original definition of "relatively split" one-motives in [3], which thus becomes:

A one-motive M/k is said to be relatively split if $W_{-2}(P(M)) = 0$.

Proof: $i) \Rightarrow ii)$ According to Theorem 2.2.5 of [6], every Hodge cycle is an absolute Hodge cycle over k^{alg}, and is thus fixed by an open subgroup of $\Pi_\ell(M)$ (via the comparison isomorphism between $TM \otimes \mathbb{Q}_\ell$ and $T_\ell(M)$, assuming σ extends τ). Therefore, the Lie algebra of $\Pi_\ell(M)$ is contained in that of $P(M) \otimes \mathbb{Q}_\ell$. In particular, $W_{-2}(\Pi_\ell(M)) \subset W_{-2}(P(M)) \otimes \mathbb{Q}_\ell$, and $i) \to ii)$ immediately follows.

$ii) \Rightarrow iii)$ Let (x,y) be an arbitrary point of $X \times Y$. The restriction of ψ to $\mathbb{Z}x \times \mathbb{Z}y$ provides a one-motive $M_{(x,y)}$ for which $W_{-2}(\Pi_\ell(M_{(x,y)}))$ again vanishes. By Proposition 3 of [3] (a symmetrized, and ℓ-adic, version of the theorem of Jacquinot and Ribet in [13]), $M_{(x,y)}$ is isotropic. Thus M is isotropic as well.

$iii) \Rightarrow i)$ Via the inclusion $W_{-2}(P) \subset \text{Hom}(X \otimes Y, \mathbb{Q}(1))$, we reduce to the case where $M = [U : \mathbb{Z} \approx \mathbb{Z}x \to G \approx y_*G]$ is of the type $M_{(x,y)}$ introduced above, i.e. where $TX = TY = \mathbb{Q}(0)$. The Cartier dual $M^V = [V : \mathbb{Z} \approx \mathbb{Z}y \to N^V \approx x_*N^V]$ of M is then of the same type, and to any such one-motive M, we may attach in a natural way a self-dual one-motive $\mathbb{M}$ as follows: denote by μ the addition law on $\mathbb{G}_m$, by δ the diagonal embedding of $\mathbb{Z}$ into $\mathbb{Z} \times \mathbb{Z}$, consider the pull-back by δ of $M \times M^V$ (viewed as an extension of $\mathbb{Z} \times \mathbb{Z}$ by $G \times N^V$), and define $\mathbb{M} = [\mathbb{U} : Z := \mathbb{Z}(x,y) \approx \mathbb{Z} \to \mathbb{G} := \mu_*(G \times N^V)]$ as the push-out by μ of $\delta^*(M \times M^V)$, viewed as an extension of $\delta^*(N \times G^V)$ by $\mathbb{G}_m \times \mathbb{G}_m$. The affine representation of $W_{-2}(P(M))$ in $\text{Hom}(\mathbb{Q}(0), \mathbb{Q}(1))$ arising from $T\mathbb{M}$ is then the sum of the standard one (given by TM) and of its contragradient (given by TM^V), so that $W_{-2}(P(M)) = 0$ if (and only if) $W_{-2}(P(\mathbb{M})) = 0$. Under our hypothesis (iii), let now B be a maximal isotropic abelian subvariety of $\mathbb{A} = A \times A^V$ containing a sufficiently large multiple b of $(u(x), v(y))$. According to [3], Lemma 1, there exists an isogeny S from $\mathbb{B} = B \times B^V$ to $\mathbb{A}$, inducing the natural injection of B, such that $S^*\mathcal{P}$ is a multiple of the Poincaré biextension on $\mathbb{B}$. Up to isogeny, we may then view $\mathbb{G}$ as the extension $B \times G_b$ of $\mathbb{B}$ by $\mathbb{G}_m$ parametrized by the point $(0, b)$ of the dual $B^V \times B$ of $\mathbb{B}$ and, analyzing the Poincaré biextension on $\mathbb{A} \times \mathbb{A}^V$, its point $\mathbb{U}(x, y)$ as $(b, 0) \in B \times G_b$. In other words, the Hodge structure $T\mathbb{M}$ is the direct sum of the extension TG_b of TB^V by $\mathbb{Q}(1)$, and of the extension of $\mathbb{Q}(0)$ by TB corresponding to the point $b \in B$. Since both present only two non-trivial consecutive steps in their weight filtrations, $W_{-2}(P(\mathbb{M}))$ must vanish, whence $W_{-2}(P(M)) = 0$.

Remarks:

i) Combined with Ribet's results on Kummer theory, André's Proposition 1 in [1] shows that the (-1)-graded pieces $\mathbb{Q}_\ell \otimes W_{-1}P(M)/W_{-2}P(M)$ and $W_{-1}\Pi_\ell(M)/W_{-2}\Pi_\ell(M)$ of $\mathbb{Q}_\ell \otimes P(M)$ and of $\Pi_\ell(M)$ coincide. Consequently, Theorem 1 implies that as soon as $rk(X) = rk(Y) = 1$, we have : $\mathbb{Q}_\ell \otimes W_{-1}P(M) = W_{-1}\Pi_\ell(M)$. It is likely that this identity holds true in all cases.

ii) Since the derived group of $W_{-1}P(M)$ is contained in $W_{-2}P(M)$ (cf. [6], [15]), Condition (i) of the theorem implies that $W_{-1}(P(M))$ is commutative. Can one characterize geometrically (i.e. in terms of the trivialization ψ, as in Condition (iii)) the one-motives M with commutative $W_{-1}P(M)$?

iii) That Condition (ii) [in sense of [3]: $U(x)$ is infinitely ℓ-divisible in $G(k(W_{-1}T_\ell(M \times M^V)))$] is a consequence of Condition (iii) can be proved geometrically as follows. Up to isogeny, we may assume that the restriction of $\mathcal{P}$ to the abelian variety $B_{x,y}$ is the trivial torsor $\mathbb{G}_m \times B_{x,y}$, and that $U(x)$ is the point $(1, (u(x), v(y)))$ on this restriction. Let then b_ℓ be a point in $B_{x,y}(k^{\text{alg}})$ such that $[\ell]_{A \times A^V}(b_\ell) = (u(x), v(y))$, let ζ_{ℓ^2} be a primitive root of unity of order ℓ^2, and consider the point (ζ_{ℓ^2}, b_ℓ) of $\mathcal{P}|_{B_{x,y}}$. Its image $U(x)_\ell$ under the partial law $[1, \ell]$ of multiplication by ℓ on $\mathcal{P}$ (above that of A^V) lies on the extension G and is defined over the field $k(W_{-1}T_\ell(M \times M^V))$. Now

$$[\ell, 1]U(x)_\ell = [\ell, 1] \circ [1, \ell](\zeta_{\ell^2}, b_\ell) = ([\ell^2]_{\mathbb{G}_m}(\zeta_{\ell^2}), [\ell]_{A \times A^V}(b_\ell)),$$

since the composition of the two partial laws $[\ell, 1] \circ [1, \ell]$ reflects the isomorphism between $([\ell]_{A \times A^V})^* \mathcal{P}$ and $([\ell^2]_{\mathbb{G}_m})_* \mathcal{P}$. Therefore,

$$[\ell]_G U(x)_\ell = [\ell, 1] U(x)_\ell = (1, (u(x), v(y))) = U(x),$$

and our claim follows by induction on the powers of ℓ. See also [5] for a related construction.

iv) As pointed out by Y. André, it would have been more natural to prove that (i) implies (iii) without using (ii), thereby avoiding any appeal to [6]. This implication probably follows from the same arguments as in [13].

v) Milne [15] says that a one-motive M is of *Kuga-type* if Condition (i): $W_{-2}(P(M)) = 0$, is satisfied. One-motives such as G, or N, are obviously of Kuga-type, but Theorem 1 shows that not all are so easy to visualize !

§2. Full local splittings of the Betti and ℓ-adic realizations

We return to the notations X, Y, G, N, attached at the beginning of §1 to our one-motive $M = [U : X \to G]$ over the number field k, and for each place v of k, we denote by k_v the completion of k at v. In [3], M was said to have *relative good reduction at v if $U(X)$ lies in the maximal compact subgroup of the Lie group $G(k_v)$*. Assuming for simplicity that A has semi-stable reduction at all finite places v, we here reinterpret this notion (which is clearly invariant under isogeny) in terms of local data attached to the Betti and ℓ-adic realizations of M, where ℓ is any given prime number.

Let thus v be archimedean (resp. finite). With a slight abuse of notation, we still denote by TM (resp. $T_\ell(M)$) the mixed $\mathbb{Q}$-Hodge structure (resp. the representation of the decomposition group at v) provided by the corresponding one-motive $M/\mathbb{C}$ (resp. M/k_v). Writing Gr_W for the direct sum of the graded pieces of a weight filtration, and tacitly requiring that isomorphisms should respect the weight filtrations, we can then state:

Lemma 1. *M has relative good reduction at the place v if and only if*

i) *[when v is finite and prime to ℓ] : as a representation of the inertia group I_v, $T_\ell(M)$ is isomorphic to $Gr_W(T_\ell(M))$.*

ii) *[when v is archimedean]* : *as a mixed $\mathbb{R}$-Hodge structure, $T(M) \otimes \mathbb{R}$ is isomorphic to $Gr_W(T(M) \otimes \mathbb{R})$.*

Proof:

i) in view of our hypotheses on A, X and Y, both extensions

$$0 \to \{W_{-2}(T_\ell(M)) = T_\ell(Y^V)\} \to \{W_{-1}(T_\ell(M)) = T_\ell(G)\} \to T_\ell(A) \to 0,$$

$$0 \to T_\ell(A) \to \{T_\ell(M)/W_{-2}(T_\ell(M)) = T_\ell(N)\} \to \{T_\ell(M)/W_{-1}(T_\ell(M)) = T_\ell(X)\} \to 0$$

split as representations of I_v, so that over $(k_v)^{nr}$, $T_\ell(M)$ is the direct sum of $T_\ell(A)$ and of an I_v-representation $T_\ell(v)$ in $\text{Ext}^1_{I_v}(T_\ell(X \otimes Y), \mathbb{Q}_\ell(1)) \cong \text{Hom}(X \otimes Y, v(k_v^*) \otimes \mathbb{Q}_\ell)$ (cf. [18] 2.5, 2.6, 3.7). When A has good reduction at v, the maximal compact subgroup of $y_*G(k_v)$ can be viewed as the kernel of a reduction map modulo v, so that $T_\ell(v)(x \otimes y) = v(y_*U(x))$ vanishes if and only if $U(x)$ lies in it, as claimed. Using the techniques of [16], we then reduce to the case where A has full multiplicative reduction, and denote by $(q_{ij})_{1 \leq i,j \leq g}$ (resp. $(q_{ij})_{0 \leq i,j \leq g}$) the period matrix of a Tate-Raynaud parametrization of A/k_v (resp. of the one-motive $M_{(x,y)}$); here, the column $(q_{0j})_{0 \leq j \leq g}$ represents the point $U(x)$ on the extension y_*G/k_v. Since the matrix $(v(q_{ij})_{1 \leq i,j \leq g}$ is non-singular, we may assume, up to an isogeny, that $v(q_{0,i}) = 0$ for $1 \leq i \leq g$. We then deduce from [16], 4.6, that $T_\ell(v)(x \otimes y) = v(q_{00})$. But $v(q_{00})$ vanishes if (and only if, under this normalization) $U(x)$ belongs to the maximal compact subgroup of $y_*G(k_v)$, and our claim follows.

ii) since a mixed $\mathbb{R}$-Hodge structure splits as soon as its weight filtration presents only two non-trivial consecutive steps, we deduce from the splitting of the "same" extensions as above that $T(M) \otimes \mathbb{R}$ is the direct sum of $T(A) \otimes \mathbb{R}$ and of an extension $T_\mathbb{R}(v)$ of $X \otimes Y \otimes \mathbb{R}$ by $\mathbb{R}(1)$) : in geometric terms (and in the notations of the begining of §1, cf. [9], Lemma 10.1.3.2), there exists a $\mathbb{C}$-linear section s of the projection π from $\text{Lie}(G)$ to $\text{Lie}(A)$, sending $H_1(A, \mathbb{R})$ into $H_1(G, \mathbb{R})$, and viewed as an element of $\text{Hom}(X \otimes Y, \mathbb{C})/\text{Hom}(X \otimes Y, \mathbb{R}(1))$ (cf. [7], 1.4 and 1.13), the extension $T_\mathbb{R}(v)$ is given by the class of $\alpha - s\pi\alpha$ in $\text{Hom}(X, \text{Hom}(Y, \mathbb{C}/\mathbb{R}(1)))$. In particular, it vanishes if and only if $\alpha(TM)$ lies in $H_1(G, \mathbb{R})$, i.e. $U(X)$ lies in the maximal compact subgroup of

$G(\mathbb{C})$. (See [10], Appendix, and [11], III.2.1.5, for further aspects of good reduction at ∞.)

Remark: When v is a real archimedean place, $T(M) \otimes \mathbb{R}$ is naturally endowed with a mixed $\mathbb{R}$-Hodge structure over $\mathbb{R}$ (cf. [11], III.1.2), and one may ask whether the splitting obtained above is compatible with the corresponding involution F_∞. In fact, this always holds in our case, in view of [18], Cor. 2.3.ii (see also J. Nekovar, PSPM AMS 55, 1994, Part 1, 537-570, 1.2.6), which implies: if an extension $T_\mathbb{R}(v)$ of $X \otimes Y \otimes \mathbb{R}$ by $\mathbb{R}(1))$ is defined over $\mathbb{R}$, it splits in that category if and only if it splits as a $\mathbb{R}$-Hodge structure over $\mathbb{C}$. In a similar vein, note that the splitting of I_v-representations we obtained for a finite place v is compatible with the corresponding Frobenius F_v, i.e. extends to the full decomposition group.

As shown by Scholl [18] (resp. Deninger and Nart [10]), combining the conditions of this lemma (and of this remark) at all finite (resp. all) places of k provides interesting subcategories $MM(S_f)$ (resp. $MM(S)$) of the category $MM(k)$ of mixed motives over k, in the sense of [14]. For our purposes, we merely need to recall the existence of a fully faithful functor Φ from the category of one-motives over k up to isogenies, to $MM(k)$, and the fact that the image $\Phi(M)$ of a one-motive M lies in $MM(S)$ if and only if all the weight filtrations considered in the lemma split as indicated. Thus, our old-fashioned definition becomes:

A one-motive M/k is said to have relative good reduction everywhere if $\Phi(M)$ lies in $MM(S)$.

Trivial examples of one-motives in $MM(S)$ are given by (the images under Φ) of $X, Y^V = \mathrm{Hom}(Y, \mathbb{Q}(1))$, the abelian variety A, and (as recalled in the proof of Lemma 1) their extensions with two consecutive weights only. Thus, *any* one-motive M/k defines two (usually non-trivial) extensions in $MM(S)$:

$$\Phi G = W_{-1}(\Phi M) \in \mathrm{Ext}^1_{MM(S)}(A, Y^V)$$
$$\Phi N = \Phi M/W_{-2}(\Phi M) \in \mathrm{Ext}^1_{MM(S)}(X, A)$$

(see Fig. 2). On the other hand:

Lemma 2. *The intersection of* $\mathrm{Ext}^1_{MM(S)}(\mathbb{Q}(0), \mathbb{Q}(1))$ *with the essential image of* Φ *is reduced to 0.*

Proof: In view of Lemma 1, this is just Kronecker's theorem: an element of k^* with vanishing height is a root of unity.

A one-motive M with vanishing $Gr_{-1}(TM) = TA$ is called a *Kummer one-motive*. According to [10], Conjecture 0.4, which we denote by (K) below, any object in $\mathrm{Ext}^1_{MM(k)}(\mathbb{Q}(0), \mathbb{Q}(1))$ should be the image under Φ of a Kummer one-motive. Thus, in view of Lemma 2, (K) would imply in particular that $\mathrm{Ext}^1_{MM(S)}(\mathbb{Q}(0), \mathbb{Q}(1))$ vanishes.

But a remarkable feature of $MM(S)$ (which is not shared by any of the other categories considered here) is the non-triviality of the Yoneda Ext-group $\mathrm{Ext}^2_{MM(S)}(\mathbb{Q}(0), \mathbb{Q}(1))$. For instance (cf. [10], 3.12, [14], 5.5), the 2-extension

$$c(M) = \Phi(G) U \Phi(N) \in Ext^2_{MM(S)}(X, Y^V) = Ext^2_{MM(S)}(X \otimes Y(0), \mathbb{Q}(1))$$

obtained by splicing the 1-extensions $\Phi(G)$ and $\Phi(N)$ at $\Phi(A)$ is in general not 0. We now combine this fact with Grothendieck's formalism of "blended extensions" [12], 9.3, to give a new characterization of relative good reduction. Recall that for any Kummer one-motive M_0 (tacitly given as an extension of X by Y^V), the one-motive $M \wedge M_0$ denotes the Baer sum of M and of the push-out of M_0 by $Y^V \to G$, viewed as extensions of X by G.

Theorem 2. *Let M/k be a one-motive, let G, N be the one-motives with two consecutive weights defined by M, and let $c(M) = \Phi(G) U \Phi(N) \in \mathrm{Ext}^2_{MM(S)}(X \otimes Y(0), \mathbb{Q}(1))$.*

i) *Suppose $\Phi(M)$ lies in $MM(S)$. Then $c(M) = 0$.*

ii) *[under (K)] Suppose $c(M) = 0$. Then there exists a unique Kummer one-motive M_0/k such that $\Phi(M \wedge M_0) \in MM(S)$*

iii) *Suppose M satisfies Conditions (i), (ii), (iii) of Theorem 1. Then $\Phi(M)$ lies in $MM(S)$.*

Proof:

i) (cf. [12], 9.3.8) If $\Phi(M)$ lies in $MM(S)$, it defines a class in $\mathrm{Ext}^1_{MM(S)}(\Phi N, Y^V)$, whose pull-back to ΦA is the element ΦG of $\mathrm{Ext}^1_{MM(S)}(\Phi A, Y^V)$. Now, $\Phi(G) U \Phi(N)$ is precisely the image of $\Phi(G)$ under the connecting morphism

$$\mathrm{Ext}^1_{MM(S)}(\Phi N, Y^V) \to \mathrm{Ext}^1_{MM(S)}(\Phi A, Y^V) \xrightarrow{\partial} \mathrm{Ext}^2_{MM(S)}(X \otimes Y(0), \mathbb{Q}(1))$$

deduced via $\text{Ext}(\ , Y^V)$ from the exact sequence:

$$0 \to \Phi A \xrightarrow{i} \Phi N \to X \otimes \mathbb{Q}(0) \to 0.$$

Thus, $\Phi(G)U\Phi(N) = \partial(i^*\Phi M) = 0$.

ii) If $\Phi(G)U\Phi(N) = 0$, the exact sequence above implies the existence of an object M' in $\text{Ext}^1_{MM(S)}(\Phi N, Y^V)$ such that $i^*(M') = \Phi(G)$. By [12], 9.3.8.b, there exists an element M_0 in $\text{Ext}^1_{MM(k)}(X, Y^V)$ (i.e., under ($\mathbb{K}$), a Kummer one-motive), such that $M' = \Phi(M \wedge M_0)$, as claimed. That M_0 is unique follows from Lemma 2.

iii) Using Condition (ii) of Theorem 1, this is Proposition 4 of [3]. It may be proved geometrically (in the style of Remark (iii) of §1), using Condition (iii).

In fact, it would be more interesting to deduce this last step of the proof directly from Condition (i) of Theorem 1, i.e., to produce an explicit Hodge tensor (in the style of [17]) responsible for the vanishing of $c(M)$. This would be a first step in the study of the main conjecture of [3], which, in the language of the present article, may now be stated as follows.

Conjecture. *Let k be a totally real number field, and let M be a one-motive over k such that $c(M) = 0$. Then there exists a Kummer one-motive M_0/k such that $W_{-2}(P(M \wedge M_0)) = 0$.*

Remarks

i) As observed in [3], §3.c, Remark (a), this statement would not hold without some reality assumption. See §3 for a bolder conjecture in that direction.

ii) Composing c with the map $cl^{-1} : \text{Ext}^2_{MM(S)}(X \otimes Y(0), \mathbb{Q}(1)) \to \text{Hom}(X \otimes Y, CH^1(S) \otimes \mathbb{Q})$ constructed in [10] provides a more arithmetic version of the conjecture. For a related approach, using Arakelov theory, see [8], 1.5.

iii) As shown in [3], the conjecture would imply that no right angle appears in the Mordell-Weil lattice of an elliptic curve over $\mathbb{Q}$.

§3. A remark on de Rham realizations

We assume from now on that the number field k is *embedded in* $\mathbb{R}$ (so that it admits at least one real embedding). When $rk(X) = rk(Y) = \dim(A) = 1$, the question of splitting the corresponding mixed $\mathbb{R}$-Hodge structure $T(M \wedge M_0) \otimes \mathbb{R}$ can be described concretely as follows. Let $\sigma, \zeta = \sigma'/\sigma$ be the usual Weierstrass sigma and zeta functions associated to a Weierstrass model of A/k, with real period and quasi-period ω, η. Denote by a, b real logarithms of the k-rational points $u(x), v(y)$ of $A \cong A^V$. After performing an isogeny, we may assume that a, b and $a + b$ do not lie in $\mathbb{Z}\omega$. In a suitable basis of $\text{Lie}(G)$, the image under the map α of the real part of the Betti realization of a one-motive of the type $M \wedge M_0$ is then represented by the subgroup of $\mathbb{R}^2$ generated by the two vectors

$$(\zeta(b)\omega - \eta b, \omega), \ (s(a,b) - \zeta(b)a - \text{Log}\gamma_0, a),$$

where γ_0 is an arbitrary element of k^* (depending only on M_0), and where we have set $s(a,b) = \text{Log}(\sigma(a+b)/\sigma(a)\sigma(b))$. Writing down their determinant, we see that if $T(M \wedge M_0) \otimes \mathbb{R}$ splits, then the number

$$(\sigma(a+b)/\sigma(a)\sigma(b))\exp(-\eta ab/\omega) = \gamma_0$$

must be algebraic. It is a natural conjecture of transcendence theory that this should happen only if a or b is rational multiple of ω, in which case one can easily construct a Kummer one-motive M_0' such that $W_{-2}P(M \wedge M_0') = 0$. From a transcendental point of view, it is therefore likely that one real archimedean condition on splitting suffices to imply the conclusion of the Conjecture, in other words : if a semi-abelian variety G defined over $k \subseteq \mathbb{R}$ contains a k-rational P lying in its maximal compact subgroup, the associated one-motive $M = [U : \mathbb{Z} \to G]$ (with $U(1) = P$) should necessarily be isotropic.

In partial support to this approach (see [19] for further motivations), we finally consider extensions of abelian varieties by vector groups, where the following generalization of Theorem 3.1 of [2] can be obtained.

Theorem 3. *Let A be an abelian variety of dimension g, defined over a real number field k, such that $\text{End}(A/k) \otimes \mathbb{Q}$ is a totally real field of degree g over $\mathbb{Q}$. Let E be the universal vector group extension of A, and let P be a point of $E(k)$ lying in the maximal compact subgroup of $E(\mathbb{R})$. Then P is a torsion point of $E(k)$.*

Proof: We shall prove that the same assertion holds for g specific linearly independent extensions of A by $\mathbb{G}_a$. This clearly implies Theorem 3. Let thus Σ be the set of embeddings of the field $F = End(A/k) \otimes \mathbb{Q}$ into $\mathbb{R}$, let $\sigma \in \Sigma$, and let Ω_σ, H_σ be k-rational differentials of the 1st and 2nd kind on A, generating the eigen-subspace of $H_{dR}(A/k)$ where F^* acts through its character σ. Modifying H_σ by an exact form, we may assume that the integrals of Ω_σ, H_σ along some path from 0 to the projection $p = \pi(P)$ of P onto A are well defined, and we denote their values by $u_\sigma, H_\sigma(u)$; the vector $u = (u_\tau, \tau \in \Sigma)$ represents a logarithm of p in the basis of Lie(A) dual to $(\Omega_\tau, \tau \in \Sigma)$, and the orbit of the vector $\omega = (\omega_\tau, \tau \in \Sigma)$ under the action of F on Lie(A) generates the $\mathbb{Q}$-vector space of real periods of A. We further set $\eta_\sigma = H_\sigma(\omega)$.

Consider the extension E_σ of A by $\mathbb{G}_a$ obtained by pushing-out E via H_σ, and let P_σ be the image of P in E_σ. Then F acts on E_σ, hence on Lie(E_σ), where σ occurs with multiplicity 2. In a basis of Lie(E_σ) formed by eigen-vectors, we see that the orbit of the vector $(\eta_\sigma; \omega)$ under the action of F on Lie(E_σ) generates the space of real periods of E_σ, and that the vector $(H_\sigma(u) - \delta, u)$ represents a logarithm of P_σ, for some element δ of k. Thus, the $(g+1) \times (g+1)$ determinant formed by this logarithm and a basis of the real periods vanishes (i.e. P_σ lies in the maximal compact subgroup of $E_\sigma(\mathbb{R})$) if and only if

$$\lambda_\sigma(u,\omega) = \delta \omega_\sigma, \quad \text{where} \quad \lambda_\sigma(u,\omega) := H_\sigma(u)\omega_\sigma - u_\sigma \eta_\sigma.$$

¿From now on, we assume that this relation holds.

Let then $\phi : A \to A^V$ be a polarization on A, whose associated Rosati involution induces the identity on F, and let G be the extension of A by $(\mathbb{G}_m)^g$ parametrized by the image by ϕ of a basis of the orbit of p under $End(A/k)$. Then, the endomorphisms of A/k lift to endomorphisms of G (cf. [2]), and F also acts on Lie(G). Furthermore, no proper algebraic subgroup of G projects onto A, unless p is a torsion point on G. After some computations, we find an eigen-basis of Lie(G) where the vector $(\lambda_\tau(u,\omega), \tau \in \Sigma; \omega)$ represents a period of G. Since the smallest algebraic subgroup of A whose Lie algebra contains $\mathbb{Q}\omega$ is A itself, we infer from Wustholz' theorem [20], and from the linear dependance of the components of this period over $\mathbb{Q}^{alg}$, that p is a torsion point, i.e. that u is in the orbit of ω under F. Then, $(H_\sigma(u) - \delta, u)$ lies in the F-orbit of $(\eta_\sigma; \omega)$ in Lie(E_σ), and P_σ is a torsion point on E_σ, as was to be proved.

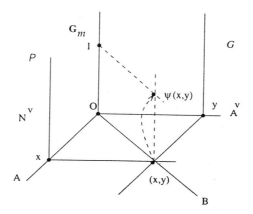

An isotropic one-motive

Figure 1

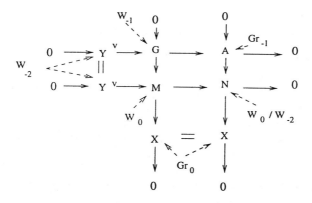

The weight filtration

Figure 2

References

[1] Y. André, Mumford-Tate groups and the theorem of the fixed part, *Compo. Math.*, **82** (1992), 1–24.

[2] D. Bertrand, Endomorphismes de groupes algébriques: applications arithmétiques, *Birkhauser Prog. Math.*, **31** (1983), 1–45.

[3] D. Bertrand, 1-motifs et relations d'orthogonalité dans les groupes de Mordell-Weil; in *Diofantovye Priblizhenya*, Fel'dman volume, *Matem. Zapiski*, **2** (1996), 7–22.

[4] D. Bertrand, Points rationnels sur les sous-groupes compacts des groupes algébriques, *Experimental Maths*, **4** (1995), 145–151.

[5] L. Breen, Biextensions alternées, *Compo. Math.*, **63** (1987), 99–122.

[6] J-L. Brylinski, 1-motifs et fomes automorphes, *Publ. math. Univ. Paris VII*, **15** (1983), 43–106.

[7] J. Carlson, The geometry of the extension class of a mixed Hodge structure, *PSPM AMS*, **46** (1987), 199–222.

[8] A. Chambert-Loir, Extensions vectorielles, périodes et gémétrie diophantienne, thèse Univ. Paris VI, Dec. 1995.

[9] P. Deligne, Théorie de Hodge III; *Publ. Math. IHES*, **44** (1975), 5–77.

[10] C. Deninger and E. Nart, On Ext^2 of motives over arithmetic curves, *American J. Math.*, **117** (1995), 602–625.

[11] J-M. Fontaine et B. Perrin-Riou, Autour des conjectures de Bloch et Kato [...], *PSPM AMS*, **55** (1994) Part 1, 599–706.

[12] A. Grothendieck, Modèles de Néron et monodromie, *SGA* VII. 1, exp. no. 9, Springer LN 288.

[13] O. Jacquinot and K. Ribet, Deficient points on extensions of abelian varieties by $\mathbb{G}_m$, *J. Number Th.*, **25** (1985), 497–507.

[14] U. Jannsen, Mixed motives, motivic cohomology and Ext-groups, *Proc. ICM Zurich* (1994), Birkhauser Prog. Math., (1995), vol. 1, 667–679.

[15] J. Milne, Canonical models of (mixed) Shimura varieties and automorphic vector bundles, *Perspectives in Maths*, **10** (1990), 284–411.

[16] M. Raynaud, Motifs et monodromie géométrique: *Astérisque* **223** (1994), exp. no 7.

[17] K. Ribet, Cohomological realization of a family of one-motives, *J. Number Th.*, **25** (1987), 152–161.

[18] A. Scholl, Height pairings and special values of L-functions, *PSPM AMS*, **55** (1994) Part 1, 571–598.

[19] M. Waldschmidt, Densité de points rationnels sur les groupes algébriques, *Experimental Maths*, **3** (1994) 329–352.

[20] G. Wustholz, Recent progress in transcendence theory, Springer LN 1068, (1983) 280–296.

D. Bertrand, Université de Paris VI, Institut de Mathématiques, T. 46, Case 247, 4, Place Jussieu; 75252 Paris Cedex 05 (France).

[e-mail : bertrand at math.jussieu.fr]

The Hasse principle in a pencil of algebraic varieties

J.-L. Colliot-Thélène

ABSTRACT. Let k be a number field, X/k a smooth, projective, geometrically connected variety and $p : X \to \mathbf{P}_k^1$ a flat morphism from X to the projective line, with (smooth) geometrically integral generic fibre. Assume that X has points in all completions of k. Does there exist a k-rational point m of $\mathbf{P}_k^1$ with smooth fibre $X_m = p^{-1}(m)$ having points in all completions of k ? This may fail, but known counterexamples can be interpreted by means of the subgroup of the Brauer group of X whose restriction to the generic fibre of p comes from the Brauer group of the function field of $\mathbf{P}_k^1$. This fibred version of the Brauer–Manin obstruction has been at the heart of recent investigations on the Hasse principle and weak approximation. Work in this area is surveyed in the present text, which develops the talk I gave at Tiruchirapalli.

1. The Hasse principle, weak approximation and the Brauer–Manin obstruction

Let k be a number field, let Ω be the set of its places and Ω_∞ the set of archimedean places. Let $\overline{k}$ denote an algebraic closure of k, and let k_v the completion of k at the place v. Let X/k be an algebraic variety, i.e. a separated k-scheme of finite type. Given an arbitrary field extension K/k, one lets $X(K) = \mathrm{Hom}_{\mathrm{Spec}(k)}(\mathrm{Spec}(K), X)$ be the set of K-rational points of the k-variety X, and one writes $X_K = X \times_k K$ and $\overline{X} = X \times_k \overline{k}$. We have obvious inclusions

$$X(k) \hookrightarrow X(\mathbb{A}_k) \subset \prod_{v \in \Omega} X(k_v),$$

the first one being the diagonal embedding into the set $X(\mathbb{A}_k)$ of adèles of X. The set $X(\mathbb{A}_k)$ is empty if and only if the product $\prod_{v \in \Omega} X(k_v)$ is empty. When X/k is proper, e.g. projective, we have $X(\mathbb{A}_k) = \prod_{v \in \Omega} X(k_v)$. If X is smooth over k and irreducible, and if $U \subset X$ is a non-empty Zariski open set of X, the conditions $X(\mathbb{A}_k) \neq \emptyset$ and $U(\mathbb{A}_k) \neq \emptyset$ are equivalent. Deciding whether the set $X(\mathbb{A}_k)$ is empty or not is a finite task. This prompts:

1991 *Mathematics Subject Classification*. Primary 14G05 Secondary 11G35, 14G25, 14D10, 14C25.

This text has benefited from numerous discussions with Alexei Skorobogatov and Peter Swinnerton-Dyer over a number of years.

DEFINITION 1. *The condition $X(\mathbb{A}_k) = \emptyset$ is the local obstruction to the existence of a rational point on X. A class $\mathcal{C}$ of algebraic varieties defined over k is said to satisfy the Hasse principle (local-to-global principle) if the local obstruction to the existence of a rational point on a variety of $\mathcal{C}$ is the only obstruction, i.e. if for X in $\mathcal{C}$ the (necessary) condition $X(\mathbb{A}_k) \neq \emptyset$ implies $X(k) \neq \emptyset$.*

A 'counterexample to the Hasse principle' is a variety X/k such that $X(\mathbb{A}_k) \neq \emptyset$ but $X(k) = \emptyset$.

Classes of varieties known to satisfy the Hasse principle include: quadrics (Legendre, Minkowski, Hasse), principal homogeneous spaces of semisimple, simply connected (linear) algebraic groups (Kneser, Harder, Tchernousov), projective varieties which are homogenous spaces under connected linear algebraic groups (a corollary of the previous case, as shown by Harder), varieties defined by a norm equation $N_{K/k}(\Xi) = c$ for $c \in k^*$ and K/k a finite, cyclic extension (Hasse). The arithmetic part of the proof of these results is encapsulated in the injection $\mathrm{Br}(k) \hookrightarrow \oplus_{v \in \Omega} \mathrm{Br}(k_v)$, which is part of the basic reciprocity sequence from class field theory:

$$(1) \qquad 0 \to \mathrm{Br}(k) \longrightarrow \bigoplus_{v \in \Omega} \mathrm{Br}(k_v) \xrightarrow{\Sigma_{v \in \Omega} \mathrm{inv}_v} \mathbb{Q}/\mathbb{Z} \to 0$$

Here $\mathrm{Br}(F)$ denotes the Brauer group of a field F, i.e. the second (continuous) cohomology group of the absolute Galois group $\mathrm{Gal}(F_s/F)$ acting on the multiplicative group F_s^* of a separable closure F_s of F, viewed as a discrete module. For each place v, the map $\mathrm{inv}_v : \mathrm{Br}(k_v) \hookrightarrow \mathbb{Q}/\mathbb{Z}$ is the embedding provided by local class field theory (an isomorphism with $\mathbb{Q}/\mathbb{Z}$ if v is a finite place, an isomorphism with $\mathbb{Z}/2$ if v is a real place, zero if v is complex).

Given a variety X over the number field k, for each place v, the set $X(k_v)$ is naturally equipped with the topology coming from the topology on k_v. Given any set $\Omega_1 \subset \Omega$ of places of k, we then have a topology on the product $\prod_{v \in \Omega_1} X(k_v)$, a basis of which is given by open sets of the shape $\prod_{v \in S} U_v \times \prod_{v \in \Omega_1 \setminus S} X(k_v)$, where S is a finite subset of Ω_1 and $U_v \subset X(k_v)$, $v \in S$ is open. If X/k is proper, hence $X(\mathbb{A}_k) = \prod_{v \in \Omega} X(k_v)$, this gives the usual topology on the set of adèles $X(\mathbb{A}_k)$. For any finite set $S \subset \Omega$, the projection map $\prod_{v \in \Omega} X(k_v) \to \prod_{v \in S} X(k_v)$ is open.

DEFINITION 2. *Let X/k be a variety over the number field k. One says that weak approximation holds for X if the diagonal map $X(k) \to \prod_{v \in \Omega} X(k_v)$ has dense image, which is the same as requiring: for any finite set S of places of k, the diagonal map $X(k) \to \prod_{v \in S} X(k_v)$ has dense image.*

If X/k is proper, this condition is also equivalent to the density of $X(k)$ in $X(\mathbb{A}_k)$ for the adèle topology (but if X/k is not proper, these conditions are far from being equivalent – think of the case where X is the additive group $\mathbf{G}_a$ or the multiplicative group $\mathbf{G}_m$). If X is smooth over k and irreducible, and if $U \subset X$ is a non-empty Zariski open set of X, then X satisfies weak approximation if and only if U does.

According to this definition, if X satisfies weak approximation, then it satisfies the Hasse principle. The reader should be aware that in many earlier papers, weak approximation is defined only under the additional assumption $X(k) \neq \emptyset$.

In parallel with the Hasse principle, one studies weak approximation, for a number of reasons:

(a) the techniques are similar ;

(b) proofs of the Hasse principle often yield a proof of weak approximation at the same stroke (taking $S = \Omega_\infty$, for specific classes of varieties, one thus gets positive answers to questions raised by Mazur [Maz2], even in cases where the group of birational automorphisms does not act transitively on rational points);

(c) weak approximation on a smooth irreducible variety implies Zariski density of rational points, as soon as there is at least one such point;

(d) proofs of the Hasse principle often rely on the weak approximation property for some auxiliary variety.

Weak approximation holds for affine space $\mathbf{A}_k^n$ and projective space $\mathbf{P}_k^n$. More generally, it holds for any smooth irreducible k-variety which is k-birational to affine space, for instance for varieties defined by a norm equation $N_{K/k}(\Xi) = 1$ when K/k is a finite, cyclic extension of k (this is a consequence of Hilbert's theorem 90). It also holds for semisimple, simply connected (linear) algebraic groups ([Pl/Ra] VII.3, Prop. 9).

Once the Hasse principle has been defined, one cannot but admit: it very rarely holds ! Indeed, (subtle) counterexamples to the Hasse principle for smooth, irreducible varieties have been exhibited among: varieties defined by a norm equation $N_{K/k}(\Xi) = c$ when K/k is not cyclic, e.g. when K/k is Galois with group $(\mathbb{Z}/2)^2$ (Hasse and Witt in the 30's), curves of genus one (Reichardt and Lind in the 40's), in particular the curve defined by the famous diagonal equation $3x^3 + 4y^3 + 5z^3 = 0$ (Selmer in the 50's), principal homogeneous spaces under some semisimple algebraic groups (Serre in the 60's, [Se1], III.4.7), smooth cubic surfaces (Swinnerton-Dyer, 1962), then the diagonal cubic surface $5x^3 + 9y^3 + 10z^3 + 12t^3 = 0$ (Cassels and Guy, 1966), conic bundles over the projective line, such as the surface given by $y^2 + z^2 = (3 - x^2)(x^2 - 2)$ (Iskovskikh, 1970), smooth intersections of two quadrics in $\mathbf{P}_k^4$ (Birch and Swinnerton-Dyer, 1975), singular intersections of two quadrics in $\mathbf{P}_k^5$.

Similarly, even under the additional assumption $X(k) \neq \emptyset$ and X smooth and irreducible, weak approximation quite often fails. Counterexamples have been found among varieties defined by a norm equation $N_{K/k}(\Xi) = 1$ when K/k is not cyclic, e.g. when K/k is Galois with group $(\mathbb{Z}/2)^2$, curves of genus one (even when rational points are Zariski dense on them), principal homogeneous spaces under some semisimple algebraic groups (Serre), smooth cubic surfaces (Swinnerton-Dyer). After Faltings' theorem (Mordell's conjecture), weak approximation clearly fails for any curve of genus at least 2 possessing at least one rational point.

As it turns out, the arguments underlying the counterexamples just listed can all be cast in a common mould, namely the Brauer-Manin obstruction to the Hasse principle, described by Manin [Ma1] in his talk at the ICM in 1970. A similar obstruction (the same indeed) accounts for the quoted counterexamples to weak approximation ([CT/San3]).

Let $\mathrm{Br}(X)$ denote the (Grothendieck) Brauer group of a scheme X, namely $H^2_{\text{ét}}(X, \mathbf{G}_m)$. If $X = \mathrm{Spec}(F)$ is the spectrum of a field F, then $\mathrm{Br}(X) = \mathrm{Br}(F)$. If X is a k-variety, F/k a field extension, and $A \in \mathrm{Br}(X)$ an element of the Brauer group, functoriality yields an evaluation map $\mathrm{ev}_A : X(F) \to \mathrm{Br}(F)$, sending the point $P \in X(F)$ to the fibre $A(P)$ of A at P.

LEMMA 1. *Let X/k be a smooth, proper, irreducible variety over the number field k, and let $A \in \mathrm{Br}(X)$.*

(i) For each place $v \in \Omega$, the evaluation map $\mathrm{ev}_A : X(k_v) \to \mathrm{Br}(k_v) \subset \mathbb{Q}/\mathbb{Z}$ is continuous and has finite image.

(ii) There exists a finite set $S_A \subset \Omega$ of places of k such that for $v \notin S_A$, the evaluation map $\mathrm{ev}_A : X(k_v) \to \mathrm{Br}(k_v) \subset \mathbb{Q}/\mathbb{Z}$ is zero.

Let X/k be a variety over k, and let $A \in \mathrm{Br}(X)$. We have the basic commutative diagramme:

$$\begin{array}{ccccc} X(k) & \hookrightarrow & X(\mathbb{A}_k) & & \\ \downarrow \mathrm{ev}_A & & \downarrow \mathrm{ev}_A & \searrow^{\theta_A} & \\ \mathrm{Br}(k) & \longrightarrow & \bigoplus_{v \in \Omega} \mathrm{Br}(k_v) & \longrightarrow & \mathbb{Q}/\mathbb{Z} \end{array}$$

where the vertical map lands in the direct sum by the lemma. That sequence (1) is exact says in particular that the bottom composite map is zero (this is a generalization of the classical quadratic reciprocity law). As indicated in the diagramme, we denote by θ_A the composite map

$$\theta_A : X(\mathbb{A}_k) \to \bigoplus_{v \in \Omega} \mathrm{Br}(k_v) \to \mathbb{Q}/\mathbb{Z}.$$

We let $\mathrm{Ker}(\theta_A) \subset X(\mathbb{A}_k)$ denote the inverse image of $0 \in \mathbb{Q}/\mathbb{Z}$. By lemma 1, this is a closed and open set of $X(\mathbb{A}_k)$.

Let $B \subset \mathrm{Br}(X)$ be a subgroup of the Brauer group. Let us define

$$X(\mathbb{A}_k)^B = \bigcap_{A \in B} \mathrm{Ker}(\theta_A) \subset X(\mathbb{A}_k)$$

and $X(\mathbb{A}_k)^{\mathrm{Br}} = X(\mathbb{A}_k)^{\mathrm{Br}(X)}$. Because of the continuity statement in Lemma 1, the closure $X(k)^{cl}$ of $X(k)$ in $X(\mathbb{A}_k)$ is contained in $X(\mathbb{A}_k)^{\mathrm{Br}}$. Let us state this as:

PROPOSITION 2. *Let X/k be a smooth, proper, irreducible variety over a number field k, and let $B \subset \mathrm{Br}(X)$ be a subgroup of the Brauer group of X. We have the natural inclusions*

$$X(k)^{cl} \subset X(\mathbb{A}_k)^{\mathrm{Br}} \subset X(\mathbb{A}_k)^B \subset X(\mathbb{A}_k)$$

of closed subsets in $X(\mathbb{A}_k) = \prod_{v \in \Omega} X(k_v)$.

Clearly, if $X(\mathbb{A}_k) \neq \emptyset$ but $X(\mathbb{A}_k)^B = \emptyset$ for some B, then we have a counterexample to the Hasse principle. What Manin [Ma1] noticed in 1970 was that this simple proposition accounts for most counterexamples to the Hasse principle hitherto known – the Cassels and Guy example awaited 1985 (work of Kanevsky, Sansuc and the author) to be fit into the Procrustean bed without damage. In these counterexamples, the rôle of sequence (1) is played by some explicit form of the reciprocity law. Clearly again, if X/k is proper and the inclusion $X(\mathbb{A}_k)^B \subset X(\mathbb{A}_k)$ is strict, i.e. if $X(\mathbb{A}_k)^B \neq X(\mathbb{A}_k)$, then weak approximation fails for X : that most known counterexamples to weak approximation can be explained in this fashion was pointed out in 1977 (see [CT/San3]).

Several remarks are in order.

(i) The set $X(\mathbb{A}_k)^B$ only depends on the image of B under the projection map $\mathrm{Br}(X) \to \mathrm{Br}(X)/\mathrm{Br}(k)$. More precisely, if $A_i \in B$, $i \in I$, generate the image of B in $\mathrm{Br}(X)/\mathrm{Br}(k)$, then $X(\mathbb{A}_k)^B = \cap_{i \in I} \mathrm{Ker}(\theta_{A_i})$. If the set I is finite, then the closed subset $X(\mathbb{A}_k)^B \subset X(\mathbb{A}_k)$ is also open. If the geometric Picard group $\mathrm{Pic}(\overline{X})$ is torsionfree, which implies that the coherent cohomology group $H^1(X, O_X)$ vanishes, and if the coherent cohomology group $H^2(X, O_X)$ also vanishes, then the quotient $\mathrm{Br}(X)/\mathrm{Br}(k)$ is finite, hence $X(\mathbb{A}_k)^{\mathrm{Br}}$ is open in $X(\mathbb{A}_k)$.

(ii) From a theoretical point of view, the best choice for $B \subset \mathrm{Br}(X)$ is $\mathrm{Br}(X)$ itself. However, for concrete computations, one prefers to use only finitely many elements in $\mathrm{Br}(X)$.

(iii) Suppose $X(\mathbb{A}_k)^B \neq X(\mathbb{A}_k)$. Then we may be more precise about the set of places where weak approximation fails. Indeed, there then exists a family $\{M_v\}_{v \in \Omega} \in X(\mathbb{A}_k)$, and an $A \in B \subset \mathrm{Br}(X)$ such that $\sum_{v \in \Omega} \mathrm{inv}_v(A(M_v)) \neq 0$. If we let S_A be as in lemma 1 (ii), then the diagonal map $X(k) \to \prod_{v \in S_A} X(k_v)$ does not have dense image.

(iv) One may formulate a version of Lemma 1 and Proposition 2 for the set of adèles $X(\mathbb{A}_k)$ of a not necessarily proper variety X/k, but until now this has been of little use. Even when one studies homogeneous spaces of a linear algebraic group, as Sansuc [San2] and Borovoi [Bo] do, the elements of the Brauer group of such a space E which turn out to play the main rôle may be shown to lie in the image of the Brauer group of smooth compactifications of E.

From now on we shall assume X/k smooth, proper, and geometrically irreducible. Let us recall some terminology. The condition $X(\mathbb{A}_k)^{\mathrm{Br}} = \emptyset$ is the Brauer–Manin obstruction to the existence of a rational point on X. The condition $X(\mathbb{A}_k)^B = \emptyset$ is the Brauer–Manin obstruction to the existence of a rational point on X attached to $B \subset \mathrm{Br}(X)$. We shall sometimes refer to it as the B-obstruction to the existence of a rational point on X. When we already know $X(\mathbb{A}_k) \neq \emptyset$, the condition $X(\mathbb{A}_k)^{\mathrm{Br}} = \emptyset$ is also referred to as the Brauer–Manin obstruction to the Hasse principle. For a class $\mathcal{C}$ of algebraic varieties over k, suppose that we have a standard way of defining a subgroup $B(X) \subset \mathrm{Br}(X)$ for any $X \in \mathcal{C}$ (e.g. $B(X) = \mathrm{Br}(X)$). We say that the Brauer–Manin obstruction to the existence of a rational point attached to B is the only obstruction for $\mathcal{C}$ if, for $X \in \mathcal{C}$, the conditions $X(k) \neq \emptyset$ and $X(\mathbb{A}_k)^{B(X)} \neq \emptyset$ are equivalent.

Similarly, the condition $X(\mathbb{A}_k)^{\mathrm{Br}} \neq X(\mathbb{A}_k)$ is the Brauer–Manin obstruction to weak approximation on X. The condition $X(\mathbb{A}_k)^B \neq X(\mathbb{A}_k)$ is the Brauer–Manin obstruction to weak approximation on X attached to B. We shall sometimes refer to it as the B-obstruction to weak approximation on X. We say that the Brauer–Manin obstruction attached to B is the only obstruction to weak approximation for X if the inclusion $X(k)^{cl} \subset X(\mathbb{A}_k)^B$ is an equality. This then implies $X(k)^{cl} = X(\mathbb{A}_k)^{\mathrm{Br}} = X(\mathbb{A}_k)^B$. It also implies that the B-obstruction to the existence of a rational point on X is the only obstruction.

Suppose that the image of B in $\mathrm{Br}(X)/\mathrm{Br}(k)$ is finite and $X(k)^{cl} = X(\mathbb{A}_k)^B$. Assume $X(k) \neq \emptyset$. Then:

(i) For any finite set S of places of k, the closure of the image of the diagonal map $X(k) \to \prod_{v \in S} X(k_v)$ is open. In particular $X(k)$ is Zariski-dense in X.

(ii) There exists a finite set of places S_0 of k such that for any finite set of places S of k with $S \cap S_0 = \emptyset$, the image of the diagonal map $X(k) \to \prod_{v \in S} X(k_v)$ is dense, i.e. weak weak approximation holds (cf. [Se2]).

Two theorems are particularly noteworthy. The first one is Manin's reinterpretation ([Ma2], VI.41.24, p. 228) of results of Cassels and Tate.

THEOREM 3 (Manin). *Let X/k be a curve of genus one. Assume that the Tate–Shafarevich group of the Jacobian of X is a finite group. Then the Brauer–Manin obstruction to the existence of a rational point on X is the only obstruction: if $X(\mathbb{A}_k)^{\mathrm{Br}} \neq \emptyset$, then $X(k) \neq \emptyset$. More precisely, let $\mathrm{Br}_0(X) \subset \mathrm{Br}(X)$ be the kernel of the map $\mathrm{Br}(X) \to \prod_{v \in \Omega} \mathrm{Br}(X_{k_v})/\mathrm{Br}(k_v)$. The $\mathrm{Br}_0(X)$-obstruction is the only obstruction: if $X(\mathbb{A}_k)^{\mathrm{Br}_0(X)} \neq \emptyset$, then $X(k) \neq \emptyset$.*

Conjecturally, the Tate–Shafarevich group of any abelian variety is a finite group (remarkable results in this direction are due to Rubin and to Kolyvagin, see [Maz1].) The theorem can be extended to cover principal homogeneous spaces of abelian varieties. Lan Wang [W] has established an analogous result for weak approximation on an abelian variety A, namely that the closure of $A(k)$ in $A(\mathbb{A}_k)$ coincides with $A(\mathbb{A}_k)^{\mathrm{Br}}$, under some additional condition. She still assumes the finiteness of the Tate–Shafarevich group of A, but there is a well-known additional difficulty here. Over the real or the complex field k_v, the Brauer group of a variety X over k_v cannot make any difference between two points in the same connected component of $X(k_v)$. Thus for an abelian variety A over a number field k, a statement like Theorem 4 below can hold only if one makes the additional assumption that the closure of the image of the diagonal map $A(k) \to \prod_{v \in \Omega_\infty} A(k_v)$ contains the connected component of identity (if A were a connected linear algebraic group, this would be automatic). This is precisely the additional assumption which L. Wang makes. Waldschmidt [Wald] has given sufficient conditions for this to hold. This problem is related to questions raised by Mazur [Maz2] [Maz3].

The second theorem builds upon work of Kneser, Harder, and Tchernousov for principal homogeneous spaces under semisimple, simply connected groups, and of Sansuc [San2], who handled the case of principal homogeneous spaces under arbitrary connected linear groups.

THEOREM 4 (Borovoi) [Bo]. *Let G be a connected linear algebraic group over k, let Y be a homogeneous space under G, and let X be a smooth, projective compactification of Y (i.e. Y is a dense open set in the smooth, projective variety X). Assume that the geometric stabilizer (isotropy group of an arbitrary $\overline{k}$-point of Y) is a connected group. Then the Brauer–Manin obstruction to weak approximation, and in particular to the existence of a rational point on X, is the only obstruction for X. More precisely, let $\mathrm{Br}_0(X) \subset \mathrm{Br}(X)$ be the kernel of the map*

$$\mathrm{Br}(X) \to \prod_{v \in \Omega} \mathrm{Br}(X_{k_v})/\mathrm{Br}(k_v).$$

The $\mathrm{Br}_0(X)$-obstruction to weak approximation, and in particular to the existence of a rational point on X, is the only obstruction.

These two theorems should not lead one to hasty generalizations: many counterexamples, such as Swinnerton-Dyer's cubic surface, or Iskovskih's conic bundle, depend on bigger subgroups of the Brauer group. This is clearly also the case for

Harari's 'transcendental example' [Ha3]: here the obstruction involves an element of $\text{Br}(X)$ which does not vanish in $\text{Br}(\overline{X})$.

These two theorems could nevertheless induce us into asking the question:

Is the Brauer–Manin obstruction the only obstruction to the existence of a rational point (resp. to weak approximation) for arbitrary smooth projective varieties ?

For arbitrary varieties, the answer to that question is presumably NO. Indeed, a positive answer would imply that the Hasse principle and weak approximation hold for smooth complete intersections of dimension at least 3 in projective space, e.g. non-singular hypersurfaces $f(x_0, x_1, x_2, x_3, x_4) = 0$ in projective space $\mathbf{P}^4$, of arbitrary degree d. The weak approximation statement would contradict the generalization of Mordell's conjecture to higher dimension, which predicts in particular that rational points on such hypersurfaces are not Zariski-dense as soon as $d \geq 6$. As for the Hasse principle, Sarnak and Wang [S/W], using the (elementary) fibration method (Theorem 6 below), have shown that this would contradict Lang's conjecture (a variation on the Mordell conjecture theme) that a smooth, projective variety $X/\mathbb{Q}$ such that $X(\mathbb{C})$ is hyperbolic has only finitely many $\mathbb{Q}$-rational points.

One should also bear in mind some examples of curves of genus at least two given by Coray and Manoil [Co/Ma]. The principle of their counterexamples to the Hasse principle is simple: they produce curves X/k equipped with a dominant k-morphism $f : X \to Y$ to a curve of genus one such that $Y(k)$ is non-empty, finite and explicitly known, $f^{-1}(Y(k)) = \emptyset$ by inspection, and $X(\mathbb{A}_k) \neq \emptyset$ (easy to check). The problem here is that the group $\text{Br}(X)/\text{Br}(k)$ is huge, and we do not have a finite algorithm for computing the Brauer–Manin obstruction.

On the basis of Theorems 3 and 4, and of the various results to be discussed later, it nevertheless makes sense to put forward:

CONJECTURE 1. *Let X/k be a smooth, projective, geometrically irreducible variety, and let $p : X \to \mathbf{P}_k^1$ be a dominant (flat) morphism. Assume:*

(α) the generic fibre of p is birational to a homogeneous space Y of a connected algebraic group G over $k(\mathbf{P}^1)$, and the geometric stabilizer of the action of G on Y is connected.

(β) For any closed point $M \in \mathbf{P}_k^1$, the fibre $X_M = p^{-1}(M)$ contains a component of multiplicity one.

Then the Brauer–Manin obstruction to the existence of a rational point on X is the only obstruction. If G is a linear group, the Brauer–Manin obstruction to weak approximation for X is the only obstruction.

The geometric stabilizer is the isotropy group of an arbitrary geometric point of the homogeneous space. The connectedness assumption for this stabilizer may be necessary. Indeed, Borovoi and Kunyavskiĭ [Bo/Ku] have recently produced a homogeneous space of a connected linear algebraic group with (non-commutative) finite geometric stabilizer, which is a counterexample to the Hasse principle, and for which it is unclear whether the Brauer–Manin obstruction holds.

Condition (β) is equivalent to condition

(β') *For any point $m \in \mathbf{P}^1(\overline{k})$, the fibre X_m contains a component of multiplicity one.*

This condition in turn is implied by condition

(γ) *The map p has a section over $\overline{k}$.*

This last condition is automatically satisfied if G in (α) is a connected linear algebraic group (indeed, in that case, the fibration p admits a section over $\overline{k}$, by a theorem of Serre). On the other hand, if one drops condition (β), and G is an elliptic curve, then it may happen that $X(k)$ is not Zariski-dense in X ([CT/Sk/SD2]).

2. Fibrations

We shall explore ways of proving the Hasse principle (and weak approximation) by fibring a variety into hopefully simpler smaller dimensional varieties. Let us first fix standard assumptions. Let X/k be a smooth, projective, geometrically irreducible variety over the field k. In this survey, we shall say that $p : X \to \mathbf{P}_k^1$ is a *fibration* if the map p is dominant (hence flat) and the generic fibre X_η over the field $k(\mathbf{P}^1)$ is smooth (automatic if char$(k) = 0$) and geometrically irreducible.

Let k be a number field and $p : X \to \mathbf{P}_k^1$ be a fibration. A naïve question would be: Assume that the Hasse principle (resp. weak approximation) holds for (smooth) fibres $X_m = p^{-1}(m)$ for $m \in \mathbf{P}^1(k)$; does it follow that the Hasse principle (resp. weak approximation) holds for X ?

Iskovskih's example, a one parameter family of conics, shows that the answer in general is in the negative: in this case, X has points in all completions of k, but for each fibre X_m over a k-point $m \in \mathbf{P}^1(k)$, there is at least one completion k_v (depending on m) such that $X_m(k_v) = \emptyset$ (otherwise, from the Hasse principle for conics, we would conclude $X(k) \neq \emptyset$). There are similar examples which give a negative answer to the question on weak approximation even when $X(k) \neq \emptyset$.

Before we raise what we feel are the relevant questions, we need a definition (see [Sk2]). Let $p : X \to \mathbf{P}_k^1$ a fibration. Let $\mathrm{Br}_{\mathrm{vert}}(X) \subset \mathrm{Br}(X)$ be the subgroup consisting of elements $A \in \mathrm{Br}(X)$ whose restriction to the generic fibre X_η lies in the image of the map $\mathrm{Br}(k(\mathbf{P}_k^1)) \to \mathrm{Br}(X_\eta)$. If the map p does not have a section (over k), this group may be bigger than the image of $\mathrm{Br}(k) = \mathrm{Br}(\mathbf{P}_k^1)$ under p^*. By 'vertical' Brauer–Manin obstruction to the existence of a rational point (resp. weak approximation), we shall mean the obstruction attached to the subgroup $\mathrm{Br}_{\mathrm{vert}}(X) \subset \mathrm{Br}(X)$. We have a fibred version of Proposition 2 :

PROPOSITION 5. *Let $p : X \to \mathbf{P}_k^1$ be a fibration. Let $\mathcal{R} \subset \mathbf{P}^1(k)$ be the set of points $m \in \mathbf{P}^1(k)$ whose fibre X_m is smooth and satisfies $X_m(\mathbb{A}_k) \neq \emptyset$. Let $\mathcal{R}_1 \subset \mathcal{R}$ be the set of points $m \in \mathbf{P}^1(k)$ whose fibre X_m is smooth and satisfies $X_m(\mathbb{A}_k)^{\mathrm{Br}(X_m)} \neq \emptyset$.*

Let $\mathcal{R}^{cl}$, resp. $\mathcal{R}_1^{cl}$, be the closure of $\mathcal{R}$, resp. $\mathcal{R}_1$, under the diagonal embedding $\mathbf{P}^1(k) \hookrightarrow \mathbf{P}^1(\mathbb{A}_k)$. We then have inclusions of closed subsets:

$$\mathcal{R}^{cl} \subset p(X(\mathbb{A}_k)^{\mathrm{Br}_{\mathrm{vert}}}) \subset p(X(\mathbb{A}_k)) \subset \mathbf{P}^1(\mathbb{A}_k)$$

and

$$\mathcal{R}_1^{cl} \subset p(X(\mathbb{A}_k)^{\mathrm{Br}}) \subset p(X(\mathbb{A}_k)) \subset \mathbf{P}^1(\mathbb{A}_k).$$

In particular, if there exists a point $m \in \mathbf{P}^1(k)$ whose fibre X_m is smooth and satisfies $X_m(\mathbb{A}_k) \neq \emptyset$, then $X(\mathbb{A}_k)^{\mathrm{Br}_{\mathrm{vert}}} \neq \emptyset$: there is no vertical Brauer–Manin obstruction to the existence of a rational point on X.

In the light of the work to be discussed in the next subsections, and of Proposition 5, it seems natural to raise the following general questions (for pencils of Severi-Brauer varieties, see also [Se1, p. 125]). We shall restrict attention to fibrations satisfying condition

(β) For any closed point $M \in \mathbf{P}_k^1$, the fibre X_M contains a component of multiplicity one.

Using the Faddeev exact sequence for the Brauer group of the function field of $k(\mathbf{P}_k^1)$, we see that, under (β), the quotient $\mathrm{Br}_{\mathrm{vert}}(X)/\mathrm{Br}(k)$ is finite and essentially computable (cf. [Sk2]). Finiteness of that quotient then implies that the (closed) set $X(\mathbb{A}_k)^{\mathrm{Br}_{\mathrm{vert}}}$ is open in $X(\mathbb{A}_k)$.

QUESTION 1. *Let k be a number field, and let $p : X \to \mathbf{P}_k^1$ be a fibration satisfying (β). Let $\mathcal{R} \subset \mathbf{P}^1(k)$ be the set of points $m \in \mathbf{P}^1(k)$ whose fibre X_m is smooth and satisfies $X_m(\mathbb{A}_k) \neq \emptyset$. Do we have $\mathcal{R}^{cl} = p(X(\mathbb{A}_k)^{\mathrm{Br}_{\mathrm{vert}}})$?*

In other words, let $\{M_v\} \in X(\mathbb{A}_k)^{\mathrm{Br}_{\mathrm{vert}}}$, and let $m_v = p(M_v)$. Is the family $\{m_v\}$ in the closure of the set T of $m \in \mathbf{P}^1(k)$ such that X_m is smooth and has points in all completions ? In particular, if $X(\mathbb{A}_k)^{\mathrm{Br}_{\mathrm{vert}}} \neq \emptyset$, i.e. if there is no vertical Brauer-Manin obstruction to the existence of a rational point on X, does there exist a rational point $m \in \mathbf{P}^1(k)$ whose fibre X_m is smooth and has points in all completions of k ?

QUESTION 2. *Let k be a number field, and let $p : X \to \mathbf{P}_k^1$ be a fibration satisfying (β). Let $\mathcal{R}_1 \subset \mathbf{P}^1(k)$ be the set of points $m \in \mathbf{P}^1(k)$ whose fibre X_m is smooth and satisfies $X_m(\mathbb{A}_k)^{\mathrm{Br}(X_m)} \neq \emptyset$. Do we have $\mathcal{R}_1^{cl} = p(X(\mathbb{A}_k)^{\mathrm{Br}})$?*

In particular, if $X(\mathbb{A}_k)^{\mathrm{Br}} \neq \emptyset$, i.e. if there is no Brauer-Manin obstruction to the existence of a rational point on X, does there then exist an $m \in \mathbf{P}^1(k)$ with smooth fibre X_m, such that $X_m(\mathbb{A}_k)^{\mathrm{Br}(X_m)} \neq \emptyset$, i.e. such that there is no Brauer-Manin obstruction to the existence of a rational point on X_m?

Giving even partial answers to these questions becomes harder and harder as the number of reducible geometric fibres of p grows. Let us be more precise. Following Skorobogatov [Sk2], given an arbitrary closed point $M \in \mathbf{P}_k^1$, let us say that the fibre X_M over the field k_M (residue field of $\mathbf{P}_k^1$ at M) is *split* if there exists at least one component Y of X_M in the divisor $X_M = p^{-1}(M) \subset X$, satisfying the two conditions:

(i) its multiplicity in X_M is one ;

(ii) the k_M-variety Y/k_M is geometrically irreducible, i.e. k_M is algebraically closed in the function field $k_M(Y)$ of Y.

As we shall now see, the difficulty to answer the above questions grows with the integer

$$\delta = \delta(p) = \sum_{M \in \mathbf{P}_k^1, X_M \text{ non-split}} [k_M : k].$$

Here M runs through the set of closed points of $\mathbf{P}_k^1$, and $[k_M : k]$ is the degree of the finite extension k_M/k. If we let $\delta_1(p)$ be the number of geometric fibres which are reducible, then we have $\delta \leq \delta_1$. In several papers, the invariant δ_1 was used. However, as pointed out by Skorobogatov, it is really δ which measures the arithmetic difficulty.

2.1 The case $\delta \leq 1$.

Let $p : X \to \mathbf{P}_k^1$ be a fibration. If $\delta(p) = 0$, all fibres are split. If $\delta(p) = 1$, there is one non-split fibre, and it lies over a k-rational point m_0 (usually taken as the point at infinity).

The following theorem admits several variants ([CT/San/SD1] p. 43; Skorobogatov [Sk1]; [CT2]). There are useful extensions of the result over $\mathbf{P}_k^n$ rather than $\mathbf{P}_k^1$, see [Sk1].

THEOREM 6. *Let k be a number field, and let $p : X \to \mathbf{P}_k^1$ be a fibration. Assume:*

(i) $\delta \leq 1$;

(ii) the map p has a section over $\bar{k}$.

Then Question 1 has a positive answer. More precisely, let $\mathcal{R} \subset \mathbf{P}^1(k)$ be the set of points $m \in \mathbf{P}^1(k)$ whose fibre X_m is smooth and satisfies $X_m(\mathbb{A}_k) \neq \emptyset$. Let $\mathcal{H} \subset \mathbf{P}^1(k)$ be a Hilbert subset of $\mathbf{P}^1(k)$. Then:

(a) The closure of $\mathcal{R} \cap \mathcal{H}$ in $\mathbf{P}^1(\mathbb{A}_k)$ coincides with $p(X(\mathbb{A}_k))$.

(b) If smooth fibres of p over $\mathcal{H} \subset \mathbf{P}^1(k)$ satisfy the Hasse principle, then X satisfies the Hasse principle: it has a k-point provided $X(\mathbb{A}_k) \neq \emptyset$; more precisely, $p(X(k))$ is then dense in $p(X(\mathbb{A}_k)) \subset \mathbf{P}^1(\mathbb{A}_k)$.

(c) If smooth fibres of p over $\mathcal{H} \subset \mathbf{P}^1(k)$ satisfy weak approximation, then X satisfies weak approximation: $X(k)$ is dense in $X(\mathbb{A}_k)$.

As a matter of fact, it was recently realized that assumption (ii) can be replaced by the weaker assumption (β) (see section 2).The same remark most certainly also holds for a number of results down below, but the (easy) details have not yet been written down. As already mentioned, some such condition is necessary (see [CT/Sk/SD2]).

In the case under consideration here, assumption (β) holds automatically for $M \neq m_0$. We could therefore replace (ii) by the simple assumption: there exists a component $Y_0 \subset X_{m_0}$ of multiplicity one (we do not require that k be algebraically closed in the function field $k(Y_0)$).

This 'fibration technique' was first used in [CT/San/SD1,2] to prove the Hasse principle and weak approximation for certain intersections of two quadrics. It was then used by Salberger and the author to prove the Hasse principle and weak approximation for certain cubic hypersurfaces, and formalized by Skorobogatov [Sk1], who studied weak approximation on certain intersections of three quadrics. The problem in these various papers is to produce fibrations such that the fibres, which are the basic building blocks, satisfy the Hasse principle (and possibly weak approximation). In [CT/San/SD1,2], the main building blocks are non-conical, integral, complete intersections of two quadrics in $\mathbf{P}_k^4$, containing a set of two skew conjugate lines. For such surfaces, the Hasse principle and weak approximation hold ([Ma2], IV.30.3.1). In my joint paper with Salberger, the basic building blocks are cubic surfaces with a set of three conjugate singular points – here the Hasse principle is a result of Skolem (1955). In [Sk1], the building blocks are the intersections of two quadrics for which the Hasse principle and weak approximation had been proved in [CT/San/SD1,2]. As is clear on these examples, when discussing concrete cases, one is soon led to consider 'fibrations' whose generic fibre need not be smooth. This can be obviated in a number of ways, for which we refer the reader to the original papers.

In the theorem above, no mention is made of the Brauer group. As a matter of fact, the assumption $\delta \leq 1$ implies $\mathrm{Br}_{\mathrm{vert}}(X)/\mathrm{Br}(k) = 0$, and the conclusion of the theorem is concerned with the existence of fibres X_m with points everywhere locally (compare Question 1).

In the next theorem, we still have $\delta \leq 1$, hence $\mathrm{Br}_{\mathrm{vert}}(X)/\mathrm{Br}(k) = 0$, but this time the whole Brauer group $\mathrm{Br}(X)$ is taken into account, and the theorem provides a positive answer to Question 2 for some fibrations. The following statement is a slight reformulation of Harari's result.

THEOREM 7 (Harari) [Ha2], [Ha4]. *Let k be a number field, and let $p : X \to \mathbf{P}^1_k$ be a fibration. Let X_{η_s} denote the geometric generic fibre of p over a separable closure of $k(\mathbf{P}^1)$. Assume:*

(i) $\delta \leq 1$;
(ii) p admits a section over $\overline{k}$;
(iii) $\mathrm{Pic}(X_{\eta_s})$ is torsionfree;
(iv) the Brauer group of X_{η_s} is finite;
Let $\mathcal{R}_1 \subset \mathbf{P}^1(k)$ be the set of points $m \in \mathbf{P}^1(k)$ whose fibre X_m is smooth and satisfies $X_m(\mathbb{A}_k)^{\mathrm{Br}} \neq \emptyset$. Let $\mathcal{H} \subset \mathbf{P}^1(k)$ be a Hilbert subset. Then:

(a) $(\mathcal{R}_1 \cap \mathcal{H})^{cl} = p(X(\mathbb{A}_k)^{\mathrm{Br}})$.

(b) If the Brauer–Manin obstruction to the existence of a rational point is the only obstruction for smooth fibres over $\mathcal{H}$, then $p(X(k))$ is dense in $p(X(\mathbb{A}_k)^{\mathrm{Br}})$. In particular the Brauer–Manin obstruction to the existence of a rational point on X is the only obstruction: X has a k-point provided $X(\mathbb{A}_k)^{\mathrm{Br}} \neq \emptyset$.

(c) If the Brauer–Manin obstruction to weak approximation on X_m for $m \in \mathcal{H}$ is the only obstruction, then the Brauer–Manin obstruction to weak approximation on X is the only obstruction: $X(k)$ is dense in $X(\mathbb{A}_k)^{\mathrm{Br}}$.

It is worth noticing that hypotheses (ii), (iii) and (iv) together imply that the group $\mathrm{Br}(X)/\mathrm{Br}(k)$ is finite.

Hypotheses (iii) and (iv) in Theorem 7 hold for instance if the generic fibre is geometrically unirational, or if it is a smooth complete intersection of dimension at least three in projective space.

In specific cases, for purely algebraic reasons, the quotient $\mathrm{Br}(X)/\mathrm{Br}(k)$ vanishes (whereas it need not vanish for the fibres X_m for general $m \in \mathbf{P}^1(k)$). In this case, to assume that there is no Brauer–Manin obstruction to the existence of a rational point on X is simply to assume that there is no local obstruction: $X(\mathbb{A}_k) \neq \emptyset$. One thus gets the corollary (a special case of which had been obtained by D. Kanevsky in 1985):

COROLLARY 8 [Ha2]. *If the Brauer–Manin obstruction to the existence of a rational point (resp. to weak approximation) is the only obstruction for smooth cubic surfaces in $\mathbf{P}^3_k$, then the Hasse principle (resp. weak approximation) holds for smooth cubic hypersurfaces in $\mathbf{P}^n_k$ for $n \geq 4$.*

The assumption made in this corollary is an open question, and the conclusion a well-known conjecture for $n \leq 7$ (using the circle method, Heath-Brown and Hooley have proved such results for $n \geq 8$).

Similarly, one obtains the Corollary (Sansuc and the author, 1986):

COROLLARY 9 (see [Ha2]). *If the Brauer–Manin obstruction to the existence of a rational point is the only obstruction for smooth complete intersections of two quadrics in $\mathbf{P}_k^4$, then the Hasse principle holds for smooth complete intersections of two quadrics in $\mathbf{P}_k^n$ for $n \geq 5$.*

The assumption made in this corollary is an open question (but see section 2.4 below). The conclusion is known to hold for $n \geq 8$ ([CT/San/SD2]). Assume $X(k) \neq \emptyset$. Then weak approximation for a smooth complete intersection X in $\mathbf{P}_k^n$ is completely under control: for $n \geq 5$, weak approximation holds by an easy application of Theorem 6 ([CT/San/SD1]), and for $n = 4$, the Brauer–Manin obstruction to weak approximation is the only obstruction. This last statement is a delicate theorem of Salberger and Skorobogatov [Sal/Sk], which builds upon [Sal1] and [CT/San3].

The proof of Theorem 7 involves a number of novel ideas. One is a systematic use of Hilbert's irreducibility theorem: one looks for points $m \in \mathbf{P}^1(k)$ such that the action of $\mathrm{Gal}(\overline{k}/k)$ on the geometric Picard group of the fibre X_m is controlled by the situation over the generic fibre. Another one is a very useful 'formal lemma' formulated by Harari ([Ha2], 2.6.1), which is a variant, for arbitrary (ramified) classes of the Brauer group of the function field $k(X)$ of X, of the Brauer–Manin condition $X(\mathbb{A}_k) \subset \mathrm{Ker}(\theta_A)$ for unramified classes $A \in \mathrm{Br}(X)$, and which elaborates on the following theorem, a kind of converse to Lemma 1 above:

THEOREM 10 [Ha2]. *Let X/k be a smooth, geometrically irreducible variety over a number field k. Let $U \subset X$ be a non-empty open set of X. If $A \in \mathrm{Br}(U)$ is not the restriction of an element of $\mathrm{Br}(X)$, then there exists infinitely many places $v \in \Omega$ such that the evaluation map $\mathrm{ev}_A : U(k_v) \to \mathrm{Br}(k_v) \subset \mathbb{Q}/\mathbb{Z}$ is not constant.*

2.2 The case where δ is small.

When the number of 'degenerate' fibres is small, a combination of the fibration method just described and of the descent method developed by Sansuc and the author ([CT/San2], [CT/San3]) has led to some general results.

THEOREM 11 [CT4] [Sal2] . *Let k be a number field, and let $p : X \to \mathbf{P}_k^1$ be a fibration. Assume*
 (i) The generic fibre X_η is a smooth conic ;
 (ii) $\delta \leq 4$.
Then the Brauer–Manin obstruction to weak approximation (hence to the existence of a rational point) on X is the only obstruction: $X(k)$ is dense in $X(\mathbb{A}_k)^{\mathrm{Br}}$.

For Châtelet surfaces, given by an affine equation $y^2 - az^2 = P(x)$, with $a \in k^*$ and $P(x) \in k[x]$ a separable polynomial of degree 4, this is a result of [CT/San/SD2]. For the general case, the proof in [CT4] is also based on descent. Salberger's results in [Sal1] (independent of the descent method) led him to an earlier proof (see [Sal2]) of the result on the existence of a rational point. As for the weak approximation statement when existence of a k-point is known, see also [Sal/Sk].

Skorobogatov used the descent technique to study pencils of 2-dimensional quadrics when the number of reducible fibres is small (general results for $\delta \leq 2$, special results for $\delta = 3$). He more recently proved a general result for $\delta = 2$, which may be reformulated as a positive answer to Question 1 in the case under study:

THEOREM 12 (Skorobogatov) [Sk2]. *Let k be a number field and $p : X \to \mathbf{P}^1_k$ a fibration. Assume:*

(i) p admits a section over $\overline{k}$;

(ii) $\delta = 2$;

(iii) if a fibre X_M/k_M over a closed point M is not split, then all its components have multiplicity one.

Let $\mathcal{R} \subset \mathbf{P}^1(k)$ be the set of points $m \in \mathbf{P}^1(k)$ whose fibre X_m is smooth and satisfies $X_m(\mathbb{A}_k) \neq \emptyset$. Let $\mathcal{H} \subset \mathbf{P}^1(k)$ be a Hilbert set. Then:

(a) The inclusion $(\mathcal{R} \cap \mathcal{H})^{cl} \subset p(X(\mathbb{A}_k)^{\text{Br}_\text{vert}})$ is an equality.

(b) If smooth fibres of p over $\mathcal{H}$ satisfy the Hasse principle, then $p(X(k))$ is dense in $p(X(\mathbb{A}_k)^{\text{Br}_\text{vert}})$. In particular, the vertical Brauer–Manin obstruction to the existence of rational points on X is the only obstruction: if $X(\mathbb{A}_k)^{\text{Br}_\text{vert}} \neq \emptyset$, then $X(k) \neq \emptyset$.

(c) If smooth fibres of p over $\mathcal{H}$ also satisfy weak approximation, then $X(k)$ is dense in $X(\mathbb{A}_k)^{\text{Br}_\text{vert}}$, which then coincides with $X(\mathbb{A}_k)^{\text{Br}}$: the vertical Brauer–Manin obstruction to weak approximation on X is the only obstruction.

When the set of closed points with non-split fibre consists of two rational points of $\mathbf{P}^1(k)$, most of the above results can also be obtained by an (unconditional) application of the technique of the next section (see [CT/Sk/SD1]) – except that assumption (iii) above has to be replaced by assumption (i) in Theorem 13. The application is unconditional, because the special case of Schinzel's hypothesis (H) it uses is Dirichlet's theorem on primes in an arithmetic progression.

It would be nice to find a common generalization of Theorem 7 and Theorem 12.

2.3 δ arbitrary (conditional results).

To prove that four-dimensional quadratic forms over a number field k have a non-trivial zero over k as soon as they have one over each k_v for $v \in \Omega$, Hasse used the result for three-dimensional quadratic forms (a consequence of exact sequence (1)) combined with the generalization for the number field k of Dirichlet's theorem on primes in an arithmetic progression. In 1979, Sansuc and the author [CT/San1] noticed that if one is willing to use a bold generalization of Dirichlet's theorem, and of the conjecture on twin primes, namely the conjecture known as Schinzel's hypothesis (H), then Hasse's technique could be pushed further. For instance, this would show that equations of the shape $y^2 - az^2 = P(x)$ with $a \in \mathbb{Q}^*$ and $P(x)$ an arbitrary *irreducible* polynomial in $\mathbb{Q}[x]$ satisfy the Hasse principle. Further developments are due to Swinnerton-Dyer [SD2], Serre (a lecture at Collège de France, see [Se1], p. 125), and Swinnerton-Dyer and the author [CT/SD].

Schinzel's hypothesis (H) claims the following. Let $P_i(t)$, $i = 1, \ldots, n$ be irreducible polynomials in $\mathbb{Z}[t]$, with positive leading coefficients. Assume that the g.c.d. of the $\prod_{i=1}^n P_i(m)$ for $m \in \mathbb{Z}$ is equal to one. Then there exist infinitely many integers $m \in \mathbb{N}$ such that each $P_i(m)$ is a prime number. In his talk, Serre formulated a convenient analogue of Schinzel's hypothesis (H) over an arbitrary number field, and showed that the hypothesis over $\mathbb{Q}$ implies this generalization over any number field (see [CT/SD]).

The most general result is:

THEOREM 13 [CT/Sk/SD1]. *Let k be a number field, and let $p: X \to \mathbf{P}_k^1$ be a fibration. Assume:*

(i) for each closed point $M \in \mathbf{P}_k^1$, there exists a multiplicity one component $Y_M \subset X_M$ such that the algebraic closure of k_M in the function field $k(Y_M)$ is abelian over k_M;

(ii) Schinzel's hypothesis (H) holds over $\mathbb{Q}$.
Let $\mathcal{R} \subset \mathbf{P}^1(k)$ be the set of points $m \in \mathbf{P}^1(k)$ whose fibre X_m is smooth and satisfies $X_m(\mathbb{A}_k) \neq \emptyset$. Then:

(a) The inclusion $\mathcal{R}^{cl} \subset p(X(\mathbb{A}_k)^{\mathrm{Br}_{\mathrm{vert}}})$ is an equality.

(b) If the Hasse principle holds for smooth fibres of p, then $p(X(k))$ is dense in $p(X(\mathbb{A}_k)^{\mathrm{Br}_{\mathrm{vert}}})$. In particular, the vertical Brauer–Manin obstruction to the existence of rational points on X is the only obstruction: if $X(\mathbb{A}_k)^{\mathrm{Br}_{\mathrm{vert}}} \neq \emptyset$, then $X(k) \neq \emptyset$.

(c) If weak approximation holds for smooth fibres of p, then $X(k)$ is dense in $X(\mathbb{A}_k)^{\mathrm{Br}_{\mathrm{vert}}}$, which then coincides with $X(\mathbb{A}_k)^{\mathrm{Br}}$: the vertical Brauer–Manin obstruction to weak approximation on X is the only obstruction.

A Severi–Brauer variety over a field F is an F-variety which becomes isomorphic to a projective space after a finite separable extension of the ground field. These varieties were studied by Severi and by F. Châtelet. One-dimensional Severi–Brauer varieties are just smooth projective conics in $\mathbf{P}_F^2$. Severi and Châtelet proved that a Severi–Brauer variety Y over F is isomorphic to projective space over F as soon as it has an F-point. Châtelet proved that Severi–Brauer varieties over a number field satisfy the Hasse principle. Weak approximation then follows. Degenerate fibres of standard models of pencils of Severi–Brauer variety have multiplicity one, and they split over a cyclic extension of the ground field. We thus have the following corollary (see [CT/SD]), which extends the original result of [CT/San1] and [SD2]:

COROLLARY 14 (Serre). *Let k be a number field, and let $p: X \to \mathbf{P}_k^1$ be a fibration. Assume that the generic fibre X_η is a Severi–Brauer variety. Under Schinzel's hypothesis (H), the vertical Brauer–Manin obstruction to weak approximation on X is the only obstruction. In particular the vertical Brauer–Manin obstruction to the existence of a rational point is the only obstruction to the existence of a rational point.*

For Y/F a Severi–Brauer variety, the map $\mathrm{Br}(F) \to \mathrm{Br}(Y)$ is surjective. Thus in the case discussed in the Corollary, the inclusion $\mathrm{Br}_{\mathrm{vert}}(X) \subset \mathrm{Br}(X)$ is an equality.

We do not see how to dispense with the abelianness requirement in assumption (i) of Theorem 13, and this is quite a nuisance. Indeed, this prevents us from producing a (conditional) extension of Theorem 7 (Harari's result, which takes into account the whole Brauer group of X) to the case where δ is arbitrary. The simplest case which we cannot handle, even under (H), is the following. Let K/k be a biquadratic extension of number fields (i.e. K/k is Galois, and $\mathrm{Gal}(K/k) = (\mathbb{Z}/2)^2$). Let $P(t) \in k[t]$ be a polynomial. Is the Brauer–Manin obstruction to the existence of a rational point the only obstruction for a smooth projective model of the variety given by the equation $N_{K/k}(\Xi) = P(t)$? The abelianness condition also prevents us from stating the analogue of Corollary 14 for fibrations whose generic fibre is an arbitrary projective homogeneous space of a connected linear algebraic group.

Before closing this subsection, let us recall how far we are from unconditional proofs. If we grant Schinzel's hypothesis, then for X as in Corollary 14, the assumption $X(k) \neq \emptyset$ implies that $p(X(k))$ is Zariski-dense in $\mathbf{P}^1$. But already for pencils of conics, there is an unconditional proof of this result only under the assumption $\delta(p) \leq 5$; in special cases, this may be extended to $\delta(p) \leq 8$, as shown by Mestre [Mes]. Upper bounds for the number of points of $p(X(k))$ of a given height have been given by Serre (see [Se1] p. 126).

2.4 Pencils of curves of genus one (Swinnerton-Dyer's recent programme).

Let k be a number field, and let $p : X \to \mathbf{P}^1_k$ be a fibration. Assume $X(\mathbb{A}_k) \neq \emptyset$. Suppose the generic fibre of p is a curve of genus one, and Hypothesis (i) of Theorem 13 is fulfilled. If all the geometric fibres are irreducible, or if one is willing to take Schinzel's hypothesis (H) for granted as in the previous section, and one assumes that there is no Brauer–Manin obstruction to the existence of a rational point on X, the previous techniques lead us to the existence of k-points $m \in \mathbf{P}^1(k)$ such that the fibre X_m/k is smooth, hence is a curve of genus one, and has points everywhere locally. Since the Hasse principle need not hold for curves of genus one (the Tate–Shafarevich group of the Jacobian of X_m is in the way), we cannot conclude that X_m has a k-point. One may however observe that the previous techniques only take into account the 'vertical' part of the Brauer group. One exception is Harari's technique, but until now that technique only applies to pencils of varieties such that the generic fibre $Y = X_\eta$ satisfies $H^1(Y, \mathcal{O}_Y) = 0$ (among other conditions), and the dimension of the vector space $H^1(Y, \mathcal{O}_Y)$ is one for a curve of genus one.

In the course of the study of a special case, taking some difficult conjectures for granted, Swinnerton-Dyer [SD3] managed to overcome the difficulty. His method has very recently been expanded in the joint work [CT/Sk/SD3], and many of the mysterious aspects of [SD3] have now been put in a more general context. In the present report, I will satisfy myself with an informal description of the results in [SD3], and with a very short indication of the progress accomplished in [CT/Sk/SD3].

In [SD3], the ground field k is the rational field $\mathbb{Q}$. Let a_i, b_i, $i = 0, \ldots, 4$ be elements of $\mathbb{Q}^*$ such that $a_i b_j - a_j b_i \neq 0$ for $i \neq j$. Let $S \subset \mathbf{P}^4_\mathbb{Q}$ be the smooth surface defined by the system of homogeneous equations

$$\sum_{i=0}^{4} a_i x_i^2 = 0, \quad \sum_{i=0}^{4} b_i x_i^2 = 0.$$

Setting $x_4 = t x_0$ defines a pencil of curves of genus one on S. A suitable blowing-up transforms the surface S into a surface X equipped with a fibration $p : X \to \mathbf{P}^1_\mathbb{Q}$, the general fibre of which is precisely the hyperplane section of Y given by $x_4 = t x_0$ (for $t \in \mathbf{A}^1(\mathbb{Q})$).

The general fibre of this fibration is thus a curve of genus one. Associated to such a fibration we have a jacobian fibration $q : \mathcal{E} \to \mathbf{P}^1_\mathbb{Q}$. For a closed point $M \in \mathbf{P}^1_\mathbb{Q}$ with smooth fibre $X_M = p^{-1}(M)$, the fibre $\mathcal{E}_M/k_M$ of the jacobian fibration is an elliptic curve, which is the jacobian of X_M. The special shape of the intersection of two quadrics (simultaneously diagonal equations) ensures that the generic fibre (hence the general fibre) of q has its 2-torsion points rational. Thus the 2-torsion subgroup of the generic fibre $\mathcal{E}_{k(\mathbf{P}^1)}$ is $(\mathbb{Z}/2)^2$, with trivial Galois action. For $m \in \mathbf{P}^1(\mathbb{Q})$ with smooth fibre X_m, there is a natural map $X_m \to \mathcal{E}_m$, which makes X_m into a 2-covering of $\mathcal{E}_m$; here we are using the classical language of descent on elliptic curves, as in Cassels' survey [Cass].

Swinnerton-Dyer throws in two hypotheses which we have already encountered:

(i) The hypothesis that the Tate–Shafarevich group of elliptic curves over $k = \mathbb{Q}$ is finite;

(ii) Schinzel's hypothesis (H).

He then makes a further algebraic assumption, referred to as (D), on the $a_i \in \mathbb{Q}^*$. Roughly speaking, these elements are supposed to be in general position (some field extensions of $\mathbb{Q}$ which depend on them are linearly independent). This condition implies the vanishing of $\mathrm{Br}(S)/\mathrm{Br}(\mathbb{Q}) = \mathrm{Br}(X)/\mathrm{Br}(\mathbb{Q})$, but it is not equivalent to the vanishing of this quotient.

Under these three assumptions, assuming that S has points in all completions of $\mathbb{Q}$, he proves that there exists at least one $m \in \mathbf{P}^1(\mathbb{Q})$ such that the fibre X_m is smooth and has points in all completions of $\mathbb{Q}$ (this is shown by the general technique described in section 2.3), and such that moreover the 2-Selmer group $\mathrm{Sel}(2, \mathcal{E}_m)$ of $\mathcal{E}_m$ has order at most 8 (the argument here is extremely elaborate). By the general theory of elliptic curves, the 2-Selmer group fits into an exact sequence

$$0 \to \mathcal{E}_m(\mathbb{Q})/2 \to \mathrm{Sel}(2, \mathcal{E}_m) \to {}_2\mathrm{III}(\mathcal{E}_m) \to 0$$

where ${}_2\mathrm{III}(\mathcal{E}_m)$ denotes the 2-torsion subgroup of the Tate–Shafarevich group of the elliptic curve $\mathcal{E}_m$. Now the order of $\mathcal{E}_m(\mathbb{Q})/2$ is at least 4, since all 2-torsion points are rational. Thus the order of ${}_2\mathrm{III}(\mathcal{E}_m)$ is at most 2. If we assume that the Tate–Shafarevich group $\mathrm{III}(\mathcal{E}_m)$ is finite, then, by a result of Cassels, this finite abelian group is equipped with a non-degenerate alternate bilinear form. In particular the order of its 2-torsion subgroup must be a square. Since this order here is at most 2, it must be 1, and ${}_2\mathrm{III}(\mathcal{E}_m) = 0$. But the 2-covering X_m has points in all completions of $\mathbb{Q}$ and defines an element of ${}_2\mathrm{III}(\mathcal{E}_m)$. This element must therefore be trivial: X_m has a $\mathbb{Q}$-rational point. The argument actually yields infinitely many such m, and it also shows that for such m the rank of $\mathcal{E}_m(\mathbb{Q})$ is one, hence $X_m(\mathbb{Q})$ is infinite. Thus $\mathbb{Q}$-rational points are Zariski-dense on X (in the case under study in [SD3], this follows from general properties of intersections of two quadrics as soon as $X(\mathbb{Q}) \neq \emptyset$, but the argument is general.)

In [CT/Sk/SD3], with input from class field theory and the duality theory of elliptic curves over local fields, we unravel some of the ad hoc computations of [SD3]. This enables us to extend the argument to arbitrary number fields, and to obtain similar conditional results (under Schinzel's hypothesis (H) and the assumption that Tate–Shafarevich groups of elliptic curves over number fields are finite) for many pencils of curves of genus one such that the (generic) jacobian has all its 2-torsion points rational. Among the surfaces controlled by these methods there are in particular some $K3$-surfaces. For such surfaces virtually nothing in the direction of the Hasse principle was known or conjectured until now.

We have a generalized condition (D). The theory of Néron minimal models and computations of Grothendieck enable us to relate condition (D) to the vanishing of the 2-torsion subgroup of $\mathrm{Br}(X)/\mathrm{Br}_{\mathrm{vert}}(X)$; however the two conditions are not exactly equivalent.

3. Zero-cycles of degree one

Let k be a field and X a k-variety. A zero-cycle on X is a finite linear combination with integral coefficients of closed points of X. It is thus an element $\sum n_M M \in \oplus_{M \in X_0} \mathbb{Z}$, where X_0 denotes the set of closed points of X. The residue field k_M of a closed point M is a finite field extension of k. One defines the degree (over k) of a zero-cycle $\sum_{M \in X_0} n_M M$ by the formula

$$\deg(\sum_{M \in X_0} n_M M) = \sum_{M \in X_0} n_M [k_M : k] \in \mathbb{Z}.$$

The following conditions are (trivially) equivalent:
 (i) There exists a zero-cycle of degree one on X.
 (ii) The greatest common divisor of the degrees of finite field extensions K/k such that $X(K) \neq \emptyset$ is equal to one.

Thus the condition that there exists a zero-cycle of degree one is a weakening of the condition that X possesses a k-rational point. For certain varieties (Severi–Brauer varieties, quadrics, curves of genus one and more generally principal homogeneous spaces of commutative algebraic groups, pencils of conics over $\mathbf{P}^1$ with at most 5 reducible geometric fibres), these conditions are equivalent, but in general they are not.

Let X/k be a smooth, projective, irreducible variety over a number field k. Using the corestriction (norm map) on Brauer groups, one may easily define a Brauer–Manin obstruction to the existence of a zero-cycle of degree one on X (see [Sai], [CT/SD]). As explained in section 1, we do not expect the Brauer–Manin obstruction to the existence of a rational point to be the only obstruction for arbitrary (smooth, projective) varieties. However, for zero-cycles of degree one, the following general conjecture still looks reasonable:

CONJECTURE 2. *Let X/k be an arbitrary smooth, projective, irreducible variety over the number field k. If there is no Brauer–Manin obstruction to the existence of a zero-cycle of degree one on X, then there exists a zero-cycle of degree one on X.*

For curves of genus one, this conjecture amounts to the conjecture on rational points, as discussed by Manin (Theorem 3 above). For rational surfaces, the special case where $\mathrm{Br}(X)/\mathrm{Br}(k) = 0$ was conjectured by Sansuc and the author in 1981. For conic bundles over $\mathbf{P}^1_k$ with at most 4 degenerate geometric fibres, it follows from the result on rational points (Theorem 11 above). In 1986, Kato and Saito put forward a very general conjecture (see [Sai]). The statement above itself is raised as a question by Saito ([Sai], §8), and as a conjecture in the survey [CT3], to which I refer for zero-cycles analogues of weak approximation.

Here is some evidence for the conjecture. The first theorem specializes to Theorem 3 when X is a curve of genus one.

THEOREM 15 (S. Saito) [Sai]. *Let X/k be a smooth, projective, geometrically irreducible curve over the number field k. Assume that the Tate–Shafarevich group of the jacobian of X is a finite group. Then the Brauer–Manin obstruction to the existence of a zero-cycle of degree one on X is the only obstruction.*

THEOREM 16 [CT/Sk/SD1]. *Let k be a number field, and let $p: X \to \mathbf{P}_k^1$ be a fibration. Assume that X has a zero-cycle of degree one over each completion k_v. Assume:*

(i) for each closed point $M \in \mathbf{P}_k^1$, there exists a multiplicity one component $Y_M \subset X_M$ such that the algebraic closure of k_M in the function field $k(Y_M)$ is abelian over k_M;

(ii) the Hasse principle for zero-cycles of degree one holds for smooth fibres X_M/k_M (where M is a closed point of $\mathbf{P}_k^1$);

(iii) there is no vertical Brauer–Manin obstruction to the existence of a zero-cycle of degree one on X.

Then there exists a zero-cycle of degree one on X.

The main technique used in the proof of Theorem 16 is due to Salberger [Sal1], who proved the theorem for conic bundles (in [Sal1], the theorem is proved explicitly under the additional assumption $\text{Br}(X)/\text{Br}(k) = 0$, in which case it reduces to a Hasse principle for zero-cycles of degree one). The theorem was then generalized to pencils of Severi–Brauer varieties and 'similar' varieties (including quadrics) in [CT/SD] and independently in [Sal2]. The above statement encompasses these previous results.

The reader will notice the striking analogy between Theorem 13, which is a statement for rational points but is conditional on Schinzel's hypothesis, and Theorem 16, which is a statement on zero-cycles of degree one, and is unconditional. As a matter of fact, the proofs of these two theorems run parallel. With hindsight, it appears that the key arithmetic idea of Salberger in [Sal1] is a substitute for Schinzel's hypothesis. Salberger's trick is explained in detail in [CT/Sk/SD1], to which I refer for a general statement. Let me here explain this simple but powerful idea on a special case, the case of twin primes.

Take $k = \mathbb{Q}$. The conjecture on twin primes predicts that there are infinitely many integers n such that n and $n+2$ are both prime numbers.

PROPOSITION 17. *For any integer $N \geq 2$, there exist infinitely many field extensions $K/\mathbb{Q}$ of degree N for which there exists an integer $\theta \in K$, prime ideals $\mathfrak{p}$ and $\mathfrak{q}$ in the ring of integers O_K of K, and prime ideals $\mathfrak{p}_2$ and $\mathfrak{q}_2$ of O_K above the prime 2, such that we have the prime decompositions*

$$(\theta) = \mathfrak{p}\mathfrak{p}_2, \ (\theta + 2) = \mathfrak{q}\mathfrak{q}_2$$

in O_K.

PROOF. Choose arbitrary prime numbers p and q. Let $R(t) \in \mathbb{Z}[t]$ be a monic polynomial of degree $N - 2$. Let $P(t) = R(t)t(t+2) + qt + p(t+2)$. For R, p and q general enough, this is an irreducible polynomial. Let $K = \mathbb{Q}[t]/P(t)$, and let $\theta \in K$ be the integer which is the class of t. Let $N_{K/\mathbb{Q}}$ denote the norm map from K to $\mathbb{Q}$. We clearly have $N_{K/\mathbb{Q}}(\theta) = \pm 2p$ and $N_{K/\mathbb{Q}}(\theta + 2) = \pm 2q$. □

In other words, in the field K, each of θ and $\theta+2$ is a prime up to multiplication by primes in a fixed bad set. The bad factors $\mathfrak{p}_2$ and $\mathfrak{q}_2$ above 2 are not a serious problem, but for the actual twin prime conjecture, we would want to take $N = 1$!

In the proof, the 2 coming in the inequality $N \geq 2$ is the sum of the degrees of the polynomials t and $t+2$. As for the second 2, the one coming in the result (the bad set), it appears because $t(t+2)$ is not separable modulo 2, and it also appears as a 'small' prime (smaller than the sum of the degrees of t and $t+2$).

In many cases, Salberger's trick replaces Schinzel's hypothesis, if one is satisfied with finding zero-cycles of degree one rather than rational points. Our next hope is that we shall be able to proceed along this way with the results of [CT/Sk/SD3]. In the original case considered by Swinnerton-Dyer [SD3], namely smooth complete intersections of two simultaneously diagonal quadrics in $\mathbf{P}_k^4$, with some algebraic conditions on the coefficients, this would ultimately lead to a proof of the existence of rational points which would 'only' depend on the finiteness assumption of Tate–Shafarevich groups.

Acknowledgements

Along with the organisers of the Tiruchirapalli meeting, I also thank the following institutions, which invited me to lecture on various aspects of the material presented here: Université de Lausanne-Dorigny; University of Southern California, Los Angeles; Université de Caen; Panjab University at Chandigarh; Tata Institute of Fundamental Research, Bombay; Matscience, Madras; Independent University, Moscow; Ohio State University; C.I.R.M., Luminy; Universités Paris 6, Paris 7 et Paris-Sud; C.U.N.Y., New York; Technion Institute, Haïfa.

References

Books and surveys

[Cass] J. W. S. Cassels, *Diophantine equations with special reference to elliptic curves*, J. London Math. Soc. **41** (1966), 193-291.

[CT1] J.-L. Colliot-Thélène, *Arithmétique des variétés rationnelles et problèmes birationnels*, Proc. ICM 1986, vol. I, Amer. Math. Soc., Berkeley, California, 1987, pp. 641-653.

[CT2] _____, *L'arithmétique des variétés rationnelles*, Ann. Fac. Sci. Toulouse (6) **I** (1992), 295-336.

[CT3] _____, *L'arithmétique du groupe de Chow des zéro-cycles*, Journal de théorie des nombres de Bordeaux **7** (1995), 51-73.

[Kol] J. Kollár, *Low degree polynomial equations : arithmetic, geometry and topology*, 2nd European Mathematical Congress, Budapest 1996, to appear.

[Ma1] Yu. I. Manin, *Le groupe de Brauer-Grothendieck en géométrie diophantienne*, Actes Congrès intern. math. Nice, 1970, tome I, Gauthier-Villars, Paris, 1971, pp. 401-411.

[Ma2] _____, *Cubic Forms : algebra, geometry, arithmetic*, 2nd engl. edition, North Holland Mathematical Library, vol. 4, Amsterdam, 1986.

[Ma/Ts] Yu. I . Manin and M. A. Tsfasman, *Rational varieties : algebra, geometry, arithmetic*, Russian math. surveys (=Uspekhi Mat. Nauk) **41** (1986), 51-116.

[Maz1] B. Mazur, *On the passage from local to global in number theory*, Bull. Amer. Math. Soc. **29** (1993), 14-50.

[Pl/R] V. P. Platonov and A. S. Rapinchuk, *Algebraic groups and number theory*, Pure and applied mathematics **139**, Academic Press, London and New York, 1994.

[San1] J.-J. Sansuc, *Principe de Hasse, surfaces cubiques et intersections de deux quadriques*, Journées arithmétiques de Besançon 1985, Astérisque **147-148**, Société mathématique de France, 1987, pp. 183-207.

[SD1] Sir Peter Swinnerton-Dyer, *Diophantine Equations : the geometric approach*, Jber. d. Dt. Math.-Verein. **98** (1996), 146-164.

[V] V. E. Voskresenskiĭ, *Algebraicheskie tory*, Nauka, Moskwa, 1977.

I refer to these books and surveys for literature prior to 1992, except for the following papers which are quoted in the text:

[CT4] J.-L. Colliot-Thélène, *Surfaces rationnelles fibrées en coniques de degré 4*, Séminaire de théorie des nombres de Paris 1988-1989, éd. C. Goldstein, Progress in Math. **91**, Birkhäuser, Boston, Basel, Berlin, 1990, pp. 43-55.

[CT/San1] J.-L. Colliot-Thélène et J.-J. Sansuc, *Sur le principe de Hasse et l'approximation faible, et sur une hypothèse de Schinzel*, Acta Arithmetica **41** (1982), 33-53.

[CT/San2] _____, *La descente sur les variétés rationnelles, I*, Journées de géométrie algébrique d'Angers, 1979, éd. A. Beauville, Sijthoff and Noordhoff, Alphen aan den Rijn, 1980, pp. 223-237.

[CT/San3] _____, *La descente sur les variétés rationnelles, II*, Duke Math. J. **54** (1987), 375-492.

[CT/San/SD1] J.-L. Colliot-Thélène, J.-J. Sansuc and Sir Peter Swinnerton-Dyer, *Intersections of two quadrics and Châtelet surfaces, I*, J. für die reine und ang. Math. (Crelle) **373** (1987), 37-107.

[CT/San/SD2] _____, *Intersections of two quadrics and Châtelet surfaces, II*, J. für die reine und ang. Math. (Crelle) **374** (1987), 72-168.

[Sai] S. Saito, *Some observations on motivic cohomology of arithmetic schemes*, Invent. math. **98** (1989), 371-414.

[Sal1] P. Salberger, *Zero-cycles on rational surfaces over number fields*, Invent. math. **91** (1988), 505-524.

[Sal/Sk] P. Salberger and A. N. Skorobogatov, *Weak approximation for surfaces defined by two quadratic forms*, Duke Math. J. **63** (1991), 517-536.

[San2] J.-J. Sansuc, *Groupe de Brauer et arithmétique des groupes algébriques linéaires sur un corps de nombres*, J. für die reine und ang. Math. (Crelle) **327** (1981), 12-80.

[Sk1] A. N. Skorobogatov, *On the fibration method for proving the Hasse principle and weak approximation*, Séminaire de théorie des nombres de Paris 1988-1989, éd. C. Goldstein, Progress in Math. **91**, Birkhäuser, Boston, Basel, Berlin, 1990, pp. 205-219.

Recent papers and books

[Bo] M. Borovoi, *The Brauer-Manin obstructions for homogeneous spaces with connected or abelian stabilizer*, J. für die reine und angew. Math. **473** (1996), 181-194.

[Bo/Ku] M. Borovoi and B. Kunyavskiĭ, *On the Hasse principle for homogeneous spaces with finite stabilizers*, Ann. Fac. Sci. Toulouse, à paraître.

[CT/SD] J.-L. Colliot-Thélène and Sir Peter Swinnerton-Dyer, *Hasse principle and weak approximation for pencils of Severi-Brauer and similar varieties*, J. für die reine und ang. Math. (Crelle) **453** (1994), 49-112.

[CT/Sk/SD1] J.-L. Colliot-Thélène, A. N. Skorobogatov and Sir Peter Swinnerton-Dyer, *Rational points and zero-cycles on fibred varieties: Schinzel's hypothesis and Salberger's device*, in preparation.

[CT/Sk/SD2] _____, *Double fibres and double covers : paucity of rational points*, Prépublication **96-44**, Université de Paris-Sud, to appear in Acta Arithmetica.

[CT/Sk/SD3] _____, *Hasse principle for pencils of curves of genus one whose Jacobians have rational 2-division points*, in preparation.

[Co/Ma] D. Coray and C. Manoil, *On large Picard groups and the Hasse principle for curves and K3 surfaces*, Acta Arithmetica **LXXVI.2** (1996), 165-189.

[Ha1] D. Harari, *Principe de Hasse et approximation faible sur certaines hypersurfaces*, Ann. Fac. Sci. Toulouse (6) **IV** (1995), 731-762.

[Ha2] _____, *Méthode des fibrations et obstruction de Manin*, Duke Math. J. **75** (1994), 221-260.

[Ha3] _____, *Obstructions de Manin "transcendantes"*, Number Theory 1993-1994, ed. S. David, Cambridge University Press, Cambridge, 1996.

[Ha4] _____, *Flèches de spécialisation en cohomologie étale et applications arithmétiques*, prépublication.

[Maz2] B. Mazur, *The Topology of Rational Points*, J. Experimental Math. **1** (1992), 35-45.

[Maz3] _____, *Speculations about the topology of rational points: an up-date*, Columbia University Number Theory Seminar 1992, Astérisque **228**, Société mathématique de France, 1995, pp. 165-181.

[Mes]   J.-F. Mestre, A series of notes, C. R. Acad. Sc. Paris : **319** (1994), 529-532; **319** (1994), 1147-1149; **322** (1996), 503-505.

[Sal2]   P. Salberger, unpublished manuscript (1993).

[S/W]   P. Sarnak and L. Wang, *Some hypersurfaces in* $\mathbf{P}^4$ *and the Hasse-principle*, C. R. Acad. Sci. Paris **321** (1995), 319-322.

[Se1]   J.-P. Serre, *Cohomologie galoisienne*, 5ème édition, Lecture Notes in Mathematics **5**, Springer-Verlag, Berlin Heidelberg New York, 1994.

[Se2]   ———, *Topics in Galois Theory*, Notes written by H. Darmon, Jones and Bartlett Publishers, Boston, 1992.

[Sk2]   A. N. Skorobogatov, *Descent on fibrations over the projective line*, Amer. J. Math. **118** (1996), 905-923.

[SD2]   Sir Peter Swinnerton-Dyer, *Rational points on pencils of conics and on pencils of quadrics*, J. London Math. Soc. (2) **50** (1994), 231-242.

[SD3]   ———, *Rational points on certain intersections of two quadrics*, Abelian Varieties, Proceedings of a conference held at Egloffstein, ed. Barth, Hulek and Lange, de Gruyter, Berlin New York, 1995, pp. 273-292.

[Wald]   M. Waldschmidt, *Densité des points rationnels sur un groupe algébrique*, J. Experimental Math. **3** (1994), 329-352.

[W]   L. Wang, *Brauer–Manin obstruction to weak approximation on abelian varieties*, Israel Journal of Mathematics **94** (1996), 189-200.

J.-L. Colliot-Thélène
C.N.R.S., U.R.A. D0752,
Mathématiques,
Bâtiment 425,
Université de Paris-Sud,
F-91405 Orsay
France

e–mail : colliot@math.u-psud.fr

Stark-Heegner points over real quadratic fields

H. Darmon

March 3, 1997

Acknowledgements: This research was partially supported by grants from NSERC and FCAR, and was carried out while the author was a member of CICMA. The article itself was written during a stay at the Mathematical Sciences Research Institute in Berkeley. It is a pleasure to thank Massimo Bertolini for the stimulating collaboration which inspired this work.

Abstract: Motivated by the conjectures of "Mazur-Tate-Teitelbaum type" formulated in [BD1] and by the main result of [BD3], we describe a conjectural construction of a global point $P_K \in E(K)$, where E is a (modular) elliptic curve over $\mathbf{Q}$ of prime conductor p, and K is a real quadratic field satisfying suitable conditions. The point P_K is constructed by applying the Tate p-adic uniformization of E to an explicit expression involving geodesic cycles on the modular curve $X_0(p)$. These geodesic cycles are a natural generalization of the modular symbols of Birch and Manin, and interpolate the special values of the Hasse-Weil L-function of E/K twisted by certain abelian characters of K. In the analogy between Heegner points and circular units, the point P_K is analogous to a Stark unit, since it has a purely conjectural definition in terms of special values of L-functions, but no natural "independent" construction of it seems to be known. We call the conjectural point P_K a "Stark-Heegner point" to emphasize this analogy.

The conjectures of section 4 are inspired by the main result of [BD3], in which the real quadratic field is replaced by an imaginary quadratic field. The methods of [BD3], which rely crucially on the theory of complex multiplication and on the Cerednik-Drinfeld theory of p-adic uniformization of Shimura curves, do not seem to extend to the real quadratic situation. One must therefore content oneself with numerical evidence for the conjectures. This evidence is summarized in the last section.

1991 *Mathematics Subject Classification*. Primary 11G05, 11G40.

1 Some motivation: Pell's equation

Mention of the so-called Pell's equation
$$x^2 - Dy^2 = 1 \qquad (D \in \mathbf{Z}, \quad D > 0)$$
can already be found in a 7th century manuscript of the Indian mathematician and astronomer Brahmagupta. (Cf. [We], I.IX.) In spite of its venerable age, Pell's equation has lost none of its fascination, and continues to be a wellspring for the most profound questions in number theory.

Here are three methods for tackling Pell's equation, arranged in increasing order of sophistication and generality:

1. The continued fraction method ([We], [HW])
Appearing in manuscripts of Jayadeva and Bhāskara dating back to the 11th and 12th centuries (and rediscovered independently much later by Fermat), it is one of the great contributions of Indian mathematics and civilization ([We], I.IX).

2. The circular unit method ([Ma], [Was])
Suppose for simplicity that $D \equiv 1 \pmod 4$ is square-free, and let $\chi_D(n) = \left(\frac{n}{D}\right)$ be the quadratic Dirichlet character. The following theorem of Gauss, intimately connected with quadratic reciprocity, is the basis for the circular unit method.

Theorem 1.1 *Every quadratic field is contained in a cyclotomic field generated by roots of unity. More precisely, the quadratic field $\mathbf{Q}(\sqrt{D})$ is contained in $\mathbf{Q}(\zeta_D)$, where ζ_D is a primitive D-th root of unity, and the homomorphism of Galois theory*
$$Gal(\mathbf{Q}(\zeta_D)/\mathbf{Q}) = (\mathbf{Z}/D\mathbf{Z})^\times \longrightarrow Gal(\mathbf{Q}(\sqrt{D})/\mathbf{Q}) = \pm 1$$
is identified with the Dirichlet character χ_D.

The importance of theorem 1.1 (for our discussion) lies in the fact that the cyclotomic field $\mathbf{Q}(\zeta_D)$ is equipped with certain natural units, the so-called *circular units*. These are algebraic integers of the form $(1 - \zeta_D^a)$ if D is not prime, and of the form $\frac{1-\zeta_D^a}{1-\zeta_D}$ if D is prime, with $a \in (\mathbf{Z}/D\mathbf{Z})^\times$. In particular, theorem 1.1 implies that the expression
$$u_D = \prod_{a=1}^{D} (1 - \zeta_D^a)^{\chi_D(a)}$$

is an element of norm 1 in the quadratic field $\mathbf{Q}(\sqrt{D})$, and in fact, in its ring of integers. This unit u_D can be used to write down an explicit solution to Pell's equation in terms of values of trigonometric functions evaluated at rational arguments.

The circular unit method appears less efficient than the continued fraction approach. Its main interest is theoretical and aesthetic (cf. the introduction of [Ma]), and also lies in its *greater generality*. For theorem 1.1 has a natural generalization, the Kronecker-Weber Theorem:

Theorem 1.2 *If K is any abelian extension of the rationals, then it is contained in a cyclotomic field $\mathbf{Q}(\zeta)$ generated by a root of unity ζ.*

Thanks to this theorem, one can construct a subgroup of the unit group of K, when K is any abelian extension of $\mathbf{Q}$, by taking the norms of circular units to K. It turns out that this subgroup is always of finite index. So the circular unit method for solving Pell's equation generalizes to a procedure for finding the unit group of an arbitrary *abelian* extension of the rationals.

3. The L-function method ([Ta], [St])
Let
$$\zeta_K(s) = \sum_{\mathcal{A}} \mathbf{N}(\mathcal{A})^{-s}, \qquad (\operatorname{Re}(s) > 1)$$
where the sum is taken over all the integral ideals of K, be the Dedekind zeta-function of the real quadratic field K. It can be shown that this function has a meromorphic continuation to the entire complex plane, and a functional equation relating its values at s and $1 - s$. The third method is based on the analytic class number formula of Dirichlet, which we state here for the special case of $\zeta_K(s)$.

Theorem 1.3 *The zeta-function $\zeta_K(s)$ has a simple zero at $s = 0$, and*
$$\zeta_K'(0) = 2h \log |u|,$$
where h is the class number of K and u is a fundamental unit in the real quadratic field K.

In particular, the expression
$$e^{\zeta_K'(0)}$$
yields a unit of K, and hence a (non-trivial) solution to Pell's equation.

From a practical and computational point of view, this third method turns out to be not very different from the second. Indeed, theorem 1.1 implies that $\zeta_K(s) = \zeta(s) L(s, \chi_D)$, and one has (cf. [Ta])

$$\zeta'_K(0) = \zeta(0) L'(0, \chi_D) = \log(\prod_{a=1}^{D} (1 - \zeta_D^a)^{\chi_D(a)}), \qquad (1)$$

so that $e^{\zeta'_K(0)} = u_D^2$, where u_D is the unit constructed from circular units following the second method.

What is important here is the change in point of view: the analytic class number formula of theorem 1.3 generalizes to an *arbitrary* number field K. In particular, when $\zeta_K(0) = 0$, the identity

$$\zeta'_K(0) = \log |u|,$$

for *some* unit u of K (not necessarily non-trivial!) continues to hold. When $\zeta_K(s)$ has a simple zero at $s = 0$, this identity gives an analytic construction of a non-trivial unit in K from special values of the Dedekind zeta-function.

Unfortunately, the only number fields for which ζ_K has a simple zero at $s = 0$ are the fields with exactly two infinite places, i.e., the real quadratic fields, the cubic fields with one real and one complex place, and the quartic fields with two complex places. This class does not even include the abelian extensions covered by the second method.

To make the method more flexible, one can enlarge the class of L-functions to include the Artin L-functions associated to irreducible representations of $\mathrm{Gal}(\bar{\mathbf{Q}}/\mathbf{Q})$. More precisely, let $\tilde{K}$ be the Galois closure of K, let G denote its Galois group, and let $H = \mathrm{Gal}(\tilde{K}/K) \subset G$. Finally, let $\rho = \mathrm{Ind}_H^G(\mathbf{1})$ be the induced representation. Then the Dedekind zeta-function $\zeta_K(s)$ is equal to the Artin L-function $L(s, \rho)$, which factorizes as a product of Artin L-series:

$$\zeta_K(s) = L(s, \rho) = \prod_i L(s, \rho_i)^{m_i}.$$

where $\sum_i m_i \rho_i$ is the decomposition of ρ as a direct sum of irreducible representations. Stark has conjectured ([St], see also [Ta]) that the leading terms of each of the factors on the right can be written down explicitly in terms of arithmetic invariants attached to K. In particular, when $L(0, \rho_i) = 0$, the first derivative $L'(0, \rho_i)$ should be expressed as an explicit combination of logarithms of units of K. In this way, one can hope to recover a unit of K from the values of $L'(0, \rho_i)$ when they are non-zero.

When K is abelian over $\mathbf{Q}$, the Artin L-series that appear in the factorization of $\zeta_K(s)$ are attached to one-dimensional characters of K, and are equal to Dirichlet L-series $L(s, \chi)$ by theorem 1.2 (Kronecker Weber). An explicit evaluation (cf. [Ta]) shows that when $L(0, \chi) = 0$, the derivative $L'(0, \chi)$ is expressed in terms of the circular units of the second method, by a formula which directly generalizes equation (1).

In general, if an irreducible representation ρ is not one-dimensional, it cuts out a non-abelian extension $K(\rho)$ of $\mathbf{Q}$, and Stark's conjecture provides a more general framework (albeit one which is *conjectural* in general) for finding units in $K(\rho)$ when $L(s, \rho)$ has a simple zero at $s = 0$. For a recent example where Stark's conjecture is used to compute the units in specific ray class fields of certain totally real cubic fields, see [DST].

2 Elliptic curves over Q

Let $E/\mathbf{Q}$ be an elliptic curve over the rationals of conductor N, given by the projective equation

$$y^2 z + a_1 xyz + a_3 yz^2 = x^3 + a_2 x^2 z + a_4 xz^2 + a_6 z^3. \qquad (2)$$

By the Mordell-Weil theorem, the Mordell-Weil group $E(\mathbf{Q})$ is a finitely generated abelian group,

$$E(\mathbf{Q}) \simeq \mathbf{Z}^r \oplus T,$$

where T is the finite torsion subgroup of $E(\mathbf{Q})$. As has been known for a long time (see for example [Ma]), there is a resonance between Pell's equation and the study of rational points on elliptic curves. In particular, each of the three methods outlined in section 1 for tackling Pell's equation has an analogue in the realm of elliptic curves.

1. The method of descent ([Si1], [Ca])
When Fermat, unaware of the Indian contributions, rediscovered the continued fraction method for solving Pell's equation, he viewed it as a "positive" application of his general method of descent, which he had used until then only to prove that certain Diophantine equations had no solutions. Unlike the continued fraction method, the descent method for computing $E(\mathbf{Q})$ is

not known to give an algorithm in general for finding $E(\mathbf{Q})$, i.e., to always terminate. Such a statement would follow if one knew that the Shafarevich-Tate group $\text{III}(E/\mathbf{Q})$ (or, even just $\text{III}(E/\mathbf{Q}) \otimes \mathbf{Z}_\ell$, for some prime ℓ which can be effectively determined) is finite.

2. The Heegner point method ([El], [Gr1], [Za])

If N is a positive integer, let $X_0(N)$ denote the *modular curve* which is the (coarse) moduli space of elliptic curves equipped with a rational subgroup of order N. This curve admits a model over the rationals.

The Heegner point method relies crucially on Wiles' theorem, formerly known as the Shimura-Taniyama conjecture. We state it here in a strong form which follows from combining the works of [Wi], [TW], and [Di].

Theorem 2.1 *Suppose that E is an elliptic curve over the rationals, having semistable reduction at 3 and 5. Then E is modular, i.e., there is a nonconstant morphism*

$$\phi_E : X_0(N) \longrightarrow E$$

defined over $\mathbf{Q}$.

Remarks:
1. The reader should note the direct analogy between theorem 2.1 and the more classical theorem 1.1. The latter is intimately connect with abelian reciprocity laws, while the former is a manifestation of a (non-abelian) reciprocity law for $\mathbf{GL}_2$.
2. Theorem 2.1 is in fact somewhat weaker than the original conjecture, which is formulated without the technical assumption of semi-stability at 3 and 5. For the rest of this paper, we will assume that E satisfies the conclusion of theorem 2.1.
3. It is customary to normalize ϕ_E so that it sends the cusp $i\infty$ on $X_0(N)$ to the identity on E, and so that the map $\phi_{E*} : J_0(N) \longrightarrow E$ induced on jacobians by covariant functoriality has connected kernel. This can always be achieved, possibly after replacing E by a curve in the same rational isogeny class.

In the same way that the cyclotomic fields $\mathbf{Q}(\zeta_D)$ are equipped with circular units, the modular curves are endowed with an explicit set of algebraic points, the so-called Heegner points. More precisely, let K be an imaginary quadratic field satisfying

$$\text{All primes } \ell | N \text{ are split in } K/\mathbf{Q}. \tag{3}$$

Let $\mathcal{O}_K$ be the ring of integers of K, and let A be an elliptic curve satisfying $\text{End}(A) \simeq \mathcal{O}_K$. (One says that A has *complex multiplication by $\mathcal{O}_K$*.) By the theory of complex multiplication, the curve A can be defined over the Hilbert class field H of K. The technical assumption (3) implies that A has a cyclic subgroup C of order N which is also defined over H. The pair (A, C) gives rise to a point $\alpha_H \in X_0(N)(H)$. Let $P_H = \phi_E(\alpha_H) \in E(H)$, where ϕ_E is the modular parametrization of theorem 2.1. Let $P_K := \text{trace}_{H/K}(P_H)$ be the trace of P_H to $E(K)$.

Unlike the solution to Pell's equation constructed from circular units, the point P_K may be trivial, so that the Heegner point method for finding a rational point in $E(K)$ does not always succeed. But (just like with circular units, cf. equation (1)), the non-triviality of the point P_K can be related to the non-vanishing of certain L-function values. More precisely, let $L(E/K, s)$ be the Hasse-Weil L-function of E over K. Thanks to theorem 2.1, it is known to have an analytic continuation and a functional equation relating its values at s and $2 - s$. (For $L(E/\mathbf{Q}, s)$, this follows from Hecke's theory. For $L(E/K, s)$, it can be proved either by exploiting the factorization $L(E/K, s) = L(E/\mathbf{Q}, s)L(E^{(K)}/\mathbf{Q}, s)$, where $E^{(K)}$ is the twist of E over K, or by using Rankin's method, as in [GZ].)

The sign in this functional equation can be written down explicitly as a product of local signs. Assumption (3) forces the sign in the functional equation of $L(E/K, s)$ to be -1, so that $L(E/K, 1) = 0$ (cf. [GZ], p. 71). Let $\Omega_{E/K} := \int_{E(\mathbf{C})} \omega \wedge i\bar{\omega}/\sqrt{d_K}$, where ω is a Néron differential on $E/\mathbf{Q}$ and d_K is the discriminant of K. The following result of Gross and Zagier [GZ] can be viewed as an elliptic curve analogue of equation (1).

Theorem 2.2 *There is an explicit non-zero rational number $\alpha \in \mathbf{Q}^\times$ such that*
$$L'(E/K, 1) = \alpha \Omega_{E/K} \langle P_K, P_K \rangle,$$
where $\langle \, , \, \rangle$ is the Néron-Tate canonical height on $E(K)$. In particular, the point P_K is of infinite order if and only if $L'(E/K, 1) \neq 0$.

A result of Kolyvagin shows that if P_K is of infinite order, then $E(K)$ has rank one. Conversely, it is expected (and it follows from the Birch and Swinnerton-Dyer conjecture) that the Heegner point method works precisely in this "rank one" situation.

3. The L-function method?

By analogy with the zeta-function method for solving Pell's equation, one might ask for a method of computing a rational point in $E(\mathbf{Q})$ from the special values of the Hasse-Weil L-function $L(E/\mathbf{Q}, s)$. The analogue of Dirichlet's analytic class number formula in this context is the Birch and Swinnerton-Dyer conjecture, which relates the arithmetic behaviour of $E/\mathbf{Q}$ to the analytic properties of $L(E/\mathbf{Q}, s)$ in the neighbourhood of $s = 1$. Recall that r is the rank of the Mordell-Weil group $E(\mathbf{Q})$, and that T is its finite torsion subgroup.

Conjecture 2.3 *The Hasse-Weil L-function $L(E/\mathbf{Q}, s)$ vanishes to order r at $s = 1$, and*

$$L^{(r)}(E/\mathbf{Q}, s) = \#\text{III}(E/\mathbf{Q}) \left(\det \left(\langle P_i, P_j \rangle \right)_{1 \leq i,j \leq r} \right) \#T^{-2} \left(\int_{E(\mathbf{R})} \omega \right) \prod_p m_p,$$

where $\text{III}(E/\mathbf{Q})$ is the (conjecturally finite) Shafarevich-Tate group of $E/\mathbf{Q}$, the points $P_1, \ldots, P_r$ are a basis for $E(\mathbf{Q})$ modulo torsion, $\langle\,,\,\rangle$ is the Néron-Tate canonical height, ω is the Néron differential on E, and m_p is the number of connected components in the Néron model of $E/\mathbf{Q}_p$.

In particular, if $L(E/\mathbf{Q}, s)$ has a simple zero at $s = 1$, then conjecture 2.3 predicts that $E(\mathbf{Q})$ has rank 1 and that

$$L'(E/\mathbf{Q}, 1) = \#T^{-2} \left(\int_{E(\mathbf{R})} \omega \right) \prod_p m_p \langle P, P \rangle,$$

where $P = \sqrt{\#\text{III}(E/\mathbf{Q})} P_0$, and P_0 is a generator for $E(\mathbf{Q})$ modulo torsion. This formula allows one to compute the Néron-Tate canonical height $h(P) = \langle P, P \rangle$ of a point $P \in E(\mathbf{Q})$ from the special value $L'(E/\mathbf{Q}, 1)$.

In [Si2], Silverman explains how the a priori knowledge of $h(P)$ can be used to assist in the calculation of P itself. Silverman's method seems quite efficient computationally – in all likelihood, alot more so than the p-adic methods we are about to describe. Still, there is no simple analytic function, analogous to the exponential in the case of Pell's equation, which would reconstruct the point P directly from $h(P)$. From this point of view the analogy with method 3 of section 1 seems to break down somewhat.

It turns out that the analogy can be pushed in another direction if one replaces the classical L-function by a p-adic avatar. Such a phenomenon was first discovered by Karl Rubin [Ru] for elliptic curves with complex multiplication. More precisely, if E is such a curve, Rubin showed (building on the formula of Gross-Zagier and on Perrin-Riou's p-adic analogue [PR]) that a global point in $E(\mathbf{Q})$ can be obtained by applying the exponential map in the formal group of $E/\mathbf{Q}_p$ to the first derivative of a certain two-variable p-adic L-function of E (which in this case interpolates the special values of a Hecke L-series with Grossencharacter).

More recently, the article [BD3] described a construction of a global point $P_K \in E(K)$ from the first derivative of a p-adic L-function, in the case when E is a (modular) elliptic curve over $\mathbf{Q}$ having a prime p of multiplicative reduction, and K is a quadratic imaginary field in which p is inert. In this formula, the role of the exponential map is played by the Tate uniformization:

$$\Phi_{\text{Tate}} : K_p^\times \longrightarrow E(K_p),$$

(where $K_p = K \otimes \mathbf{Q}_p$ is the completion of K at p). The next section recalls the formula of [BD3].

3 p-adic L-functions and rational points

Assume as before that E is a (modular) elliptic curve of conductor N. Let K be a quadratic imaginary field of discriminant D relatively prime to N. Furthermore, suppose that

1. The curve E has good or multiplicative reduction at all primes which are inert in $K/\mathbf{Q}$.

2. There is at least one prime, p, which is inert in K and for which E has multiplicative reduction.

3. The sign in the functional equation for $L(E/K, s)$ is -1.

Write $N = N^+N^-p$, where N^+, resp. N^- is divisible only by primes which are split, resp. inert in K. Note that by assumptions 1 and 2, N^- is square-free and not divisible by p.

Let H be the Hilbert class field of K, and let H_n the ring class field of conductor p^n. We write $H_\infty = \bigcup H_n$, and set

$$G_n := \text{Gal}(H_n/H), \quad \tilde{G}_n := \text{Gal}(H_n/K),$$

$$G_\infty := \mathrm{Gal}(H_\infty/H), \quad \tilde{G}_\infty := \mathrm{Gal}(H_\infty/K), \quad \Delta := \mathrm{Gal}(H/K).$$

There is an exact sequence of Galois groups

$$0 \longrightarrow G_\infty \longrightarrow \tilde{G}_\infty \longrightarrow \Delta \longrightarrow 0,$$

and, by class field theory, G_∞ is canonically isomorphic to $K_p^\times/\mathbf{Q}_p^\times$, which can be identified with the group $(K_p)_1^\times$ of elements of norm 1 in $K_p^\times$ (by sending z to $\frac{z}{\bar{z}}$). The completed integral group rings $\mathbf{Z}[\![G_\infty]\!]$ and $\mathbf{Z}[\![\tilde{G}_\infty]\!]$ are defined as the inverse limits of the integral group rings $\mathbf{Z}[G_n]$ and $\mathbf{Z}[\tilde{G}_n]$ under the natural projection maps.

Let

$$\Omega_E := \iint_{E(\mathbf{C})} \omega \wedge i\bar{\omega}$$

be the complex period (or Parshin-Faltings height) of E, where ω is a Néron differential on E. Write d for the discriminant of the order $\mathcal{O}$ of conductor c and u for one half the order of the group of units of $\mathcal{O}$.

Theorem 3.1 *There exists an element $\mathcal{L}_p(E/K) \in \mathbf{Z}[\![\tilde{G}_\infty]\!]$ such that*

$$|\chi(\mathcal{L}_p(E/K))|^2 = \frac{L(E/K, \chi, 1)}{\Omega_E} \sqrt{d} \cdot u^2,$$

for all finite order characters χ of $\tilde{G}_\infty$.

Remark. The interpolation property of theorem 3.1 determines $\mathcal{L}_p(E/K)$ uniquely, up to right multiplication by elements in $\tilde{G}_\infty$, if it exists. The existence amounts to a statement of rationality and integrality for the special values $L(E/K, \chi, 1)$. The construction of $\mathcal{L}_p(E/K)$, which is based on work of Gross [Gr2] and Daghigh [Dag], is explained in chapter 2 of [BD3].

If χ is the trivial character (or, more generally, any character of $\tilde{G}_\infty$ which is unramified at p, i.e., factors through Δ) then the interpolation property of theorem 3.1 implies that

$$\chi(\mathcal{L}_p(E/K)) = 0. \tag{4}$$

In particular, $\mathcal{L}_p(E/K)$ belongs to the augmentation ideal $\tilde{I}$ of $\mathbf{Z}[\![\tilde{G}_\infty]\!]$. Let $\mathcal{L}'_p(E/K)$ denote the image of $\mathcal{L}_p(E/K)$ in $\tilde{I}/\tilde{I}^2 = \tilde{G}_\infty$. The reader should view $\mathcal{L}'_p(E/K) \in \tilde{G}_\infty$ as the first derivative of $\mathcal{L}_p(E/K)$ evaluated at the central point.

Lemma 3.2 *The element $\mathcal{L}'_p(E/K)$ belongs to $G_\infty \subset \tilde{G}_\infty$.*

Proof: Formula (4) implies that $\mathcal{L}_p(E/K)$ belongs to the kernel of the natural projection $\mathbf{Z}[\![\tilde{G}_\infty]\!] \longrightarrow \mathbf{Z}[\Delta]$. This implies that $\mathcal{L}'_p(E/K)$ belongs to the kernel of the map $\tilde{G}_\infty \longrightarrow \Delta$.

Thanks to lemma 3.2, the element $\mathcal{L}'_p(E/K)$ can (and will) be viewed as an element of $K_p^\times$ of norm 1.

The main formula of [BD3] is:

Theorem 3.3 *The local point $\Phi_{\text{Tate}}(\mathcal{L}'_p(E/K)) \in E(K_p)$ is a global point in $E(K)$.*

Crucial to the proof of theorem 3.3 is the fact that the global point P_K constructed from special values of L-functions has an *alternate construction*. This construction relies on two basic ingredients: the theory of complex multiplication, and the Cerednik-Drinfeld theory of p-adic uniformization of Shimura curves associated to indefinite quaternion algebras.

Before making this more precise, we record the following lemma:

Lemma 3.4 *The integer N^- is the product of an odd number of primes.*

Proof: By page 71 of [GZ], the sign in the functional equation of the complex L-function $L(E/K, s)$ is $(-1)^{\#\{\ell | N^- p\}+1}$. The result follows.

Let B be the indefinite quaternion algebra which is ramified exactly at the primes dividing pN^-. Such a B exists, by Hilbert's reciprocity law and lemma 3.4. Choose a maximal order R in B, and an Eichler order $R(N^+) \subset R$ of level N^+, defined as in [BD3]. Let Γ be the subgroup of $R(N^+)^\times$ of elements of reduced norm 1. By fixing an embedding of $B \otimes \mathbf{R}$ into $M_2(\mathbf{R})$, the group Γ acts on the standard complex upper half plane by Mobius transformations. The complex-analytic quotient $X = \mathcal{H}/\Gamma$ is a complex model for a curve X. Shimura showed that X has a model over $\mathbf{Q}$ by identifying it with a (coarse) moduli space for polarized abelian surfaces with endomorphisms by R and an appropriate level N^+ structure. For details, see [Ro] or [BD1] for example.

The curve X is endowed with a set of *Heegner points* corresponding to quaternionic surfaces A with complex multiplication by the maximal order $\mathcal{O}_K$ of K. (By complex multiplication by $\mathcal{O}_K$, one means that the ring of endomorphisms of A which commute with the quaternionic multiplications and respect the level N^+-structure is isomorphic to $\mathcal{O}_K$.) These points are defined over the Hilbert class field H of K, and are permuted transitively by the group $\text{Gal}(H/K) \times W$, where W is the group of exponent 2 generated by all the Atkin-Lehner involutions on X (cf. [BD1]). Let $\alpha_1, \ldots, \alpha_h$ ($h = [H:K]$) be a $\text{Gal}(H/K)$-orbit of Heegner points, and let $\alpha'_j = w_{N/p}\alpha_j$, where $w_{N/p}$ is the product of the Atkin-Lehner involutions w_ℓ over all primes $\ell | N/p$. Note that $\alpha'_1, \ldots, \alpha'_h$ is another $\text{Gal}(H/K)$-orbit of Heegner points, so that the effective divisor $(\alpha'_1) + \cdots + (\alpha'_h)$ is K-rational.

The Jacobian J of X is an abelian variety over $\mathbf{Q}$. By a theorem of Jacquet-Langlands [JL], it is isogenous to the quotient of $J_0(N^+N^-p)$ corresponding to cusp forms which are new at N^-p. Hence, the modularity of E (theorem 2.1) implies the existence of a generically surjective map

$$\phi_E : J \longrightarrow E.$$

The degree 0 divisor on X:

$$D = (\alpha_1) + \cdots + (\alpha_h) - (\alpha'_1) - \cdots - (\alpha'_h) \qquad (5)$$

is defined over K, and hence gives rise to a canonical "Heegner element" in $J(K)$, which depends on the choice of the α_i only up to the action of the Atkin-Lehner involutions. Let

$$P_K = \phi_E(D) \in E(K)$$

be its image in $E(K)$. Note that P_K depends only up to sign on the choice of the K-rational effective divisor $(\alpha_1) + \cdots + (\alpha_h)$.

The more precise form of theorem 3.3 proved in [BD3] states that the local point $\Phi_{\text{Tate}}(\mathcal{L}'_p(E/K))$ is equal to the global point P_K, up to a sign and a simple fudge factor. The proof exploits a p-adic analytic construction of P_K supplied by the Cerednik-Drinfeld theory of p-adic uniformization of X. For more details, see [BD3].

Again, the formula relating P_K to $\mathcal{L}'_p(E/K)$ is analogous to formula (1) expressing circular units in terms of derivatives of abelian L-series. Circular units (and, in our situation, Heegner points on Shimura curves) lead to examples of what Kolyvagin has called an "Euler system". When they can be constructed, such Euler systems provide powerful insights into the associated L-function values. However, there are many instances where no Euler system is known to exist. The units defined conjecturally from derivatives of non-abelian Artin L-functions – whose construction would supply a key to the Stark conjectures – are a case in point. Motivated by the discussion in section 1, one might ask whether the (p-adic) L-function methods of the present section suggest (conjectural) constructions of global points on elliptic curves, in situations where there is no (known) Euler system construction. Or, put more succintly: "Are there Stark-Heegner points"?

The remainder of this article describes a fragment of experimental mathematics suggesting that the answer to this question is "yes".

4 A real quadratic analogue

We will restrict ourselves for simplicity to the case where the elliptic curve E has prime conductor $N = p$. This simplifying assumption allows us, in particular, to avoid Shimura curves and formulate our construction entirely within the setting of classical modular curves – a luxury which was not available to us in section 3.

Let K be a real quadratic field in which p is inert. (This corresponds to assumptions 1 and 2 of section 3.) Let $L(E/K, s)$ be the Hasse-Weil L-function for E/K.

Lemma 4.1 *If $N = p$ is inert in K, then the sign in the functional equation for $L(E/K, s)$ is -1.*

In particular, $L(E/K, 1) = 0$, and the conjecture of Birch and Swinnerton-Dyer leads us to expect that $E(K)$ is infinite. We will now describe a conjectural method for constructing a global point $P_K \in E(K)$ (or rather, a p-adic approximation of it) using modular symbols and Tate's p-adic analytic theory. We caution the reader that, just as with all the methods covered previously, we expect that this method yields a non-trivial global point precisely when $E(K)$ has rank 1.

4.1 Modular symbols

Let $\mathcal{O}_K = \mathbf{Z}[\omega]$ be the ring of integers of K. An order in K is a subring of K which is finitely generated as a $\mathbf{Z}$-module. Every order in K is contained in $\mathcal{O}_K$, and is of the form $\mathbf{Z}[c\omega]$, for a (unique) positive integer c, called the *conductor* of the order.

A *lattice* in K is a **Z**-submodule of K which is free of rank 2. If I is any lattice, the set of elements
$$\text{End}(I) := \{x \in K | xI \subset I\}$$
is an order in $\mathcal{O}_K$, which we also call the *order associated to I*. By abuse of notation, we will sometimes say that a lattice has conductor c if its associated order is of conductor c.

Let c be an integer which is prime to p, and let I be a lattice of conductor cp^n with $n \geq 1$.

Lemma 4.2 *There is a unique sublattice $I_0 \subset I$ contained in I with index p having conductor cp^{n-1}.*

Proof: Let $\mathbf{P}_1(I)$ be the set of all index p sublattices of I. It is in bijection with $\mathbf{P}_1(\mathbf{F}_p)$ and hence has cardinality $p+1$. If $\mathcal{O} = \text{End}(I)$ is the order associated to I, then we have (non-canonical) isomorphisms of groups:
$$(\mathcal{O}/p\mathcal{O})^\times/\mathbf{F}_p^\times \simeq (\mathbf{F}_p[\epsilon]/(\epsilon^2))^\times/\mathbf{F}_p^\times \simeq \mathbf{Z}/p\mathbf{Z}.$$

The first isomorphism sends $a + bcp^n\omega$ ($a, b \in \mathbf{Z}$) to $a + b\epsilon$, and the second isomorphism sends $a + b\epsilon \in \mathbf{F}_p[\epsilon]^\times$ to b/a. The group $(\mathcal{O}/p\mathcal{O})^\times/\mathbf{F}_p^\times$ acts on $\mathbf{P}_1(I)$ in the natural way, and has exactly one fixed point. This fixed point corresponds to the sublattice I_0. The p remaining sublattices have conductor cp^{n+1}.

Choose a real embedding of K, and say that an element of K is positive (resp. negative) if its image by this embedding is positive (resp. negative).

Definition 4.3 *A basis ω_1, ω_2 for $K/\mathbf{Q}$ is called* positive *if*
$$\det\begin{pmatrix} \omega_1 & \omega_2 \\ \bar{\omega}_1 & \bar{\omega}_2 \end{pmatrix} > 0.$$

Now, choose a **Z**-basis (ω_1, ω_2) for I satisfying

1. ω_1 belongs to I_0.

2. The basis (ω_1, ω_2) is positive.

Note that I_0 is the set of elements in I of the form $a\omega_1 + b\omega_2$ with $p|b$. Let now u be any unit in $\mathcal{O}^\times$ of norm 1. Multiplication by u gives an endomorphism of I. Since the sublattice I_0 is stable under this endomorphism (indeed, it is stable under multiplication by the order of conductor cp^{n-1}) it follows that the matrix describing the multiplication by u in the basis (ω_1, ω_2) belongs to $\Gamma_0(p)$. Let $m_u(I)$ denote this matrix.

Lemma 4.4 *The matrix $m_u(I)$ is well-defined up to conjugation in $\Gamma_0(p)$.*

Proof: Any two choices of bases for I satisfying the conditions 1 and 2 above differ by an element of $\Gamma_0(p)$.

Two lattices I and J are said to be *equivalent* if there exists an element $\alpha \in K^\times$ of positive norm such that

$$J = \alpha I.$$

Of course, equivalent lattices have the same associated order. Furthermore:

Lemma 4.5 *If I and J are equivalent, then the matrices $m_u(I)$ and $m_u(J)$ are conjugate in $\Gamma_0(p)$.*

Proof: If (ω_1, ω_2) is a basis for I satisfying conditions 1 and 2 above, and $J = \alpha I$, then $(\alpha\omega_1, \alpha\omega_2)$ is a basis for J satisfying the same conditions. Relative to these bases, the matrices expressing the multiplication by u are in fact equal. The lemma follows.

An element of $\Gamma = \Gamma_0(p)$ (well-defined up to conjugation) gives rise in the usual way to a class in the integral homology $H_1(X_0(p), \mathbf{Z})$ which is a quotient (by the torsion and parabolic elements) of the commutator factor group $\Gamma/[\Gamma, \Gamma]$. Let $\mathrm{Pic}(\mathcal{O})$ be the set of equivalence classes of lattices of conductor cp^n. The map m_u sets up an assignment, which we denote by γ_u to emphasize the dependence on u:

$$\gamma_u : \mathrm{Pic}(\mathcal{O}) \longrightarrow H_1(X_0(p), \mathbf{Z}).$$

Definition 4.6 *We call $\gamma_u(I)$ the modular symbol attached to the lattice I of conductor cp^n ($n \geq 1$) and to the unit $u \in \mathcal{O}^\times$.*

Remarks:
1. To an equivalence class of lattices I of conductor cp^n one can associate the primitive binary quadratic form of discriminant $\mathrm{Disc}(K)c^2p^{2n}$

$$F(x, y) = \mathrm{norm}(x\omega_1 + y\omega_2)g^{-1} = Ax^2 + Bxy + Cy^2,$$

where ω_1, ω_2 is a basis for I chosen as above and g is the unique rational number such that $A, B, C \in \mathbf{Z}$ and $\gcd(A, B, C) = 1$. Note that $p^2|A$ and $p|B$. The roots of the polynomial $F(X, 1)$ are two elements of $K \subset \mathbf{R}$ which are Galois conjugate. Consider the geodesic in the Poincaré upper half plane which joins these two roots on the real line. This geodesic maps to an infinitely repeating periodic path on $X_0(p)$. If u is a fundamental unit of norm 1 in $\mathcal{O}^\times$, then the element $\gamma_u(I)$ is the basic period in this cycle, viewed as a homology class of $X_0(p)$. For this reason, it is sometimes called the *geodesic cycle* on $X_0(p)$ associated to the binary quadratic form $F(x, y)$.

2. Like the modular symbols of Birch and Manin, the geodesic cycles on $X_0(p)$ encode special values of L-functions. More precisely, they interpolate the special values of $L(E/K, s)$ at $s = 1$, twisted by ring class characters of K of conductor dividing cp^n. They can be used, just as in [MT], to construct "theta-elements" which are adèlic analogues "at finite level" of the more familiar p-adic L-functions. This point of view, which forms the basis for the present article, is developped in [Dar].

3. In the same way that the modular symbols of Birch and Manin are calculated efficiently by computing the continued fraction expansion of certain rational numbers, the geodesic cycles attached to a binary quadratic form can be calculated from the (periodic) continued fraction expansion of certain real quadratic irrationalities.

4.2 The tree associated to a lattice

In this subsection, let I be a sublattice of conductor c prime to p, and let $\mathcal{O} = \text{End}(I)$ be its associated order. Let $\mathcal{T}(I)$ be the graph whose vertices correspond to homothety classes of sublattices of I which are contained in I with index p^n for some n. The edges of $\mathcal{T}(I)$ join vertices which correspond to lattices which are contained one inside the other with index p.

The graph $\mathcal{T}(I)$ is a homogenous tree of weight $p + 1$, equipped with a distinguished vertex v_0 which corresponds to the homothety class of I. One defines a distance function on $\mathcal{T}(I)$ in the natural way. If v is a vertex of $\mathcal{T}(I)$ corresponding to a lattice I_v, then the order $\text{End}(I_v)$ has conductor cp^m, where m is the distance from v_0 to v. The vertex v is then said to be of *level* m.

Let
$$G_m = (\mathcal{O}_K \otimes \mathbf{Z}/p^m\mathbf{Z})^\times / (\mathbf{Z}/p^m\mathbf{Z})^\times = (\mathcal{O}_K \otimes \mathbf{Z}/p^m\mathbf{Z})_1^\times,$$
where the isomorphism between these two descriptions sends z to $\frac{z}{\bar{z}}$. The group
$$G_\infty = K_p^\times / \mathbf{Q}_p^\times = (\mathcal{O}_K \otimes \mathbf{Z}_p)^\times / \mathbf{Z}_p^\times \simeq K_{p,1}^\times = \varprojlim G_n$$
acts naturally on $\mathcal{T}(I)$, leaving v_0 fixed and permuting transitively the vertices of a given level m. The isotropy group of a vertex of level m is the group of elements which are congruent to a scalar modulo p^m, and hence, G_m acts simply transitively on the set of vertices of level m.

Let
$$u = a + b\omega \in \mathcal{O}^\times$$
be a unit of norm 1 in $\mathcal{O}^\times$, and let n be the largest integer such that $p^n | b$. The modular symbol γ_u defined in the previous section gives rise to a function (which we denote also by γ_u by abuse of notation) on the set of all vertices of $\mathcal{T}(I)$ satisfying
$$0 < \text{level}(v) \leq n.$$
Extending the domain of definition of γ_u slightly, we define $\gamma_u(v_0) := 0$.

Let v be any vertex of level $m < n$, and let $v_1, \ldots, v_p$ be the p vertices of level $m+1$ which are adjacent to it. Recall the Atkin-Lehner involution w_p which acts on $X_0(p)$ and hence on the homology $H_1(X_0(p), \mathbf{Z})$.

Lemma 4.7 *The function γ_u satisfies the relation*
$$\gamma_u(v_1) + \cdots + \gamma_u(v_p) = -w_p \gamma_u(v).$$

Proof: The homology class
$$\gamma_u(v_1) + \cdots + \gamma_u(v_p) + w_p \gamma_u(v)$$
is in the image of the map $H_1(X_0(1), \mathbf{Z}) \longrightarrow H_1(X_0(p), \mathbf{Z})$ induced by the natural degeneracy maps $X_0(p) \longrightarrow X_0(1)$. Since $H_1(X_0(1), \mathbf{Z}) = 0$, the lemma follows.

4.3 The element $\mathcal{L}'_p(I)$

In this section, let I be again a (fixed) lattice of conductor prime to p, and $\mathcal{O}$ its associated order. Let

$$u = a + b\omega \in \mathcal{O}^\times$$

be the fundamental unit of norm 1 in $\mathcal{O}^\times$, and let n be the largest integer such that $p^n | b$.

Let $v_0, v_1, \ldots, v_n$ be a sequence of vertices of level $0, 1, \ldots, n$ such that v_m is adjacent to v_{m+1}, and let

$$\mathcal{L}_{p,m}(I) := \sum_{\sigma \in G_m} (-w_p)^m \gamma_u(\sigma v_m) \cdot \sigma^{-1} \in H_1(X_0(p), \mathbf{Z}) \otimes \mathbf{Z}[G_m].$$

This element is analogous to the theta-elements introduced in [MT]. It is closely related to the special values of the partial L-function

$$L(f, I, s) = \sum_n a_n(f) r_I(n) n^{-s},$$

at $s = 1$ twisted by ring class characters of conductor p^n. Here, f is a cusp form of weight 2 on $\Gamma_0(p)$, $a_n(f)$ is its n-th Fourier coefficient, and $r_I(n)$ is the number of lattices of norm n which are equivalent to I. For more details, see for example [GKZ] or [Ko].

It follows from lemma 4.7 that the elements $\mathcal{L}_{p,m}(I)$ are compatible under the natural projection maps $\mathbf{Z}[G_m] \longrightarrow \mathbf{Z}[G_{m-1}]$, and that $\mathcal{L}_{p,m}(I)$ belongs to $H_1(X_0(p), \mathbf{Z}) \otimes I_m$, where I_m is the augmentation ideal of $\mathbf{Z}[G_m]$. Let $\mathcal{L}'_{p,m}(I)$ be the image of $\mathcal{L}_{p,m}(I)$ by the natural projection to $H_1(X_0(p), \mathbf{Z}) \otimes (I_m/I_m^2)$. After identifying I_m/I_m^2 with G_m in the usual way, we have:

$$\mathcal{L}'_{p,m}(I) := \sum_{\sigma \in G_m} (-w_p)^m \gamma_u(\sigma v_m) \cdot \sigma^{-1} \in H_1(X_0(p), \mathbf{Z}) \otimes G_m.$$

Since $\mathcal{L}_{p,m}(I)$ depends on the choice of the vertex v_m of level m only up to right multiplication by an element of G_m, and since the induced action of G_m on I_m/I_m^2 is trivial, the elements $\mathcal{L}'_{p,m}(I)$ ($m \le n$) do not depend on the choice of the vertex v_m, and they are compatible under the natural projection maps from G_m to G_{m-1}.

To lighten notations, set

$$\mathcal{L}'_p(I) := \mathcal{L}'_{p,n}(I).$$

It is a canonical element in $H_1(X_0(p), \mathbf{Z}) \otimes G_n$ associated to I, well-defined up to a sign and the action of the Atkin-Lehner involution w_p.

Let $H_1(X_0(p), \mathbf{Z})^+ \subset H_1(X_0(p), \mathbf{Z})$ be the subgroup of the homology which is fixed under the action of complex conjugation.

Lemma 4.8 *The element $\mathcal{L}'_p(I)$ belongs to $H_1(X_0(p), \mathbf{Z})^+ \otimes G_n$.*

The proof of this lemma, which we leave to the reader, follows by comparing the action of the Atkin-Lehner involution w_p on the modular symbols with the action of complex conjugation on $H_1(X_0(N), \mathbf{Z})$.

4.4 Local points

Let
$$\mathcal{L}'_p(c) := \sum_I \mathcal{L}'_p(I),$$
where the sum is taken over $\text{Pic}(\mathcal{O})$, the set of equivalence classes of lattices of conductor c.

The basic intuition is that the element $\mathcal{L}'_p(c)$ should encode the position of a special point in $J_0(p)(K)$, analogous to the Heegner divisor of equation (5) except that the role of the imaginary quadratic field is now played by a real quadratic field.

To make this precise, fix now an elliptic curve E of conductor p. Let f be the modular form on $X_0(p)$ which is associated to E by Wiles' theorem, and let w be the sign of the Atkin-Lehner involution w_p acting on f. Let $E^w(K)$ (resp. $E^w(K_p)$) be the subgroup of $E(K)$ (resp. $E(K_p)$) on which complex conjugation acts like w_p. We note the following two properties of $E^w(K)$ and $E^w(K_p)$ (the first global, and the second local):

1. The sign in the functional equation for $L(E/\mathbf{Q}, s)$ is $-w$. Hence, it follows from the Birch and Swinnerton-Dyer conjecture that $E^w(K)$ has odd rank and that $E^{-w}(K)$ has even rank.

2. The curve E has split (resp. non-split) multiplicative reduction at p if and only if $w = -1$ (resp. $w = 1$). In particular, the group $E^w(K_p)$ is contained in the group $E_{ns}(K_p)$ of points having non-singular reduction, and the Tate uniformization Φ_{Tate} induces an isomorphism

$$\Phi_{\text{Tate}} : (K_p^\times)_1 \longrightarrow E^w(K_p).$$

The compact group $(K_p^\times)_1$ is equipped with a canonical filtration

$$(K_p^\times)_1 \supset (K_p^\times)_1^{(1)} \supset (K_p^\times)_1^{(2)} \supset \cdots$$

with the property that $G_n = (K_p^\times)_1/(K_p^\times)_1^{(n)}$. Likewise, the group $E^w(K_p)$ is equipped with the canonical p-adic filtration defined in [Si2]

$$E^w(K_p) \supset E^w(K_p)^{(1)} \supset E^w(K_p)^{(2)} \supset \cdots$$

and the isomorphism between $(K_p^\times)_1$ and $E^w(K_p)$ given by the Tate uniformization respects these filtrations. In particular, by passing to the quotient one has isomorphisms

$$\Phi_{\text{Tate},n} : G_n \longrightarrow E^w(K_p)/E^w(K_p)^{(n)}.$$

Now, let
$$\phi_E : X_0(p) \longrightarrow E$$
be the modular parametrization of theorem 2.1. It induces a surjection on the real homology:

$$\phi_{E*} : H_1(X_0(p), \mathbf{Z})^+ \longrightarrow H_1(E, \mathbf{Z})^+ \simeq \mathbf{Z},$$

and a corresponding map $H_1(X_0(p), \mathbf{Z})^+ \otimes G_n \longrightarrow G_n$, also denoted ϕ_{E*} by abuse of notation.

By lemma 4.8, the element $\mathcal{L}_p'(c)$ belongs to $H_1(X_0(p), \mathbf{Z})^+ \otimes G_n$. Let

$$\mathcal{L}_p'(c, E) := \phi_{E*}(\mathcal{L}_p'(c)) \in G_n.$$

Now, define the local point

$$P_K(c) := \Phi_{\text{Tate},n}(\mathcal{L}_p'(c, E)) \in E^w(K_p)/E^w(K_p)^{(n)}.$$

This point is well-defined as a function of E, K, and c, up to an ambiguity of sign, and can be viewed as an approximation to a point in $E(K_p)$, with a p-adic accuracy of p^{-n}. Let t denote the order of the torsion subgroup of $E(K)$. Suppose that c is square-free and relatively prime to $p\text{Disc}(K)$; let c^+ (resp. c^-) denote the product of the primes which are split (resp. inert) in $K/\mathbf{Q}$.

Conjecture 4.9 *The point $P_K(c)$ is trivial if the rank of $E(K)$ is greater than 1. Otherwise, there is a global point $P \in E(K)$ (not deepending on c) such that:*

$$t \cdot P_K(c) \equiv \pm \prod_{\ell \mid c^+}(a_\ell - 2) \prod_{\ell \mid c^-} a_\ell \cdot \sqrt{\#\text{III}(E/K)} \cdot (P + w\bar{P}) \quad (\text{mod } E^w(K_p)^{(n)}),$$

where $\bar{P}$ is the complex conjugate point.

Furthermore, if the modular parametrization used to define $P_K(c)$ is a strong parametrization, then P is a generator for $E(K)$ modulo torsion.

Frequently, t is relatively prime to $(p+1)p$. In this case conjecture 4.9 can be written

$$P_K(c) \equiv \pm \prod_{\ell \mid c^+}(a_\ell - 2) \prod_{\ell \mid c^-} a_\ell \cdot \sqrt{\#\text{III}(E/K)} t^{-1} \cdot (P + w\bar{P}) \quad (\text{mod } E^w(K_p)^{(n)}),$$

where the inverse of t is taken modulo $(p+1)p^{n-1}$.

Remarks:

1. Conjecture 4.9 predicts that there is a global point $P_K \in E(K)$ such that

$$t \cdot P_K(c) = \pm \prod_{\ell \mid c^+}(a_\ell - 2) \prod_{\ell \mid c^-} a_\ell \cdot P_K \quad (\text{mod } E^w(K_p)^{(n)}), \qquad (6)$$

for all positive integers c which are relatively prime to p. Given any $n \geq 0$, it is possible to find c so that the fundamental unit of norm 1 in the order of conductor c also belongs to the order of conductor p^n. Hence equation (6) defines the point P_K uniquely up to sign, and gives a p-adic recipe for computing it. We call P_K the *Stark-Heegner point* associated to E over the real quadratic field K.

2. The appearance of the factor

$$\prod_{\ell \mid c^+}(a_\ell - 2) \prod_{\ell \mid c^-} a_\ell$$

may seem unnatural. This factor plays the role of a product of Euler factors at the primes ℓ dividing c. By replacing the element $\mathcal{L}_p(c)$ with the "regularized element" $\sum_{d \mid c} \epsilon(d)\mu(c/d)\mathcal{L}_p(c/d)$, where ϵ is the Dirichlet character associated to K, and μ is the Mobius function, one would obtain a slightly different construction of a point $P_K(c)$, involving the more natural factor

$$\prod_{\ell | c^+} (\ell + 1 - a_\ell) \prod_{\ell | c^-} (\ell + 1 + a_\ell)$$

in the conjecture.

3. *A caveat*: The precise form of conjecture 4.9 was suggested by the analogy with the formula in [BD3], as well as by the numerical experiments of section 5. Note that some of the fudge factors one might expect to find in a Birch and Swinnerton-Dyer type formula, such as the order m_p of the group of connected components of the Néron model of $E/\mathbf{Q}_p$, do not appear in our formula. It would be of interest to formulate a precise conjecture along the lines of conjecture 4.9 for curves of arbitrary conductor, where we expect some of the integers m_ℓ, with $\ell \neq p$, to appear. We felt that our numerical evidence was too scant, and our conceptual understanding too incomplete, to make confident predictions about the precise fudge factors which would appear in general.

5 Experimental evidence

5.1 Experiments with $X_0(11)$

Let E be the elliptic curve $X_0(11)$ with minimal Weierstrass equation

$$y^2 + y = x^3 - x^2 - 10x - 20.$$

Its Mordell-Weil group over $\mathbf{Q}$ is finite, of order $t = 5$. The Atkin-Lehner involution w_{11} acts on E by -1, and hence $w = -1$.

Let $K = \mathbf{Q}(\sqrt{2})$. The prime 11 is inert in this real quadratic field, and the maximal order $\mathcal{O}_K = \mathbf{Z}[\sqrt{2}]$ has class number 1 and fundamental unit equal to $1 + \sqrt{2}$.

The Mordell-Weil group of E over K is of rank 1, and is generated (modulo torsion) by the point in $E^w(K)$

$$P = (9/2, \frac{-2 + 7\sqrt{2}}{4}).$$

Let
$$\Phi_{\text{Tate}} : K_{11}^\times \longrightarrow E(K_{11})$$
be the Tate 11-adic uniformization, and let $\tilde{P}$ be a lift of P to the group of units in $\mathcal{O}_{11}^\times$. Since

$$\Phi_{\text{Tate}}(40612 + 94673\sqrt{2}) \equiv (9/2, \frac{-2 + 7\sqrt{2}}{4}) \pmod{11^4},$$

it follows that $\tilde{P}$ is equal to $40612 + 94673\sqrt{2}$ for this degree of 11-adic accuracy.

In order to try out the conjectures of the previous section, we need to exploit some non-maximal orders whose fundamental unit $a + b\sqrt{2}$ satisfies $11^n | b$, for some $n > 0$.

For simplicity, we will work only with orders of prime conductor ℓ. One sees directly that, if the fundamental unit $u = a + b\omega$ of $\mathcal{O}$ satisfies $11^n | b$, then necessarily we have either:

1. ℓ is split in $\mathbf{Q}(\sqrt{11})$ and $12 \cdot 11^{n-1}$ divides $\ell - 1$, or

2. ℓ is inert in $\mathbf{Q}(\sqrt{11})$ and $12 \cdot 11^{n-1}$ divides $\ell + 1$.

Here is a small table of the first few primes ℓ satisfying these conditions, together with their narrow class numbers and fundamental units: (Here $u = 1 + \sqrt{2}$ is a fundamental unit of K.)

ℓ	h	Unit	ℓ	h	Unit	ℓ	h	Unit
73	4	u^{36}	347	2	u^{348}	673	4	u^{336}
83	2	u^{84}	433	4	u^{216}	683	6	u^{228}
97	4	u^{48}	467	2	u^{468}	769	4	u^{384}
107	2	u^{108}	491	2	u^{492}	827	6	u^{276}
*131	2	u^{132}	563	2	u^{564}	937	4	u^{468}
179	10	u^{36}	587	2	u^{588}	947	2	u^{948}
193	4	u^{96}	601	20	u^{60}	971	2	u^{972}
251	6	u^{84}	*659	2	u^{660}	1009	28	u^{72}

Note that in all cases, the fundamental unit of $\mathcal{O}$ listed in the table is a power of u^{12}, and hence belongs to the order of conductor 11 in $\mathbf{Z}[\sqrt{2}]$. Note also that in the two cases marked with a *, the fundamental unit is actually a power of $u^{12\cdot 11}$, and hence belongs to the order of conductor 11^2 of K. In these two cases, the constructions of the previous section will allow us to construct an approximation to a global point in $E^w(\mathcal{O}_K \otimes (\mathbf{Z}/121\mathbf{Z}))$, and not just in $E^w(\mathcal{O}_K \otimes (\mathbf{Z}/11\mathbf{Z}))$.

Suppose first that the 11 divides b exactly. Then conjecture 4.9, assuming that $\#\text{III}(E/K) = 1$, predicts that

$$\mathcal{L}'_{11}(\ell, E) \equiv \tilde{P}^{\pm 2\frac{\ell+1-a_\ell}{5}} \pmod{11} = \begin{cases} -1 & \text{if } 2 \nmid \ell+1-a_\ell, \\ +1 & \text{if } 2 | \ell+1-a_\ell. \end{cases}$$

We indeed checked that this was true on the 24 discriminants listed in the table.

For further verifications, we carried out the calculations modulo 11^2 with the orders of conductor 131 and 659. Here, we obtained a canonical element in $(\mathcal{O}/11^2\mathcal{O})^\times$. For example, in the case of $\ell = 131$, there are two narrow ideal classes I_1 and I_2. A calculation shows that

$$\mathcal{L}'_{11}(I_1) = \mathcal{L}'_{11}(I_2) = (120 + 77\sqrt{2}) \otimes \omega,$$

where ω is a real period of E. Hence,

$$\mathcal{L}'_{11}(131, E) \equiv (120 + 77\sqrt{2})^2 \equiv 1 + 88\sqrt{2} \pmod{11^2}.$$

On the other hand, the 131st Fourier coefficient for $X_0(11)$ is $a_{131} = -18$, so that the predicted right hand side on the conjecture is

$$\tilde{P}^{2(131+1-a_{131})/5} \equiv \tilde{P}^{60} \equiv 1 + 88\sqrt{2} \pmod{11^2},$$

confirming the conjecture.

Likewise, in the case of $\ell = 659$, we found that

$$\mathcal{L}'_{11}(659, E) \equiv 1 + 99\sqrt{2} \pmod{11^2},$$

and that the right hand side (noting that $(\ell+1-a_\ell)/5 = 130$) is

$$\tilde{P}^{260} \equiv 1 + 99\sqrt{2} \pmod{11^2}.$$

We have performed similar verifications with the orders of conductor 23, 43, and 89, which have the property that if u denotes their fundamental unit, then u^6 belongs to the order of conductor 11^2. Hence, by working with the modular symbols γ_{u^6} instead of γ_u, one could obtain an approximation to $\tilde{P}^{12(\ell+1-a_\ell)/5}$. The results are listed in the following table:

ℓ	h^+	a_ℓ	$\mathcal{L}'_{11}(\ell, E)^6$	$\tilde{P}^{12(\ell+1-a_\ell)/5}$
23	2	-1	$1+88\sqrt{2}$	$1+88\sqrt{2}$
43	2	-6	$1+55\sqrt{2}$	$1+55\sqrt{2}$
89	4	15	$1+22\sqrt{2}$	$1+22\sqrt{2}$

We could have pushed the calculation further, using primes ℓ for which the fundamental unit of the order of conductor ℓ also belongs to the unit of conductor 11^3. This would give an approximation to the generator of $E(K)$ with an 11-adic accuracy of 11^{-3}, provided that 11 does not divide $\ell+1-a_\ell$. The first prime with this property is $\ell = 727$. Unfortunately, the calculation of $\mathcal{L}'_{11}(727, E)$ with our implementation of the algorithm seemed to require more computer time than we were willing to devote (about 3 hours). It is likely that the algorithm could be improved, and it would be interesting to explore issues of computational efficiency more carefully.

Needless to say, one could in principle construct a global point to an arbitrary level of 11-adic accuracy, by using these conjectures and exploiting primes ℓ for which

1. The fundamental unit in the order of conductor ℓ also belongs to the order of conductor 11^n;

2. 11 does not divide $\ell+1-a_\ell$.

It can be shown, using the Chebotarev density theorem, that there are infinitely many primes ℓ with these properties, for any given n. (For instance, with $n = 4$ the smallest prime satisfying the above conditions is $\ell = 2663$.)

5.2 Experiments with $X_0(37)$

In this section, let E be the elliptic curve $X_0(37)^+$ with minimal Weierstrass equation

$$y^2 + y = x^3 - x.$$

Here the sign of w_{37} is 1, and hence $w = 1$. The Mordell-Weil group $E^w(K) = E(\mathbf{Q})$ is isomorphic to $\mathbf{Z}$ and is generated by the point $P = (0, 0)$.

This time, we varied the real quadratic field, seeking fields whose fundamental unit also belongs to the order of conductor 37. It is likely that there are infinitely many such fields, although proving such a statement appears to be difficult. In the range $D \leq 9,000$, we found 12 such fields with narrow class number 1. (Restricting to fields of narrow class number 1 allowed for some simplication in the calculations, but was not really necessary.)

For each of these fields, we computed the element $\mathcal{L}'_{37}(1, E)$ associated to the maximal order of K. We could verify that

$$\Phi_{\text{Tate}}(\mathcal{L}'_{37}(1, E)) = \pm 2\sqrt{\#\text{III}(E/K)^?} \cdot (0,0), \quad \text{in} \quad E(\mathbf{F}_{37}).$$

Here

$$\#\text{III}(E/K)^? = L(E^K, 1)/\Omega_{E^K},$$

where E^K is the twist of E over K, and Ω_{E^K} is its associated period. Thus, $s^2 = \#\text{III}(E/K)^?$ is the order of $\text{III}(E/K)$ that is predicted by the Birch and Swinnerton-Dyer conjecture, at least when it is non-zero. The results are summarized in the table below.

D	$\mathcal{L}'_{37}(1, E)$	$\Phi_{\text{Tate}}(\mathcal{L}'_{37}(1, E))$	s	$2 \cdot s \cdot (0,0)$
1277	$17 + 13\sqrt{1277}$	$(30, 13)$	12	$(30, 23)$
1609	$28 + 5\sqrt{1609}$	$(2, 2)$	2	$(2, 34)$
1613	$22 + 8\sqrt{1613}$	$(17, 33)$	14	$(17, 33)$
2333	$28 + 15\sqrt{2333}$	$(2, 2)$	2	$(2, 34)$
2437	$4 + 16\sqrt{2437}$	$(15, 31)$	8	$(15, 31)$
4993	$28 + 16\sqrt{4993}$	$(2, 2)$	2	$(2, 34)$
5009	$28 + 23\sqrt{5009}$	$(2, 34)$	2	$(2, 34)$
5869	1	∞	0	∞
7349	$13 + 5\sqrt{7349}$	$(26, 3)$	4	$(26, 3)$
7369	$13 + 18\sqrt{7369}$	$(26, 3)$	4	$(26, 3)$
7793	$4 + 18\sqrt{7793}$	$(15, 5)$	8	$(15, 31)$
8677	$17 + 13\sqrt{8677}$	$(30, 13)$	12	$(30, 23)$

Note that when $D = 5869$, one has $s = 0$, suggesting that the curve E^K has infinite Mordell-Weil group and that $\text{rank}(E(K)) > 1$. In this case, conjecture 4.9 predicts that the points $P_K(c) \in E^w(K_p)/E^w(K_p)^{(n)}$ are trivial for all c, so that the Stark-Heegner point $P_K \in E(K)$ is trivial.

References

[BD1] M. Bertolini and H. Darmon, *Heegner points on Mumford-Tate curves*, CICMA preprint; Invent. Math., to appear.

[BD2] M. Bertolini and H. Darmon (with an appendix by B. Edixhoven) *A rigid analytic Gross-Zagier formula and arithmetic applications*, CICMA preprint; Annals of Math., to appear.

[BD3] M. Bertolini and H. Darmon, *Heegner points, p-adic L-functions and the Cerednik-Drinfeld uniformization*, CICMA preprint; submitted.

[Ca] J.W.S. Cassels, Lectures on elliptic curves, London Mathematical Society Student Texts **24**, Cambridge University Press, Cambridge 1991.

[Dag] H. Daghigh, McGill PhD. Thesis, in progress.

[Dar] H. Darmon, *Heegner points, Heegner cycles, and congruences*, in "Elliptic curves and related topics", CRM proceedings and lecture notes vol. **4**, H. Kisilevsky and M. Ram Murty eds. (1992) pp. 45-60.

[Di] F. Diamond, *On deformation rings and Hecke rings*, to appear in Annals of Math.

[DST] D.S. Dummit, J.W. Sands, and B.A. Tangedal, *Computing Stark units for totally real cubic fields*, preprint.

[El] N. Elkies, *Heegner point computations*, Algorithmic number theory (Ithaca, NY, 1994) 122–133, Lecture Notes in Computer Science **877** Springer, Berlin, 1994.

[Gr1] B.H. Gross, *Heegner points on $X_0(N)$*, in Modular Forms, R.A. Rankin ed., p. 87-107, Ellis Horwood Ltd., 1984.

[Gr2] B.H. Gross, *Heights and the special values of L-series*, Number theory (Montreal, Que., 1985), 115–187, CMS Conf. Proc., **7**, Amer. Math. Soc., Providence, RI, 1987.

[GKZ] B.H. Gross, W. Kohnen, D. Zagier, *Heegner points and derivatives of L-series. II.* Math. Ann. **278** (1987), no. 1-4, 497–562.

[GZ] B.H. Gross, D.B. Zagier, *Heegner points and derivatives of L-series*, Inv. Math. **84** (1986), no. 2, 225–320.

[HW] G.H. Hardy, E.M. Wright, An introduction to the theory of numbers, fifth ed., the Clarendon Press, Oxford University Press, New York, 1979

[JL] H. Jacquet, R.P. Langlands, *Automorphic forms on* **GL(2)**, Springer Lecture Notes, **114**, (1970).

[Ko] W. Kohnen, *Modular forms and real quadratic fields.* Automorphic functions and their applications (Khabarovsk, 1988), 126–134, Acad. Sci. USSR, Inst. Appl. Math., Khabarovsk, 1990.

[Ma] B. Mazur, *Modular curves and arithmetic*, Proceedings of the Int. Congress of Math., (1983), Warszawa, pp. 185-209.

[MT] B. Mazur, J. Tate, *Refined conjectures of the "Birch and Swinnerton-Dyer type"*, Duke Math. J. **54** (1987), no. 2, 711–750.

[MTT] B. Mazur, J. Tate, and J. Teitelbaum, *On p-adic analogues of the conjectures of Birch and Swinnerton-Dyer*, Inv. Math. **84** 1-48 (1986).

[PR] B. Perrin-Riou, *Points de Heegner et dérivées de fonctions L p-adiques*, Invent. Math. **89** 1987, no. 3, 455-510.

[Ro] D. Roberts, Shimura curves analogous to $X_0(N)$, Harvard PhD. Thesis, 1989.

[Ru] K. Rubin, *p-adic L-functions and rational points on elliptic curves with complex multiplication*, Invent. Math. **107**, 323-350 (1992).

[Si1] J.H. Silverman, The arithmetic of elliptic curves, GTM **106**, Springer-Verlag, New York 1986.

[Si2] J.H. Silverman, *Computing rational points on rank 1 elliptic curves via L-series and canonical heights*, (preprint).

[St] H.M. Stark, *Values of L-functions at s = 1, I, II, III, and IV*, Advances in Math. **7** (1971), 301–343; **17** (1975), 60–92; **22** (1976), 64–84; **35** (1980), 197–235.

[Ta] J. Tate, Les conjectures de Stark sur les fonctions L d'Artin en $s = 0$. Birkhäuser, Boston, 1984.

[TW] R. Taylor and A. Wiles, *Ring theoretic properties of certain Hecke algebras*, Annals of Math. **141**, No. 3, 1995, pp. 553-572.

[Wal] J-L. Waldspurger, *Sur les valeurs de certaines fonctions L automorphes en leur centre de symétrie*, Compos. Math. **54** (1985) no. 2, 173–242.

[Was] L. Washington, Introduction to Cyclotomic Fields, GTM **83**, Springer-Verlag, 1982.

[We] A. Weil, Number Theory: An approach through history, from Hammurapi to Legendre. Birkhäuser, 1984.

[Wi] A. Wiles, *Modular elliptic curves and Fermat's last theorem*, Annals of Math. **141**, No. 3, 1995, pp. 443-551.

[Za] D.B. Zagier, *Modular points, modular curves, modular surfaces, and modular forms* Workshop Bonn 1984 (Bonn 1984) 225–248, Lecture Notes in Math. **1111**, Springer Berlin,New York 1985

On the Hecke action on the cohomology of Hilbert-Blumenthal surfaces

Fred Diamond

1. Introduction

In this paper we consider the problem of determining the multiplicities of certain Galois representations occurring in finite subquotients of the étale cohomology of Hilbert modular surfaces. We prove a "multiplicity one" result and apply it to obtain information about the action of the Hecke algebra on spaces of Hilbert modular forms.

Before giving a more precise formulation, we describe the problem first addressed by Mazur [**M**, II.14a] (and by Ribet [**R1**, Section 5]) in the classical setting of cusp forms of weight two on $\Gamma_0(N)$. Let **T** denote the ring generated by the Hecke operators T_n on the space S of such forms. Let **m** be a maximal ideal of **T** of residue characteristic ℓ. Then **m** is the kernel of a homomorphism defined by $T_n \mapsto a_n \mod \lambda$ (for all $n \geq 1$), where a_n is the eigenvalue of T_n acting on a common eigenform f in S, and λ is a prime in the field generated by the a_n. Using the theory of Eichler and Shimura, one can attach to f a certain representation $\rho_f : \text{Gal}(\overline{\mathbf{Q}}/\mathbf{Q}) \to \mathbf{GL}_2(\mathcal{O}_E)$. Forming the semi-simplification of the reduction of $\rho_f \mod \lambda$, one associates to **m** a representation

$$\rho_{\mathbf{m}} : \text{Gal}(\overline{\mathbf{Q}}/\mathbf{Q}) \longrightarrow \mathbf{GL}_2(\mathbf{T}/\mathbf{m})$$

which is unramified at primes q not dividing $N\ell$ and such that $\rho_{\mathbf{m}}(\text{Frob}_q)$ has characteristic polynomial $X^2 - T_q X + q \pmod{\mathbf{m}}$. We suppose that the representation $\rho_{\mathbf{m}}$ is irreducible. Consider now $J = J_0(N)$, the Jacobian of the modular curve $X = X_0(N)$, endowed with the natural action of the Hecke algebra **T**. The kernel of **m** in J, denoted $J[\mathbf{m}]$, is a $(\mathbf{T}/\mathbf{m})[\text{Gal}(\overline{\mathbf{Q}}/\mathbf{Q})]$-module containing a model for $\rho_{\mathbf{m}}$. (In fact $J[\mathbf{m}]$ is isomorphic to a direct sum of copies of $\rho_{\mathbf{m}}$ [**BLR**].) Mazur's argument in [**M**] (see [**R1**, Section 5]) shows that if ℓ does not divide $2N$, then $J[\mathbf{m}]$ is isomorphic to $\rho_{\mathbf{m}}$. The idea is to use Dieudonné theory to relate the kernel of ℓ in $J_{/\mathbf{F}_\ell}$ to the deRham cohomology of $X_{/\mathbf{F}_\ell}$, and then to use the q-expansion principle to bound dimensions. This has been generalized by Tilouine [**Ti**], Mazur-Ribet [**MR**], Edixhoven [**E**] and Wiles [**W**] to certain cases in which ℓ divides N, by Faltings-Jordan [**FJ**] in the context of modular forms of higher weight, and by Ribet [**R2**] and Yang [**Y**] for modular forms on Shimura curves.

1991 *Mathematics Subject Classification.* Primary 11F41; Secondary 11F33, 11G18, 14G35.

This research was conducted while the author was a Ritt Assistant Professor at Columbia University.

As a corollary, $\mathrm{Ta}_\ell(J)_\mathbf{m}$, the localization at $\mathbf{m}$ of the Tate module, is free of rank two over the completion $\mathbf{T_m}$ [**M**, II.15]. Using auto-duality of the Jacobian, one can produce an isomorphism of $\mathbf{T_m}$-modules $\mathbf{T_m} \cong \mathrm{Hom}_{\mathbf{Z}_p}(\mathbf{T_m}, \mathbf{Z}_\ell)$ and deduce that $\mathbf{T_m}$ is a Gorenstein ring.

In the context of Hilbert modular forms associated to a real quadratic field F, we consider the action of a Hecke algebra $\mathbf{T}$ on the space of Hilbert modular cusp forms of weight two and level $\mathfrak{n}$, where $\mathfrak{n}$ is an ideal in the ring of integers $\mathcal{O}$ of F. These forms are certain holomorphic two-forms on a Hilbert modular surface X, and $\mathbf{T}$ is generated by the operators $T_\mathfrak{a}$ for non-zero ideals $\mathfrak{a}$ of $\mathcal{O}$ (see [**H**] and Section 2 below for definitions). Let $\mathbf{m}$ be a maximal ideal of $\mathbf{T}$. (We assume throughout that the residue characteristic ℓ of $\mathbf{m}$ does not divide $6D\mathbf{N}(\mathfrak{n})$, where D is the discriminant of F over $\mathbf{Q}$ and $\mathbf{N}$ is the norm, and we replace $\mathbf{T}$ with the Hecke algebra over $\mathbf{Z}_\ell$.) One can again associate to $\mathbf{m}$ a Galois representation

$$\rho_\mathbf{m} : \mathrm{Gal}(\overline{F}/F) \longrightarrow \mathbf{GL}_2(\mathbf{T}/\mathbf{m})$$

with properties analogous to those above. In this generality, the existence of the representation $\rho_\mathbf{m}$ is due to Taylor [**Ta**]. To construct it, we can replace the eigenform f as above with a "congruent" eigenform g of weight two and level $\mathfrak{n}\mathfrak{q}$ (for some $\mathfrak{q}$ not dividing $\ell\mathfrak{n}$) so that the automorphic representation corresponding to g is special at $\mathfrak{q}$. The representation ρ_g can be constructed using the Jacquet-Langlands correspondence and the Jacobian of a certain Shimura curve [**Ca**].

It is now the intersection cohomology of X, in degree two and relative to the Baily-Borel-Satake compactification, which arises naturally in connection with the weight two cuspidal automorphic representations (see [**BL**]). In Section 2 we use intersection cohomology to define a $\mathbf{Q}_\ell$-vector space B with an action of $\mathbb{T}[\mathrm{Gal}(\overline{\mathbf{Q}}/\mathbf{Q})]$, where $\mathbb{T}$ is the polynomial ring over $\mathbf{Z}_\ell$ generated by the variables $T_\mathfrak{a}$. Using the toroidal compactifications constructed by Rapoport [**Rp**] and the étale-crystalline comparison theorem of Fontaine-Messing [**FM**], we establish a relationship between B and the space of Hilbert modular cusp forms of weight two and level $\mathfrak{n}$. More precisely, we use an integral version of the comparison theorem due to Faltings [**Fa**] to relate a stable lattice M in B to the $\mathbf{T}$-module of such forms with Fourier coefficents in $\mathbf{Z}_\ell$. We express this relationship in Proposition 1 using a finite version of a Dieudonné module functor constructed by Fontaine [**F2**]. (We use Faltings' variant of the one defined by Fontaine and Laffaille [**FL**].) We remark that Proposition 1 uses Faltings' comparison theorem only in the case of varieties with good reduction. We also assume that Rapoport's toroidal compactifications are schemes, not just algebraic spaces. (This is stated in the literature without proof in [**HLR**]; an indication of the method of proof is given in [**Rm**].)

We can now hope to apply the q-expansion principle to deduce multiplicity one results, but the representations occurring as Jordan-Holder constituents of $M/\ell M$ are not the representations $\rho_\mathbf{m}$ described above. In Section 3, we consider the tensor-induction $\sigma_\mathbf{m}$ of $\rho_\mathbf{m}$ to $\mathrm{Gal}(\overline{\mathbf{Q}}/\mathbf{Q})$ (where we have replaced $\rho_\mathbf{m}$ with its dual as we are working with cohomology). The representation $\sigma_\mathbf{m}$ is four-dimensional, and its restriction to $\mathrm{Gal}(\overline{F}/F)$ is the tensor product of $\rho_\mathbf{m}$ with its conjugate under $\mathrm{Gal}(F/\mathbf{Q})$. It follows from the work of Brylinski-Labesse [**BL**] that its Jordan-Holder constituents are the ones of interest in $(M/\ell M)[\mathfrak{m}]$. Here $\mathfrak{m}$ is the inverse image of $\mathbf{m}$ in $\mathbb{T}$, and we will write $\mathbf{F}$ for $\mathbf{T}/\mathbf{m} = \mathbb{T}/\mathfrak{m}$. Using our finite Dieudonné functor, we define a Jordan-Holder constituent $\sigma_{\mathbf{m},2}$ of $\sigma_\mathbf{m} \otimes_\mathbf{F} \overline{\mathbf{F}}$ which, as a consequence of Proposition 1 and the q-expansion principle, occurs with multiplicity one in $(M/\ell M)[\mathbf{m}] \otimes_{\mathbf{F}} \overline{\mathbf{F}}$. We would also like some type of dual result, namely that a certain constituent $\sigma_{\mathbf{m},0}$ also occurs with multiplicity one, and so we are led to analyze the possible reducibility of $\sigma_{\mathbf{m}}$. (It is this analysis of the Jordan-Holder constituents which is our main reason for restricting to the case of real quadratic F.) We find that if $\rho_{\mathbf{m}}$ is irreducible, then $\sigma_{\mathbf{m},0} = \sigma_{\mathbf{m},2}$ and thus occurs with multiplicity one. The multiplicity results are collected in Theorem 1.

In Section 4 we combine Theorem 1 with Poincaré duality in order to show that $\mathbf{T}_{\mathbf{m}}$ is Gorenstein.

THEOREM 2. *Let F be a real quadratic field, and let $\mathfrak{n}$ be a non-zero ideal in $\mathcal{O}_F$. Let ℓ be a rational prime not dividing $6D_{F/\mathbf{Q}}\mathbf{N}_{F/\mathbf{Q}}(\mathfrak{n})$. Let $\mathbf{T}$ be the $\mathbf{Z}_\ell$-algebra of endomorphisms of $S_2(V_1(\mathfrak{n}), \mathbf{Z}_\ell)$ generated by all the Hecke operators $T_\mathfrak{a}$, and let $\mathbf{m}$ be a maximal ideal of $\mathbf{T}$. Suppose that the associated Galois representation $\rho_{\mathbf{m}} : \mathrm{Gal}(\overline{F}/F) \to \mathbf{GL}_2(\mathbf{T}/\mathbf{m})$ is irreducible. Then $\mathbf{T}_{\mathbf{m}}$ is Gorenstein.*

The main motivation for this result was initially provided by Wiles' use of the Gorenstein property for certain Hecke algebras for classical modular forms in his work on the Shimura-Taniyama-Weil conjecture. The research described in this paper was done in 1993 with a view to extending his methods to the setting of Hilbert modular forms. That project was set aside until Wiles' proof of the conjecture in the semistable case [**W**] was completed by his work with Taylor in [**TW**]. Their methods actually made use of mod ℓ multiplicity one results as well as the Gorenstein property. Upon returning to the project for Hilbert modular forms and considering the obstacles to improving Theorem 2, the author [**Di**] found a modification of the methods of [**W**] and [**TW**] which did not use mod ℓ multiplicity one and the Gorenstein property; in fact, they became corollaries. It was also discovered independently by K. Fujiwara [**Fu**] that these were not needed as ingredients to obtain the main results of [**W**] and [**TW**] in the case of "minimal level." He generalizes this to the setting of Hilbert modular forms, and this also yields a proof that certain Hecke algebras acting on Hilbert modular forms are Gorenstein. The results described here are not subsumed by ones expected from the approach of [**Di**] and [**Fu**], which require that the Hecke algebra correspond to a certain type of deformation problem. The results in [**Fu**] for Hilbert modular forms are also subject to technical hypotheses, including that ℓ not divide the class number of F. More work is needed to remove those hypotheses, and also to obtain mod ℓ multiplicity one results in that context from the new approach.

Finally, we remark on generality and limitations of the methods of this paper: In the case $\ell = 3$, Theorem 2 may be accessible provided T_3 is in $\mathbf{m}$ (the "non-ordinary" case), but the arguments will require a great deal more care. It seems likely that our methods work for modular forms of higher weight, provided ℓ is accordingly large. Our restriction to primes ℓ not dividing $2D\mathbf{N}(\mathfrak{n})$ is more serious. For totally real fields of higher degree the main obstacle is understanding the constituents of $\sigma_{\mathbf{m}}$, and of course we must restrict to $\ell > [F : \mathbf{Q}] + 1$ (and perhaps some cases of equality).

The author is grateful to Ching-Li Chai, George Pappas, Richard Taylor and Andrew Wiles for helpful conversations.

2. Hilbert modular forms

We fix a real quadratic field F and a non-zero ideal $\mathfrak{n}$ in the ring of integers $\mathcal{O}$ of F. We fix a rational prime $\ell > 3$, and we always assume that ℓ does not divide

$D\mathbf{N}(\mathfrak{n})$, where $D = D_{F/\mathbf{Q}}$ is the discriminant and $\mathbf{N}$ is the norm $\mathbf{N}_{F/\mathbf{Q}}$. We will write $\mathbf{A}$ for the ring of adeles over $\mathbf{Q}$ and $\mathbf{A_f}$ for the subring of finite adeles. We use the definitions and conventions of Hida [**H**] with regard to Hilbert modular forms.

In this section we use the étale-crystalline comparison theorem of Fontaine and Messing [**FM**] (and its integral version due to Faltings [**Fa**]) to establish a relationship between the Hilbert modular cusp forms of level $\mathfrak{n}$ and the étale intersection cohomology of a Hilbert modular surface.

2.1. Hilbert modular surfaces. We first review the results we will be using about Hilbert modular surfaces and their compactifications.

Let I denote the set of embeddings $\tau : F \hookrightarrow \mathbf{R}$, and let K_∞ denote the subgroup $\prod_\tau K_\tau$ of $\mathbf{GL}_2(F \otimes \mathbf{R}) \cong \prod_\tau \mathbf{GL}_2(F_\tau)$ where

$$K_\tau = \left\{ \begin{pmatrix} a & b \\ -b & a \end{pmatrix} \Big| a, b \in F_\tau \right\}.$$

For each open compact subgroup U of $\mathbf{GL}_2(F \otimes \mathbf{A_f})$, we let X_U denote the surface over $\mathbf{Q}$ with complex points

$$\mathbf{GL}_2(F) \backslash \mathbf{GL}_2(F \otimes \mathbf{A}) / K_\infty U,$$

obtained from Shimura's theory of canonical models (see [**S1**] and [**De**]). The surface X_U is a coarse moduli space for two-dimensional abelian schemes with $\mathcal{O}$-action and level U-structure. We let $\overline{X}_U$ denote the Baily-Borel-Satake compactification of X_U. Thus for open compact subgroups U and V, and $g \in \mathbf{GL}_2(F \otimes \mathbf{A_f})$ satisfying $g^{-1}Ug \subset V$, we have a finite morphism $J(g) : X_U \to X_V$ extending to $\overline{J}(g) : \overline{X}_U \to \overline{X}_V$. The morphisms $\overline{J}(g)$ satisfy the usual compatibilities, and on complex points of X_U, $J(g)$ corresponds to right multiplication by g. For sufficiently small V (contained, for example, in a principal congruence subgroup of level at least three), the surface X_V is smooth over $\mathbf{Q}$ and $J(g)$ is étale.

Suppose that V is sufficiently small and that $V_\ell = \mathbf{GL}_2(\mathcal{O}_\ell)$. Then $X_V = \mathcal{X}_V \times_{\mathbf{Z}_{(\ell)}} \mathbf{Q}$ for a smooth scheme $\mathcal{X}_V$ over $\mathbf{Z}_{(\ell)}$. A construction of Rapoport [**Rp**] (see also [**Ch**]) provides a toroidal compactification $\tilde{\mathcal{X}}_V$ of $\mathcal{X}_V$ which is a smooth proper scheme over $\mathbf{Z}_{(\ell)}$. (The context under consideration in [**Rp**] is slightly different from ours, but [**HLR**] and [**Rm**] describe the modifications needed to apply the construction in our case.) We recall that this compactification is not canonical, but depends on a choice of certain polyhedral cone decompositions. However if U and V are sufficiently small with $U_\ell = V_\ell = \mathbf{GL}_2(\mathcal{O}_\ell)$ and $g \in \mathbf{GL}_2(F \otimes \mathbf{A_f})$ is such that $g^{-1}Ug \subset V$, then we can choose compatible polyhedral cone decompositions so that $J(g)$ extends to a morphism $\tilde{J}(g) : \mathcal{X}_U \to \mathcal{X}_V$.

Let $V_1(\mathfrak{n})$ denote the open compact subgroup of $\prod_v \mathbf{GL}_2(\mathcal{O}_v)$ consisting of matrices congruent to $\begin{pmatrix} * & * \\ 0 & 1 \end{pmatrix}$ mod $\mathfrak{n}$. Choose a rational prime $q \geq 3$ with $q\mathcal{O}$ prime to $\ell D\mathfrak{n}$ satisfying $q \equiv 1 \mod D$ and $q \not\equiv \pm 1 \mod \ell$. Let W be the intersection of $V_1(\mathfrak{n})$ with the principal congruence subgroup of level q. Then W is normal in $V_1(\mathfrak{n})$ of index prime to ℓ, and in this case we will omit the subscript W and write simply X, $\overline{X}$, $\mathcal{X}$ and $\tilde{\mathcal{X}}$. The natural action of $G = V_1(\mathfrak{n})/W$ on X extends to $\mathcal{X}$.

2.2. The q-expansion principle. For each open compact subgroup U of $\mathbf{GL}_2(F \otimes \mathbf{A_f})$, we let $S_2(U, \mathbf{C})$ denote the space of Hilbert modular cusp forms of weight two and level U (see [**H**] and [**S2**]). The Koecher principle allows us to identify $S_2(W, \mathbf{C})$ with the image of $H^0(\tilde{\mathcal{X}} \times \mathbf{C}, \Omega^2_{\tilde{\mathcal{X}}/\mathbf{C}}) \hookrightarrow H^0(\mathcal{X} \times \mathbf{C}, \Omega^2_{\mathcal{X}/\mathbf{C}})$ (see

[**Fr**, II.4] for example). The image is thus independent of the choice of toroidal compactification and its intersection with $H^0(\mathcal{X} \times \mathbf{C}, \Omega^2_{\mathcal{X}/\mathbf{C}})^G$ is $S_2(V_1(\mathfrak{n}), \mathbf{C})$. We see also that
$$H^0(\tilde{\mathcal{X}}, \Omega^2_{\tilde{\mathcal{X}}/\mathbf{Z}_{(\ell)}}) \hookrightarrow H^0(\mathcal{X}, \Omega^2_{\mathcal{X}/\mathbf{Z}_{(\ell)}})$$
has torsion-free cokernel, and the image is independent of the choice of compactification. We write $S_2(V_1(\mathfrak{n}), \mathbf{Z}_{(\ell)})$ for the intersection of the image with $H^0(\mathcal{X}, \Omega^2_{\mathcal{X}/\mathbf{Z}_{(\ell)}})^G$.

If A is a $\mathbf{Z}_{(\ell)}$-algebra and C is a set of cusps of $\overline{X}$ stable under the action of $\text{Gal}(\overline{\mathbf{Q}}/\mathbf{Q})$, then we have a natural q-expansion homomorphism
$$\phi_{A,C} : H^0(\tilde{\mathcal{X}} \times A, \Omega^2_{\tilde{\mathcal{X}}/A}) \to H^0(\mathcal{T}(C) \times A, \Omega^2_{\mathcal{T}(C)/A}),$$
where $\mathcal{T}(C)$ denotes the completion of $\tilde{\mathcal{X}}$ along the components of $\tilde{\mathcal{X}} - \mathcal{X}$ corresponding to cusps in C. In the notation of [**HLR**], $H^0(\mathcal{T}(s), \Omega^2_{\mathcal{T}(s)/R})$ is canonically isomorphic to $H^0(\mathcal{T}(s), \mathcal{I}(s)) \otimes \text{Hom}(\wedge^2 N(s), \mathbf{Z})$, where $\mathcal{I}(s)$ is the largest ideal of definition of $\mathcal{T}(s)$ (so $H^0(\mathcal{T}(s), \mathcal{I}(s))$ is the kernel of the map $H^0(\mathcal{T}(s), \mathcal{O}_{\mathcal{T}(s)}) \to R$ defined by evaluation at s). This isomorphism provides a description of $H^0(\mathcal{T}(C) \times \mathbf{C}, \Omega^2_{\mathcal{T}(C)/\mathbf{C}})$ by which $\phi_{\mathbf{C},C}$ becomes compatible with the classical notion of the Fourier expansions of the cusp form. We also find that $H^0(\mathcal{T}(C), \Omega^2_{\mathcal{T}(C)/\mathbf{Z}_{(\ell)}})$ is flat over $\mathbf{Z}_{(\ell)}$ and that $H^0(\mathcal{T}(C), \Omega^2_{\mathcal{T}(C)/\mathbf{Z}_{(\ell)}}) \otimes A \cong H^0(\mathcal{T}(C) \times A, \Omega^2_{\mathcal{T}(C)/A})$. If C meets every component of $\overline{X}$, then we have the following q-expansion principle (see [**Rp**, Theorem 6.7]):

- $\phi_{A,C}$ is injective
- Suppose that B is a subalgebra of A and that $f \in H^0(\tilde{\mathcal{X}} \times A, \Omega^2_{\tilde{\mathcal{X}}/A})$. Then f is in the image of $H^0(\tilde{\mathcal{X}} \times B, \Omega^2_{\tilde{\mathcal{X}}/B})$ if and only if $\phi_{A,C}(f)$ is in the image of $H^0(\mathcal{T}(C) \times B, \Omega^2_{\mathcal{T}(C)/B})$.

In particular, we can identify $S_2(V_1(\mathfrak{n}), \mathbf{Z}_{(\ell)})$ with the set of cusp forms in $S_2(V_1(\mathfrak{n}), \mathbf{C})$ whose q-expansions have coefficients in $\mathbf{Z}_{(\ell)}$ (in the sense of [**H**, 4.5a]).

For each non-zero ideal $\mathfrak{a}$ of $\mathcal{O}$, we can define an action of the Hecke operator $T_\mathfrak{a}$ on $S_2(V_1(\mathfrak{n}); \mathbf{C})$ (see [**H**, Section 2]). These operators commute and $S_2(V_1(\mathfrak{n}); \mathbf{Z}_{(\ell)})$ is stable under their action ([**H**, Section 4]). Let $\mathbb{T}$ denote the polynomial ring over $\mathbf{Z}_\ell$ with variables $T_\mathfrak{a}$ and write $\mathbf{T}$ for its image in $\text{End}(S)$, where $S = S_2(V_1(\mathfrak{n}), \mathbf{Z}_\ell) = S_2(V_1(\mathfrak{n}); \mathbf{Z}_{(\ell)}) \otimes \mathbf{Z}_\ell$. Then the $\mathbf{Z}_\ell$-algebra $\mathbf{T}$ is free and finitely generated as a $\mathbf{Z}_\ell$-module. Moreover $S \cong \text{Hom}_{\mathbf{Z}_\ell}(\mathbf{T}, \mathbf{Z}_\ell)$ as a $\mathbf{T}$-module ([**H**, Section 5]).

2.3. Crystalline representations. We now summarize results of Fontaine and Lafaille [**FL**], Fontaine and Messing [**FM**] and Faltings [**Fa**] on crystalline representations. We refer the reader to [**F1**], [**F2**] and [**I**] for more fundamental definitions and results.

Let K be a number field, and $\mathfrak{l}$ a prime of $\mathcal{O}_K$ unramified over ℓ. Let R denote $\mathcal{O}_{K,\mathfrak{l}}$ and fix an embedding $\overline{K} \hookrightarrow \overline{K}_\mathfrak{l}$. Recall that an ℓ-adic representation V of $\text{Gal}(\overline{K}_\mathfrak{l}/K_\mathfrak{l})$ is called crystalline if $\dim_{\mathbf{Q}_\ell} V = \dim_K \mathbb{D}_{\text{cris}}(V)$, where $\mathbb{D}_{\text{cris}}(V)$ denotes the filtered Dieudonné module $(B_{\text{cris}} \otimes_{\mathbf{Q}_\ell} V)^{\text{Gal}(\overline{K}_\mathfrak{l}/K_\mathfrak{l})}$. The functor $\mathbb{D}_{\text{cris}}$ establishes an equivalence between the category of crystalline representations and the category of admissible filtered Dieudonné modules. If Δ is a filtered Dieudonné module with $\text{Fil}^\ell D = 0$ and $\text{Fil}^0 \Delta = \Delta$, then Δ is admissible if and only if it contains an adapted lattice (see [**FL**]). We will call an ℓ-adic representation of

$\mathrm{Gal}(\overline{K}/K)$ crystalline (at $\mathfrak{l}$) when the local representation (at $\mathfrak{l}$) is crystalline. In particular $V = \mathbf{Q}_\ell(-1)$ is crystalline with $\mathbb{D}_{\mathrm{cris}}(V) = \mathrm{gr}^1(\mathbb{D}_{\mathrm{cris}}(V)) \cong K_\mathfrak{l}$.

Fontaine and Lafaille define an abelian category which (following [**Fa**]) we denote $\mathfrak{MF}_{[a,b]}(R)$. The objects of $\mathfrak{MF}_{[a,b]}(R)$ are pairs (M, ϕ) where M is a filtered R-module of finite length with $\mathrm{Fil}^{b+1} M = 0$ and $\mathrm{Fil}^a M = M$, and ϕ is an isomorphism $\overline{M} \otimes_{R,\Phi} R \to M$. Here $\Phi : R \to R$ is the Frobenius and $\overline{M}$ denotes the inductive limit of the diagram

$$\cdots \hookrightarrow \mathrm{Fil}^{i+1}(M) \xleftarrow{\ell} \mathrm{Fil}^{i+1}(M) \hookrightarrow \mathrm{Fil}^i(M) \xleftarrow{\ell} \mathrm{Fil}^i(M) \hookrightarrow \cdots.$$

Morphisms are homomorphisms of filtered R-modules respecting the maps ϕ. Since the morphisms are strict for filtrations [**FL**, Section 1], Fil^i and gr^i are exact. If $\Lambda_2 \subset \Lambda_1$ are adapted lattices in D with $\mathrm{Fil}^{\ell-1} D = 0$ and $\mathrm{Fil}^0 D = D$, then Λ_1/Λ_2 is an object of $\mathfrak{MF}_{[0,\ell-1]}(R)$. Following a construction of [**FL**], Faltings considers a functor $\mathbf{D}$ which establishes an anti-equivalence between $\mathfrak{MF}_{[0,\ell-1]}(R)$ and a full subcategory of finite length $\mathbf{Z}_\ell[\mathrm{Gal}(\overline{K}_\mathfrak{l}/K_\mathfrak{l})]$-modules. We write $\mathbb{V}$ for the Pontrjagin dual of $\mathbf{D}$. The essential image of $\mathbb{V}$ is closed under subobjects and quotients (see [**Fa**, Theorem 2.6]). We write $\mathbb{D}$ for the quasi-inverse of $\mathbb{V}$ and $\mathbb{D}^i$ for $\mathrm{gr}^i \circ \mathbb{D}$.

We will also regard $\mathbb{D} = \mathbb{D}_E(K, \mathfrak{l})$ and $\mathbb{D}^i = \mathbb{D}^i_E(K, \mathfrak{l})$ in the obvious way as functors on a full subcategory $\mathfrak{CR} = \mathfrak{CR}_E(K, \mathfrak{l})$ of $E[\mathrm{Gal}(\overline{K}/K)]$-modules for any $\mathbf{Z}_\ell$-algebra E. Thus $\mathbb{D}$ takes values in a category whose objects are pairs (M, ϕ), where M is a filtered $R \otimes E$-module and ϕ is $R \otimes E$-linear. The functor $\mathbb{D}$ respects tensor products in the following sense: If A and B are in $\mathfrak{CR}$ with $\mathbb{D}(A)$ in $\mathfrak{MF}_{[a_1,a_2]}(R)$, $\mathbb{D}(B)$ in $\mathfrak{MF}_{[b_1,b_2]}(R)$ and $a_2 + b_2 < \ell - 1$, then $A \otimes_E B$ is in C and $\mathbb{D}(A \otimes_E B) \cong \mathbb{D}(A) \otimes_{E \otimes R} \mathbb{D}(B)$ (where the filtration and ϕ on the latter tensor product are defined in the obvious way). A similar description can be given for $\mathrm{Hom}_E(A, B)$ (requiring $b_1 \geq a_2$ instead of $a_2 + b_2 < \ell - 1$). We recall also that the $\mathbf{Z}_\ell$-length of A coincides with the R-length of $\mathbb{D}(A)$.

Suppose K', $\mathfrak{l}'$ and R' are as above with $K \hookrightarrow K'$ and $\mathfrak{l}'$ lying over $\mathfrak{l}$. An ℓ-adic representation V of $\mathrm{Gal}(\overline{K}/K)$ is crystalline at $\mathfrak{l}$ if and only if its restriction to $\mathrm{Gal}(\overline{K}'/K')$ is crystalline at $\mathfrak{l}'$. In this case there is a natural isomorphism $\mathbb{D}_{\mathrm{cris}}(\mathrm{res}\, V) \cong \mathbb{D}_{\mathrm{cris}}(V) \otimes_{K_\mathfrak{l}} K'_{\mathfrak{l}'}$. A similar statement holds for objects of $\mathfrak{CR}_E(K, \mathfrak{l})$ (cf. [**FL**, Section 3]).

If V is crystalline and satisfies $\mathrm{Fil}^{\ell-1} \mathbb{D}_{\mathrm{cris}}(V) = 0$ and $\mathrm{Fil}^0 \mathbb{D}_{\mathrm{cris}}(V) = V$, then any $\mathbf{Z}_\ell[\mathrm{Gal}(\overline{K}_\mathfrak{l}/K_\mathfrak{l})]$-subquotient of V of finite length is in the essential image of $\mathbb{V}$. If L is a Galois stable lattice in V, let L_n denote $L/\ell^{n+1}L$. Then

$$\Lambda = \varprojlim_n \mathbb{D}(L_n)$$

is an adapted lattice in the filtered Dieudonné module $\mathbb{D}_{\mathrm{cris}}(V) \cong \Lambda \otimes \mathbf{Q}_\ell$.

Suppose that $\mathcal{Y}$ is a smooth proper scheme over R_0 of dimension $d < \ell - 1$, where R_0 is the localization of $\mathcal{O}_K$ at $\mathfrak{l}$. Let L denote the image of $H^b_{\mathrm{et}}(\mathcal{Y} \times \overline{K}, \mathbf{Z}_\ell)$ in $V = H^b_{\mathrm{et}}(\mathcal{Y} \times \overline{K}, \mathbf{Q}_\ell)$, and let Λ denote the image of $H^b_{\mathrm{cris}}(\mathcal{Y}_0/R) \cong H^b_{\mathrm{DR}}(\mathcal{Y} \times R/R)$ in $\Delta = H^b_{\mathrm{DR}}(\mathcal{Y} \times K_\mathfrak{l}/K_\mathfrak{l})$ (where $\mathcal{Y}_0 = \mathcal{Y} \times \mathcal{O}_K/\mathfrak{l}$). Then by theorems of Fontaine and Messing [**FM**] and Faltings [**Fa**], V is crystalline (at $\mathfrak{l}$), $\mathbb{D}_{\mathrm{cris}}(V)$ is naturally isomorphic to Δ, Λ is an adapted lattice in Δ, and $\mathbb{D}(L_n)$ is naturally isomorphic to $\Lambda/\ell^{n+1}\Lambda$ (in $\mathfrak{MF}_{[0,a]}(R)$ where $a = \min\{b, d\}$). In particular when $b = d$ we obtain natural isomorphisms $\mathrm{gr}^d(\mathbb{D}_{\mathrm{cris}}(V)) \cong H^0(\mathcal{Y}, \Omega^d_{\mathcal{Y}/R_0}) \otimes_{R_0} K_\mathfrak{l}$ and

$$\mathbb{D}^d(L_n) \cong H^0(\mathcal{Y}, \Omega^d_{\mathcal{Y}/R_0}) \otimes_{R_0} R_0/\ell^{n+1}R_0.$$

2.4. Intersection cohomology.

Now we apply these results to the intersection cohomology of the Hilbert modular surface. For sufficiently small open compact U, consider the inclusion $j_U : X_U \to \overline{X}_U$. We define the intersection complex $\mathcal{C}_U^\bullet = \tau_{\leq 1}(Rj_*\mathbf{Q}_\ell)$ of ℓ-adic sheaves on $\overline{X}_U \times \overline{\mathbf{Q}}$ (see [**BL**]). Then $A_U = H^2(\overline{X}_U \times \overline{\mathbf{Q}}, \mathcal{C}_U^\bullet)$ is isomorphic to the image of $H^2_c(X_U \times \overline{\mathbf{Q}}, \mathbf{Q}_\ell)$ in $H^2(X_U \times \overline{\mathbf{Q}}, \mathbf{Q}_\ell)$. We let L_U denote the image of $H^2_c(X_U \times \overline{\mathbf{Q}}, \mathbf{Z}_\ell)$ in A_U. Then L_U and A_U are naturally $\mathrm{Gal}(\overline{\mathbf{Q}}/\mathbf{Q})$-modules. We omit the subscripts when $U = W$ (a fixed "nice" normal subgroup of $V_1(\mathfrak{n})$).

For g and U such that $g^{-1}Ug \subset V$, the adjunction $\mathbf{Z}_\ell \to J(g)_*J(g)^*\mathbf{Z}_\ell$ and trace $J(g)_*J(g)^*\mathbf{Z}_\ell \to \mathbf{Z}_\ell$ induce homomorphisms $g_* : L_V \to L_U$ and $g^* : L_U \to L_V$ satisfying the usual compatibilities. In particular the maps g_* define an action of G on L (and on A), and we let $M = L^G$ and $B = A^G$. Moreover note that the maps g_* make $\mathcal{A} = \varinjlim A_U$ an admissible $\mathbf{GL}_2(F \otimes \mathbf{A_f})$-module and that the inclusion $A \hookrightarrow \mathcal{A}$ induces $B \cong \mathcal{A}^{V_1(\mathfrak{n})}$. We thus obtain an action of the Hecke algebra $\mathbb{T}$ on B commuting with the action of $\mathrm{Gal}(\overline{\mathbf{Q}}/\mathbf{Q})$. One checks easily that the lattice M is stable under the action of $\mathbb{T}$. (Note that the image of M in $\mathcal{A}^U$ is independent of the choice of q in the definition of W. Thus we may assume that $\mathfrak{a}$ is prime to q and express $T_\mathfrak{a}$ in terms of restrictions of homomorphisms g^*1_* on A.)

Let N denote the image of $H^2(\tilde{\mathcal{X}} \times \overline{\mathbf{Q}}, \mathbf{Z}_\ell)$ in $C = H^2(\tilde{\mathcal{X}} \times \overline{\mathbf{Q}}, \mathbf{Q}_\ell)$. Then by [**HLR**, Proposition 5.3] we have a natural inclusion $A \hookrightarrow C$, so A and hence B are crystalline at ℓ. The results of [**HLR**, Section 5] in fact yield an exact sequence

$$0 \longrightarrow L \longrightarrow N \longrightarrow H^2(X^\infty \times \overline{\mathbf{Q}}, \mathbf{Z}_\ell)$$

where X^∞ (the complement of X in $\tilde{\mathcal{X}} \times \mathbf{Q}$) is a relative divisor with normal crossings. Thus the cokernel of the above inclusion $L \hookrightarrow N$ is isomorphic to a direct sum of copies of $\mathbf{Z}_\ell(-1)$ (as a module over $\mathrm{Gal}(\overline{\mathbf{Q}}/K)$ where K, the field of definition of the cusps of $\overline{X}$, is a finite abelian extension of $\mathbf{Q}$ unramified at ℓ). We record the following immediate consequence.

LEMMA 1. *The inclusion $L \hookrightarrow N$ induces an isomorphism $\mathbb{D}^i(L_n) \cong \mathbb{D}^i(N_n)$ for $i \neq 1$.*

Composing with the isomorphism defined by the étale-crystalline comparison theorem, we obtain

$$\mathbb{D}^2(L_n) \cong H^0(\tilde{\mathcal{X}}, \Omega^2_{\tilde{\mathcal{X}}/\mathbf{Z}_{(\ell)}}) \otimes \mathbf{Z}/\ell^{n+1}\mathbf{Z}.$$

This isomorphism is independent of the choice of compactification (this follows from choosing a common refinement of the cone decompositions and applying naturality of the comparison isomorphism). Furthermore, restricting to $\mathcal{X}$ defines an inclusion

$$\mathbb{D}^2(L_n) \hookrightarrow H^0(\mathcal{X}, \Omega^2_{\mathcal{X}/\mathbf{Z}_{(\ell)}}) \otimes \mathbf{Z}/\ell^{n+1}\mathbf{Z}$$

which is compatible with the natural action of G. Composing with $\mathbb{D}^2$ of the inclusion $M_n = L_n^G \hookrightarrow L_n$, we get

$$\mathbb{D}^2(M_n) \hookrightarrow H^0(\mathcal{X}, \Omega^2_{\mathcal{X}/\mathbf{Z}_{(\ell)}}) \otimes \mathbf{Z}/\ell^{n+1}\mathbf{Z},$$

and the image coincides with that of $S/\ell^{n+1}S$ under the natural inclusion. Again using naturality of the comparison isomorphism, we find that the isomorphism $\mathbb{D}^2(M_n) \cong S/\ell^{n+1}S$ just defined is even independent of the choice of the auxiliary prime q.

PROPOSITION 1. $\mathbb{D}^2(M_n) \cong S/\ell^{n+1}S$ as a $\mathbb{T}$-module.

PROOF. We have already constructed an isomorphism of $\mathbf{Z}_\ell$-modules, and we now sketch a proof of its compatibility with the Hecke action. It suffices to prove that $\mathrm{gr}^2(\mathbb{D}_{\mathrm{cris}}(B \otimes K)) \to S \otimes K$ is $\mathbb{T} \otimes K$-linear for a finite extension K of $\mathbf{Q}_\ell$. But since $I = \mathrm{Ann}(M \otimes K)$ annihilates $S \otimes K$ (see [**HLR**, Satz 1.9], for example), it suffices to consider the action of $(\mathbb{T} \otimes K)/I$. For sufficiently large K, it follows from strong multiplicity one that this ring is generated by the Hecke operators $T_\mathfrak{a}$ with $\mathfrak{a}$ prime to ℓq (since ℓq is prime to $\mathfrak{n}$). Therefore it suffices to prove that $\mathrm{gr}^2(\mathbb{D}_{\mathrm{cris}}(B)) \to S \otimes \mathbf{Q}_\ell$ commutes with such $T_\mathfrak{a}$. But for an open compact subgroup U such that $U_\ell = \mathbf{GL}_2(\mathcal{O}_\ell)$, we can use the above method to define a map $\mathrm{gr}^2(\mathbb{D}_{\mathrm{cris}}(A_U)) \hookrightarrow H^0(X_U \times \mathbf{Q}_\ell, \Omega^2_{X_U/\mathbf{Q}_\ell})$ (which is also independent of the compactification). If moreover $g^{-1}Ug \subset W$, then we get a commutative diagram

$$\begin{array}{ccc} \mathrm{gr}^2(\mathbb{D}_{\mathrm{cris}}(A)) & \longrightarrow & \mathrm{gr}^2(\mathbb{D}_{\mathrm{cris}}(A_U)) \\ \downarrow & & \downarrow \\ H^0(X \times \mathbf{Q}_\ell, \Omega^2_{X/\mathbf{Q}_\ell}) & \longrightarrow & H^0(X_U \times \mathbf{Q}_\ell, \Omega^2_{X_U/\mathbf{Q}_\ell}), \end{array}$$

where the top inclusion is $\mathrm{gr}^2(\mathbb{D}_{\mathrm{cris}}(g_*))$ and the bottom one is $J(g)^*$. The desired compatibility now follows from the definitions of the Hecke operators. $\square$

3. Multiplicity one

Suppose that $f \in S \otimes \overline{\mathbf{Q}}_\ell$ is an eigenform of all the Hecke operators $T_\mathfrak{a}$. Then $f|T = \theta(T)f$ defines a homomorphism $\theta : \mathbb{T} \to \mathcal{O}_E$, where E is the field generated by the eigenvalues of the $T_\mathfrak{a}$. A theorem of Taylor [**Ta**] associates to such an f a continuous representation $\rho_f : \mathrm{Gal}(\overline{F}/F) \longrightarrow \mathbf{GL}_2(\mathcal{O}_E)$ satisfying

- ρ_f is unramified outside $\ell\mathfrak{n}$.
- For any prime ideal $\mathfrak{q}$ of F not dividing $\ell\mathfrak{n}$, $\rho_f(\mathrm{Frob}_\mathfrak{q})$ has trace $\theta(T_\mathfrak{q})$ and determinant $\theta(S_\mathfrak{q})\mathbf{N}\mathfrak{q}$.

(The operator $S_\mathfrak{q}$ on S satisfies $S_\mathfrak{q} \mathbf{N}\mathfrak{q} = T_\mathfrak{q}^2 - T_{\mathfrak{q}^2}$ and is thus in $\mathbb{T}$.) It will be more convenient to replace ρ_f by its dual and let $\mathrm{Frob}_\mathfrak{q}$ denote the *geometric* Frobenius, so $\det \rho_f \cong \chi(-1)$ for a character χ of $\mathrm{Gal}(\overline{F}/F)$ of conductor dividing $\mathfrak{n}$. Let λ denote the maximal ideal of $\mathcal{O}_E$. The semi-simplification of the reduction of ρ_f modulo λ then defines a representation $\overline{\rho}_f : \mathrm{Gal}(\overline{F}/F) \longrightarrow \mathbf{GL}_2(\mathcal{O}_E/\lambda)$. We recall that if $\overline{\rho}_f$ is irreducible, then it is absolutely irreducible (since the determinant of a complex conjugation is -1).

Now fix a maximal ideal $\mathbf{m}$ of $\mathbb{T}$, and let $\mathfrak{m}$ denote the inverse image of $\mathbf{m}$ in $\mathbb{T}$. Let $\mathbb{T}'$ denote the $\mathbf{Z}_\ell$-subalgebra of $\mathbb{T}$ generated by the Hecke operators $T_\mathfrak{a}$ with $\mathfrak{a}$ prime to $\mathfrak{n}$. Let $\mathfrak{m}' = \mathbb{T}' \cap \mathfrak{m}$, and let $\mathbf{F}$ be a finite field containing $\mathbb{T}'/\mathfrak{m}'$. Choosing f as above so that $\mathbf{m}$ is the kernel of $\overline{\theta} : \mathbb{T} \to \mathcal{O}_E/\lambda$, we can construct a semi-simple representation

$$\rho_\mathbf{m} : \mathrm{Gal}(\overline{F}/F) \longrightarrow \mathbf{GL}_2(\mathbf{F})$$

characterized by

- $\rho_\mathbf{m}$ is unramified outside $\ell\mathfrak{n}$.
- For any prime ideal $\mathfrak{q}$ of F not dividing $\ell\mathfrak{n}$, $\rho_\mathbf{m}(\mathrm{Frob}_\mathfrak{q})$ has trace $T_\mathfrak{q}$ and determinant $S_\mathfrak{q}\mathbf{N}\mathfrak{q}$.

We will also write $\rho = \rho_\mathbf{m}$ for the corresponding $\mathbf{F}[\mathrm{Gal}(\overline{F}/F)]$-module.

Let $\mathfrak{l}$ be a prime dividing ℓ in F, and let $k = \mathcal{O}/\mathfrak{l}$. From [**Ta**, Theorem 1] (see also [**Ca**]), we see that the Jordan-Holder constituents of ρ arise as $\mathbf{F}[\mathrm{Gal}(\overline{F}/F)]$-subquotients of $H^1(Y, \mathbf{F})$, where Y is a Shimura curve defined over F having good reduction at ℓ. Thus ρ is in $\mathfrak{CR}_\mathbf{F}(F, \mathfrak{l})$ with $\mathbb{D}(\rho)$ in $\mathfrak{MF}_{[0,1]}(\mathcal{O}_\mathfrak{l})$. Considering $\det(\rho)$, we see that $\mathrm{gr}^i(\Lambda^2_{\mathbf{F} \otimes k} \mathbb{D}(\rho)) = 0$ for $i \neq 1$. Thus $\mathbb{D}^i(\rho)$ is free of rank one over $\mathbf{F} \otimes k$ for $i = 0, 1$.

Let τ be the non-trivial automorphism of F. Let $\sigma = \sigma_\mathbf{m}$ denote the tensor-induction of ρ to $\mathrm{Gal}(\overline{\mathbf{Q}}/\mathbf{Q})$ (see [**CR**, Section 13], for example). The restriction of σ to $\mathrm{Gal}(\overline{F}/F)$ is isomorphic to $\rho \otimes_\mathbf{F} \rho^\tau$, where ρ^τ denotes the composition of ρ with conjugation by a lift of τ. Thus ρ^τ is in $\mathfrak{CR}_\mathbf{F}(F, \mathfrak{l})$, and $\mathbb{D}(\rho^\tau)$ is isomorphic to $\mathbb{D}_\mathbf{F}(F, \mathfrak{l}^\tau)(\rho)$ if ℓ splits and to $\mathbb{D}(\rho)^\Phi$ if ℓ is inert. Thus σ is an object of $\mathfrak{CR}_\mathbf{F}(\mathbf{Q}, \ell)$ satisfying

$$\dim_\mathbf{F} \mathbb{D}^i(\sigma) = \begin{cases} 1 & \text{if } i = 0 \text{ or } 2, \\ 2 & \text{if } i = 1, \\ 0 & \text{otherwise.} \end{cases}$$

The key observation (and our main reason for assuming that F is quadratic) is the following:

LEMMA 2. *Suppose that ρ is irreducible. Then $\dim_\mathbf{F} \mathbb{D}^0(\psi) = \dim_\mathbf{F} \mathbb{D}^2(\psi)$ for any $\mathbf{F}[\mathrm{Gal}(\overline{\mathbf{Q}}/\mathbf{Q})]$ Jordan-Holder constituent ψ of σ.*

PROOF. First suppose that χ is an $\mathbf{F}[\mathrm{Gal}(\overline{F}/F)]$-submodule of $\rho \otimes_\mathbf{F} \rho$ which is one-dimensional over $\mathbf{F}$. Since ρ is irreducible, we see that $\rho \otimes_\mathbf{F} \chi \cong \rho \otimes_\mathbf{F} \det \rho$. Therefore $\mathbb{D}^1(\chi) \cong k \otimes \mathbf{F}$. This applies also if χ is a one-dimensional quotient, for then $\mathrm{Hom}_\mathbf{F}(\chi, (\det \rho)^2)$ is a submodule. It follows that if χ is *any* Jordan-Holder constituent of $\rho \otimes_\mathbf{F} \rho \cong \det \rho \oplus \mathrm{Symm}^2 \rho$ (as an $\mathbf{F}[\mathrm{Gal}(\overline{F}/F)]$-module), then the $\mathbb{D}^i(\chi)$ are free over $k \otimes \mathbf{F}$ and satisfy $\dim \mathbb{D}^i(\chi) = \dim \mathbb{D}^{2-i}(\chi)$.

Now we turn our attention to $\rho \otimes \rho^\tau$. Suppose this has a one-dimensional constituent χ. Then we can assume χ is a submodule. (If χ is a quotient, then $\mathrm{Hom}_\mathbf{F}(\chi, \det \rho \det \rho^\tau)$ is isomorphic to a submodule.) We then have $\rho^\tau \otimes_\mathbf{F} \det \rho \cong \chi \otimes_\mathbf{F} \rho$, and therefore

$$\rho \otimes_\mathbf{F} \rho^\tau \cong \chi' \otimes_\mathbf{F} \rho \otimes_\mathbf{F} \rho$$

where $\chi' = \mathrm{Hom}_\mathbf{F}(\det \rho, \chi)$. But we can form $\mathbb{D}(\chi')$ and observe that $\mathbb{D}^0(\chi') \cong k \otimes \mathbf{F}$ and $\mathbb{D}^i(\chi') = 0$ for $i > 0$. It follows in this case that we have $\dim \mathbb{D}^i(\psi) = \dim \mathbb{D}^{2-i}(\psi)$ for any Jordan-Holder constituent ψ of $\rho \otimes \rho^\tau$ as an $\mathbf{F}[\mathrm{Gal}(\overline{F}/F)]$-module.

We are now left with the case that σ is reducible but has no one-dimensional constituent (as a module for $\mathbf{F}[\mathrm{Gal}(\overline{\mathbf{Q}}/\mathbf{Q})]$). It suffices to prove the lemma for the two-dimensional constituent ψ with $\mathbb{D}^2(\psi) = 0$. We have $\det \psi$ as a one-dimensional $\mathbf{F}[\mathrm{Gal}(\overline{F}/F)]$ constituent of

$$\Lambda^2_\mathbf{F}(\rho \otimes_\mathbf{F} \rho^\tau) \cong \left(\mathrm{Symm}^2(\rho) \otimes_\mathbf{F} \det \rho^\tau\right) \oplus \left(\mathrm{Symm}^2(\rho^\tau) \otimes_\mathbf{F} \det \rho\right).$$

From our analysis of $\rho \otimes_\mathbf{F} \rho$ we see that $\mathbb{D}^i(\det \psi) = 0$ for $i \neq 2$. So $\mathrm{gr}^i \Lambda^2_\mathbf{F} \mathbb{D}(\psi) = 0$ for $i \neq 2$, and consequently $\mathbb{D}^0(\psi) = 0$. The lemma follows. □

We remark that the same argument applies to $\mathbf{F}[\mathrm{Gal}(\overline{F}/F)]$ constituents (at least if $\ell > 5$, or if $\ell = 5$ and splits in F).

Now we examine the consequences for the intersection cohomology of the Hilbert modular surface.

Recall that the $\mathbf{GL}_2(F \otimes \mathbf{A}_f)[\mathrm{Gal}(\overline{\mathbf{Q}}/\mathbf{Q})]$-module $\mathcal{A} = \lim_{\to} A_U$ has a well-known decomposition indexed by automorphic representations of $\mathbf{GL}_2(F \otimes \mathbf{A})$. The results of Brylinski-Labesse and Harder-Langlands-Rapaport (see [**BL**], [**HLR**] and [**Rm**]) provide the information we need about the corresponding components of the $\mathbb{T}'[\mathrm{Gal}(\overline{\mathbf{Q}}/\mathbf{Q})]$-module

$$B \otimes_{\mathbf{Q}_\ell} \overline{\mathbf{Q}}_\ell \cong \bigoplus_\pi V_\pi^{m(\pi)}$$

(where B denotes $\mathcal{A}^{V_1(\mathfrak{n})}$). Two types of automorphic representations π appear in the sum with $m(\pi) > 0$. If π is one-dimensional, then $\pi = \nu \circ \det$ for a character ν of trivial conductor, in which case $V_\pi \cong \iota(-1) \oplus \epsilon(-1)$ where ι is trivial and ϵ is the non-trivial character of $\mathrm{Gal}(F/\mathbf{Q})$. Otherwise π is cuspidal of weight two and conductor dividing $\mathfrak{n}$, and V_π is four-dimensional. In this case, one can associate to π an eigenform f as above (not necessarily unique) so that $\mathbb{T}'$ acts on V_π via the composition $\mathbb{T}' \hookrightarrow \mathbb{T} \to \mathcal{O}_E \hookrightarrow \overline{\mathbf{Q}}_\ell$ (where the second map is θ_f). Moreover, viewed as representations of $\mathrm{Gal}(\overline{\mathbf{Q}}/\mathbf{Q})$, the semi-simplification of V_π is isomorphic to that of the tensor-induction of $\rho_f \otimes_{\mathcal{O}_E} \overline{\mathbf{Q}}_\ell$ from $\mathrm{Gal}(\overline{F}/F)$. (This follows from the fact that the characteristic polynomials of Frobenius coincide for all but finitely primes.)

LEMMA 3. *Let P be a $\mathbb{T}[\mathrm{Gal}(\overline{\mathbf{Q}}/\mathbf{Q})]$-subquotient of B annihilated by $\mathfrak{m}$, and let $\mathbf{F} = \mathbb{T}/\mathfrak{m}$. Let Q be a Jordan-Holder constituent of $P \otimes_{\mathbf{F}} \overline{\mathbf{F}}$. Then Q is isomorphic to $\iota(-1)$, to $\epsilon(-1)$ or to a Jordan-Holder constituent of $\sigma_{\mathfrak{m}} \otimes_{\mathbf{F}} \overline{\mathbf{F}}$.*

PROOF. We can regard P as a $\mathbb{T}'[\mathrm{Gal}(\overline{\mathbf{Q}}/\mathbf{Q})]$ subquotient of $B \otimes_{\mathbf{Q}_\ell} \overline{\mathbf{Q}}_\ell$ annihilated by $\mathfrak{m}' = \mathbb{T}' \cap \mathfrak{m}$. Thus it suffices to consider such subquotients of the V_π. If π is one-dimensional then Q is isomorphic to $\iota(-1)$ or $\epsilon(-1)$. Otherwise choose f as in the description of V_π, and note that $P = 0$ unless $\theta_f(\mathfrak{m}')$ is contained in the maximal ideal λ of $\mathcal{O}_E$. Thus the $\overline{\mathbf{F}}[\mathrm{Gal}(\overline{\mathbf{Q}}/\mathbf{Q})]$-constituents of $P \otimes_{\mathbb{T}'/\mathfrak{m}'} \overline{\mathbf{F}}$ are constituents of $\sigma_{\mathfrak{m}} \otimes_{\mathbf{F}} \overline{\mathbf{F}}$. □

Before stating our multiplicity one result, we recall our hypotheses and notation. We have fixed a real quadratic field F and a non-zero ideal $\mathfrak{n}$ in $\mathcal{O}_F$. We assume that the rational prime ℓ does not divide $6D_{F/\mathbf{Q}}\mathbf{N}(\mathfrak{n})$. We let $\mathbb{T}$ be the polynomial algebra over $\mathbf{Z}_\ell$ with variables $T_{\mathfrak{a}}$, where $\mathfrak{a}$ runs over the non-zero ideals of $\mathcal{O}_F$. Let $\mathbf{T}$ denote the image of $\mathbb{T}$ in the ring of endomorphisms of $S = S_2(V_1(\mathfrak{n}), \mathbf{Z}_\ell)$, the Hilbert modular cusp forms of weight two and level $\mathfrak{n}$ with coefficients in $\mathbf{Z}_\ell$ (Section 2.2). We have also defined an action of $\mathbb{T}$ on a lattice M in $A^{V_1(\mathfrak{n})}$ where A is the (ℓ-adic) intersection cohomology of a Hilbert modular surface (Section 2.4). For a maximal ideal $\mathbf{m}$ of $\mathbf{T}$, let $\mathfrak{m}$ denote its inverse image in $\mathbb{T}$ and let $\mathbf{F} = \mathbf{T}/\mathbf{m} = \mathbb{T}/\mathfrak{m}$. We have associated to $\mathbf{m}$ certain Galois representations $\rho_{\mathbf{m}} : \mathrm{Gal}(\overline{F}/F) \to \mathbf{GL}_2(\mathbf{F})$ and $\sigma_{\mathbf{m}} : \mathrm{Gal}(\overline{\mathbf{Q}}/\mathbf{Q}) \to \mathbf{GL}_4(\mathbf{F})$. The functors $\mathbb{D}^i$ were defined in Section 2.3.

We make another definition before stating the theorem. For any $\mathbf{F}'$ containing $\mathbf{F}$ there is a unique Jordan-Holder constituent ψ of $\sigma_{\mathbf{m}} \otimes_{\mathbf{F}} \mathbf{F}'$ such that $\mathbb{D}^2(\psi)$ is one-dimensional. Choose $\mathbf{F}'$ large enough so that ψ is absolutely irreducible, and we let $\sigma_{\mathbf{m},2} = \psi \otimes_{\mathbf{F}'} \overline{\mathbf{F}}$. Similarly define $\sigma_{\mathbf{m},0}$.

THEOREM 1. *With the above hypotheses and notation, $\dim_{\mathbf{F}} \mathbb{D}^2(M_0[\mathbf{m}]) = 1$, and $\sigma_{\mathbf{m},2}$ occurs with multiplicity one in the semi-simplification of $M_0[\mathbf{m}] \otimes_{\mathbf{F}} \overline{\mathbf{F}}$. If $\rho_{\mathbf{m}}$ is irreducible, then $\sigma_{\mathbf{m},0} = \sigma_{\mathbf{m},2}$ and $\dim_{\mathbf{F}} \mathbb{D}^0(M_0[\mathbf{m}]) = 1$.*

PROOF. Proposition 1 shows that $\mathbb{D}^2(M_0[\mathbf{m}]) \cong (S/\ell S)[\mathbf{m}]$ as a module for $\mathbb{T}/\mathbf{m}$. But since $S \cong \mathrm{Hom}_{\mathbf{Z}_\ell}(\mathbf{T}, \mathbf{Z}_\ell)$ (Section 2.2), we see that
$$(S/\ell S)[\mathbf{m}] \cong \mathrm{Hom}_{\mathbf{F}_\ell}(\mathbf{T}/\mathbf{m}, \mathbf{F}_\ell)$$
is one-dimensional over $\mathbf{T}/\mathbf{m}$. The assertion regarding $\sigma_{\mathbf{m},2}$ follows from Lemma 3, and the rest follows from Lemma 2. □

Note that more generally, if $\rho_{\mathbf{m}}$ is irreducible and P is a finite $\mathbb{T}[\mathrm{Gal}(\overline{\mathbf{Q}}/\mathbf{Q})]$-subquotient of $B_{\mathbf{m}}$, then $\mathbb{D}^2(P)$ and $\mathbb{D}^0(P)$ have the same length.

4. Poincaré duality

We will now appeal to Poincaré duality to show that under the hypotheses of the second part of Theorem 1, $S/\mathbf{m}S$ is one-dimensional over $\mathbf{T}/\mathbf{m}$. It will then follow that $\mathrm{Hom}_{\mathbf{Z}_\ell}(\mathbf{T}_\mathbf{m}, \mathbf{Z}_\ell) \cong S_\mathbf{m}$ is free of rank one over $\mathbf{T}_\mathbf{m}$, and consequently that $\mathbf{T}_\mathbf{m}$ is Gorenstein.

THEOREM 2. *Let F be a real quadratic field, and let $\mathfrak{n}$ be a non-zero ideal in $\mathcal{O}_F$. Let ℓ be a rational prime not dividing $6D_{F/\mathbf{Q}}\mathbf{N}_{F/\mathbf{Q}}(\mathfrak{n})$. Let $\mathbf{T}$ be the $\mathbf{Z}_\ell$-algebra of endomorphisms of $S_2(V_1(\mathfrak{n}), \mathbf{Z}_\ell)$ generated by all the Hecke operators $T_\mathfrak{a}$, and let $\mathbf{m}$ be a maximal ideal of $\mathbf{T}$. Suppose that the associated Galois representation $\rho_\mathbf{m} : \mathrm{Gal}(\overline{F}/F) \to \mathbf{GL}_2(\mathbf{T}/\mathbf{m})$ is irreducible. Then $\mathbf{T}_\mathbf{m}$ is Gorenstein.*

PROOF. By the Poincaré Duality Theorem, the natural symmetric pairing $C \otimes_{\mathbf{Q}_\ell} C \to \mathbf{Q}_\ell(-2)$ is non-degenerate (where $C = H^2(\tilde{\mathcal{X}} \times \overline{\mathbf{Q}}, \mathbf{Q}_\ell))$. Moreover the pairings $A_U \otimes_{\mathbf{Q}_\ell} A_U \to \mathbf{Q}_\ell(-2)$ are non-degenerate (where A_U is the image of $H^2_c(X_U \times \overline{\mathbf{Q}}, \mathbf{Q}_\ell)$ in $H^2(X_U \times \overline{\mathbf{Q}}, \mathbf{Q}_\ell))$, and the pairings on $A = A_W$ and C are compatible with the inclusion with $A \hookrightarrow C$. Since $g_* : A \to A$ is adjoint to $g_*^{-1} = g^*$ for $g \in V_1(\mathfrak{n})$, we see in fact that the pairing on A restricts to a non-degenerate pairing $B \otimes_{\mathbf{Q}_\ell} B$ where $B = A^G$. We thus obtain isomorphisms of $\mathrm{Gal}(\overline{\mathbf{Q}}/\mathbf{Q})$-modules
$$\alpha_V : V \to \mathrm{Hom}_{\mathbf{Q}_\ell}(V, \mathbf{Q}_\ell(-2))$$
for $V = A$, B and C. The above pairings restrict to non-degenerate pairings on the corresponding lattices and thus define injections
$$\alpha_P : P \to \mathrm{Hom}_{\mathbf{Z}_\ell}(P, \mathbf{Z}_\ell(-2))$$
for $P = N$, L and M. This is an isomorphism for $P = N$, but for $P = L$ or M, it only follows that α_P has finite cokernel. Tensoring with $\mathbf{F}_\ell$, we obtain homomorphisms
$$\alpha_{P,0} : P_0 \to \mathrm{Hom}_{\mathbf{F}_\ell}(P_0, \mu_\ell^{\otimes (-2)}).$$
Again this is an isomorphism if $P = N$, but not necessarily for $P = L$ or M. However we have the following immediate consequence of Lemma 1.

LEMMA 4. $\mathbb{D}^2(\alpha_{M,0})$ *is an isomorphism.*

The homomorphism $\alpha_{M,0}$ is not equivariant for the Hecke operators, but we can adjust for this by composing with the homomorphism induced by a certain involution of the Hilbert modular surface. We begin by observing that the involution $x \mapsto x^{-\iota} = x \det x^{-1}$ of $\mathbf{GL}_2(F \otimes \mathbf{A}_\mathbf{f})$ defines an isomorphism $t_U : X_U(\mathbf{C}) \to X_{U^\iota}(\mathbf{C})$ (where ι denotes the main involution of $\mathbf{GL}_2$). If U contains the principal congruence subgroup of level $N\mathcal{O}$, then t_U can be defined as a map on the coarse moduli spaces
$$X_U \times \mathbf{Q}(\mu_N) \to X_{U^\iota} \times \mathbf{Q}(\mu_N).$$

Indeed if V is an abelian variety with $\mathcal{O}$ multiplication and level U structure, then the corresponding point is sent to that of $V^t \otimes_{\mathcal{O}} \mathfrak{d}$ (where V^t is the dual abelian variety and $\mathfrak{d}$ is the different of F) with level U^ι structure induced by the Weil pairing and a choice of ζ_N.

Now fix an element $b = \begin{pmatrix} 0 & 1 \\ \varpi_{\mathfrak{n}} & 0 \end{pmatrix}$ of $\mathbf{GL}_2(F \otimes \mathbf{A_f})$ where $\varpi_{\mathfrak{n}} \prod_v \mathcal{O}_v = \prod_v \mathfrak{n}\mathcal{O}_v$. Then $b^{-1}W^\iota b = W$, so we have $J(b) : X_{W^\iota} \to X_W$, and $J(b) \circ t_W$ defines an involution $X_W \times \mathbf{Q}(\mu_n)$. This induces an involution of A (and of L) as a module for $\mathrm{Gal}(\overline{\mathbf{Q}}/\mathbf{Q}(\mu_{nq}))$, which we denote $w_{\mathfrak{n}}$. One checks easily that $M = L^G$ is stable under $w_{\mathfrak{n}}$ and that the restriction to M is independent of the choice of q in the definition of W. We have thus defined an involution of M (and of $B = A^G$) as a module for $\mathrm{Gal}(\overline{\mathbf{Q}}/\mathbf{Q}(\mu_n))$. We also find that $w_{\mathfrak{n}} T w_{\mathfrak{n}}$ and T are adjoint under the pairing $B \otimes B \to \mathbf{Q}_\ell(-2)$. (This is true of $T_{\mathfrak{a}}$ on A for $\mathfrak{a}$ prime to q, and the pairing on B is independent of q up to multiplication by a unit.) Therefore the composition $\alpha_B \circ w_{\mathfrak{n}} : B \to \mathrm{Hom}_{\mathbf{Q}_\ell}(B, \mathbf{Q}_\ell(-2))$ is $\mathbb{T}$-linear. The same is true of $\alpha_{M,0} \circ w_{\mathfrak{n},0}$, where $w_{\mathfrak{n},0}$ is the involution induced by $w_{\mathfrak{n}}$ on M_0. We can then apply $\mathbb{D} = \mathbb{D}_{\mathbb{T}}(\mathbf{Q}(\mu_n), \mathfrak{l})$ for a prime $\mathfrak{l}$ above ℓ, and it follows from the preceding lemma that $\mathbb{D}^2(\alpha_{M,0} \circ w_{\mathfrak{n},0})$ defines an isomorphism

$$\mathbb{D}^2(M_0) \otimes R \to \mathbb{D}^2(\mathrm{Hom}(M_0, \mu_\ell^{\otimes(-2)})) \otimes R$$

of $\mathbb{T} \otimes R$-modules, where $R \cong \mathbf{F}_\ell(\mu_n)$ is the residue field of $\mathfrak{l}$. We have now proved

LEMMA 5. $\mathbb{D}^2(M_0) \otimes R \cong \mathrm{Hom}_{\mathbf{Z}_\ell}(\mathbb{D}^0(M_0), \mathbf{F}_\ell) \otimes R$ as $\mathbb{T} \otimes R$-modules.

Tensoring over $\mathbb{T}$ with $\mathbf{F} = \mathbb{T}/\mathfrak{m}$, we conclude that

$$\dim_{\mathbf{F}}(\mathbb{D}^2(M_0)/\mathfrak{m}\mathbb{D}^2(M_0)) = \dim_{\mathbf{F}}(\mathbb{D}^0(M_0)[\mathfrak{m}]).$$

But if $\rho_{\mathfrak{m}}$ is irreducible, then this dimension is one, and therefore so is that of $S/\mathfrak{m}S \cong \mathbb{D}^2(M_0)/\mathfrak{m}\mathbb{D}^2(M_0)$. This concludes the proof of the theorem. □

References

[BLR] N. Boston, H. W. Lenstra and K. Ribet, *Quotients of group rings arising from two-dimensional representations*, C. R. Acad. Sci. Paris, Série I **312** (1991), 323–328.

[BL] J.-L. Brylinski and J.-P. Labesse, *Cohomologie d'intersection et fonctions L de certaines variétés de Shimura*, Ann. Sci. Ec. Norm. Super. **17** (1984), 361–412.

[Ca] H. Carayol, *Sur les représentations ℓ-adiques associées aux formes modulaires de Hilbert*, Ann. Sci. Ec. Norm. Super. **19** (1986), 409–468.

[Ch] C.-L. Chai, *Arithmetic minimal compactification of the Hilbert-Blumenthal moduli spaces*, Annals of Math. **131** (1990), 541–554.

[CR] C. Curtis and I. Reiner, *Methods of representation theory*, vol. I, Wiley, New York, 1981.

[De] P. Deligne, *Travaux de Shimura*, Sém. Bourbaki, 23e année, n° 389, (1970/71), 123–165.

[Di] F. Diamond, *The Taylor-Wiles construction and multiplicity one*, to appear in Inv. Math.

[E] B. Edixhoven, *The weight in Serre's conjectures on modular forms*, Invent. Math. **109** (1992), 563–594.

[Fa] G. Faltings, *Crystalline cohomology and p-adic Galois representations*, Algebraic Analysis, Geometry and Number Theory, Proc. JAMI Inaugural Conference, Johns-Hopkins Univ. Press (1989), 25-79.

[FJ] G. Faltings and B. Jordan, *Crystalline cohomology and* $\mathbf{GL}(2, \mathbf{Q})$, Israel J. Math. **90** (1995), 1–66.

[F1] J.-M. Fontaine, *Modules galoisiens, modules filtrés et anneaux de Barsotti-Tate*, Journées de Géometrie Algébrique de Rennes, Astérisque **65** (1979), 3–80.

[F2] J.-M. Fontaine, *Sur certains types de représentations p-adiques du groupe de Galois d'un corps local; construction d'un anneau de Barsotti-Tate*, Ann. of Math. **115** (1982), 529–577.

[FL] J.-M. Fontaine and G. Lafaille, *Constructions de représentations p-adiques*, Ann. Sci. Ec. Norm. Super. **15** (1982), 547-608.

[FM] J.-M. Fontaine and W. Messing, *p-adic periods and p-adic étale cohomology*, Comtemp. Math. **67** (1987), 179–207.

[Fr] E. Freitag, *Hilbert modular forms*, Springer-Verlag, 1990.

[Fu] K. Fujiwara, *Deformation rings and Hecke algebras in the totally real case*, preprint.

[HLR] G. Harder, R. P. Langlands, M. Rapoport, *Algebraische Zyklen auf Hilbert-Blumenthal-Flachen*, J. Reine Angew. Math. **366** (1986), 53–120.

[H] H. Hida, *On p-adic Hecke algebras for GL_2 over totally real fields*, Ann. of Math. **128** (1988), 295–384.

[I] L. Illusie, *Cohomologie de de Rham et cohomologie étale p-adiques*, Sém. Bourbaki, 42e année, n° 726, (1989/90), 325–374.

[M] B. Mazur, *Modular curves and the Eisenstein ideal*, Publ. Math. IHES **47** (1977), 33–186.

[MR] B. Mazur and K. Ribet, *Two-dimensional representations in the arithmetic of modular curves*, Astérisque **196-197** (1991), 215–255.

[Rm] D. Ramakrishnan, *Arithmetic of Hilbert-Blumenthal Surfaces*, Montreal Conference in Number Theory, CMS Conf. Proc. vol. 7 (1985), 285–370.

[Rp] M. Rapoport, *Compactifications de l'espace de modules de Hilbert-Blumenthal*, Comp. Math. **36** (1978), 255–335.

[R1] K. Ribet, *On modular representations of $\mathrm{Gal}(\overline{\mathbf{Q}}/\mathbf{Q})$ arising from modular forms*, Invent. Math. **100** (1990), 431–476.

[R2] K. Ribet, *Multiplicities of Galois representations in Jacobians of Shimura curves*, in Festschrift of I. I. Piatetski-Shapiro (Part II), Israel Math. Conf. Proc. **3** (1990), 221–236.

[S1] G. Shimura, *On canonical models of arithmetic quotients of bounded symmetric domains*, Ann. of Math. **91** (1970), 144–222.

[S2] G. Shimura, *The special values of the zeta functions associated with Hilbert modular forms*, Duke Math. J. **45** (1978), 637–679.

[Ta] R. Taylor, *On Galois representations associated to Hilbert modular forms*, Invent. Math. **98** (1989), 265–280.

[TW] R. Taylor and A. Wiles, *Ring theoretic properties of certain Hecke algebras*, Annals of Math. **141** (1995), 553–572.

[Ti] J. Tilouine, *Un sous-groupe p-divisible de la jacobienne de $X_1(Np^r)$ comme module sur l'algèbre de Hecke*, Bull. Soc. Math. France **115** (1987), 329–360.

[W] A. Wiles, *Modular elliptic curves and Fermat's Last Theorem*, Annals of Math. **141** (1995), 443–551.

[Y] L. Yang, *Multiplicities for Galois representations in the higher weight sheaf cohomology associated to Shimura curves*, Ph. D. Thesis, The City University of New York, 1996.

DEPARTMENT OF MATHEMATICS, MASSACHUSETTS INSTITUTE OF TECHNOLOGY, CAMBRIDGE, MA 02139

E-mail address: `fdiamond@math.mit.edu`

THE ABC CONJECTURE
AND
EXPONENTS OF CLASS GROUPS OF QUADRATIC FIELDS

M. Ram Murty[1]

1. Introduction.

Jayadeva, in a commentary of the 11th century, describes a method he calls *cakravāla* (from the Sanskrit word *cakra* which means 'wheel') to determine all solutions of the equation

(1) $$x^2 - dy^2 = m$$

with $d > 0$. (He however does not prove that he has all of the solutions.) This equation, usually called Pell's equation, had been studied much earlier by several Indian mathematicians. In the work of Brahmagupta, dating back to the 7th century, we find a whole section devoted to such equations. His 'bhavana' (meaning production in Sanskrit) rules gave an algorithm for producing many solutions. Later in the 12th century, Bhaskara had given an algorithm for finding all solutions and it was believed, until the work of Jayadeva, that Bhaskara's work was the oldest detailed treatment of such equations. See Weil [W, p. 17-24] for an illuminating discussion on this matter.

Not to upset tradition, we will continue to refer to (1) as Pell's equation. We shall use the 'bhavana' and 'cakravala' methods to study exponents of ideal class groups of real quadratic fields. The corresponding results for imaginary quadratic fields are also established and are much stronger.

To be precise, the problem we wish to investigate in this paper concerns quadratic extensions of the rational number field. Given a natural number g, we wish to count the number of imaginary quadratic fields $\mathbb{Q}(\sqrt{-d})$ with $0 < d < x$ whose order of the class group is divisible by g. We investigate the identical question for real quadratic fields.

There have been many papers written showing that the number of such fields is infinite. First, if g is a power of 2, then genus theory gives a very precise answer. The number of such fields with absolute value of the discriminant less than x is asymptotic to cx for some constant c which of course depends on the power of 2.

When $g = 3$, Davenport and Heilbronn [DH] have obtained precise results. If $r_3(d)$ denotes the rank of the 3-part of the ideal class group of $\mathbb{Q}(\sqrt{-d})$, they proved

$$\sum_{d<x} 3^{r_3(d)} \sim 2x.$$

If $R_3(d)$ is the 3-rank of the class group of $\mathbb{Q}(\sqrt{d})$ they showed

$$\sum_{d<x} 3^{R_3(d)} \sim \frac{4}{3}x.$$

---

[1] *Research partially supported by NSERC, FCAR and CICMA.*

It is not clear if one can deduce that a positive proportion of such fields have class number divisible by 3.

It is however, possible to deduce some quantitative estimate from a related result of Davenport and Heilbronn [DH]. Indeed, they show that the number of cubic fields of positive discriminant less than x is asymptotically $1/12\zeta(3)$. By class field theory, the class number of $\mathbb{Q}(\sqrt{d})$ is divisible by 3 if and only if there exists a cubic field of discriminant d. It is now immediate that the number of real quadratic fields whose discriminant is $< x$ and whose class number is divisible by 3 is $\gg x^{1/3-\epsilon}$. An identical result can be deduced in the imaginary quadratic case.

Many authors, such as Nagell [Na], Humbert [Hu], Ankeny and Chowla [AC] and Kuroda [K], have shown that for a given number g, there are infinitely many imaginary quadratic fields whose class number is divisible by g. For real quadratic fields, infinitude has been shown by Yamamoto [Y] and Weinberger [We] independently.

In either the imaginary quadratic or the real quadratic case, nothing quantitative is known, for $g \geq 5$. Cohen and Lenstra [CL] formulated the following conjecture when $g = p$ is a prime. In the imaginary quadratic case, they conjecture that the probability p divides the class number is

$$1 - \prod_{i=1}^{\infty}\left(1 - \frac{1}{p^i}\right).$$

In the real quadratic case, they conjecture

$$1 - \prod_{i=2}^{\infty}\left(1 - \frac{1}{p^i}\right)$$

as the probability that p divides the class number.

Assuming the ABC conjecture, we will prove that the number of imaginary quadratic fields whose disriminant is $< x$ with exponent of the class group divisible by g is $\gg x^{\frac{1}{g}-\epsilon}$. In the real quadratic case, we obtain $\gg x^{\frac{1}{2g}-\epsilon}$. These we state as:

Theorem 13. *Let $g \geq 3$ be odd and assume the ABC conjecture. The number of imaginary quadratic fields $\mathbb{Q}(\sqrt{-d})$ with $0 < d < x$ and whose class group has exponent divisible by g is $\gg x^{\frac{1}{g}-\epsilon}$ for any $\epsilon > 0$. (Here the implied constant will depend on ϵ.)*

Theorem 14. *Let $g \geq 3$ be odd and assume the ABC conjecture. The number of real quadratic fields $\mathbb{Q}(\sqrt{d})$ with $0 < d < x$ and whose class group has exponent divisible by g is $\gg x^{1/2g-\epsilon}$ for any $\epsilon > 0$.*

We also deduce in the last section an interesting result about exponents of the class group of certain real quadratic fields.

2. The imaginary quadratic case.

We begin with the following theorem.

Theorem 1. *Suppose that $n^g - 1 = d$ is squarefree with n odd and ≥ 5. Then the class group of $\mathbb{Q}(\sqrt{-d})$ has an element of order g.*

Proof. We have the ideal factorization in $\mathbb{Q}(\sqrt{-d})$:

$$(n)^g = (1 + \sqrt{-d})(1 - \sqrt{-d}).$$

Since n is odd, the ideals $(1 + \sqrt{-d})$ and $(1 - \sqrt{-d})$ are coprime. Thus, they must each be the g-th power of an ideal. Therefore,

$$\mathfrak{A}^g = (1 + \sqrt{-d}) \quad \text{and} \quad (\mathfrak{A}')^g = (1 - \sqrt{-d})$$

where $\mathfrak{A}$ and $\mathfrak{A}'$ are coprime and $N\mathfrak{A} = n$. If for some $m \leq g-1$, we have that $\mathfrak{A}^m$ is principal, then for some $u, v \in \mathbb{Z}$,

$$\mathfrak{A}^m = (u + v\sqrt{-d}) \quad \text{or} \quad (\frac{u + v\sqrt{-d}}{2})$$

depending on whether $-d \equiv 2, 3 \pmod{4}$ or $-d \equiv 1 \pmod{4}$ respectively. We must have $v \neq 0$ for otherwise $\mathfrak{A}$ and $\mathfrak{A}'$ would have a common factor. Thus, taking norms of the above equation, we find in case $-d \equiv 2, 3 \pmod{4}$:

$$n^{g-1} \geq n^m = u^2 + dv^2 \geq d = n^g - 1$$

so that $1 \geq n^{g-1}(n-1) \geq 4 \cdot 5^{g-1}$, a contradiction. In the case $-d \equiv 1 \pmod{4}$, we find:

$$n^{g-1} \geq n^m \geq \frac{u^2 + dv^2}{4} \geq \frac{d}{4} = \frac{n^g - 1}{4}$$

so that $1 \geq n^{g-1}(n-4) \geq 5^{g-1}$, a contradiction. This completes the proof of the theorem.

Theorem 2. *Let $g > 2$. Suppose that $n^g - 1 = p^2 d$, with d squarefree, n odd and ≥ 5. If $p < n^{g/4}/2\sqrt{2}$, then the class group of $\mathbb{Q}(\sqrt{-d})$ has an element of order g.*

Proof. We proceed as before:

$$(n)^g = (1 + p\sqrt{-d})(1 - p\sqrt{-d}).$$

Since n is odd, $(1 + p\sqrt{-d})$ and $(1 - p\sqrt{-d})$ are coprime. Thus, they must each be the g-th power of an ideal. Therefore,

$$\mathfrak{A}^g = (1 + p\sqrt{-d}) \quad \text{and} \quad (\mathfrak{A}')^g = (1 - p\sqrt{-d})$$

where $\mathfrak{A}$ and $\mathfrak{A}'$ are coprime with $N\mathfrak{A} = N\mathfrak{A}' = n$. Thus, the order of the ideal class of $\mathfrak{A}$ divides g. Suppose $\mathfrak{A}^m$ is principal for some $m \leq g/2$. Then, as before

$$\mathfrak{A}^m = (u + v\sqrt{-d}) \quad \text{or} \quad (\frac{u + v\sqrt{-d}}{2})$$

according as the residue class d belongs to mod 4. Thus, as before,

$$n^{g/2} \geq n^m = \frac{u^2 + dv^2}{4} \geq \frac{d}{4} = \frac{n^g - 1}{4p^2} \geq \frac{n^g}{8p^2}$$

so that $p \geq n^{g/4}/2\sqrt{2}$ contrary to hypothesis. This completes the proof.

3. The real quadratic case.

We begin by reviewing some classical material about continued fractions.

Lemma 3. *If N and d are integers with $d > 0$ and $|N| < \sqrt{d}$ and d is not a square, then all positive solutions of the Pell's equation*

$$x^2 - dy^2 = N$$

are such that x/y is a convergent of $\sqrt{d}$.

Proof. See LeVeque [L, p. 181, Theorem 9-8].

We apply this as follows:

Lemma 4. *The continued fraction of $\sqrt{d}$ with $d = a^2 + 1$ is*

$$[a, 2a, 2a, ...].$$

Moreover, if $|u^2 - dv^2| \neq 0$ or 1, then

$$|u^2 - dv^2| > \sqrt{d}.$$

Proof. We find from the continued fraction algorithm that the convergents p_k/q_k always satisfy

$$p_k^2 - dq_k^2 = \pm 1$$

so that the result easily follows from Lemma 3.

Theorem 5. *Suppose that n is odd, $n \geq 5$ and $n^{2g} + 1 = d$ is squarefree. Then the class group of $\mathbb{Q}(\sqrt{d})$ has an element of order g.*

Remark. This theorem is essentially due to Ankeny and Chowla [AC]. Since they did not supply a complete proof, we do so here. This will have its use since we will also need to modify their result later on.

Proof. As before,
$$(n)^{2g} = (-1 + \sqrt{d})(1 + \sqrt{d}).$$

Since n is odd, each of the ideals $(-1 + \sqrt{d})$ and $(1 + \sqrt{d})$ must be coprime. Hence

$$\mathfrak{A}^{2g} = (-1 + \sqrt{d}) \quad \text{and} \quad (\mathfrak{A}')^{2g} = (1 + \sqrt{d}).$$

If $\mathfrak{A}^m = (u + v\sqrt{d})$ or $(\frac{u+v\sqrt{d}}{2})$ according as $d \equiv 2, 3 \pmod 4$ or $d \equiv 1 \pmod 4$, then
$$n^m = |u^2 - dv^2| \quad \text{or} \quad |\frac{u^2 - dv^2}{4}|.$$
In either case, we deduce from Lemma 4 that
$$n^m \geq |\frac{u^2 - dv^2}{4}| \geq \frac{n^g}{4}.$$

If $m \leq g - 1$, we deduce $n \leq 4$, a contradiction. Therefore, $\mathfrak{A}$ has order $\geq g$. Since $\mathfrak{A}^{2g}$ is principal, $\mathfrak{A}$ has order g or $2g$. In either case, the class group has an element of order divisible by g.

We modify Theorem 5 in the following way:

Theorem 6. Let g be odd and ≥ 5. Suppose that n is odd, $n \geq 5$ and $n^{2g} + 1 = p^2 d$ with $1 < d$ squarefree, p satisfying $p < n^{3/2}/2$. Then the class group of $\mathbb{Q}(\sqrt{d})$ has an element of order g.

Proof. We proceed as before to find that $(-1 + p\sqrt{d})$ and $(1 + p\sqrt{d})$ must be $2g$-th powers of coprime ideals (since n is odd). Thus,
$$\mathfrak{A}^{2g} = (-1 + p\sqrt{d})$$
and if
$$\mathfrak{A}^m = (u + v\sqrt{d}) \quad \text{or} \quad (\frac{u + v\sqrt{d}}{2})$$
with $m \leq g - 3$ we find by Lemma 4 that
$$n^{g-3} \geq n^m \geq |\frac{u^2 - dv^2}{4}| > |\frac{(pu)^2 - p^2 dv^2}{4p^2}| > \frac{n^g}{4p^2}$$
so that $2p > n^{3/2}$, contrary to hypothesis. Thus, if $\mathfrak{A}^m$ is principal, then $m \geq g - 1$ and must divide $2g$, since g is odd. Thus, $m = g$ or $2g$ and in either case, we are done.

4. Application of the ABC conjecture.

One can use the ABC conjecture to determine if any large prime factors divide $n^g - 1$. Recall that the ABC conjecture states that if
$$A + B = C$$
and A, B, C are three integers mutually coprime, then
$$\max(|A|, |B|, |C|) \ll \prod_{p | ABC} p^{1+\epsilon}$$
for any $\epsilon > 0$ and the implied constant depends on ϵ. Now let $0 < \delta < g$. Thus, if $p^2 | n^g - 1$ and $p > n^{1+\delta}$, then considering the ABC equation $1 + (n^g - 1) = n^g$, the ABC conjecture implies
$$n^g \ll n^{(1+\epsilon)(g-\delta)}.$$

If we choose $\epsilon = \delta/(g - \delta)$, this is a contradiction for n sufficiently large. We conclude that if $p^2 | n^g - 1$ then $p < n^{1+\delta}$. This we state as:

Proposition 7. *Let $0 < \delta < g$ and assume the ABC conjecture. If p is prime and $p^2 | n^g - 1$, then $p < n^{1+\delta}$ for all n sufficiently large.*

We can state a similar result for $n^{2g} + 1$. Assuming the ABC conjecture, Granville [G] has recently established that for any polynomial f, the number of $n < x$ such that a given polynomial $f(n)$ is squarefree is asymptotically $c_f x$ for some constant c_f. In out context, we need such a result for the polynomial $f(n) = n^g - 1$ and n odd. One can modify the method of Granville to yield such a result and from it deduce Theorems 13, with exponent $1/g$ and Theorem 14 with exponent $1/2g$. Granville's application of the ABC invokes the deep theorem of Belyi. For the sake of independent interest, we will adopt here a slightly different route. We begin by stating the analog of Proposition 7 tailored to be applied in the real quadratic case.

Proposition 8. *Let $0 < \delta < 2g$ and assume the ABC conjecture. If p is prime and $p^2 | n^{2g} + 1$, then $p < n^{1+\delta}$.*

We record, for later calculation, the following corollaries:

Corollary 9. *Let $0 < \delta < g$ and assume the ABC conjecture. Then the number of $n \leq x$ such that there is a prime $p > n^{1+\delta}$ and $p^2 | (n^g - 1)$ is bounded.*

Corollary 10. *Let $0 < \delta < 2g$ and assume the ABC conjecture. Then, the number of $n \leq x$ such that there is a prime $p > n^{1+\delta}$ and $p^2 | (n^{2g} + 1)$ is bounded.*

5. The simple asymptotic sieve.

We now proceed as in Hooley [H]. We use the identity

$$\sum_{d^2 | n} \mu(d) = \begin{cases} 1 & \text{if } n \text{ is squarefree} \\ 0 & \text{otherwise.} \end{cases}$$

Let $0 < \delta < g$. We will write $n \sim x$ to mean that there exist positive constants a and b so that $ax < n < bx$. Let $N(x)$ be the number of $n \sim x$ such that $f(n) = n^g - 1$ is squarefree or has a prime divisor p satisfying $n < p < n^{1+\delta}$. Let P_z be the product of the primes $\leq z$. Clearly, if $z = c \log x$ for sufficiently small c, then

$$N(x) \leq N(x, z)$$

where

$$N(x, z) = \sum_{n \sim x} \sum_{d^2 | (f(n), P_z^2)} \mu(d).$$

By Corollary 9, the number of $n \leq x$ such that there is a prime $p > n^{1+\delta}$ and $p^2 | (n^g - 1)$ is bounded. Hence,

$$N(x) \geq N(x, z) - \sum_{z < p < x^{1+\delta}} N_p(x) + O(1)$$

where $N_p(x)$ is the number of $n \sim x$ such that $p^2|f(n)$. We will choose z sufficiently small so that the error terms are controlled. Indeed,

$$N(x, z) = \sum_{d|P_z} \mu(d) N_d(x)$$

where $N_d(x)$ is the number of $n \sim x$ such that $d^2|f(n)$. Letting $\rho(p^2)$ be the number of solutions of the congruence $f(m) \equiv 0 \pmod{p^2}$, we find that choosing $z = \frac{1}{2 \log 2g} \log x$ gives

$$N(x, z) = x \prod_{p<z} \left(1 - \frac{\rho(p^2)}{p^2}\right) + O(x^{1/2}).$$

Now,

$$\sum_{z<p<x} N_p(x) \ll \frac{x}{\log x}$$

since $\rho(p^2) \leq g$ for p coprime to g by a simple calculation via Hensel's lemma. This leads to:

Theorem 11. *Let g be odd, $0 < \delta < g$ and assume the ABC conjecture. Let $\tilde{N}(x)$ be the number of odd $n \sim x$ such that $f(n) = n^g - 1$ is either squarefree or has a prime divisor p satisfying $p^2|(n^g - 1)$ with $n < p < n^{1+\delta}$. Then*

$$\tilde{N}(x) = \frac{x}{4} \prod_{p>2} \left(1 - \frac{\rho(p^2)}{p^2}\right) + O\left(\frac{x(\log \log x)^{g-1}}{\log x}\right).$$

Proof. We begin by observing that by an argument similar to the above, the number of even $n \sim x$ such that $n^g - 1$ has no squared prime factor $p < x$ is

$$= \frac{x}{2} \prod_{2<p<z} \left(1 - \frac{\rho(p^2)}{p^2}\right) + O\left(\frac{x}{\log x}\right).$$

Since g is odd, $\rho(4) = 1$ so that

$$N(x) = \frac{3}{4} x \prod_{2<p<z} \left(1 - \frac{\rho(p^2)}{p^2}\right) + O\left(\frac{x}{\log x}\right).$$

Thus,

$$\tilde{N}(x) = \frac{1}{4} x \prod_{2<p<z} \left(1 - \frac{\rho(p^2)}{p^2}\right) + O\left(\frac{x}{\log x}\right).$$

In the above calculation, it remains to estimate

$$\prod_{p} \left(1 - \frac{\rho(p^2)}{p^2}\right) - \prod_{p<z} \left(1 - \frac{\rho(p^2)}{p^2}\right) \ll \sum_{d>z} \frac{g^\nu(d)}{d^2},$$

where $\nu(d)$ denotes the number of distinct prime factors of d. Since

$$\sum_{n \leq x} g^{\nu(d)} \ll x(\log x)^{g-1}$$

we have by partial summation

$$\sum_{d>z} \frac{g^{\nu(d)}}{d^2} \ll \int_z^\infty \frac{(\log t)^{g-1} dt}{t^2} \ll \frac{(\log z)^{g-1}}{z}$$

which completes the proof since $z = \frac{1}{2 \log 2g} \log x$.

In an identical manner, we prove

Theorem 12. *Let $0 < \delta < 2g$. Assume the ABC conjecture. Let $N_1(x)$ be the number of $n \sim x$ such that $f(n) = n^{2g} + 1$ is either squarefree or has $p^2 | (n^{2g} + 1)$ with $n < p < n^{1+\delta}$. Then,*

$$N_1(x) = x \prod_p \left(1 - \frac{\rho(p^2)}{p^2}\right) + O\left(\frac{x(\log \log x)^{2g-1}}{\log x}\right)$$

where $\rho(p^2)$ is the number of solutions of the congruence $n^{2g} + 1 \equiv 0 (\bmod p^2)$.

6. The main theorems.

We can now prove:

Theorem 13. *Let $g \geq 3$ be odd and assume the ABC conjecture. The number of imaginary quadratic fields $\mathbb{Q}(\sqrt{-d})$ with $0 < d < x$ whose class group has an element of order g is $\gg x^{1/g - \epsilon}$ for any $\epsilon > 0$.*

Proof. Let $x^{1/g}/2 \leq n \leq x^{1/g}$ and $0 < \delta < g$. The number of n in this range, for which $n^g - 1$ is either squarefree or has a prime divisor satisfying $p^2 | n^g - 1$ and $n < p < n^{1+\delta}$ is by Theorem 11 equal to

$$c_1 x^{1/g} + O\left(\frac{x^{1/g}(\log \log x)^{g-1}}{\log x}\right)$$

where

$$c_1 = \prod_p \left(1 - \frac{\rho(p^2)}{p^2}\right).$$

Observe that if p is coprime to g, then $\rho(p^2) \leq g$ (by a simple application of Hensel's lemma) so that each factor in the above product for $p > \sqrt{g}$ is non-zero. However the congruence $n^g \equiv 1 (\bmod p)$ has $(g, p-1)$ solutions and these lift to $(g, p-1)$ solutions mod p^2 if p is coprime to g. If $p | g$ the number of solutions is $(g, p(p-1))$ which is clearly less than p^2 and hence $c_1 \neq 0$. We now assert that for each n enumerated by $\tilde{N}(x^{1/g})$ in Theorem 11, either $n^g - 1$ is squarefree or $n^g - 1 = p^2 d$ where d is squarefree and p is a prime satisfying $n < p < n^{1+\delta}$. Indeed, if $n^g - 1$

has two prime factors p_1 and p_2 satisfying $n < p_1 < n^{1+\delta}$, $n < p_2 < n^{1+\delta}$ and $p_1^2 p_2^2 | n^g - 1$, then by the ABC conjecture,

$$n^g \ll \left(\frac{n^{g+1}}{p_1 p_2}\right)^{1+\epsilon}$$

which implies

$$n^{2+2\epsilon} < (p_1 p_2)^{1+\epsilon} \ll n^{1+\epsilon(g+1)}.$$

Thus, the number of such n's is bounded. Therefore, the n's enumerated by $\tilde{N}(x^{1/g})$ can be put into one of two sets S_1 or S_2 (say) according as $n^g - 1$ is squarefree or $n^g - 1 = p^2 d$ with d squarefree and p a prime satisfying $n < p < n^{1+\delta}$. Since $\tilde{N}(x^{1/g}) \gg x^{1/g}$, the size of at least one of these sets is $\gg x^{1/g}$. If S_1 has size $\gg x^{1/g}$ then we are done by Theorem 1. If S_2 has size $\gg x^{1/g}$, then for each $n \in S_2$, $n^g - 1 = p^2 d$. We enumerate the number of distinct d's that can arise. Since $(n-1)|(n^g - 1)$ and $p > n$, we deduce $(n-1)|d$. Hence any given d can be repeated at most $O(d^\epsilon)$ times. Thus, the number of distinct quadratic fields arising from S_2 is $\gg x^{1/g-\epsilon}$. The result now follows from Theorems 1 and 2.

In an identical fashion, we derive:

Theorem 14. *Let $g \geq 3$ be odd and assume the ABC conjecture. The number of real quadratic fields $\mathbb{Q}(\sqrt{d})$ $0 < d < x$ whose class group has an element of order g is $\gg x^{1/2g-\epsilon}$ for any $\epsilon > 0$.*

Proof. We proceed as in the proof of Theorem 13 with $x > (4g)^g$. The corresponding set S_1 enumerating $n^{2g} + 1$ squarefree presents no problem. We must deal with S_2. Since g is odd, $(n^2+1)|(n^{2g}+1)$. For $n \in S_2$, $n^{2g}+1 = p^2 d$ with d squarefree. If $p|(n^2+1)$ and $p|(\frac{n^{2g}+1}{n^2+1})$, then p divides the discriminant of $n^{2g} + 1$. This means $p|2g$, which is a contradiction since $p > n > 2g$. Therefore, either $p^2|(n^2+1)$ or $p^2|(\frac{n^{2g}+1}{n^2+1})$. The former case leads to $p \leq n$, which is a contradiction. Thus, $(n^2+1)|d$. Now the argument continues as before. Any given d can be repeated at most $O(d^\epsilon)$ times in the enumeration of S_2. This completes the proof.

7. Exponents of class groups of certain real quadratic fields.

Lemma 4 allows us to derive a result on exponents of class groups of quadratic fields. In the imaginary quadratic case, Boyd and Kisilevsky [BK] proved assuming the generalised Riemann hypothesis (GRH) that the exponent of the class group of the imaginary quadratic field $\mathbb{Q}(\sqrt{-d})$, $d > 0$ satisfies

$$\gg \frac{\log d}{\log \log d}.$$

One cannot, of course, expect such a result to hold in the real quadratic case since extensive numerical calculations suggest that there are infinitely many real quadratic fields of class number 1. However, the real quadratic fields $\mathbb{Q}(\sqrt{a^2+1})$ have large class groups and the identical analogue of the theorem of [BK] seems to be:

Theorem 15. *Suppose $a^2 + 1$ is squarefree and let $e(a)$ be the exponent of the class group of $\mathbb{Q}(\sqrt{a^2+1})$. Then, assuming GRH,*

$$e(a) \gg \frac{\log a}{\log \log a}.$$

Proof. Let d be the discriminant of $k = \mathbb{Q}(\sqrt{a^2+1})$. By the Chebotarev density theorem and the GRH (see [MMS]) we know there is always a prime ideal $\mathfrak{p}$ satisfying

$$\mathbb{N}_{k/\mathbb{Q}}(\mathfrak{p}) \leq (\log d)^2$$

in any given ideal class. Let $\mathfrak{p}$ be non-principal with exponent $e(a)$. Then $\mathfrak{p}^{e(a)}$ is principal:

$$\mathfrak{p}^{e(a)} = (u + v\sqrt{d})$$

(say). Then, taking norms, and applying Lemma 4, we deduce

$$e(a) \log \log d \gg \log d.$$

Since $d = a^2 + 1$ or $4(a^2 + 1)$, the result is now immediate.

REFERENCES

[AC] N. Ankeny and S. Chowla, On the divisibility of the class number of quadratic fields, Pacific Journal of Math., 5 (1955) p. 321 - 324.

[BK] D. Boyd and H. Kisilevsky, On the exponent of the ideal class groups of complex quadratic fields, Proc. Amer. Math. Soc., 31 (1972) 433 - 436.

[CL] H. Cohen and H.W. Lenstra Jr., Heuristics on class groups of number fields, Springer Lecture Notes, 1068, in Number Theory Noordwijkerhout 1983 Proceedings.

[DH] H. Davenport and H. Heilbronn, On the density of discriminants of cubic fields II, Proc. Royal Soc., A 322 (1971) p. 405 - 420.

[G] A. Granville, ABC allows us to count squarefrees, preprint.

[H] C. Hooley, Applications of sieve methods, Cambridge Tracts in Mathematics, 1976.

[Hu] P. Humbert, Sur les nombres de classes de certains corps quadratiques, *Comment. Math. Helv.* **12** (1939/40) 233-245; also **13** (1940/41) 67.

[K] S. Kuroda, On the class number of imaginary quadratic fields, *Proc. Japan Acad.*, **40** (1964) 365-367.

[L] W. LeVeque, Topics in Number Theory, Vol. 1, Addison-Wesley, Reading. Mass., 1956.

[Na] T. Nagell, Uber die klassenzahl imaginar-quadratischer Zahlkorper, *Abh. Math. Sem. Univ. Hamburg* **1** (1922) 140 - 150.

[MMS] M. Ram Murty, V. Kumar Murty and N. Saradha, Modular forms and the Chebotarev density theorem, *American Journal of Mathematics*, **110** (1988) 253-281.

[Y] Y. Yamamoto, On unramified Galois extensions of quadratic number fields, *Osaka J. Math.*, **7** (1970) 57-76.

[W] A. Weil, Number Theory, An approach through history, From Hammurapi to Legendre, Birkhäuser, Boston, Basel, Stuttgart, 1984.

[We] P. Weinberger, Real Quadratic Fields with Class Number Divisible by n, *Journal of Number Theory*, **5** (1973) 237-241.

Department of Mathematics,
Queen's University,
Kingston, Ontario
K7L 3N6, Canada
murty@mast.queensu.ca

Heights of Points on Subvarieties of $\mathbb{G}_m^n$

WOLFGANG SCHMIDT

Let $h(x)$ be the absolute logarithmic height of a nonzero algebraic number x. Then $h(x) \geqq 0$, with equality precisely when x is a root of 1, i.e., x is a torsion point of $\mathbb{G}_m = \mathbb{G}_m(\mathbb{C})$. Lehmer's conjecture, that $h(x) > c_0/d$, for every x of degree d which is not a root of 1, and where $c_0 > 0$ is absolute, is still unproved. It was therefore a surprise when Zhang [7] proved that if $x + y = 1$ where x, y are nonzero and not primitive 6-th roots of 1, then $h(x) + h(y) > c_1 > 0$. Zagier [6] determined the best value for c_1; namely $c_1 = \frac{1}{2} \log \eta$, where $\eta = (1 + \sqrt{5})/2$ is the Golden Ratio.

When $\mathbf{x} = (x_1, \ldots, x_n) \in \mathcal{A}_m^n$, where $\mathcal{A}_m$ is the multiplicative group of algebraic numbers, set

$$h(\mathbf{x}) = \sum_{i=1}^n h(x_i).$$

We have $h(\mathbf{x}) \geqq 0$, with equality precisely when $\mathbf{x}$ lies in the torsion group $\mathcal{T}$ of $\mathcal{A}_m^n$. Soon after the work of Zagier, I showed [4] that if $F(\mathbf{X}) \in \mathbb{Z}[X_1, \ldots, X_n]$ and $F(\mathbf{x}) = 0$, $F(\mathbf{x}^{-1}) \neq 0$, then $h(\mathbf{x}) > c_2(F) > 0$, where $c_2(F)$ may be taken to be $1/(2^{4f+2n}H)$, with f the total degree and H the maximum modulus of the coefficients of F. Recently Beukers and Zagier [1] established a much better value for c_2.

When $\mathbf{x}, \mathbf{y} \in \mathcal{A}_m^n$, set

$$\delta(\mathbf{x}, \mathbf{y}) = h(\mathbf{xy}^{-1}).$$

Then

(i) $\delta(\mathbf{x}, \mathbf{y}) \geqq 0$, with equality precisely when $\mathbf{xy}^{-1}$ is in the torsion group $\mathcal{T}$,

(ii) $\delta(\mathbf{x}, \mathbf{y}) = \delta(\mathbf{y}, \mathbf{x})$,

(iii) $\delta(\mathbf{x}, \mathbf{z}) \leqq \delta(\mathbf{x}, \mathbf{y}) + \delta(\mathbf{y}, \mathbf{z})$.

Thus δ is a semidistance on $\mathcal{A}_m^n$, and induces a distance on the factor group $\mathcal{A}_m^n/\mathcal{T}$. I call $\mathbf{x}$, $\mathbf{y}$ *neighbors of distance* $< \varepsilon$ if $\delta(\mathbf{x}, \mathbf{y}) < \varepsilon$. I showed [4] that for a wide class of curves $\mathcal{C} \subset \mathbb{C}^2$, a point $\mathbf{x} \in \mathcal{A}_m^2$ has at most q neighbors of distance $< \varepsilon$ on $\mathcal{C}$, where $q = q(\mathcal{C})$, $\varepsilon = \varepsilon(\mathcal{C})$ depend on $\mathcal{C}$ in a simple way.

Soon after this work, Bombieri and Zannier [3] affected a wide generalization. Let $V \subset \mathbb{G}_m^n$ be an algebraic variety. They defined a certain subvariety V^u and showed that when $\mathbf{x} \in V \backslash V^u$, then $h(\mathbf{x}) > c_3(V) > 0$. The subvariety V^u is defined as follows. An algebraic subgroup of $\mathbb{G}_m^n$

is a subgroup which is an algebraic variety. By *coset* we will understand a coset $\mathbf{g}H$ where H is an algebraic subgroup. A *torsion coset* is a coset $\mathbf{u}H$ where $\mathbf{u} \in \mathcal{T}$. Then V^u is the union of all torsion cosets contained in V. The structure of V^u is elucidated by the following two theorems. Let $V^u(H)$ be the union of all torsion cosets $\mathbf{u}H \subset V$, and $V_1^u(H)$ the union of all maximal torsion cosets $\mathbf{u}H \subset V$, i.e., cosets such that there is no coset $\mathbf{u}H'$ with $\mathbf{u}H \subsetneq \mathbf{u}H' \subset V$.

Theorem 1. *For given d, there are $m < (2d)^{n^2}$ algebraic subgroups $H_1, \ldots, H_m$ depending only on n and d, such that when V is defined by polynomial equations of total degree $\leq d$, then*

$$V^u = \bigcup_{i=1}^{m} V^u(H_i) = \bigcup_{i=1}^{m} V_1^u(H_i).$$

Theorem 2. *When V is defined by polynomial equations of total degree $\leq d$ with rational coefficients, then each $V_1^u(H)$ is the union of fewer than $\exp(3N^{3/2} \log N)$ torsion cosets, where*

$$N = \binom{n+d}{d}.$$

When f is a polynomial, let $H'(f)$ be the sum of the moduli of its coefficients.

Theorem 3. *Suppose V is defined by polynomial equations $f_1 = \cdots = f_\ell = 0$, where each f_i has total degree $\leq d$, coefficients in $\mathbb{Z}$ and $H'(f_i) \leq M$. Then every $\mathbf{x} \in V \backslash V^u$ has*

$$h(\mathbf{x}) > 1/(dN \cdot e^{3M}).$$

This gives an explicit value for the constant $c_3(V)$ of Bombieri and Zannier. A direct application of their method would have given a value which is n times exponential in M. An essential tool of Bombieri and Zannier is the following easily proved lemma.

Let f be a polynomial in $\mathbb{Z}[\mathbf{X}]$ with partial degrees $\leq d$ and $H'(f) \leq M$. Let $p > M$ be a prime number. Then $f(\mathbf{x}) = 0$, $f(\mathbf{x}^p) \neq 0$ implies

$$h(\mathbf{x}) \geq \frac{1}{pd} \log(p/M).$$

In applications, using one prime p at a time leads to weak estimates. It would help if one had long arithmetic progressions of primes. However, the above lemma has an almost trivial extension as follows.

Let f be as above. Let m be an integer all of whose prime factors are $\geq e^\varepsilon M$. Then $f(\mathbf{x}) = 0$, $f(\mathbf{x}^m) \neq 0$ implies

$$h(\mathbf{x}) \geq \varepsilon/pd.$$

It is not hard to construct suitable arithmetic progressions of numbers m all of whose prime factors are large.

Let V^a be the union of all cosets gH of *positive dimension* contained in V. There are theorems playing the role of Theorems 1, 2 for V^a. Bombieri and Zannier showed that a point $\mathbf{x} \in \mathcal{A}_m^n$ has at most q neighbors of distance $< \varepsilon$ in $V \setminus V^a$, where $q = q(V)$, $\varepsilon = \varepsilon(V)$. We make this explicit, as follows.

Theorem 4. *Suppose V is defined by polynomial equations of total degree $\leq d$. Then a point $\mathbf{x} \in \mathcal{A}_m^n$ has at most q neighbors of distance $1/q$ in $V \setminus V^a$, where*

$$q = \exp((4n)^{2dN}).$$

A theorem of this type has applications to diophantine equations of exponential type, e.g., equations $a_1 x_1 + \cdots + a_n x_n = 1$ where $\mathbf{x} = (x_1, \ldots, x_n)$ lies in a given group $\Gamma \subset \mathbb{G}_m^n$ of finite rank r. See, e.g., the work of Schlickewei [3] which antedates the results described here, except for [6], [7].

References

[1] F. Beukers and D. Zagier. Lower bounds of heights of points on hypersurfaces, preprint. Univ. Utrecht.

[2] E. Bombieri and U. Zannier. Algebraic Points on Subvarieties of $\mathbb{G}_m^n$, International Mathematics Research Notices 7 (1995), 333–347.

[3] H. P. Schlickewei. Equations $ax + by = 1$, Annals of Math. (to appear).

[4] W. M. Schmidt. Heights of Algebraic Points Lying on Curves or Hypersurfaces, Proc. A.M.S. (to appear).

[5] W. M. Schmidt. Heights of Points on Subvarieties of $\mathbb{G}_m^n$, Sem. Number Theory, Paris, 93–94.

[6] D. Zagier. Algebraic Numbers Close to Both 0 and 1, Math. Computation 61 (1993), 485–491.

[7] S. Zhang. Positive line bundles on arithmetic surfaces, Annals of Math. 136 (1992), 569–587.

A note on a geometric analogue of Ankeny-Artin-Chowla's conjecture

Jing Yu and Jiu-Kang Yu

ABSTRACT. A well-known conjecture of Ankeny-Artin-Chowla states that if L is a real quadratic number field with prime discriminant p, then the conductor of the fundamental unit of L is not divisible by p. In this note we show that in the function field case, the analogous conjecture is true in arbitrary characteristic, without any hypothesis on the discriminant. The geometric statements proved can be reformulated in terms of the non-vanishing of relative differentials of fundamental units. An reexamination of the original conjecture from this light is also recorded.

1.

In this section, k is a field of characteristic $\neq 2$. Let T be an indeterminate and $K = k(T)$ be the field of rational functions in T. Let L/K be a quadratic extension in which the unique pole ∞ of T splits into two distinct places ∞^+ and ∞^- of L. Such an extension is called a *real quadratic extension* of K.

The extension L can be generated by a quadratic irrationality α over K. It is well-known that we can normalize the choice of α in the following way: we require that α is a square root of a monic, square free polynomial D of even degree $2g+2$ in $k[T]$ ($g \geq 0$). It follows that the integral closure of $A = k[T]$ in L is $B = A[\sqrt{D}]$ and the discriminant of B/A is simply D.

THEOREM 1. *Let $u = r + s\sqrt{D}$ be a unit of $k[T, \sqrt{D}]$, with $r, s \in k[T]$, $s \neq 0$. Suppose that u is separable as a morphism $C \to \mathbb{A}^1$, where C is the affine hyperelliptic curve defined by $y^2 = D(x)$. Then $\deg \gcd(D, s) \leq g$.*

FIRST PROOF. Without lose of generality, we may assume that k is algebraically closed. Write $D' = \gcd(D, s)$, $s = D's'$, $\deg D' = g'$, $\deg s' = m$, then $\deg r = m + g + g' + 1$. Let $a = r^2$, $b = -D(D's')^2$, $c = N_{L/K}(u) \in k^*$. Then $a + b = c$ and $\gcd(a, b, c) = 1$. The separability assumption on u implies that not all three of a, b, c are p^{th} powers if k is of characteristic $p \neq 0$. By Mason's theorem [**M, p.156**],

$$\max \deg\{a, b, c\} \leq n_0(abc) - 1,$$

1991 *Mathematics Subject Classification.* 14H05; 11R11.

where $n_0(abc)$ is the number of distinct zeros of abc, and is $\leq \deg(rDs') = 2m + 3g + g' + 3$ in our case. Clearly, $\max \deg\{a,b,c\} = 2m + 2g + 2g' + 2$. Now Mason's theorem tells us $g' \leq g$, which is what we want to prove. ∎

SECOND PROOF. Let $D' = \gcd(D, s)$ and $g' = \deg D'$. Let $\bar{C}/k$ be the smooth compactification of C. Then $\bar{C}$ is of genus g and $\bar{C} \setminus C$ consists of two points ∞^+, ∞^-. The divisor $\text{div}(u)$ is supported on $\{\infty^+, \infty^-\}$. Thus $\text{div}(u)$ equals to $n(\infty^+ - \infty^-)$ for some integer n.

Let $D' = \prod(T - t_i)$ and for each i, let P_i be the point on C lying above $(T - t_i)$. From $r^2 = N_{L/K}(u) + Ds^2$, we see that $\text{ord}_{P_i}(r^2 - r^2(P_i)) \geq 6$. Therefore $\text{ord}_{P_i}(r - r(P_i)) \geq 6$ as well. It follows that we have $\text{ord}_P(u - u(P_i)) \geq 3$.

Consider the algebraic differential $\omega = du$. By the separability assumption, ω is not zero and represents the canonical divisor class. In particular, $\deg \omega = 2g - 2$. However, we have

$$\text{ord}_P(\omega) \geq \begin{cases} n - 1, & \text{if } P = \infty^+ \\ -n - 1, & \text{if } P = \infty^- \\ 2, & \text{if } P = P_i \text{ for some } i \\ 0, & \text{otherwise.} \end{cases}$$

Consequently, we have $\deg \omega \geq 2g' - 2$. It follows that $g' \leq g$. ∎

REMARK 1. The inequality in the theorem bounding $\gcd(D, s)$ by genus is sharp. This can be seen by examining elliptic curves over $\mathbb{F}_5$. Here u are easily computed via continued fractions in the field $\mathbb{F}_5((1/T))$.

REMARK 2. The condition $s \neq 0$ in the theorem implies that s is not a constant. That is, $s \notin k$. A non-constant unit exists if and only if the divisor class $(\infty^+ - \infty^-)$ is a torsion point on the Jacobian of $\bar{C}$. It is certainly the case when k is a finite field. When there is a non-constant unit, B^*/k^* is a free abelian group of rank 1, and a unit u such that uk^* generates B^*/k^* is called a *fundamental unit*.

COROLLARY. *Let k be a finite field of odd characteristic, $D \in A = k[T]$ be monic, square free, of even degree ≥ 2. Let $u = r + s\sqrt{D}$ be a fundamental unit of $B = A[\sqrt{D}]$. Then D does not divide s. In other words, the discriminant D does not divide the conductor of the order $A[u]$.*

PROOF. The function u is separable. Otherwise the rational function field $k(u)$ is contained in $k(T, \sqrt{D})^p$ so that u is the pth power of another unit (p being the characteristic of k). The corollary then follows from the theorem. ∎

This corollary is clearly valid for any perfect field k, provided that we assume that there exists a non-constant unit in B.

2.

In this section, k is a field of characteristic 2 and $\bar{k}$ is an algebraic closure of k. Let $K = k(T)$ and L/K be a separable quadratic extension in which the unique pole ∞ of T splits into two distinct places ∞^+ and ∞^- of L. Such an extension is again called a *real quadratic extension* of K.

The extension L can be generated by a quadratic irrationality α over K. We may assume that α satisfies an Artin-Schreier equation

$$\wp(\alpha) = a,$$

where $\wp(x) = x^2 + x$. Replacing a by $a + \wp(b)$ ($b \in K$) does not change the extension L. We are going to normalize the choice of a following Hasse. The rational function a has a unique partial fraction expansion

$$a = c + f_\infty(T) + \sum_{t \in \bar{k}} f_t\left(\frac{1}{T-t}\right),$$

where $c \in k$, and each f_t ($t \in \mathbb{P}^1(\bar{k})$) is a polynomial over $\bar{k}$ whose constant term is 0. Let $\pi = 1/T$. Then a is in $\wp(K_\infty) = \wp(k[T] \oplus \pi k[[\pi]]) = \wp(k[T]) \oplus \pi k[[\pi]]$. Observe that $\sum_{t \in \bar{k}} f_t(1/(T-t))$ is in $\pi k[[\pi]]$. Therefore $c + f_\infty(T)$ is in $\wp(k[T])$ and we may in fact assume that $c + f_\infty(T) = 0$. Next we may choose a so that the degree of the denominator of a is smallest possible. In this case, each non-zero f_t is necessarily of odd degree, because if certain f_t has a leading term cT^{2k}, by replacing a with $a+$ the sum of conjugates of $\wp(c^{1/2}(T-t)^{-k})$, we can lower the degree of the denominator of a.

To summarize, we may, and do, assume that a is of the form $W/(DV^2)$, where D is monic and square free, V divides a high power of D, and W is relatively prime to D and of degree $< \deg(DV^2)$. The quadratic irrationality $\beta = DV\alpha$ also generates the extension L, and satisfies the equation

$$\beta^2 + DV\beta + DW = 0.$$

It is easily shown that the integral closure of $A = k[T]$ in L is $B = A[\beta]$, and the discriminant of B/A is $(DV)^2$.

THEOREM 2. *Let $u = r + s\beta$ be a unit of B, with $r, s \in A$. Then V divides s. But if u is separable as a morphism $\operatorname{Spec} B \to \mathbb{A}^1$, and DV divides s, then u is a constant.*

PROOF. Without loss of generality, we may do a base field extension and assume that k is algebraically closed.

First we show that V divides s. We have $N_{L/K}(u) = r^2 + DVrs + DWs^2 = c \in k^*$. Let P be a prime dividing D. Then $\operatorname{ord}_P(DVrs + DWs^2) = \operatorname{ord}_P(r + c^{1/2})^2$ is positive and even. But $\operatorname{ord}_P(DWs^2)$ is clearly odd. So we must have $\operatorname{ord}_P(DVrs) \leq \operatorname{ord}_P(DWs^2)$, from which we deduce immediately $\operatorname{ord}_P(V) \leq \operatorname{ord}_P(s)$. Since this is valid for all primes P dividing V, we have proved that V divides s.

To prove the second part of the theorem, suppose on the contrary that u is separable, DV divides s, and u is not a constant. Let $\bar{C}/k$ be the smooth compactification of C. By the Riemann-Hurwitz relation [**H, p.301**], the genus g of $\bar{C}$ is determined by the relation $2g + 2 = \deg(DV)^2$.

Then $\bar{C} \setminus C$ consists of two points ∞^+, ∞^-. The divisor $\operatorname{div}(u)$ is supported on $\{\infty^+, \infty^-\}$. Therefore, $\operatorname{div}(u) = n(\infty^+ - \infty^-)$ for some integer n.

Let $D = \prod(T - t_i)$, $DV = \prod(T - t_i)^{e_i}$. For each i, let P_i be the closed point of C lying above $(T - t_i)$. Write $s = DVt$. Then we have the following relation

$$(r + \sqrt{c})^2 = (DV)^2 rt + D(DV)^2 Wt^2.$$

It follows that $\text{ord}_{P_i}(r+\sqrt{c}) \geq \text{ord}_{P_i}(DV) = 2e_i$. If $\text{ord}_{P_i}(r+\sqrt{c}) > 2e_i$, we surely have $\text{ord}_{P_i}(dr) \geq \text{ord}_{P_i}(r+\sqrt{c}) - 1 \geq 2e_i$. If $\text{ord}_{P_i}(r+\sqrt{c})$ is exactly $2e_i$, we still have $\text{ord}_{P_i}(dr) \geq 2e_i$ because we are working in characteristic 2. We also observe that $\text{ord}_{P_i}(d(DVt\beta)) \geq \text{ord}_{P_i}(DVt\beta) - 1 \geq 2e_i$.

By the separability assumption, $du = dr + d(DVt\beta)$ is not zero and hence represents the canonical divisor class. By the above discussions,

$$\text{ord}_P(du) \geq \begin{cases} n-1, & \text{if } P = \infty^+ \\ -n-1, & \text{if } P = \infty^- \\ 2e_i, & \text{if } P = P_i \text{ for some } i \\ 0, & \text{otherwise}. \end{cases}$$

Consequently, we have $\deg(\text{div}(du)) \geq -2 + \sum 2e_i = 2g$, contradicting the fact that $\deg(\text{div}(du)) = 2g - 2$. ∎

COROLLARY. *Suppose that k is a finite field of characteristic 2, and u is a fundamental unit of $B = A[\beta]$. Write $u = r + s\beta$ with $r, s \in A$. Then V divides s but DV does not divide s.*

3.

We are going to reformulate the results in sections 1 and 2 in a uniform way by using the formalism of relative differentials.

Let k be a finite field of *any* characteristic. Let $K = k(T)$, $A = k[T]$. Consider a real quadratic extension L of K and the integral closure B of A in L. Let $d: B \to \Omega_{B/A}$ be the canonical map from B into the module of relative differentials of B over A. Structure of $\Omega_{B/A}$ can be given explicitly and we deduce from it the following

THEOREM 3. *Let u be a fundamental unit of B, then $du \neq 0$ in $\Omega_{B/A}$.*

PROOF. We know that $B = A \oplus A.\beta$, where β satisfies a minimal equation $\beta^2 + b\beta + c = 0$, $b, c \in A$. The different of B/A is generated by $\sqrt{b^2 - 4c} \in B$, and $\Omega_{B/A}$ is isomorphic to the cyclic module $B/B.\sqrt{b^2 - 4c}$, and is generated by $d\beta$ [**S, p.59**].

Let $u = r + s\beta$. Then $du = s \cdot d\beta$. Therefore Theorem 3 simply asserts that $s \notin B.\sqrt{b^2 - 4c}$. If the characteristic of k is odd, $b^2 - 4c = D$ and the assertion is equivalent to that D does not divide s (we are using the notations of section 1). If the characteristic of k is 2, $b^2 - 4c = (DV)^2$ and the assertion is equivalent to that DV does not divides s (we are using the notations of section 2).

Thus Theorem 3 follows from the corollaries to Theorem 1 and Theorem 2. ∎

In view of the analogy between number fields and function fields, it seems reasonable to question whether the following two equivalent statements are true.

(a) Let B be the maximal order in a real quadratic extension L of $\mathbb{Q}$, and $u \in B$ be a fundamental unit. Then $du \neq 0$ in $\Omega_{B/\mathbb{Z}}$.
(b) Let d be a square free positive integer, $D = d$ or $4d$ according as $d \equiv 1 \pmod 4$ or not. Let $\beta = (D + \sqrt{D})/2$ and $u = r + s\beta$ be a fundamental unit in $\mathbb{Z}[\beta]$. Then d or $2d$ does not divide s according as $d \equiv 1 \pmod 4$ or not. Note that $|s|$ is the conductor of the order $\mathbb{Z}[u]$ hence also being called the conductor of the fundamental unit.

Ankeny, Artin and Chowla [1] proposed the special case where d is a prime $\equiv 1$ (mod 4) of the above question, which was the inspiration of the present work. A computer search by the authors found no counterexamples to this original form of Ankeny-Artin-Chowla's conjecture in which $d < 10^8$. However, Washington [**W, p.83**] has pointed out that such an evidence is not convincing enough.

Question (b) stated in the above generality is however false. We have found 6 counterexamples in which $d < 10^8$, namely $d = 2 \cdot 23$, $23 \cdot 79$, $2 \cdot 3 \cdot 7 \cdot 19 \cdot 73$, $3 \cdot 69997$, $41 \cdot 79 \cdot 541$, $2 \cdot 1562159$. It worths mentioning that in all 6 cases, we have $N_{\mathbb{Q}(\sqrt{d})/\mathbb{Q}}(u) = +1$ but in Ankeny-Artin-Chowla's original problem, d is a prime $\equiv 1$ (mod 4) and hence $N_{\mathbb{Q}(\sqrt{d})/\mathbb{Q}}(u) = -1$.

Acknowledgement. This work has benefited from a conversation between the second named author and Chia-Fu Yu. We would like to thank him, and our respective institutions for their supports.

References

[A] N. C. Ankeny, E. Artin, S.Chowla, *The class-number of real quadratic number fields*, Annals of Math. **56** (1953), 479–492.
[H] R. Hartshorne, *Algebraic geometry*, Springer-Verlag, New York, 1977.
[M] R.C.Mason, *Equations over function fields*, Number Theory, Noordwijkerhout 1983, Lecture Notes in Mathematics 1068, Springer-Verlag, New York, 1984.
[S] J.-P. Serre, *Local fields*, Springer-Verlag, New York, 1979.
[W] L. Washington, *Introduction to Cyclotomic Fields*, Springer-Verlag, New York, 1982.

INSTITUTE OF MATHEMATICS, ACADEMIA SINICA, TAIPEI 11529, TAIWAN, R.O.C.
E-mail address: yu@math.sinica.edu.tw

DEPARTMENT OF MATHEMATICS, PRINCETON UNIVERSITY, PRINCETON, NJ 0854, U.S.A.
E-mail address: yu@math.princeton.edu

Part II

AUTOMORPHIC FORMS

Determinant of representations of division algebras of prime degree over local fields.

S.Jayasree and R.Tandon

Deligne considered the question of when special orthogonal representations of Weil-Deligne groups can be lifted to the spin group. In a recent paper D.Prasad and D.Ramakrishnan [DP-DR] answer this question for special orthogonal representations of D^*/F^* for a quaternion division algebra D over F with odd residue characteristic. In the process the authors prove that representations of D^*/F^* are orthogonal and calculate their determinant. It seems a natural question to calculate the determinant of irreducible representations of D^*/F^* for general division algebra. In this paper we consider a division algebra D of prime degree over a finite extension F of $\mathbb{Q}_p$ and evaluate the determinant of representations of D^*/F^*. To evaluate the determinant of a representation with trivial central character we use the explicit construction of minimal irreducible representations as given in [Ho]. We show that if the degree of D over F is odd or if D is a quaternion division algebra with even residue characteristic, then the irreducible representations of D^*/F^* of sufficiently large conductors have trivial determinant with the exception of tamely ramified division algebras over F. In the case of tamely ramified division algebras, if the degree of the extension of $\overline{F}$ got by adjoining a primitive n th root of unity is even, then the determinant could be nontrivial even if the conductor is very large.

---

[0]1991 Mathematics Subject Classification. Primary 11F85, 11F70

The first author thanks the Council of Scientific and Industrial Research, India, for research support. The second author thanks the University Grants Commission, India, for its Special Assistance Programme under the Departmental Research Support scheme.

1 Notations and Conventions

By a minimal irreducible representation ρ of D^* we mean a continuous representation of dimension more than 1. Let D be a division algebra of prime degree n over a finite extension F of $\mathbb{Q}_p$. We say that D is **tamely ramified** over F if the degree of D over F is coprime to the residue charateristic of F and D is **wildly ramified** over F if the degree of D over F is equal to the residue characteristic of F. Let e, f denote respectively the ramification index and the modular degree of F over $\mathbb{Q}_p$. For a division algebra D and a field L let D^*, L^* denote the multiplicative groups of non-zero elements in D, L respectively. Let R_D and R_L be the rings of integers in D and L with prime ideals P_D and P_L and residue fields $\overline{D}$ and $\overline{L}$ respectively and let U_D denote the group of units in D^*. Let v_D and v_L denote respectively the valuation of D and L and let $|\ |_D$ denote the absolute value in D. If q' is the order of $\overline{D}$, and q that of $\overline{F}$, then $q' = q^n$ and $q = p^f$. Let F_n be an unramified extension of F of degree n in D. Let ν be the reduced norm and τ the reduced trace on D. For an extention L of F, let $Tr_{L/F}$, $N_{L/F}$ denote the trace and norm on L respectively. Let $e_{L/F}$, $f_{L/F}$ denote respectively the ramification index and the modular degree of L over F. Let $d_{L/F}$ denote the differential exponent of L over F. If L is a quadratic extension of F, let $w_{L/F}$ be the quadratic character associated to F^* by local class field theory. For χ, a character of F^*, let χ_ν denote $\chi \circ \nu$. For $i \in \mathbb{N}$, let $G_i = 1 + P_D^i$. Let π be a uniformising element of D and ζ a $(q^n - 1)$th root of unity in D^* such that π^n is a uniformising element of F and such that conjugation by π generates the Galois group of $F(\zeta)$ over F. We sometimes write π^n as π_F. Let $l = \dfrac{q^n - 1}{q - 1}$. Then ζ^l generates the group of $(q-1)$th roots of unity in F^*. Let $\sigma : \langle \zeta \rangle \to \langle \zeta \rangle$ be given by $\zeta^\sigma = \pi \zeta \pi^{-1} = \zeta^{q^m}$ for some m, $1 \leq m < n$.

On the additive group $(F, +)$ we fix a character ψ_F, of conductor $n(\psi_F) = 0$. This means ψ_F is trivial on R_F and nontrivial on $\pi_F^{-1} R_F$. For a group G let G^c denote the derived subgroup of G. If H is a subgroup of G, we let $t_{G \setminus H}$ denote the transfer map from G/G^c to H/H^c. Moreover, in the transfer calculations by a set of coset representatives of H in G, we mean a set of right coset representatives of H in G. Let h be an element of D^* such that h is in general position [Ho] and $v_D(h) = -k - n$. If k is coprime to n, then h is conjugate to $\lambda \pi^{-k-n} u$ for some $\lambda \in \langle \zeta^\ell \rangle$ and $u \in G_1$. Hence if $F' = F(h)$ is a totally ramified extension, then F'

is conjugate to $F(\pi u)$ for some $u \in G_1$. Let F' be an extension of F such that F'^*G_j is a subgroup of D^*. Let $1, \pi, \pi^2 \ldots \pi^{(n-1)}$ (respectively $1, \zeta, \zeta^2 \ldots \zeta^{\ell-1}$) be a set of right coset representaives of F'^*G_1 in D^* and if F' is unramified (respectively totally ramified). If D is wildly ramified over F and F' is totally ramified, let $\{1 + \zeta_r \pi^j, \ r = 1, 2, \ldots, q^{(p-1)}\}$ be a set of coset representatives of F'^*G_{j+1} in F'^*G_j such that for $q \geq 3$, $\sum_{r=1}^{q^{(p-1)}} \bar{\zeta}_r = \bar{0}$ in the residue field. If $q = 2$ let ζ be a cube root of unity.

The statements of the three main theorems that we prove are as follows:

Theorem A(Tame case):

Let D be a division algebra such that the degree n of D over F is not equal to the residue characteristic p of F. Let m' be the degree of the extension of the residue field of F obtained by adjoining a primitive nth root of unity. Let Π' be an irreducible representation of D^* such that its central character $w_{\Pi'}$ is trivial. Then Π' is minimal and Π' is either induced from a representation ρ_1 of F'^*G_1 where F' is an extension of degree n of F in D or Π' is an extension of a representation ρ_1 of $F_n^*G_1$ [Ho].

1. Let n be odd. Then,

$$\det \Pi' \equiv \begin{cases} 1 & \text{if } m' \text{ is odd} \\ \mu & \text{if } m' \text{ is even} \end{cases}$$

   where μ is a character of D^*/U_D into the group of nth roots of unity in $\mathbb{C}^*$.

2. Let $n = 2$. Then,

$$\det \Pi' = \begin{cases} w_{F'/F} \circ \nu & \text{if } F' \text{ is unramified} \\ w_{L/F} \circ \nu & \text{if } F' \text{ is totally ramified} \end{cases}$$

   where $L = F'$ if $q \equiv 1 \bmod 4$ and L is the other totally ramified extension if $q \equiv 3 \bmod 4$.

Remarks: In statement (1) of the above theorem, μ could be the trivial character. In fact given a character μ of D^*/U_D there exists an extension Π' of ρ_1 such that $\det \Pi' \equiv \mu$. Statement (2) of the above theorem is a result of [DP-DR].

For the next two theorems we assume that the degree n of D over F is equal to the residue characteristic p of F.

Theorem B (The Wild Case - F' Unramified):
Let ρ be a minimal irreducible representation of D^* of conductor $k+2$ such that k is a multiple of p. Then ρ is induced from a representation of $F_p^* G_1$ [Ho]. Suppose there exists a character χ of F^* such that $\Pi' = \rho \otimes \chi$ has trivial central character $w_{\Pi'}$. If p is an odd prime, then $\det \Pi'$ is the trivial character of D^*. If $p = 2$, then for $k > 0$, $\det \Pi'$ is the trivial character of D^* and for $k = 0$, $\det \Pi' = w_{F_2/F} \circ \nu$.

Theorem C (The Wild Case - F' Totally Ramified):
Let ρ be a minimal irreducible representation of D^* of conductor $k+2$ where k is coprime to p. Then there exists an extension F' totally ramified over F such that ρ is induced from a character $\tilde{\phi}$ of $F'^* G_j$ or $F'^* G'_{j-1}$ for a 'maximal isotropic' subgroup G'_{j-1}, $G_j \subset G'_{j-1} \subset G_{j-1}$ [Ho]. Suppose there exists a character χ of F^* such that $\Pi' = \rho \otimes \chi$ has trivial central character $w_{\Pi'}$.

1. Let the conductor of ρ be 3.
   This implies, χ is trivial on $(1+P_F)^p$. Let χ' be a character of F^* such that $\chi' = \chi$ on $1 + P_F$, $\chi'_\nu(\zeta) = 1$ and

   $$\chi'_\nu(\pi) = \begin{cases} \tilde{\phi}\chi_\nu(\pi) & \text{if } p \text{ is odd} \\ (-1)^{q/2}\tilde{\phi}\chi_\nu(\pi) & \text{if } p = 2 \end{cases}$$

   where π is any uniformising element of D such that $\pi^p \in F$ and $\pi \in F'^* G_1$. Then χ' is a character of F^* of order p and $\det \Pi' = \chi'_\nu$. If $p = 2$ and $K_{\chi'}$ is a quadratic extension of F determined by χ', then $\det \Pi' = w_{K_{\chi'}/F} \circ \nu$

2. Let the conductor of ρ be more than 3.

   (a) If $q \geq 3$, then $\det \Pi' \equiv 1$ except if $q = 4$, $j = 2$ and $d_{F'/F} > 2$ in which case, $\det \Pi'$ could be either the trivial character or the unramified character. Both cases occur.

   (b) If $q = 2$ and $e \geq 2$, then $\det \Pi' \equiv 1$ for $j \geq 5$ and in fact if $d_{F'/F} = 2$ for $j \geq 3$. If $d_{F'/F} = 2$ and $j = 2$, then $\det \Pi' \equiv (w_{F_2/F}) \circ \nu$. If $d_{F'/F} > 2$, then for $j = 2$ and $j = 3$ there exists a ramified quadratic extension K of F such that $\det \Pi'$ is either $w_{K/F} \circ \nu$ or $(w_{F_2/F} w_{K/F}) \circ \nu$ and both cases occur. If $j = 4$, then $\det \Pi'$ is the trivial character if $d_{F'/F} > 4$ and the unramified character if $d_{F'/F} = 4$.

(c) If $q = 2$ and $e = 1$, then $F = \mathbb{Q}_2$. Let π be a uniformising element such that $\pi^2 = 2$ and $\pi \in F'^* G_2$.
(i) If $j = 2$, then $1 + \pi \in F'^* G_2$ and $\det \Pi'$ is either $w_{\mathbb{Q}_2(\sqrt{3})/\mathbb{Q}_2} \circ \nu$ or $w_{\mathbb{Q}_2(\sqrt{-1})/\mathbb{Q}_2} \circ \nu$ and both cases occur.

(ii) If $j \geq 3$, then

$$\det \Pi' = \begin{cases} 1 & \text{if } d_{F'/F} = 2 \text{ or if } j \geq 4 \\ w_{\mathbb{Q}_2(\sqrt{-3})/\mathbb{Q}_2} \circ \nu & \text{if } d_{F'/F} > 2 \ \& \ j = 3. \end{cases}$$

Remark: In statements 1 and 2(c) of the above theorem one can show that there exists a uniformising element π satisfying the conditions of the theorem. In statament 1, $\det \Pi'$ given in terms of χ'_ν is indeed well defined. In statement 2(c) if π' is another uniformising element satisfying the same conditions then $\tilde{\phi}(1 + \pi') = \tilde{\phi}(1 + \pi)$.

An outline of the proof:

Minimal irreducible representations of D^* are obtained from characters of G_i by repeated extension and induction [Ho]; moreover, by the following general result about the determinant of induced representations, evaluation of the determinant reduces to evaluating the transfer map of the appropriate groups.

Lemma[De 1]: Let G be a group and H a subgroup of finite index in G. Let $t : G^{ab} \to H^{ab}$ denote the transfer map. For a finite dimensional representation η of H we have,

$$\det \Phi(x) = \epsilon(x)^{\dim \eta} \det \eta(t(x)) \tag{1}$$

where $\Phi = \operatorname{Ind}_H^G \eta$ and ϵ is the determinant of the permutation representation of G on G/H.

Observe that there exists a character χ of F^* such that $\Pi' = \rho \otimes \chi$ has trivial central character if and only if ρ is trivial on the pth roots of unity in F^*. We know from a theorem of Nakayama and Matsushima [Wa] that the derived subgroup D' of D^* equals the subgroup D_1 of norm 1 elements of D^*. This, together with $w_{\Pi'} = 1$ implies $D'F^* \subseteq \ker \det \Pi'$. But, $D^*/D'F^* \cong F^*/F^{*n}$ by the norm map ν. Thus, $\det \Pi'$ is a character of order n and if we know $\det \Pi'$ on F^*/F^{*n}, we know it on the whole of D^*. If D is a tamely ramified division algebra over F, we show that $G_1 \subset D_1 F^* \subseteq \ker \det \Pi'$. Thus if we evaluate $\det \Pi'$ on a uniformising element π of D and a primitive $(q^n - 1)$th root of unity ζ, we know it on the whole of D^*.

If D is a wildly ramified division algebra over F, let ρ be a minimal irreducible representation of D^* such that $\Pi' = \rho \otimes \chi$ has trivial central character. Then since $p\, (= n)$ is coprime to $(q^p - 1)$, $\det \Pi'$ is trivial on ζ. Therefore, to know $\det \Pi'$ on D^*, we need to know it on π and on a set of coset representatives of $D'F^*$ in D^*. If F' is unramified we choose a suitable set of coset representatives and show that $\det \Pi'$ is trivial except if $p = 2$ and the conductor of ρ is 2. If F' is totally ramified the proof is more involved. We show that if $q \geq 3$, or if $q = 2$ but $j \geq 4$, tensoring by χ_ν does not affect the value of $\det \Pi'$ on G_1 (cf. Lemma 4.21) and the transfer $t_{D^* \backslash F'^* G_j}(x)$ for $x \in G_j$ lies in the kernel of the inducing character $\tilde{\phi}$. Finally we show that for sufficiently large values of j, $t_{D^* \backslash F'^* G_j}(\pi) = \pi^{\dim \rho} \mod \ker \tilde{\phi}$.

2 A Proof of Theorem A

Let D be a tamely ramified division algebra over F.

Lemma 2.1 For an irreducible representation Π' of D^* such that $w_{\Pi'}$ is trivial, the assumption $(n, p) = 1$ implies that Π' is minimal.

Proof: Let Π' be of conductor $k+1$. Suppose Π' is not minimal. For a character χ of F^* $\Pi' \otimes \chi$ has smaller conductor. This implies χ_ν is trivial on G_k. For $i \in \mathbb{N}$ let $H(i) = G_i \cap F$. Then ν maps G_i onto $H(i)$. As χ is non-trivial on $H(k-1)$ and trivial on $H(k)$, we have $H(k-1) \neq H(k)$ and χ is a nontrivial character of the p-group $H(k-1)/H(k)$. But then $\Pi' \otimes \chi$ is of smaller conductor than Π' implies $w_{\Pi' \otimes \chi} = \chi^n$ is trivial on $H(k-1)$ which is impossible because $(n,p) = 1$. Therefore Π' must be minimal. $\square$

Lemma 2.2 If Π' is an irreducible representation of D^* such that $w_{\Pi'}$ is trivial, then $G_1 \subset D_1 F^* \subseteq \ker \det \Pi'$ and $\det \Pi'$ takes values in the group of nth roots of unity in $\mathbb{C}^*$.

Proof: As $F^* D_1 \subseteq \ker \det \Pi'$, $\det \Pi'$ can be treated as a character of F^*/F^{*n} from which it follows that $\det \Pi'$ takes values in the group of nth roots of unity in $\mathbb{C}^*$. If $z \in G_1$, there exists $y \in 1 + P_F$ such that $\nu(z) = y^n = \nu(y)$ and hence $G_1 \subseteq D_1 F^* \subseteq \ker \det \Pi'$. $\square$

From Lemma 2.2 it follows that to know $\det \Pi'$ on D^* it is enough to evaluate it on the two elements π and ζ of D^*. Π' is either induced from a representation ρ_1 of F'^*G_1 where F' is an extension of degree n of F in D or Π' is an extension of a representation ρ_1 of $F_n^*G_1$ [Ho]. For the next three propositions Π' denotes an irreducible representation of D^* with conductor $k+2$ and trivial central character.

Proposition 2.3 If $\Pi' = \operatorname{Ind}_{F'^*G_1}^{D^*} \rho_1$ where F' is the unramified extension of F in D, then $\det \Pi'(\zeta) = 1$ and

$$\det \Pi'(\pi) = \begin{cases} 1 & \text{if n is odd} \\ -1 & \text{if n = 2} \end{cases}$$

Proof: The index of F'^*G_1 in D^* is n and $\dim \rho_1 = q^{(n-1)k/2}$. Also,

$$t_{D^* \backslash F'^* G_1}(\zeta) = \prod_{r=0}^{n-1} \pi^r \zeta \pi^{-r} = \zeta^{\sum_{r=0}^{n-1} q^{rm}} \in F$$

and hence by Lemma 1, $\det \Pi'(\zeta) = \epsilon(\zeta)^{\dim \rho_1} \det \rho_1(t(\zeta)) = 1$. Next, $t_{D^* \backslash F'^* G_1}(\pi) = \pi^n$. Also, $\epsilon(\pi)$ is the sign of the permutation $(1, 2, \ldots, n)$. Therefore, $\det \Pi'(\pi) = (-1)^{q^{(n-1)k/2}} = -1$. Hence the proposition. □

Proposition 2.4 If $\Pi' = \operatorname{Ind}_{F'^*G_1}^{D^*} \rho_1$ where F' is totally ramified, then

$$\det \Pi'(\zeta) = \begin{cases} 1 & \text{if n is odd} \\ -1 & \text{if n = 2} \end{cases}$$

and

$$\det \Pi'(\pi) = \begin{cases} 1 & \text{if n is odd or if n=2 and } q \equiv 1 \bmod 4 \\ -1 & \text{if n = 2 and } q \equiv 3 \bmod 4 \end{cases}$$

Proof: Without loss of generality we may assume that $\pi \in F'^*G_1$. Then $t_{D^* \backslash F'^* G_1}(\zeta) = \zeta^\ell$ and $\epsilon(\zeta)$ the sign of the permutation $(1, 2, \ldots, \ell)$ is 1 if and only if n is odd. Therefore, $\epsilon(\zeta)^{\dim \rho_1} = 1$ if and only if n is odd. Now by Lemma 1, it follows that

$$\det \Pi'(\zeta) = \epsilon^{\dim \rho_1}(\zeta) \det \rho_1(t_{D^* \backslash F'^* G_1}(\zeta)) = 1$$

if n is odd and -1 otherwise. Next, $t_{D^* \backslash F'^* G_1}(\pi) = \pi \prod_{i=1}^{\ell-1} \zeta^{\ell r_i} \pi = \pi^\ell \zeta^{\ell \sum_{i=1}^{\ell-1} r_i} \in F$. Since the orbits under the action of π on $1, \zeta, \zeta^2, \ldots, \zeta^{\ell-1}$ are of length 1 or n, $\epsilon(\pi) = 1$ if n is odd. If $n = 2$, then 1 and $\zeta^{(q+1)/2}$

are fixed under the action of π and the remaining elements break up as $(q-1)/2$ transpositions. Therefore if $n=2$, we have $\epsilon(\pi) = (-1)^{(q-1)/2}$ and since q is odd $\epsilon(\pi)^{\dim \rho_1} = 1$ if $q \equiv 1 \mod 4$ and -1 otherwise. Now by equation (1)

$$\det \Pi'(\pi) = \epsilon^{\dim \rho_1}(\pi) \det \rho_1(t_{D^*\backslash F'^*G_1}(\pi))$$
$$= \begin{cases} 1 & \text{if n is odd or if n=2 and } q \equiv 1 \mod 4 \\ -1 & \text{if n =2 and } q \equiv 3 \mod 4 \end{cases}$$

Hence the proposition. □

Lemma 2.5 Suppose Π' is an extension of an irreducible representation ρ of $F_n^*G_1$ where $\rho = \text{Ind}_{F^*G_1}^{F_n^*G_1} \rho'$ for some representation ρ' of F^*G_1. Then $\det \Pi' \equiv \mu$ where μ is a character of D^*/U_D into the group of nth roots of unity in $\mathbb{C}^*$.

Proof: Π' is not induced from F'^*G_1 implies m is even [Ho]. Therefore $n \nmid (q-1)$. Since n divides $q^n - 1$ if and only if n divides $q-1$, n does not divide ℓ. Let $r, s \in \mathbb{Z}$ be such that $rn + s\ell = 1$. Then $\det \Pi'(\zeta) = \det \Pi'(\zeta^{(rn+s\ell)}) = \det \Pi'(\zeta^n)^r \det \Pi'(\zeta^\ell)^s = 1$. As Π' is an extension of $\rho = \text{Ind}_{F^*G_1}^{F_n^*G_1} \rho'$, $\Pi'(\pi) = \lambda A$ where λ is an nth root of unity and A is an intertwining operator of $\text{Ind}_{F^*G_1}^{F_n^*G_1} \rho'$ and $(\text{Ind}_{F^*G_1}^{F_n^*G_1} \rho')^\pi$ such that $A^n = \text{Id}$. Thus, $\det \Pi'(\pi) = \det(\lambda A) = \lambda^{q^{\dim \rho}} \det A$. If μ is the character of D^*/U_D given by $\mu(\pi) = \lambda^{q^{\dim \rho}} \det A$, then $\det \Pi' \equiv \mu$. Thus for the n different extensions of ρ corresponding to $\Pi'(\pi) = \lambda^i A$, $i = 0, 1, \ldots, n-1$, we get n distinct characters $\det \Pi'$ of D^* into the group of nth roots of unity in $\mathbb{C}^*$. □

Theorem A now follows from Propositions 2.3, 2.4 and Lemma 2.5.

3 A Proof of Theorem B

Without loss of generality we can assume that $F' = F(\zeta)$. Since $\det \pi'(\zeta) = 1$, to know $\det \Pi'$, we need to know it only on π and on a set of elements $c_1, c_2 \ldots c_r$ such that $\nu(c_i) = b_i$ where b_i are a set of coset representatives of $(1+P_F)^p$ in $1+P_F$. Clearly, $t_{D^*\backslash F'^*G_1}(\pi) = \pi^p \in F^*$ and $\epsilon(\pi)^{\dim \tilde{\rho}} = 1$ if $k > 0$ and $\epsilon(\pi) = -1$ if $k = 0$. Therefore,

$$\det \Pi'(\pi) = \epsilon(\pi)^{\dim \tilde{\rho}} \det \tilde{\rho}\chi_\nu \mid_{F'^*G_1}(\pi^p)$$

$$= \begin{cases} 1 & \text{if } p \text{ is odd or if } p = 2 \text{ and } k > 0 \\ -1 & \text{if } p = 2 \text{ and } k = 0 \end{cases}$$

Further

$$t_{D^* \setminus F'^* G_1}(c_i) = \prod_{r=0}^{p-1} \pi^r c_i \pi^{-r} = \prod_{r=0}^{p-1} c_i^{\sigma^r} = N_{F'/F}(c_i) \in F^*$$

where $\sigma : x \to \pi x \pi^{-1}$ generates $\text{Gal}(F'/F)$. As $\epsilon(x) = 1$ for $x \in G_1$, we have

$$\det \Pi'(c_i) = \det \tilde{\rho} \chi_\nu \big|_{F'^* G_1} (N_{F'/F}(c_i)) = 1$$

for all i. Thus, if p is odd or if $p = 2$ and $k \geq 2$, then $\det \Pi' \equiv 1$. If $p = 2$ and $k = 0$, then $\det \Pi' \equiv w_{F_2/F} \circ \nu$. Hence the theorem. □

We conclude this section with the following result from Tunnell's paper [Tu, page 1300].

Lemma 3.1 Let D be a quaternion division algebra over a finite extension F of $\mathbb{Q}_2$. Let ρ be a minimal irreducible representation of D^* of even conductor $k + 2 > 2$. Then the central character w_ρ of ρ cannot be trivial.

4 A Proof of Theorem C

This section is divided into three subsections. In the first subsection we find conditions under which $(1 + P_F)^{\dim \Pi} \subset (1 + P_F^t)^p \subseteq \ker \chi$ and in the second one we calculate the tranfer $t_{D^* \setminus F'^* G_j}(x)$ for $x \in G_1$ and $x = \pi$ where F' is totally ramified and finally in the third subsection we prove Theorem C

4.1 Some Results on F

Given a positive integer k coprime to p, let t be such that $k = (t-1)p + s$, $0 < s < p$ and let j be $(k/2) + 1$ if k is even and $(k+1)/2$ if k is odd. Note that $j = t$ if $p = 2$ and $j > t$ otherwise.

The following generalisation of a result in [Me, page 163] will be used repeatedly to prove some of the results in this subsection.

Lemma 4.1 If F is a non-archimedean local field and P_F is the maximal ideal in F, then for any positive integer r and prime p (p not necessarily the residue characteristic of F) such that $P_F^r \subseteq pP_F$, we have $(1+P_F^r)^p = 1+pP_F^r$. In particular, if F has residue characteristic p, then for any $r \geq e+1$, $e = v_F(p)$, we have $(1+P_F^r)^p = 1+P_F^{r+e}$.

The next Lemma can be proved by induction.

Lemma 4.2 Given a positive integer s, suppose $n \in \mathbb{N}$ is such that $sp^{n-1} \leq e < sp^n$. Then for any r such that $0 \leq r \leq n$,
$(1+P_F^s)^{p^r(p-1)} \subseteq 1+P_F^{sp^r}$.

If $q = 2$, we have a slightly stronger result.

Lemma 4.3 Let $q = 2$. For any positive integer s
(i) if $s < e$, then $(1+P_F^s)^2 \subseteq 1+P_F^{2s}$
(ii) if $s = e$, then $(1+P_F^s)^2 \subseteq 1+P_F^{2s+1}$
(iii) if $s > e$, then $(1+P_F^s)^2 = 1+P_F^{s+e}$.

Proof: (i) and (iii) are straight forward. To prove (ii) suppose $x \in (1+P_F^s)$. Then $x = 1 + \alpha_s \pi_F^s + \alpha_{s+1} \pi_F^{s+1} + ...$ where $\alpha_i = 0$ or 1. Therefore, $x^2 = 1 + 2\alpha_s \pi_F^s + 2\alpha_{s+1} \pi_F^{s+1} + ... + (2\alpha_{2s} + \alpha_s^2) \pi_F^{2s} + ...$ Since $q = 2$, $(\alpha_s + \alpha_s^2) \in P_F$ and hence $x^2 \in 1+P_F^{2s+1}$. Hence the lemma. $\square$

Lemma 4.4 Given e, if there exists a positive integer n such that $sp^{n-1} \leq e < sp^n$, then for any $r > n$, $(1+P_F^s)^{p^r(p-1)} \subseteq 1+P_F^{sp^n+(r-n)(p-1)e}$. Otherwise, $s > e$ and $(1+P_F^s)^{p^r} = 1+P_F^{s+re}$.

Proof: If $s > e$, then by Lemma 4.1, $(1+P_F^s)^{p^r} = 1+P_F^{s+re}$. Suppose $s \leq e$. Then, there exists a positive integer n such that $sp^{n-1} \leq e < sp^n$ and by Lemma 4.2 for any $r > n$, $(1+P_F^s)^{p^r(p-1)} \subseteq 1+P_F^{sp^n+(r-n)(p-1)e}$ as $sp^n > e$ proving the lemma. $\square$

Proposition 4.5 Let $q \geq 3$. Then, for $k \geq 2$, we have,

$$(1+P_F)^{q^{(k-1)(p-1)/2}} \subseteq (1+P_F^t)^p$$

Proof: To prove the proposition, it is enough to show that

$$(1+P_F)^{p^{(f(k-1)(p-1)/2)-1}} \subseteq 1+P_F^t. \tag{2}$$

For $j \geq 3$ we will in fact show that

$$(1+P_F)^{q^{(j-2)(p-1)}} \subseteq 1+P_F^j. \tag{3}$$

As $(f(k-1)(p-1)/2) - 1 \geq f(j-2)(p-1)$ we have, $(1+P_F)^{q^{(k-1)(p-1)/2}} = (1+P_F)^{p^{f(k-1)(p-1)/2}} \subseteq (1+P_F)^{p^{f(j-2)(p-1)}}$. Also, $1+P_F^j \subseteq 1+P_F^t$ and hence (2) follows from (3). Let $j \geq 3$. Suppose $f(j-2) \leq n$. By Lemma 4.2, $(1+P_F)^{p^{f(j-2)(p-1)}} \subseteq 1 + P_F^{p^{f(j-2)}}$. For $j = 3$, $p^{f(j-2)} = p^f = q \geq 3 = j$ and by induction we are through. Next, suppose $f(j-2) > n$. Then by Lemma 4.4 $(1+P_F)^{p^{f(j-2)(p-1)}} \subseteq 1 + P_F^{p^n+(f(j-2)-n)(p-1)e}$. As before, to prove (3) we need to show that $p^n + (f(j-2) - n)(p-1)e \geq j$. Since, $p^n + (f(j+1-2) - n)(p-1)e \geq p^n + (f(j-2) - n)(p-1)e + 1$ it is enough to show that $p^n + (f(j-2) - n)(p-1)e \geq j$ for some j. Let $j = 3$. Then $p^n + (f(j-2) - n)(p-1)e = p^n + (f-n)(p-1)e \geq p^n + (f-n)(p-1) \geq p+1 \geq 3 = j$ as $f(j-2) = f > n$, proving (3) for $f(j-2) > n$. Finally let $j = 2$. This means k is either 2 or 3. If p is odd, $t = 1$. and hence (2) is trivially true. Let $p = 2$. Then $k = 3$ and $t = 2$. Hence $(1+P_F)^{p^{(f(k-1)(p-1)/2)-1}} = (1+P_F)^{2^{f-1}}$. If $(f-1) \leq n$, by Lemma 4.2, $(1+P_F)^{2^{f-1}} \subseteq 1 + P_F^{2^{f-1}} \subseteq 1 + P_F^2 = 1 + P_F^t$. If $(f-1) > n$, by Lemma 4.4, $(1+P_F)^{2^{f-1}} \subseteq 1 + P_F^{2^n + (f-1-n)e}$. But, $2^n + (f-1-n)e \geq 2^n + (f-1-n) \geq 2$. This completes the proof of the proposition. $\square$

Proposition 4.6 Let $q = 2$. Then, for $j \geq 4$, $(1+P_F)^{q^{j-1}} \subseteq (1+P_F^j)^2$.

Proof: Let n be the positive integer such that $2^{n-1} \leq e < 2^n$. Suppose $(j-2) \leq n$. Then, by Lemma 4.2, $(1+P_F)^{2^{j-2}} \subseteq 1 + P_F^{2^{j-2}}$ and as $j \geq 4$, we have $1 + P_F^{2^{j-2}} \subseteq 1 + P_F^j$. Next let $(j-2) > n$. Then $j \geq 4$ and by Lemma 4.4 $(1+P_F)^{2^{j-2}} \subseteq 1 + P_F^{2^n+(j-2-n)e}$. Now if $n \geq 2$, then $2^n + (j-2-n)e \geq j$. Therefore, $(1+P_F)^{q^{j-1}} = (1+P_F)^{2^{j-1}} \subseteq (1+P_F^j)^2$. If $n = 1$ by Lemma 4.3, $(1+P_F)^2 \subseteq 1+P_F^3$. By Lemma 4.4, $(1+P_F)^{q^{j-1}} \subseteq (1+P_F^3)^{2^{j-2}} = 1 + P_F^{3+(j-2)e} = 1 + P_F^{j+1} = (1+P_F^j)^2 = 1+P_F^{j+1}$. $\square$

Proposition 4.7 Let $q = 2$. Then,
(i) For $j = 2$ and $e = 1$, $(1+P_F)^{q^{j-1}} = (1+P_F^j)^2$, and for $j = 2$ and $e \geq 2$, $(1+P_F)^{q^{j-1}} \not\subseteq (1+P_F^j)^2$.
(ii) For $j = 3$ and $e = 1$ or $e = 2$, $(1+P_F)^{q^{j-1}} \subseteq (1+P_F^j)^2$, and for $j = 3$ and $e \geq 3$,
$(1+P_F)^{q^{j-1}} \not\subseteq (1+P_F^j)^2$ if $e \geq 3$.

Proof: Let $j = 2$. If $e = 1$, the claim follows from Lemma 4.3 and 4.4. If $e \geq 2$, $(1+P_F^j)^2 \subseteq 1 + P_F^4$ whereas $(1+\pi_F)^2 = 1 + 2\pi_F + \pi_F^2 \in (1+P_F^2) - (1+P_F^3)$ proving (i). By a similar argument (ii) follows.. $\square$
However, for $j \geq (1+P_F^2)^{q^{(k-1)(p-1)/2}} \subseteq (1+P_F^t)^p$.

4.2 Some results on Transfer

In this section, we calculate the transfer $t_{D^*\setminus F'^*G_j}(x)$ for $x \in G_1$ and $x = \pi$. In order to avoid cumbersome notation we write $t_{D^*\setminus F'^*G_j}(x)$ as an element of D^* though it is in fact a coset of an appropriate subgroup in D^*. Observe that, for any root of unity η and any positive integers r and k,

$$(1+\eta\pi^r)(1+\alpha_k\pi^k+\alpha_{k+1}\pi^{k+1}+\ldots)(1+\eta\pi^r)^{-1} =$$
$$1 + \alpha_k\pi^k + \alpha_{k+1}\pi^{k+1} + \ldots + (\alpha_{k+r} + (\eta\alpha_k^{\sigma^r} - \alpha_k\eta^{\sigma^k}))\pi^{k+r}$$
$$+(\alpha_{k+r+1} + (\eta\alpha_{k+1}^{\sigma^r} - \alpha_{k+1}\eta^{\sigma^{k+1}}))\pi^{k+r+1} + \ldots \quad (4)$$
$$+(\alpha_{k+2r-1} + (\eta\alpha_{k+r-1}^{\sigma^r} - \alpha_{k+r-1}\eta^{\sigma^{k+r-1}}))\pi^{k+2r-1} + \ldots$$

Lemma 4.8 Let $q \geq 3$, $x \in G_k$. Let $1 \leq r < k$ or $r = k$ and $x \in F'^*G_{k+1}$ and let $s = \min\{(p-1)v_D(q)+k,\ k+r+p,\ k+2r\}$. Then, $t_{F'^*G_r\setminus F'^*G_{r+1}}(x) \in G_s$.

Proof: Let $x = 1 + \alpha_k\pi^k + \alpha_{k+1}\pi^{k+1} + \ldots$. Since G_k is normal in D^*, by (4) we have,

$$t_{F'^*G_r\setminus F'^*G_{r+1}}(x) = \prod_{i=1}^{q^{(p-1)}}(1+\zeta_i\pi^r)x(1+\zeta_i\pi^r)^{-1}$$

$$\prod_{i=1}^{q^{(p-1)}}(1+\alpha_k\pi^k+\alpha_{k+1}\pi^{k+1}+\ldots+(\alpha_{k+r}+(\zeta_i\alpha_k^{\sigma^r}-\alpha_k\zeta_i^{\sigma^k}))\pi^{k+r}$$
$$+(\alpha_{k+r+1}+(\zeta_i\alpha_{k+1}^{\sigma^r}-\alpha_{k+1}\zeta_i^{\sigma^{k+1}}))\pi^{k+r+1}+\ldots$$
$$+(\alpha_{k+2r-1}+(\zeta_i\alpha_{k+r-1}^{\sigma^r}-\alpha_{k+r-1}\zeta_i^{\sigma^{k+r-1}}))\pi^{k+2r-1}+\ldots)$$

Suppose $k \geq 2r$. Then the above equation becomes,

$$t_{F'^*G_r\setminus F'^*G_{r+1}}(x) =$$
$$1 + q^{(p-1)}\alpha_k\pi^k + q^{(p-1)}\alpha_{k+1}\pi^{k+1} + \ldots + (q^{(p-1)}\alpha_{k+r}$$
$$+(\alpha_k^{\sigma^r}\sum_{i=1}^{q^{(p-1)}}\zeta_i - \alpha_k\sum_{i=1}^{q^{(p-1)}}\zeta_i^{\sigma^k}))\pi^{k+r}$$
$$+(q^{(p-1)}\alpha_{k+r+1}+(\alpha_{k+1}^{\sigma^r}\sum_{i=1}^{q^{(p-1)}}\zeta_i-\alpha_{k+1}\sum_{i=1}^{q^{(p-1)}}\zeta_i^{\sigma^{k+1}}))\pi^{k+r+1}+\ldots$$
$$+(q^{(p-1)}\alpha_{k+2r-1}+(\alpha_{k+r-1}^{\sigma^r}\sum_{i=1}^{q^{(p-1)}}\zeta_i-\alpha_{k+r-1}\sum_{i=1}^{q^{(p-1)}}\zeta_i^{\sigma^{k+r-1}}))\pi^{k+2r-1}+\ldots$$

But, $\sum_{i=1}^{q^{(p-1)}} \bar{\zeta}_i = \sum_{i=1}^{q^{(p-1)}} \bar{\zeta}_i^{\sigma^t} = \bar{0}$ since $q \geq 3$. Therefore $t_{F'^*G_r \setminus F'^*G_{r+1}}(x) \in G_s$ where, $s = \min\{(p-1)v_D(q) + k,\ k+r+p,\ k+2r\}$. Next, let $k < 2r$. Then, in the expression of $t_{F'^*G_r \setminus F'^*G_{r+1}}(x)$ the coefficient of π^{2k} is

$$q^{(p-1)}\alpha_{2k} + (\alpha_{2k-r}^{\sigma^r} \sum_{i=1}^{q^{(p-1)}} \zeta_i - \alpha_{2k-r} \sum_{i=1}^{q^{(p-1)}} \zeta_i^{\sigma^{2k-r}}) + \frac{1}{2}q^{(p-1)}(q^{(p-1)}-1)\alpha_k \alpha_k^{\sigma^k}$$

and that of π^{2k+t} for $1 \leq t \leq 2r-k-1$, is

$$q^{(p-1)}\alpha_{2k+t} + (\alpha_{2k+t-r}^{\sigma^r} \sum_{i=1}^{q^{(p-1)}} \zeta_i - \alpha_{2k+t-r} \sum_{i=1}^{q^{(p-1)}} \zeta_i^{\sigma^{2k+t-r}})$$
$$+ \frac{1}{2}q^{(p-1)}(q^{(p-1)}-1) \sum_{t_1 < t_2} (\alpha_{k+t_1}\alpha_{k+t_2}^{\sigma^{k+t_1}} + \alpha_{k+t_2}\alpha_{k+t_1}^{\sigma^{k+t_2}}) + \Delta$$

where t_1 and t_2 are such that $t = t_1 + t_2$ and

$$\Delta = \begin{cases} \frac{1}{2}q^{(p-1)}(q^{(p-1)}-1)\alpha_{k+t_1}\alpha_{k+t_1}^{\sigma^{k+t_1}} & \text{if } t = 2t_1 \\ 0 & \text{if } t \text{ is odd} \end{cases}$$

Now, if q is odd, then 2 does not divide $q^{(p-1)}$ and since $2k+p \geq k+r+p$, as before $t_{F'^*G_r \setminus F'^*G_{r+1}}(x) \in G_s$ where, $s = \min\{(p-1)v_D(q)+k,\ k+r+p,\ k+2r\}$. If $q = 2^f$, then 2 divides $q^{(p-1)}$. To prove the lemma it suffices to show that $v_D(q) + 2k - 2e \geq k+r+2$. But, $v_D(q) + 2k - 2e = 2e(f-1) + 2k$ and since $f \geq 2$ and $k \geq r$ we have $2e(f-1) + 2k \geq k+r+2$. Hence the lemma. . □

The next lemma is a consequence of the fact that G_1 is normal in D^* and equation (4).

Lemma 4.9 Let $q=2$ and let $x \in G_{2j}$. Then for $j \geq 2$, $t_{F'^*G_j \setminus F'^*G_{j+1}}(x) \in G_{2(j+1)}$.

Proposition 4.10 If $x \in G_1$, then $t_{D^* \setminus F'^*G_1}(x) \in G_2$.

Proof: Let $x = 1 + \alpha_1 \pi + \alpha_2 \pi^2 + \ldots$ and let $\pi \zeta \pi^{-1} = \zeta^{q^m}$ for some m, $1 \leq m < p$. Then $t_{D^* \setminus F'^*G_1}(x) = \prod_{i=0}^{\ell-1} \zeta^{-i} x \zeta^i = 1 + \alpha_1 \sum_{i=0}^{\ell-1} \zeta^{(q^m-1)i} \pi \mod P_D^2 = 1 \mod P_D^2$ since $\zeta^{(q^m-1)}$ is an ℓth root of unity . □

The next theorem follows easily from Lemma 4.8 and Proposition 4.10. In fact a much stronger result is true. However the version given here is adequate for our purpose.

Theorem 4.11 Let $q \geq 3$ and $x \in G_1$. Then for $j \geq 1$, $t_{D^* \setminus F'^* G_j}(x) \in G_{2j}$. In fact for $j \geq 4$, $t_{D^* \setminus F'^* G_{j-1}}(x) \in G_{2j-1}$.

Proof: Clear.

Proposition 4.12 Let $q = 2$. If $x \in G_1$, then for $j = 1$, or for $j \geq 4$, $t_{D^* \setminus F'^* G_j}(x) \in G_{2j}$. Moreover, $t_{D^* \setminus F'^* G_2}(x)$ is

$$\begin{cases} 1 + \zeta \pi^3 \mod P_D^4 & \text{if } x = 1 + \delta \pi + \pi^2, 1 + \delta \pi, 1 + \zeta \pi^2, 1 + \zeta^2 \pi^2 \mod P_D^3 \\ 1 \mod P_D^4 & \text{otherwise} \end{cases}$$

and $t_{D^* \setminus F'^* G_3}(x)$ is

$$\begin{cases} 1 + \zeta \pi^5 \mod P_D^6 & \text{if } t_{D^* \setminus F'^* G_2}(x) = 1 + \zeta \pi^3 \mod P_D^4 \text{ and } e \geq 2 \\ 1 \mod P_D^6 & \text{otherwise} \end{cases}$$

where $\delta \in \langle \zeta \rangle$.

Proof: Let $x = 1 + \alpha_1 \pi + \alpha_2 \pi^2 + \ldots$, $\alpha_i \in \langle \zeta \rangle \cup \{0\}$. Then $t_{D^* \setminus F'^* G_1}(x) = x \, \zeta x \zeta^2 \, \zeta^2 x \zeta = 1 + (\alpha_2 + \alpha_1^3 \zeta^2) \pi^2 \mod P_D^3$. Let $\beta_2 = (\alpha_2 + \alpha_1^3 \zeta^2) \mod P_D$. Then

$$\beta_2 = \begin{cases} 0 & \text{if } \alpha_1 = \alpha_2 = 0 \text{ or } \alpha_1 \neq 0, \alpha_2 = \zeta^2 \\ 1 & \text{if } \alpha_2 = \zeta, \alpha_1 \neq 0 \\ \zeta \text{ or } \zeta^2 & \text{if } \alpha_2 = 0 \text{ or } 1 \text{ \&} \alpha_1 \neq 0 \text{ or if } \alpha_1 = 0 \text{ \&} \alpha_2 = \zeta \text{ or } \zeta^2 \end{cases}$$

$t_{D^* \setminus F'^* G_1}(x) = 1 + \beta_2 \pi^2 \mod P_D^3 \in G_2$. Next,

$$t_{D^* \setminus F'^* G_2}(x) = 1 + \zeta(\beta_2^2 - \beta_2) \pi^3 \mod P_D^4$$

$$= \begin{cases} 1 + \zeta \pi^3 \mod P_D^4 & \text{if } \alpha_2 = 0 \text{ or } 1 \text{ and } \alpha_1 \neq 0 \\ & \text{or if } \alpha_1 = 0 \text{ and } \alpha_2 = \zeta \text{ or } \zeta^2 \\ 1 \mod P_D^4 & \text{otherwise} \end{cases}$$

If $t_{D^* \setminus F'^* G_2}(x) \in G_4$, then by Lemma 4.9, $t_{D^* \setminus F'^* G_j}(x) \in G_{2j}$ for all $j \geq 3$ and if $t_{D^* \setminus F'^* G_2}(x) = 1 + \zeta \pi^3 \mod P_D^4$. Then,

$$t_{D^* \setminus F'^* G_3}(x) = \begin{cases} 1 + \zeta \pi^5 \mod P_D^6 & \text{if } e \geq 2 \\ 1 \mod P_D^6 & \text{if } e = 1 \end{cases}$$

Now it is easy to see that if $x \in G_1$ and $j \geq 4$, then $t_{D^* \setminus F'^* G_j}(x) \in G_{2j}$ □

Proposition 4.13 $t_{D^*\setminus F'^*G_1}(\pi) = \pi^\ell$

Proof: Note that $\zeta^i \pi = \pi \zeta^{iq^{(p-m)}}$. Let $iq^{(p-m)} = \ell r_i + j_i$ where $1 \leq j_i < \ell$. Since $\pi \in F'^*G_1 = F(\pi)^*G_1$, and $\zeta^{\ell \sum_{i=1}^{\ell-1} r_i} = 1$, $t_{D^*\setminus F'^*G_1}(\pi) = \pi \prod_{i=1}^{\ell-1} \zeta^{\ell r_i}\pi = \pi^\ell \zeta^{\ell \sum_{i=1}^{\ell-1} r_i} = \pi^\ell$. □

We need the next two general lemmas to proceed with the calculation of $t_{D^*\setminus F'^*G_j}(\pi)$ for $j \geq 2$.

Lemma 4.14 Let $F' = F(\pi')$ where $\pi' = \pi(1 + \beta\pi + \beta_2\pi^2 + \ldots)$. Then $d_{F'/F} = p$ if and only if $Tr_{F(\zeta)/F}(\beta)$ is in U_D.

Proof: Let $f_{\pi'}(X) = X^p + a_1 X^{p-1} + \ldots + a_{p-1}X + a_p$ be the minimal polynomial for π' over F. Then $p \leq d_{F'/F} \leq v_{F'}(p) + p - 1$ and

$$d_{F'/F} = \min_{i=0,1,2,\ldots,p-1}\{v_{F'}[(p-i)a_i\pi'^{p-i-1}]\}.$$

$d_{F'/F} = p$ implies $a_{p-1} = \pi^p u_{p-1}$. The coefficient of π^{p+1} in $f_{\pi'}(\pi') = Tr_{F(\zeta)/F}(\beta) + u_{p-1} = 0$ which is equivalent to saying that $Tr_{F(\zeta)/F}\beta$ is in U_D. Conversely, if $Tr_{F(\zeta)/F}(\beta)$ is in U_D, then the coefficient of π^{p+1} in the expression of $f_{\pi'}(\pi')$ is zero only if $v_{F'}(a_{p-1}) = 1$ or equivalently, $d_{F'/F} = p$. Hence the lemma. □

The condition that $Tr_{F(\zeta)/F}(\beta)$ is in U_D is equivalent to saying that $\bar{\beta}$ is not a trace zero element in $\overline{D}$. If $p = 2$, this is equivalent to saying $\bar{\beta} \notin \overline{F}$ or, equivalently that $F'^*G_2 \neq F(\pi)^*G_2$.

We state the following lemma without proof.

Lemma 4.15 Let $F'^*G_2 = F(\pi(1 + \beta\pi))^*G_2$. Then,

1. $\pi(1 + \gamma\pi) \in F'^*G_2$ if and only if $\bar{\gamma} = \bar{\beta} + \bar{\alpha}$ for some $\alpha \in \langle \zeta^\ell \rangle \cup \{0\}$.

2. $(1 + \zeta_i\pi)\pi(1 + \zeta_i\pi)^{-1} \in F'^*G_2$ for some i if and only if $d_{F'/F} > p$.

3. If for some i, $(1 + \zeta_i\pi)\pi(1 + \zeta_i\pi)^{-1} \notin F'^*G_2$, then there exist distinct indices $i_1, i_2, \ldots, i_p$ with $i_1 = i_{p+1} = i$ such that for $1 \leq j \leq p$, $(1 + \zeta_{i_j}\pi)\pi = x_j(1 + \zeta_{i_{j+1}}\pi)$ where $x_j \in F'^*G_2$.

4. There exists a uniformising element $\pi' \in F'^*G_2$ and a $(q^p - 1)$th root of unity ζ' in D^* such that $\pi'^p \in F$ and $\pi'\zeta'\pi'^{-1} = (\zeta')^{q^m}$ for some m, $1 \leq m < n$ if and only if $\bar{\beta}$ is a trace zero element in $\overline{D}$.

Theorem 4.16 Let $d_{F'/F} = p$. Then, $t_{D^* \backslash F'^* G_j}(\pi) = \pi^{\ell q^{(j-1)(p-1)}}$

Theorem 4.17 Let $q \geq 3$. Then, $t_{D^* \backslash F'^* G_1}(\pi) = \pi^\ell$, $t_{D^* \backslash F'^* G_2}(\pi) = \pi^{\ell q^{p-1}}(1+\beta_2\pi^2+\beta_3\pi^3) \bmod P_D^4$ and for $j \geq 3$, $t_{D^* \backslash F'^* G_j}(\pi) = \pi^{\ell q^{(p-1)(j-1)}} \bmod P_D^{2j}$ where,

$$\beta_2 = \begin{cases} \eta + \delta_1 & \text{if } q = 3 \\ \delta_1 & \text{if } q \geq 4 \end{cases}$$

and $\beta_3 = \begin{cases} \alpha\eta + \delta_2 & \text{if } q \text{ is odd} \\ \delta_2 & \text{if } q = 2^f,\ f \geq 3 \\ -\eta^2 + \delta_2 & \text{if } q = 4 \end{cases}$

for α, δ_i in $\langle \zeta^\ell \rangle \cup \{0\}$

Proof: Let η be a $(q^p - 1)$th root of unity such that $\overline{\eta^{\sigma^{-1}} - \eta} = \bar{1}$. Then $\eta^{\sigma i} = \eta - i \bmod P_D^p$. Let $\zeta_i = \alpha_i \eta$ for $1 \leq i \leq q$ where $\alpha_i \in \langle \zeta^\ell \rangle \cup \{0\}$. $1 + \zeta_i \pi$ can be completed to a full set of coset representatives $F(\pi)^* G_2$ in $F(\pi)^* G_1$ such that $\sum_{i=1}^{q^{(p-1)}} \bar{\zeta}_i = 0$. Let $\eta^{\sigma^{-1}} - \eta = 1 + \gamma_p \pi^p + \gamma_{2p} \pi^{2p} + \ldots$. Then

$$t_{F'^* G_1 \backslash F'^* G_2}(\pi) = \pi^{q^{p-2}} \prod_{i=1}^{q}(1+\zeta_i\pi)\pi(1+\zeta_i\pi)^{-1}$$

$$= \pi^{q^{p-1}} \prod_{i=1}^{q}(1 + \alpha_i\pi - \alpha_i^2(\eta - i - 1))\pi^2 + (\alpha_i^3(\eta - i - 1)(\eta - i - 2)$$

$$+ \alpha_i \lambda^{\sigma^i})\pi^3) \bmod P_D^4$$

where $\lambda = \gamma_2$ if $p = 2$ and zero otherwise. Hence the coefficient of π and π^2 are as claimed in the theorem. The Coefficient of π^3 is

$$\eta^2 \sum_{i=1}^q \alpha_i^3 - 2\eta \sum_{i=1}^q \alpha_i^3 - 3\eta \sum_{i=1}^q \alpha_i^3 + \sum_{i=1}^q \alpha_i \lambda^{\sigma^i} - \eta \sum_{i<j}(\alpha_i^2 \alpha_j + \alpha_i \alpha_j^2) + \delta$$

for some $\delta \in \langle \zeta^\ell \rangle \cup \{0\}$. Since $\sum_{i<j}(\alpha_i^2\alpha_j + \alpha_i\alpha_j^2) = -\sum \alpha_i^3 \bmod P_F$, the coefficient of π^3 is as claimed in the theorem. Let β_2, β_3 denote the coefficient of π^2, π^3 respectively. Then $t_{D^* \backslash F'^* G_3}(\pi) = \pi^{\ell q^{2(p-1)}} t_{F'^* G_2 \backslash F'^* G_3}(1+ \beta_2\pi^2 + \beta_3\pi^3)$ and if $q \geq 4$, $(1 + \beta_2\pi^2 + \beta_3\pi^3) \in F'^* G_3$ and hence $t_{F'^* G_2 \backslash F'^* G_3}(1 + \beta_2\pi^2 + \beta_3\pi^3) = 1 \bmod P_D^6$. If $q = 3$, $(1 + \zeta_i\pi^2)(1+$

$\beta_2\pi^2 + \beta_3\pi^3)(1 + \zeta_i\pi^2)^{-1} \notin F'^*G_3$ for any i. Moreover $(1 + \beta_2\pi^2 + \beta_3\pi^3)^3 \in F'^*G_3$. Therefore, $t_{F'^*G_2 \setminus F'^*G_3}(1 + \beta_2\pi^2 + \beta_3\pi^3) = (1 + \beta_2\pi^2 + \beta_3\pi^3)^9 = 1 \mod P_D^6$ which implies $j \geq 3$ we have, $t_{D^* \setminus F'^*G_j}(\pi) = \pi^{\ell q(p-1)(j-1)} \mod P_D^{2j}$. Hence the theorem. □

Finally, let $p = q = 2$. The next three theorems give the calculation of transfer in this case for $d_{F'/F} > 2$.

Theorem 4.18 Let $q = 2$. Suppose $e = 1$. Then,

$$t_{D^* \setminus F'^*G_1}(\pi) = \pi^3$$
$$t_{D^* \setminus F'^*G_2}(\pi) = \pi^6(1 + \pi + \zeta^2\pi^2 + \zeta\pi^3) \mod P_D^4$$
$$t_{D^* \setminus F'^*G_3}(\pi) = \pi^{12}(1 + \pi^2) \mod P_D^6 \text{ and}$$
$$t_{D^* \setminus F'^*G_j}(\pi) = \pi^{(q+1)q^{j-1}}(1 + \pi^{2j-2}) \mod P_D^{2j} \text{ for } j \geq 4$$

Theorem 4.19 Let $q = 2$. Suppose $e = 2$. Then,

$$t_{D^* \setminus F'^*G_1}(\pi) = \pi^3$$
$$t_{D^* \setminus F'^*G_2}(\pi) = \pi^6(1 + \pi + \zeta^2\pi^2 + \pi^3) \mod P_D^4$$
$$t_{D^* \setminus F'^*G_3}(\pi) = \pi^{12}(1 + \pi^2)(1 + \pi^3 + \zeta\pi^4 + \pi^5) \mod P_D^6$$
$$t_{D^* \setminus F'^*G_4}(\pi) = \pi^{24}(1 + \pi^4 + \pi^6)(1 + \zeta\pi^7) \mod P_D^8 \text{ and}$$
$$t_{D^* \setminus F'^*G_j}(\pi) = \pi^{(q+1)q^{j-1}} \mod P_D^{2j} \text{ for } j \geq 5$$

Theorem 4.20 Let $q = 2$ and let $e \geq 3$. Then,

$$t_{D^* \setminus F'^*G_1}(\pi) = \pi^3$$
$$t_{D^* \setminus F'^*G_2}(\pi) = \pi^6(1 + \pi + \zeta^2\pi^2 + \pi^3) \mod P_D^4$$
$$t_{D^* \setminus F'^*G_3}(\pi) = \pi^{12}(1 + \pi^2)(1 + \pi^3 + \zeta\pi^4) \mod P_D^6$$
$$t_{D^* \setminus F'^*G_4}(\pi) = \pi^{24}(1 + \pi^4)(1 + \zeta^2\pi^7) \mod P_D^8$$
$$t_{D^* \setminus F'^*G_5}(\pi) = \pi^{48}(1 + \pi^8) \mod P_D^{10} \text{ and}$$
$$t_{D^* \setminus F'^*G_j}(\pi) = \pi^{(q+1)q^{j-1}} \mod P_D^{2j} \text{ for } j \geq 6$$

4.3 Proof of the Theorem

In this subsection we give the proof of Theorem C by breaking it into several cases. If ρ is a minimal irreducible representation of D^* of conductor $k+2$, then there exists a character ϕ of G_j/G_{j+1} such that ρ lies over ϕ and if ψ_F is an additive character of F with $n(\psi_F) = 0$, then there exists $h = \lambda \pi^{-k-p}(1 + \alpha_1 \pi + \alpha_2 \pi^2 + \ldots) \in D^*$ such that $\phi(1+d) = \psi_F \circ \tau(hd)$ for $d \in P_D^j$. Let $F' = F(h)$. This implies F' is totally ramified. If k is odd we can extend ϕ to $\tilde{\phi}$ on $F'^* G_j$ and $\rho = \mathrm{Ind}_{F'^* G_j}^{D^*} \tilde{\phi}$ ([Ho]). If k is even, p must be odd and ϕ induces a non degenerate skew symmetric bilinear form on $G_k/G_{k+1} \cong \overline{D}$ and hence there exists a maximal isotropic subgroup G'_{j-1} such that $G_j \subset G'_{j-1} \subset G_{j-1}$ and ϕ can be extended to $\tilde{\phi}$ on $F'^* G'_{j-1}$ and $\rho = \mathrm{Ind}_{F'^* G'_{j-1}}^{D^*} \tilde{\phi}$ ([Ho]). Note that since F' is conjugate to $F(\pi u)$ for some $u \in G_1$ and in view of the fact that $\mathrm{Ind}_{zHz^{-1}}^G \phi \cong \mathrm{Ind}_H^G \phi^z$, we can assume $h = \lambda \pi^{-k-p} u$ where $\lambda \in \langle \zeta^\ell \rangle$ and $u \in G_1$.

Lemma 4.21 Let ρ be a minimal irreducible representation of D^* of conductor $k+2$ such that k is coprime to p. Suppose there exists a character χ of F^* such that $\Pi' = \rho \otimes \chi$ has trivial central character. If $q \geq 3$, then for $k \geq 2$, $(1 + P_F)^{q^{(k-1)(p-1)/2}} \subseteq \ker \chi$. If $q = 2$, then for $k \geq 7$, $(1 + P_F)^{q^{j-1}} \subseteq \ker \chi$. If $k = 3$ (resp. $k = 5$) $(1 + P_F)^{q^{j-1}} \subseteq \ker \chi$ only if $e = 1$ (resp. $e = 1$ or 2.)

Proof: Since ϕ is trivial on G_{k+1}, for $x \in G_{k+1} \cap F^* = 1 + P_F^t$, $1 = w_{\Pi'}(x) = w_{\rho \otimes \chi}(x) = \phi(x)\chi^p(x) = \chi^p(x)$. Therefore, $(1 + P_F^t)^p \subseteq \ker \chi$. Now by Propositions 4.5, 4.6 and 4.7 the lemma follows. $\square$

We evaluate $\det \Pi'$ by expressing it in terms of the inducing character $\tilde{\phi}$. Since $\Pi' = \rho \otimes \chi$ and since $\dim \rho = \ell q^{(p-1)(k-1)/2}$ we have $\det \Pi' = \det \rho \otimes \chi = \chi_\nu^{\dim \rho} \det \rho = \chi_\nu^{\ell q^{(p-1)(k-1)/2}} \det \rho$. Also if G''_j is G'_{j-1} or G_j according as k is even or odd, then for $x \in D^*$, by equation (1) $\det \rho(x) = \epsilon(x)\tilde{\phi}(t_{D^* \setminus F'^* G''_j}(x))$ where ϵ is the determinant of the permutation representation of D^* on $D^*/F'^* G''_j$. Therefore, we have

$$\det \Pi' = \epsilon(x)\chi_\nu^{\ell q^{(p-1)(k-1)/2}}(x)\tilde{\phi}(t_{D^* \setminus F'^* G''_j}(x)) \tag{5}$$

Lemma 4.22 Let ϵ be the determinant of the permutation representation of D^* on $D^*/F'^* G''_j$ for $j \geq 1$. Then $\epsilon(x) = 1$ if $x \in G_1$. Moreover,

$\epsilon(\pi) = 1$ if $q \geq 3$, if $q = 2$ and $j \geq 4$, if $q = 2$, $j = 2$ and $d_{F'/F} > 2$ and if $q = 2$, $j = 3$ and $d_{F'/F} = 2$. $\epsilon(\pi) = -1$ otherwise.

Proposition 4.23 Let ρ be a minimal irreducible representation of D^* of conductor 3. Suppose there exists a character χ of F^* such that $\Pi' = \rho \otimes \chi$ has trivial central character. Then, χ is trivial on $(1 + P_F)^p$ and if χ' is a character of F^* such that $\chi' = \chi$ on $1 + P_F$, $\chi'_\nu(\zeta) = 1$ and

$$\chi'_\nu(\pi) = \begin{cases} \phi\chi_\nu(\pi) & \text{if } p \text{ is odd} \\ (-1)^{q/2}\phi\chi_\nu(\pi) & \text{if } p = 2 \end{cases}$$

where π is any uniformising element of D such that $\pi^p \in F$ and $\pi \in F'^* G_1$. Then χ' is a character of F^* of order p and $\det \Pi' = \chi'_\nu$. If $p = 2$ and $K_{\chi'}$ is a quadratic extension of F determined by χ', then $\det \Pi' = w_{K_{\chi'}/F} \circ \nu$

Proof: Conductor of $\rho = k + 2 = 3$ implies χ is a character of $(1 + P_F)/(1 + P_F)^p$. If $x \in G_1$, by Proposition 4.10 $\tilde{\phi}(t_{D^* \setminus F'^* G_1}(x)) = 1$. Also, $\chi_\nu(x^{\ell-1}) = 1$ and hence $\det \Pi' = \chi_\nu$ on G_1. Next, by Proposition 4.13 $\det \Pi'(\pi) = \epsilon(\pi)\tilde{\phi}(\pi^\ell)\chi_\nu(\pi^\ell)$. Since $\pi^{\ell-1} \in F^*$ and $\epsilon(\pi) = (-1)^{q/2}$ if $p = 2$ and $\epsilon(\pi) = 1$ otherwise we have $\det \Pi'(\pi) = \tilde{\phi}\chi_\nu(\pi)$. Let χ' be a character of F^* defined as in the statement of the proposition. Then, $\det \Pi' = \chi'_\nu$. □

Theorem 4.24 Let $q \geq 3$ and let ρ be a minimal irreducible representation of D^* of conductor $k + 2 \geq 4$. Suppose there exists a character χ of F^* such that $\Pi' = \rho \otimes \chi$ has trivial central character. Then, $\det \Pi' \equiv 1$ except if $q = 4$, $j = 2$ and $d_{F'/F} > p$ in which case, $\det \Pi'$ could be either the trivial character or the unramified character.

Proof: Let $x \in G_1$. Then $\epsilon(x) = 1$ and by Proposition 4.5 if $q \geq 3$, $(1 + P_F)^{q(p-1)(k-1)/2} \subseteq (1 + P_F^t)^p$. Since the conductor of ρ is $k + 2$ $\chi(1 + P_F^t)^p = 1$. Also since $\nu(G_1) = 1 + P_F$, $\chi_\nu(G_1^{\dim \rho}) = \chi(1 + P_F)^{q(p-1)(k-1)/2} \subseteq \chi((1 + P_F^t)^p) = 1$. Let k be odd or let k be even and $j \geq 4$.. Then by theorem 4.11, $\tilde{\phi}(t_{D^* \setminus F'^* G_j}(x)) = 1$. Let k be even and $j = 2$ or 3. Then $t_{D^* \setminus F'^* G'_{j-1}}(x) = 1 \mod P_D^{2j-1}$ on which $\tilde{\phi}$ is trivial and hence $\det \Pi' \equiv 1$ on G_1.

To evaluate $\det \Pi'(\pi)$ first let k be odd. Then if $j \geq 3$ or if $j = 2$ and $d_{F'/F} = p$, $\det \Pi'(\pi) = \epsilon(\pi)\chi_\nu\tilde{\phi}(\pi^{\ell q(j-1)(p-1)}) = 1$. Now suppose $d_{F'/F} > p$ and $j = 2$. Then k is odd and $q \geq 4$. If $q \neq 4$, $h = \lambda \pi^{-k-p}(1 + \alpha_1 \pi + \alpha_2 \pi^2 + \ldots)$ for some $\lambda \in \langle \zeta^\ell \rangle \cup \{0\}$, the condition $d_{F'/F} > p$ implies $\tau(\alpha_1) = Tr_{F(\zeta)/F}(\alpha_1) \in P_F$. Therefore, $\det \Pi'(\pi) = \psi_F(\tau((\beta_2 \pi^2 + \beta_3 \pi^3)\lambda \pi^{-3-p}(1 + \alpha_1 \pi + \alpha_2 \pi^2 + \ldots))) = \psi_F(\lambda \tau(\beta_3) \pi^p)$. If q is odd then $\tau(\eta) = p\eta + \frac{p(p-1)}{2} \mod P_F$ which implies $\psi_F(\lambda \alpha \tau(\eta) \pi^{-p}) = 1$. If $q = 2^f$, $f \geq 3$, then $\beta_3 \in F$ and hence $\psi_F(\lambda \tau(\beta_3) \pi^p) = 1$. Next let $q = 4$. Then, $\det \Pi'(\pi) = \tilde{\phi}(1 + \beta_2 \pi^2 + \beta_3 \pi^3) = \psi_F(-\lambda(\eta^2 + (\eta^2)^\sigma)\pi^{-2})$ since $\alpha_1 \in F$. But, $(\eta^2)^\sigma = \eta^2 + 2\eta + 1 \mod P_F$ and hence, $\eta^2 + (\eta^2)^\sigma = 1 \mod P_F$. Therefore, $\det \Pi'(\pi) = \psi_F(\lambda \pi^{-2})$. As $\pi^{-2} R_F / R_F \cong \mathbb{Z}_2 \times \mathbb{Z}_2$, and λ is a cube root of unity the value of $\det \Pi'(\pi)$ depends on λ or in other words the extension F'. We remark that both cases, namely $\det \Pi'(\pi) = 1$ and $\det \Pi'(\pi) = -1$ occur.

Next let k be even. Then p is odd. Suppose $k = 2$. Then $j = 2$ and by Proposition 4.13 $t_{D^* \setminus F'^* G_1}(\pi) = \pi^\ell$. G_1' is generated by $1 + \alpha_i \eta^r \pi$, $1 \leq i \leq q$ and $1 \leq r \leq (p-1)/2$ where $\alpha_i \in \langle \zeta^\ell \rangle \cup \{0\}$. and $t_{D^* \setminus F'^* G_1'}(\pi) = \pi^{\ell q(p-1)/2}$. This implies $\det \Pi'(\pi) = \chi_\nu \tilde{\phi}(\pi^\ell q^{(p-1)/2}) = 1$. Now if $d_{F'/F} = p$ and $k \geq 4$, by Theorem 4.16 $t_{D^* \setminus F'^* G_{j-1}'}(\pi) = \pi^{\ell q(k-1)(p-1)/2}$ and again $\det \Pi'(\pi) = 1$. Suppose $d_{F'/F} > p$. If $k = 4$, then $j = 3$ and by Theorem 4.17 $t_{D^* \setminus F'^* G_2}(\pi) = \pi^{\ell q^{p-1}}(1 + \beta_2 \pi^2 + \beta_3 \pi^3) \mod P_D^4$. This implies $t_{D^* \setminus F'^* G_2'}(\pi) = \pi^{\ell q 3(p-1)/2} \prod_{i=1}^{q(p-1)/2}(1 + \zeta_i \pi^2)(1 + \beta_2 \pi^2 + \beta_3 \pi^3)(1 + \zeta_i \pi^2)^{-1} = \pi^{\ell q 3(p-1)/2} \mod P_D^5$ which gives $\det \Pi'(\pi) = \chi_\nu \tilde{\phi}(\pi^\ell q^{3(p-1)/2}) = 1$. Now if $k \geq 6$, then by Theorem 4.17 $t_{D^* \setminus F'^* G_j}(\pi) = \pi^{\ell q(j-1)(p-1)} \mod P_D^{2j}$ and hence $t_{D^* \setminus F'^* G_{j-1}'}(\pi) = \pi^{\ell q(k-1)(p-1)/2} \mod P_D^{2j-1}$ and we have again $\det \Pi'(\pi) = 1$. Hence the theorem. $\square$

In the next three theorems we consider the case $q = p = 2$. Note that $q = 2$ implies every element in F has a power series expansion interms of π^2 with coefficients either zero or 1. In particular, $h = \pi^{k-2} u$ and $2 = \pi^{2e} u_1$ where $u \in G_1$ and $u_1 \in U_F$. One can show that if ρ is a minimal irreducible representation of D^* of conductor $k+2$ induced from a character $\tilde{\phi}$ of $F'^* G_j$ and if $d_{F'/F} = 2$, and $k = 2e+1$, there does not exist a character χ of F^* such that $\Pi' = \rho \otimes \chi$ has trivial central character.

Theorem 4.25 Let $q = 2$ and $e = 1$. Let ρ be a minimal irreducible representation of D^* of conductor $k + 2 \geq 4$. Suppose there exists a character χ of F^* such that $\Pi' = \rho \otimes \chi$ has trivial central character.

1. If $d_{F'/F} = 2$, then $\det \Pi' \equiv 1$ for $j \geq 2$.

2. If $d_{F'/F} > 2$, then

    (a) for $j = 2$, $1 + \pi \in F'^* G_2$ and $\det \Pi'$ is either $w_{Q_2(\sqrt{3})/Q_2} \circ \nu$ or $w_{Q_2(\sqrt{-1})/Q_2} \circ \nu$ and both cases occur.

    (b) for $j = 3$, $\det \Pi' \equiv w_{Q_2(\sqrt{(-3)})/Q_2} \circ \nu$

    (c) for $j \geq 4$, $\det \Pi' \equiv 1$.

Proof: In this case $F = Q_2$ and we choose π such that $\pi^2 = 2$. The group F^*/F^{*2} is of order eight and is generated by $-\pi^2 F^{*2}, (1+\pi^2)F^{*2}$ and $(1+\pi^2+\pi^4)F^{*2}$. Therefore to know $\det \Pi'$ on D^* we need to know it only on $\pi, (1+\zeta\pi)$ and $(1+\zeta\pi^2)$. It is easy to prove that if $d_{F'/F} = 2$, $h = \pi^{-k-2} u$ where $u = 1 + \alpha_1 \pi + \alpha_2 \pi^2 + \ldots$ such that α_1 is 0 or 1 and for $i \geq 2$, $\alpha_i \in \langle \zeta \rangle \cup \{0\}$.

(i) Let $j = 2$. This means $k = 3$. By Proposition 4.12 we have $t_{D^* \backslash F'^* G_2}(x) = 1 + \zeta\pi^3$ for $x = 1 + \zeta\pi$ and $x = 1 + \zeta\pi^2$. Therefore, $\det \Pi'(1+\zeta\pi) = \det \Pi'(1+\zeta\pi^2) = \tilde{\phi}(1+\zeta\pi^3) = \psi_F(\tau(\zeta\pi^3 \pi^{-5}(1 + \alpha_1\pi + \alpha_2\pi^2 \ldots))) = \psi_F(-\pi^{-2}) = 1$. Now $\det \Pi'(\pi) = \chi_\nu \tilde{\phi}(\pi^6)\tilde{\phi}(1+\pi)\tilde{\phi}(1+\zeta\pi^2)$. But,

$$\tilde{\phi}(1+\zeta\pi^2) = \psi_F(\tau(\zeta^2\pi^2 h)) = \begin{cases} 1 & \text{if } \alpha_1 = 0 \\ -1 & \text{if } \alpha_1 = 1 \end{cases}$$

Also since $(1+\pi)^2 = 1+\pi^2+\pi^3$, $\tilde{\phi}(1+\pi) = \pm 1$. Therefore,

$$\det \Pi'(\pi) = \tilde{\phi}(1+\pi)\tilde{\phi}(1+\zeta^2\pi^2) = \begin{cases} 1 & \text{if } \alpha_1 = 0 \& \tilde{\phi}(1+\pi) = 1 \text{ or if } \alpha_1 = 1 \& \tilde{\phi}(1+\pi) = -1 \\ -1 & \text{otherwise} \end{cases}$$

which proves the claim of the theorem.

(ii) Let $j = 3$. This implies $k = 5$ and $\dim \rho = 12$. Clearly $\det \Pi' \equiv 1$ on G_1. As $t_{D^* \backslash F'^* G_2}(\pi) \notin F'^* G_3$, $\epsilon(\pi) = -1$. Also, $t_{D^* \backslash F'^* G_3}(\pi) = \pi^{12}(1+\pi^2)$. Therefore, $\det \Pi'(\pi) = \epsilon(\pi)\chi_\nu \tilde{\phi}(\pi^{12})\tilde{\phi}(1+\pi^2) = -1$ proving that $\det \Pi' \equiv w_{Q_2(\sqrt{(-3)})/Q_2} \circ \nu$

(iii) Let $j \geq 4$. Then it is easy to see that $\det \Pi' \equiv 1$ on G_1 and

$\epsilon(\pi) = 1$. Now $\det \Pi'(\pi) = \chi_\nu \tilde{\phi}(\pi^{(q+1)q^{j-1}})\tilde{\phi}(1+\pi^{2j-2}) = \tilde{\phi}(1+\pi^{2j-2}) = \psi_F(\tau(\alpha_1 \pi^{-2})) = 1$. Thus $\det \Pi'(\pi) \equiv 1$ for $j \geq 4$. Hence the theorem
. $\square$

Theorem 4.26 Let $q = 2$ and $e \geq 2$. Let ρ be a minimal irreducible representation of D^* of conductor $k + 2 \geq 4$. Suppose there exists a character χ of F^* such that $\Pi' = \rho \otimes \chi$ has trivial central character.

1. If $d_{F'/F} = 2$, then $\det \Pi' \equiv w_{F_2/F} \circ \nu$ for $j = 2$ and $\det \Pi' \equiv 1$ for $j \geq 3$.

2. If $d_{F'/F} > 2$, then

    (a) for $j = 2$ and $j = 3$ there exists a ramified quadratic extension K of F such that $\det \Pi'$ is either $w_{K/F} \circ \nu$ or $(w_{F_2/F} w_{K/F}) \circ \nu$ and both cases occur.

    (b) for $j = 4$
    $$\det \Pi' \equiv \begin{cases} 1 & \text{if } d_{F'/F} > 4 \\ w_{F_2/F} \circ \nu & \text{if } d_{F'/F} = 4 \end{cases}$$

    (c) for $j \geq 5$ $\det \Pi' \equiv 1$.

Proof: First let $d_{F'/F} = 2$. This implies $h = \pi^{-k-2} u$ where $u = 1 + \alpha_1 \pi + \alpha_2 \pi^2 + \ldots$ such that α_1 is ζ or ζ^2 and $\alpha_i \in \langle \zeta \rangle \cup \{0\}$ for $i \geq 2$. For $j = 2$ or 3 if $x \in G_1$, then $\tilde{\phi}(t_{D^* \setminus F'^* G_j}(x)) = -1$ if and only if $\chi_\nu^{(q+1)q^{j-1}}(x) = -1$. Also $\epsilon(x) = 1$ Therefore for $j = 2$ and $j = 3$, $\det \Pi' \equiv 1$ on G_1. If $j \geq 4$, then by Propositions 4.6, 4.12 and Lemma 4.22, we have $\det \Pi' \equiv 1$ on G_1. Next, to calculate $\det \Pi'(\pi)$, note that since $d_{F'/F} = 2$, from Lemma 4.22 $\epsilon(\pi) = -1$ if $j = 2$ and $\epsilon(\pi) = 1$ for $j \geq 3$. Therefore,

$$\det \Pi'(\pi) = \epsilon(\pi)\chi_\nu \tilde{\phi}(\pi^{(q+1)q^{j-1}}) = \begin{cases} -1 & \text{if } j = 2 \\ 1 & \text{otherwise.} \end{cases}$$

Thus we have proved that $\det \Pi' \equiv w_{F_2/F} \circ \nu$ for $j = 2$ and $\det \Pi' \equiv 1$ for $j \geq 3$.

Next let $d_{F'/F} > 2$ then $h = \pi^{-k-2} u$ where $u = 1 + \alpha_1 \pi + \alpha_2 \pi^2 + \ldots$ such that α_1 is zero or 1 and $\alpha_i \in \langle \zeta \rangle \cup \{0\}$ for $i \geq 2$. In this case

$\chi_\nu^{(q+1)q^{j-1}}$ is trivial on G_1 for $j \geq 2$. For $x \in G_1$ we have $\det \Pi' = \tilde{\phi}(t_{D^* \backslash F'^* G_j}(x))$. This implies for $j \geq 4$ $\det \Pi' \equiv 1$ on G_1. For $j = 2$ or 3, if x is of the form $1 + \beta\pi + \delta\pi^2$, $1 + \zeta\pi^2$ or $1 + \zeta^2\pi^2$ with $\beta \in \langle \zeta \rangle$ and $\delta = 0$ or 1, then $\tilde{\phi}(t_{D^* \backslash F'^* G_j}(x)) = -1$. Therefore, for $j = 2$ or 3 $\det \Pi'$ is nontrivial on G_1. To evaluate $\det \Pi'(\pi)$ for $j = 2$, $j = 3$, $j = 4$ and $j \geq 5$. Now $\det \Pi'(\pi) = \chi_\nu^{(q+1)q^{j-1}}\tilde{\phi}(t_{D^* \backslash F'^* G_2}(\pi)) = \chi_\nu\tilde{\phi}(\pi^6)\tilde{\phi}(1 + \pi)\tilde{\phi}((1 + \zeta^2\pi^2 + \zeta^2\pi^3)$. Now

$$\tilde{\phi}(1 + \zeta^2\pi^2 + \zeta^2\pi^3) = \psi_F(\tau((\zeta^2\pi^2 + \zeta^2\pi^3)h))$$
$$= \begin{cases} -1 & \text{if } \alpha_1 = 0 \\ 1 & \text{if } \alpha_1 = 1 \end{cases}$$

Moreover, since $(1+\pi)^2 = 1+\pi^2 \mod P_D^4$, $\tilde{\phi}((1+\pi)^2) = \psi_F(\tau(\alpha_1\pi^{-2})) = 1$ we have $\tilde{\phi}(1 + \pi) = \pm 1$. Therefore if $j = 2$,

$$\det \Pi'(\pi) = \begin{cases} 1 & \text{if } \alpha_1 = 0 \& \tilde{\phi}(1+\pi) = -1 \text{ or if } \alpha_1 = 1 \& \tilde{\phi}(1+\pi) = 1 \\ -1 & \text{otherwise} \end{cases}$$

Next suppose $j = 3$. Then we have $\epsilon(\pi) = -1$ and from Theorems 4.19 and 4.20.

$$t_{D^* \backslash F'^* G_3}(\pi) = \begin{cases} \pi^{12}(1+\pi^2)(1+\pi^3+\zeta\pi^4+\pi^5) \mod P_D^6 & \text{if } e = 2 \\ \pi^{12}(1+\pi^2)(1+\pi^3+\zeta\pi^4) \mod P_D^6 & \text{if } e \geq 3 \end{cases}$$

and therefore

$$\det \Pi'(\pi) = \epsilon(\pi)\chi_\nu\tilde{\phi}(\pi^{12})\tilde{\phi}(1+\pi^3+\zeta\pi^4)$$
$$= -(\psi_F(2\pi^{-4} + (\alpha_2 + \alpha_2^2)\pi^{-2})\psi_F(\alpha_1(\zeta + \zeta^2)\pi^{-2}))$$
$$= \begin{cases} -1 & \text{if } \alpha_2 \text{ is 0 or 1 and } \alpha_1 = 0 \\ & \text{or if } \alpha_2 \text{ is } \zeta \text{ or } \zeta^2 \text{ and } \alpha_1 = 1 \\ 1 & \text{otherwise} \end{cases}$$

Next let $j \geq 4$. $\epsilon(\pi) = 1$ for $j \geq 4$ we have

$$\det \Pi'(\pi) = \chi_\nu\tilde{\phi}(\pi^24)\tilde{\phi}(1+\pi^4)$$
$$= \begin{cases} 1 & \text{if } \alpha_3 \text{ is 0 or 1} \\ -1 & \text{if } \alpha_3 \text{ is } \zeta \text{ or } \zeta^2 \end{cases}$$

Note that since we already have $d_{F'/F} > 2$ the condition α_3 is 0 or 1 is equivalent ot saying that $d_{F'/F} \geq 4$. Therefore in this case the value of det $\Pi'(\pi)$ depends on α_3 or equivalently on $d_{F'/F}$ and

$$\det \Pi' \equiv \begin{cases} 1 & \text{if } d_{F'/F} > 4 \\ w_{F_2/F} \circ \nu & \text{if } d_{F'/F} = 4 \end{cases}$$

Finally if $j \geq 5$,

$$t_{D^* \backslash F'^* G_3}(\pi) = \begin{cases} \pi^{48}(1+\pi^8) \mod P_D^{10} & \text{if } e \geq 3 \text{ and } j = 5 \\ \pi^{q^{j-1}(q+1)} \mod P_D^{2j} & \text{otherwise} \end{cases}$$

But if $j = 5$, one can easily see that $\tilde{\phi}(1+\pi^8) = 1$. Therefore, for $j \geq 5$, it follows that $\det \Pi'(\pi) = \chi_\nu \tilde{\phi}(\pi^{q^{j-1}(q+1)}) = 1$ and hence $\det \Pi' \equiv 1$ proving the theorem. □

Now Theorem C follows from Proposition 4.23, Theorem 4.11 and Theorems 4.24, 4.25 and 4.26.

References

[De 1] P.Deligne, Les constentes des equations fonctionelles des fonctions L, in *Modular Functions of One Variable-II*, Lecture Notes in Mathematics, Vol. 349 Springer-Verlag, New York (1973), 501-595

[De 2] P.Deligne, Les constentes locales de l'équation fonctionelle de la fonction L d'Artin d'une representation orthogonale, Inv. Math. **35** (1976), 299-316.

[DP-DR] D.Prasad and D.Ramakrishnan, Lifting orthogonal representations to spin groups and local root numbers, Proc. Indian Acad. Sci. (Math.Sci.)., Vol 105,**3** August (1995) 259-267.

[Ho] Roger E.Howe, Kirillov theory for compact p-adic groups, Pacific Journal of Mathematics, Vol. 73, **2**, (1977).

[Me] O.T.O'Meara, Introduction to Quadratic Forms, Grund.math.Wiss., Springer-Verlag, New York, (1973).

[Se 2] J.P.Serre, Local fields, Graduate texts in Mathematics **67**, Springer-Verlag (1979).

[Ta] J.Tate, Number theoretic background in *Aotomorphic Forms, Representations and L-functions*, Proc. Symp.Pure Math. **33** vol.2 (1979) 3-26. AMS, Providence, R.I.

[Tu] J.Tunnell, Local epsilon factors and characters of GL(2), American Journal of Math., (1983) 1277-1307.

[Wa] S.Wang, Amer. Jour. of Maths **72**, (1950) 323-334.

[WeA] A.Weil, Basic Number Theory, Grund.math.Wiss., 3rd edit. Springer-Verlag, New York (1974).

Department of Mathematics & Statistics,
University of Hyderabad,
Hyderabad - 500 046, India.
E-mail address: rtsm@uohyd.ernet.in

Mod p modular forms

Chandrashekhar Khare

1. Introduction

In this note we study the space of mod p modular forms of Serre and Swinnerton-Dyer, especially with a view to understanding the structure of the mod p Hecke algebra. We rely upon the paper of Jochnowitz, cf. [J], to obtain some information about local components of the mod p Hecke algebra. The new ingredient is the fact that it is now known from the deformation theory of Mazur, cf. [M], that the local components are Noetherian. This allows one to strengthen some of the results of [J] (see §2).

We also include in the §3 of this note speculation about certain sequences of Hecke modules which we believe hold the key to understanding local components better. In §4 we study modular forms arising from varying p-power levels.

2. Varying the weight

Fix an integer $N \geq 5$ and a prime $p > 3$ such that $(N, p) = 1$. Fix embeddings $\iota_p : \overline{\mathbb{Q}} \hookrightarrow \overline{\mathbb{Q}}_p$ and $\iota_\infty : \overline{\mathbb{Q}} \hookrightarrow \mathbb{C}$. We consider the congruence subgroup $\Gamma := \Gamma_1(N)$ of $SL_2(\mathbb{Z})$, and for any integer $k \geq 2$, denote by $M'_k(\Gamma)$ the space of weight k modular forms for Γ such that the coefficients of the Fourier expansion at infinity are in $\overline{\mathbb{Q}}$, and are integral for the valuation induced by ι_p.

We identify M'_k with its image in $\overline{\mathbb{Q}}[[q]]$ given by the q-expansion map, where hereafter we eliminate the mention of Γ as we will consider modular forms only for this fixed Γ. We define $M_k := M'_k \otimes \overline{\mathbb{Z}}_p$ (in this is implicit the embedding ι_p). We denote by $\widetilde{M_k}$ the reduction of M_k modulo the maximal ideal of $\overline{\mathbb{Z}}_p$.

We have inclusions $\widetilde{M_k} \subset \widetilde{M_{k+p-1}}$, given by multiplication by the Eisenstein series E_{p-1} of weight $p-1$ (see [Ka]; note that we are assuming $p > 3$).

We denote by $\widetilde{M^a}$, where $a \in \mathbb{Z}/(p-1)$, the union of the $\widetilde{M_k}$'s for $k \geq 2$ and $k \equiv a \bmod (p-1)$. On this space we may define a filtration w, by defining $w(f)$ to be

---

1991 *Mathematics Subject Classification.* Primary: 11F, 11R.

the least k such that f is in the image of M_k under the reduction map. As usual, we have the Hecke action on the space of modular forms. We only note here that the action of the p th Hecke operator, which we denote by U, is

$$\sum a_n q^n \to \sum a_{np} q^n.$$

We denote by $\widetilde{M}$ the space $\bigoplus_a \widetilde{M}^a$. This is for us the space of mod p modular forms for the group Γ which comes equipped with a grading mod $p-1$. On the space of mod p forms one has several interesting operators, chief amongst these being the Θ operator and V which are defined by their effect on q expansions by:
$\Theta(\sum a_n q^n) = \sum n a_n q^n$ and $V(\sum a_n q^n) = \sum a_n q^{np}$. We also define $\Phi = \Theta^{p-1}$.

These preserve the space of mod p modular forms. Their effect on the filtration is given by $w(\Theta(f)) \leq w(f) + p + 1$ and $w(V(f)) = pw(f)$. These results follow from [Ka1]. Note that although our definition of $\widetilde{M}_k$ is different from the one used by Katz, the base-change results of §1.7 of [Ka2], ensure that the definitions agree. Briefly, our assumption $k \geq 2$, ensures that $H^1(X_1(N)_{\mathbb{Z}_p}, \underline{\omega}^k_{\overline{\mathbb{F}}_p})$ vanishes because of the Riemann-Roch theorem.

It is the existence of the above operators which makes the theory of mod p modular forms more tractable than that in characteristic 0, as was discovered by Serre, Swinnerton-Dyer and Tate.

We note the basic relations between the operators U, V and Θ considered as operators on the space of mod p modular forms. $ker(\Theta) = Im(V)$, $ker(U) = Im(\Theta)$, V is injective and U is surjective. These facts follow from the basic relation $\Phi + VU = UV = 1$. We also take heed of the following results on filtrations which follow from [Ka1] (see also [J] and [J1]):

1. $w(U(f)) < w(f)$ if $w(f) > p + 1$ (Lemma 1.9 of [J1]; this was earlier proved when $\Gamma = SL_2(\mathbb{Z})$ by Serre in [S]). For the convenience of the reader, we rewrite the proof of the lemma of Jochnowitz. Namely, write $f = \Phi(f) + V(g)$. Note that $U(f) = g$. We also have that $w(g) \leq \frac{w(f) + p^2 - 1}{p}$. From this the result follows.

2. If p does not divide $w(f)$, then we have $w(\Phi(f)) \geq w(f)$.

We note that the operator Φ is idempotent. Observe that any $f \epsilon \widetilde{M}$ has the decomposition:

$$f = \sum f_i$$

where $f_i \epsilon \overline{\mathbb{F}}_p[[q]]$ is such that only exponents of q of the form q^{np^i}, where $(n, p) = 1$, occur with non-zero coefficient. As f_i is checked to be given by $f_i = V^i \Phi U^i(f)$, we see that it is a mod p modular form. From the results above about filtrations, we deduce that the filtration of f_i satisfies, $w(f_i) \leq w(f) + p^2 - 1$.

In particular the filtrations of the f_i's are bounded. This suggests that the image of $\widetilde{M}$ inside $M = \widehat{\bigoplus} V^n \Phi(\widetilde{M})$ is small. Here $\widehat{\bigoplus}$ denotes direct product.

Proposition 1. $\widetilde{M} / \bigoplus V^n \Phi(\widetilde{M})$ *is a finite dimensional* $\overline{\mathbb{F}}_p$ *vector space.*

Proof. We denote the vector space in the Proposition by S. We note that U acts bijectively on S. Note that by the lemma of Jochnowitz that we have quoted (Lemma 1.9 of [J1]), it easily follows that any subspace or quotient of $\widetilde{M}$, which is U-stable and on which U acts injectively, is finite dimensional (in fact, even more is true, in that such a space has bounded dimension). As S is a quotient on which U acts bijectively, we deduce that S is finite dimensional.

Remark. The proposition is a mere rephrasing of the fact that the subspace of $\widetilde{M}$ which belongs to components of the Hecke algebra (see below), such that U is not in the maximal ideal, is finite-dimensional. Thus the components which are more mysterious are the U-nilpotent, or nonordinary, components.

Now we study local components of the mod p Hecke algebra. We set up some notation. The inclusion

$$\widetilde{M_k} \subset \widetilde{M_{k+p-1}}$$

induces a map of the Hecke algebras $h_{k+p-1} \to h_k$ where h_r is the image of the Hecke operators in the endomorphisms of $\widetilde{M_r}$. We define the limit Hecke algebra H as the algebra obtained as the inverse limit of the h_r's taken with respect to these maps. H is a semi-local ring with finitely many local components, cf. [J] (though [J] focusses on the case $\Gamma = SL_2(\mathbb{Z})$, the resullts from [J] that we use are valid for the Γ we consider).

We focus on a local component R and denote its maximal ideal by m_R and assume that U belongs to m_R (see remark above).

In [J] it is proven that local components for which the operator U is nilpotent (i.e., U belongs to the maximal ideal), have Zariski tangent space dimension at least 1. Though in [J], the case of only $SL_2(\mathbb{Z})$ is considered, all the arguments go through on noting the Proposition 7.2 in [E] which says that, using notation in [J1], that for k large enough, W_{k+p^2-1} is isomorphic to W_k. Here for any $r \geq p+1$, $W_r = \widetilde{M}_r/\widetilde{M}_{r-p+1}$. Further this isomorphism is given by Φ, and hence is Hecke compatible for operators T_ℓ ($\ell \neq p$), and preserves components on which U is nilpotent. This replaces Theorem 2.4 of [J1] (due to Serre-Tate) which is the main ingredient in the arguments of [J].

In fact using her arguments, which prove that any local component R on which U is nilpotent has the property that $R/(U)$ is infinite dimensional over $\overline{\mathbb{F}_p}$, and the fact which is known by Mazur's theory of deformations that any local component is Noetherian, we see that all local components for which U is nilpotent, have Krull dimension ≥ 2. We now give an argument for this after stating it as a proposition.

As notation for a local component R, we denote by $R^{(p)}$ the closed subalgebra generated by the operators T_n, for $(n,p) = 1$. We denote by $\tilde{\rho}$ the Galois representation $G_{\mathbb{Q}} \to GL_2(R/m_R)$ attached to R. It is characterised, up to semisimplification (recall that we are assuming p odd), by the property that for a prime r such that $(r, Np) = 1$, $\tilde{\rho}$ is unramified at r, and the image of the Frobenius at r has trace the image of T_r in R/m_R.

Proposition 2. *If R is a nonordinary local component of H, then R is a power series ring in U over $R^{(p)}$. Further, if we assume that the mod p Galois representation $\tilde{\rho}$ above is irreducible, then the Krull dimension of R is at least 2.*

Proof. The first part is proven as Theorem 6.3 of [J], but we recall the main point of the proof. We need to check that no expression of the form $\sum c_i U^i$ is 0 in R where the c_i's belong to $R^{(p)}$. We assume that n is the least i for which c_i is not 0.

Then there is a g in $\widetilde{M}$ on which m_R acts nilpotently, and such that c_n does not kill g. We may assume that $\Theta(g)$ is not 0 by applying an appropriate power of U to g. Then $\Phi V^n(g)$ is not in the kernel of U^n, but is in the kernel of U^m for all $m > n$. This shows what we wanted.

Now we prove that the Krull dimension of R is at least 2 under the assumptions in the proposition. We denote by $\widetilde{M_R}$ the subspace of $\widetilde{M}$ on which m_R is locally nilpotent. By Corollary 6.6 of [J] it follows that $\widetilde{M_R}/V(\widetilde{M_R})$ is an infinite dimensional vector space over $\overline{\mathbb{F}}_p$. From this it follows, upon using the nondegenerate pairing

$$R^{(p)} \times \widetilde{M_R}/V(\widetilde{M_R}) \to \overline{\mathbb{F}}_p,$$

induced by $(f, T_n) = a_1(f|T_n)$, that $R^{(p)}$ is infinite dimensional as a $\overline{\mathbb{F}}_p$ algebra.

Now we will explain how it follows from the deformation theory of [M] that R, and hence $R^{(p)}$, is Noetherian. Using notions and notations as in [M], observe that the universal ring corresponding to the mod p representation associated to R is Noetherian, and surjects onto $R^{(Np)}$, where N is the level of Γ and as usual, by this notation, we mean R deprived of operators T_n which are not coprime to Np. This surjection is a consequence of the construction in [H1] of a representation $\rho : G_{\mathbb{Q}} \to GL_2(R)$ which deforms the mod p representation $\tilde{\rho}$ above. Note that the situation in [H1] is slightly different, but the method of proof there works here, as we are assuming $p > 3$ and that the mod p representation is absolutely irreducible.

By Eichler-Shimura relations, we see that the deformation ring, which Mazur in [M] attaches to $\tilde{\rho}$, surjects onto $R^{(Np)}$ (which denotes the subalgebra of R generated by T_n such that $(n, Np) = 1$).

¿From the two facts:

1. $R^{(p)}$ is infinite-dimensional as a $\overline{\mathbb{F}}_p$ algebra

2. $R^{(p)}$ is Noetherian

it follows that $R^{(p)}$ has Krull dimension at least 1. Then by the first line of the proposition, that we have already proved, it follows that R has Krull dimension at least 2.

Remark. Its also proven in Theorem 8.1 of [J] that, in fact, most local components have Zariski tangent space dimension ≥ 3.

3. A comparison of filtrations on mod p forms

In studying local components of Hecke algebras acting on the space of mod p modular forms for the group $\Gamma_1(N)$) and all weights ≥ 2, as in §2, a question naturally comes up, about the relationship between two naturally occuring filtrations on the space of mod p forms. One filtration is the classical one due to Serre and Swinnerton-Dyer which comes from the least weight in which the q-expansion of a mod p form occurs; we have discussed this in §2. The other arises from looking at the space $\tilde{M}$ of mod p modular forms for $\Gamma_1(N)$. Write this as the direct sum of spaces on which the (finitely many) maximal ideals of the Hecke algebra acting on it are respectively nilpotent. Then the other filtration comes from the degree of nilpotence of the modular form lying in the space corresponding to the local component R. More precisely, the filtration of a form f belonging to the R-component, for a local component R, is defined to be the least n such that m^n kills the form where m is the maximal ideal of the corresponding local component R. We shall always assume that R is non-Eisenstein (i.e., the mod p representation attached to R is irreducible) in what follows.

Thus it is natural to ask what is the relation between these two filtrations. This question is of interest in determining algebraic properties of local components which have proven to be elusive. As far as the author knows, local components have been essentially studied in detail only in [J], though in the particular case of ordinary components, there is much more known even without reducing mod p due to the work of Hida, cf. [H].

There was little known about local components till Mazur introduced his idea of deformations. From his theory, as remarked before, it follows that the local components are Noetherian. It is somewhat surprising that this fact is not known without using deformations. The Noetherian property of local components has interesting consequences for the structure of mod p forms. Below we will use it to study some exact sequences of modular forms which arise from the peculiar property of characteristic p that there are inclusions between spaces of modular forms of different weights, on identifying forms with their q expansion.

We are, in particular, interested in studying the following exact sequence:

$$0 \to \widetilde{M_k} \to \widetilde{M_{k+p-1}} \to SS_k \to 0 \qquad (1)$$

as a sequence of Hecke modules, where SS_k is defined by the exactness of (1). We say that (1) is *essential* if $\widetilde{M_k}$ is maximal amongst the Hecke submodules N of $\widetilde{M_{k+p-1}}$ with the property that N contains $\widetilde{M_k}$, and the resulting exact sequence

$$0 \to \widetilde{M_k} \to N \to N' \to 0$$

(N' defined by the exactness of the above sequence) is split as a Hecke module.

Proposition 3. *The sequence (1) is an essential extension of Hecke modules for k large enough.*

Proof. For the proof, note that by the results of Serre-Tate (see [J] and [J1], particularly Theorem 2.4 of the latter, or Proposition 7.2 of [E]), the Hecke action on $\widetilde{M}$ has only finitely many eigenforms (up to scalar multiplication). Choose k larger than the filtrations of all these eigenforms. Then for such a k, (1) above is essential, as otherwise we would deduce that there is an eigenform of filtration $k+p-1$.

The study of the exact sequence (1) seems to hold the key to understanding the structure of local components of Hecke algebras. SS_k was defined by the exactness of the sequence. But there is another more direct interpretation of SS_k which depends on an idea of Serre which interprets this, at least for k large enough, as functions on supersingular elliptic curves, cf. [E]. The mere existence of such a sequence as (1) produces a relationship between the filtrations we alluded to above. It is an immediate consequence of the fact that SS_k has bounded dimension, which follows from the dimension formula for $S_k(\Gamma_1(N))$, that the part of $\widetilde{M_k}$ on which m is nilpotent is contained in $\widetilde{M}[\mathrm{m}^{ak+b}]$, for some constants a and b. We can make a and b explicit. It can, for instance, be easily checked from (1), that a maybe chosen to be any integer larger than the dimensions of the SS_k's ($2 \leq k \leq p^2 + 1$), and b can be taken to be the dimension of $\widetilde{M}_k$, where k is the weight filtration of the eigenform corresponding to m, which is unique upto scalar. Henceforth (in a change of notation from §2) by $\widetilde{M}_k$ we will mean those mod p forms of weight k on which m is locally nilpotent, for the local component R and the corresponding maximal ideal m that we have fixed (so we are dropping the subscript R with which this space was adorned in §2). To get information about local components we need also to be able to control the terms of the nilpotency filtration in terms of the weight filtration. It is unrealistic to expect that there is a similar linear relation in the other direction as it has been shown in §2 that the dimension of nonordinary local components is at least 2 which makes this impossible. But we may modify our space $\widetilde{M}_k$ a little and look only at the part which is killed by the Atkin operator U. Then we may ask the following question. By $\mathcal{M}$ below we mean the subspace of $\widetilde{M}$ which is given by the intersection of the kernel of U and the submodule of $\widetilde{M}$ on which m is nilpotent.

Question 1. *Does there exist a "nice function" $f(n)$ of n such that $\mathcal{M}[\mathrm{m}^n] \hookrightarrow \mathcal{M}_{f(n)}$?*

Remark. There is a strong connection between the nature of the function $f(n)$ asked for in Question 1 and the Krull dimension of R.

To study this question we return now to a modified version of (1) and study some of its properties in greater detail. For this, we denote by $\mathcal{M}_k$ the subspace of $\widetilde{M}_k$ on which m is nilpotent, and which is killed by U. We have to study the exact sequence:

$$0 \to \mathcal{M}_k \to \mathcal{M}_{k+p-1} \to \mathcal{SS}_k \to 0 \tag{2},$$

where $\mathcal{SS}_k$ is defined by (2). This exact sequence results from (1) by applying the idempotent corresponding to R and looking at the kernel of U, as a sequence of $\mathcal{R}$ modules. We can ask a related question to the one stated above about this exact sequence.

Question 2. *Does there exist a "nice function" $g(r)$ such that*

$$0 \to \frac{\mathcal{M}_k}{\mathcal{M}_r} \to \frac{\mathcal{M}}{\mathcal{M}_r} \to \mathcal{C}_k \to 0 \tag{3}$$

is an essential extension of $\mathcal{R}$ modules for $k \geq g(r)$? Here $\mathcal{C}_{g(r)}$ is defined by (3).

We state and prove two propositions related to these questions.

Proposition 4. *With notation as in Question 2, for a fixed r, (3) is an essential extension for $g(r)$ large enough.*

Remark. This shows that there exists at least some function as asked for in Question 3, which may not be evident *a priori*.

Proof. We note that there exists a n such that $\mathcal{M}_r$ is contained in $\mathcal{M}[\mathrm{m}^n]$. Then the weight filtration being exhaustive, we note that as $\mathcal{R}$ is Noetherian, there exists a k_r such that $\mathcal{M}[\mathrm{m}^{n+1}]$ is contained in $\mathcal{M}_{k_r}$. Then we claim that setting $g(r) = k_r$ works. As if (3) were not essential with this k_r, there would exist a S strictly containing $\mathcal{M}_{k_r}$ such that the sequence corresponding to (3), made with S instead of $\mathcal{M}$, splits as a Hecke module. Then, as $\mathcal{R}$ is commutative and as its maximal ideal m acts nilpotently on $\mathcal{M}$, we get that there is a non-zero f which is not in $\mathcal{M}_{k_r}$ but which is killed by m^{n+1}. This contradicts our choice of k_r.

In order to indicate the relatedness of the above questions, we prove the following proposition.

Proposition 5. *If the function $g(r)$ can be taken to be of the form $r + x$ for some constant x, then $f(n)$ can be taken to be of the form $\alpha n + \beta$, for some constants α and β.*

Proof. We assume that $g(r)$ can to be taken as in the proposition. We use induction on the n which appears in the proposition. Thus let us assume that the eigenform corresponding to m has weight filtration y. Thus $\mathcal{M}[\mathrm{m}] \hookrightarrow \mathcal{M}_y$. Now we assume that $\mathcal{M}[\mathrm{m}^n] \hookrightarrow \mathcal{M}_{xn+y}$ for x as in the proposition. Then our hypothesis implies that the sequence:

$$0 \to \frac{\mathcal{M}_{x(n+1)+y}}{\mathcal{M}_{xn+y}} \to \frac{\mathcal{M}}{\mathcal{M}_{xn+y}} \to \mathcal{C}_{x(n+1)+y} \to 0$$

is an essential extension. But from this we claim that $\mathcal{M}[\mathrm{m}^{n+1}] \hookrightarrow \mathcal{M}_{x(n+1)+y}$. As, if $f \in \mathcal{M}[\mathrm{m}^{n+1}]$, then the one dimensional vector space spanned by the image of f in $\frac{\mathcal{M}}{\mathcal{M}_{xn+y}}$ is stable under the Hecke action. This contradicts the property of the above sequence of being essential, after using our inductive hypothesis that $\mathcal{M}[\mathrm{m}^n] \hookrightarrow \mathcal{M}_{xn+y}$. This completes the induction, with the constants α and β being set equal to x and y respectively.

Remark. It is clear from the proof above that knowledge of the function g of Question 2, enables us to determine some f (in terms of g) which satisfies the property asked of in Question 1.

To make *real* progress with the questions above, we essentially have to make quantitatively precise the fact that the action of $\mathcal{R}$ on $\mathcal{M}$ is highly non semisimple.

4. Varying the level

Instead of varying the weight as above, we now fix the weight to be 2, but vary the p power level. In the same way as above, we can consider the space of modular forms mod p of weight 2 for the 'group' $(\Gamma)' = \Gamma_1(N) \cap \Gamma_1(p^\infty)$ (i.e., we consider the union of $S_2(\Gamma_1(N) \cap \Gamma_1(p^r))$, for all integers $r \geq 1$) identifying forms with their q expansion, reducing modulo p etc. We are assuming that $(N,p) = 1$.

This time we have a filtration with respect to the p-power level the form comes from, i.e., we define $\ell(f) = r$ if the minimal p power level of a lift of f to a characteristic 0 form of weight 2 for the group Γ' is p^r (note that in contrast to the previous section, this filtration makes sense in "characteristic 0".)

Just as before we have the operators U, V, Θ, Φ, and though their meaning is different, the effect on q-expansions is the same. For example, the Θ operator in our context may be seen as the twisting action of the Teichmuller character (again this oprator exists in characteristic 0). We have that $ker(\Phi) = Im(V)$, $ker(U) = Im(\Phi)$, V is injective and U is surjective. We may note, to make the analogy between varying levels and weights explicit, V always increases the filtration by 1, Θ increases the filtration by at most 2, etc. These are standard results. For instance, that V increases the filtration by 1, is Lemma 16 in [AL]. All this is in good analogy with what we saw in section 1.

We can study nonordinary local components of the Hecke algebra in this situation. So fix such a local component, say R. Note that a modular form f for Γ' has the representation $f = \sum f_i$, with $f_i = V^i \Phi U^i(f)$. Using this, one can imitate the argument of Jochnowitz in Theorem 6.3 of [J] (see also first paragraph of the proof of Proposition 2 above) and prove that R is a power series ring in U over the subalgebra generated by the Hecke operators outside p.

In this situation too, we may think about similar questions as in §3, and investigate the relationship between the two natural filtrations, i.e., one coming from the level filtration and the other from the degree of nilpotence with respect to the maximal ideal of a local component. We see that the local component being Noetherian immediately implies that the level filtration grows much faster, upon using the dimension formula for the space of cusp forms of weight 2 and given level.

The fact analogous to Proposition 2, that the dimension of nonordinary local components is at least 2, follows from Theorem 5 of [K]. Thus Theorem 5 of [K] may in fact be viewed as a qualitative analog of the theorem of Serre and Tate (which is for varying weights) quoted as Theorem 2.4 in [J1], in the context of varying levels.

Under the rubric of the title, we study another issue which is part of the gestalt of the study of modular forms in the context of varying p-power levels.

Consider an irreducible, mod p representation ρ, which arises from a local component of the Hecke algebra of the type we have been considering, i.e., we assume that ρ arises from $S_2(\Gamma_1(Np^s))$ for some $s \geq 1$.

We would like to consider the question of how many lifts there are of ρ to modular p-adic representations which arise from $S_2(\Gamma_1(Np^r))$ for varying r, assuming that ρ has at least one such lift. This has been qualitatively studied in [K], in the language of deformations, where it is proven that the "ring of modular deformations" of ρ (for irreducible ρ) has Krull dimension at least 4. But one may ask more exactly about the asymptotics of the number of eigencuspforms in $S_2(\Gamma_1(Np^r))$ which give rise to ρ, as r varies. As the number of isomorphism types of the mod p representations which arise from $S_2(\Gamma_1(Np^r))$ is finite (ρ as above will arise also from $S_2(\Gamma_1(Np^2))$, as follows from 2.2 and 3.2 of [R]), one may believe that given a mod p representation ρ which arises from $S_2(\Gamma_1(Np^s))$ for some $s \geq 1$, the number of eigencuspforms (here by eigenform we mean eigenvector for almost all Hecke operators T_q^* where q runs through the primes) which give rise to ρ is asymptotically of the form $c.p^{2r}$ where c is a non-zero positive constant. This belief arises from the fact that the genus of the curve $X_1(Np^r)$ is asymptotic to $C.p^{2r}$ for non-zero positive constant C.

We consider in fact a different but related question here which is connected to the arithmetic of the Jacobians $J_1(Np^r)$ of modular curves $X_1(Np^r)$ for varying r. If we consider a nonordinary maximal ideal $\mathfrak{m}$ (at p, i.e., $U_p \in \mathfrak{m}$ where we assume that $r \geq 1$) of the Hecke algebra $\mathbb{T}_{Np^r}$ acting on $J_1(Np^r)$, then we conjecture that the dimension $\dim_{\mathbb{T}_{Np^r}/\mathfrak{m}} J_1(Np^r)[\mathfrak{m}]$ grows linearly with r. At least in the case that ρ is irreducible, we have proven (see [K] and [K1], and especially the latter, for more exact statements) that it grows at least linearly with r. At the moment no reasonable upper bound seems to be available.

We now consider the restricted Hecke algebra $\mathbb{T}'_r := \mathbb{T}_{Np^r}^{(Np)}$ generated by the Hecke operators T_q, for primes q coprime to Np, in the ring of endomorphisms of $J_1(Np^r)$. Then again we know that the number of maximal ideals $\mathfrak{m}$ of $\mathbb{T}'_r$ of residue characteristic p is bounded independently of r. Thus we would again believe that the length of the semisimplification of $J_1(Np^s)[p]_\mathfrak{m}$ as a $\mathbb{T}'_r$-module asymptotically grows like $c.p^{2r}$, for a positive constant c. Here, as usual, we are pulling back a maximal ideal $\mathfrak{m}$ of $\mathbb{T}'_r$ to $\mathbb{T}'_s$ for any $s \geq r$ and using the same symbol for both these ideals. The way this belief plays off with our conjecture in [K1] that the dimension $\dim_{\mathbb{T}_{Np^r}/\mathfrak{m}} J_1(Np^r)[\mathfrak{m}]$ grows linearly with r, is that we expect the Hecke action on $J_1(Np^r)[p]$ to be "highly nonsemisimple".

We again modify the question to make it simpler for us to answer it in the form of the proposition below, whose proof we only sketch. We hope to address these issues, relating to growth of multiplicities in p-towers, in greater detail in future work.

Proposition 6. *Consider $H^1(\Gamma(Np^r), \overline{\mathbb{F}}_p)$ with the natural Hecke action of T_q for $(q.Np) = 1$ on it. Let $\mathfrak{m}$ be a maximal ideal of the algebra generated by these operators over $\overline{\mathbb{F}}_p$ which we denote by h_r. Then the length of the semisimplification of $H^1(\Gamma(Np^r), \overline{\mathbb{F}}_p)_{\mathfrak{m}}$ as a h_r module is $c.p^{3r}$ for some positive constant c.*

Sketch of proof. The proof is a simple application of Shapiro's lemma. By this, $H^1(\Gamma(Np^r), \overline{\mathbb{F}}_p)$ is isomorphic to $H^1(\Gamma(N), \overline{\mathbb{F}}_p[SL_2(\mathbb{Z}/p^r)])$ even as Hecke modules (for the definition of Hecke action see [AS]). The latter carries a natural action of $SL_2(\mathbb{Z}/p^r)$ which acts on the coefficients of the cohomology group. The irreducible representations of $SL_2(\mathbb{Z}/p^r)$ are given by $\mathbb{L}^n(\overline{\mathbb{F}}_p)$ (for $n = 0, \cdots, p-1$), where this is the nth symmetric power of the standard 2-dimensional representation. It will be enough to prove that the semisimplification of $\overline{\mathbb{F}}_p[SL_2(\mathbb{Z}/p^r)]$, as a module over itself, will contain the $\mathbb{L}^n$'s with roughly the same multiplicity. Namely, if $\mathbb{L}^n$ occurs with multiplicity c in the semisimplification of $\overline{\mathbb{F}}_p[SL_2(\mathbb{Z}/p)]$ as a module over itself, then $\mathbb{L}^n$ will occur with multiplicity $c.p^{3(r-1)}$ in the semisimplification of $\overline{\mathbb{F}}_p[SL_2(\mathbb{Z}/p^r)]$, as a module over itself. This follows from:

1. The kernel of $SL_2(\mathbb{Z}/p^r) \to SL_2(\mathbb{Z}/p)$ is a normal p-subgroup, and that the only irreducible module of a p-group in characteristic p is the trivial module.

2. A semisimple representation of $SL_2(\mathbb{Z}/p^r)$ remains semisimple when restricted to this kernel.

This proves the proposition, as the cohomology group in the proposition, up to semisimplification, breaks up as

$$\bigoplus_{n=0}^{p-1} H^1(\Gamma(N), \mathbb{L}^n(\overline{\mathbb{F}}_p))^{c_n p^{3(r-1)}}$$

for some constants c_n.

Remarks.

1. The proposition is motivated by an e-mail message of Barry Mazur, in which he had proposed that the "packets" corresponding to different modular mod p representations ρ occur with roughly the same multiplicity in the Jordan-Holder series of $J_1(Np^r)$, considered as a Galois module, as r varies.

2. The referee has remarked that the c_n's which appear towards the end of the proof above are as follows: $c_0 = c_{p-1} = p$ and $c_i = 2p$ for $0 < i < p-1$.

Acknowledgements. The author thanks Haruzo Hida for many helpful conversations on the mathematics of this note, and the referee for a careful reading of the manuscript.

References

[AL] A. Atkin, J. Lehner, *Hecke operators on* $\Gamma_0(M)$, Math. Ann. 185 (1970), 134-160.

[AS] A. Ash, G. Stevens, *Modular forms in characteristic ℓ and special values of their L-functions*, Duke Math. J. 53 (1986), 849-868.

[E] B. Edixhoven, *The weight in Serre's conjectures on modular forms*, Inv. Math. 109 (1992), 563-594.

[H] H. Hida, *Galois representations into $GL_2(\mathbb{Z}_p[[X]])$ attached to ordinary cusp forms*, Inv. Math. 85 (1986), 545-613.

[H 1] H. Hida, *Modular p-adic L functions and p-adic Hecke algebras*, to appear in Sugaku expositions.

[J] N. Jochnowitz, *A study of the local components of the Hecke algebra mod ℓ*, Trans. AMS 270 (1982), 253-267.

[J1] N. Jochnowitz, *Congruences between systems of eigenvalues of modular forms*, Trans. AMS 270 (1982), 269-285.

[K] C. Khare, *Congruences between cusp forms: the (p,p) case*, Duke Math. J. 80 (1995), 32-67.

[K1] C. Khare, *Multiplicities of mod p Galois representations*, preprint.

a] N. Katz, *Higher congruences between modular forms*, Ann. of Math. 101 (1975), 332-367.

[Ka1] N. Katz, *A result on modular forms in characteristic p*, SLNM vol. 601 (1977), 52-61.

[Ka2] N. Katz, *p-adic properties of modular schemes and modular forms*, SLNM vol. 350 (1973), 69-190.

[M] B. Mazur, *Deforming Galois representations*, in Galois Groups over $\mathbb{Q}$, eds. Y. Ihara, K. Ribet, J-P. Serre, MSRI Publ. 16, Springer-Verlag (1989), 385-437.

[R] K. Ribet, *Report on mod ℓ representations of* $\mathrm{Gal}(\overline{\mathbb{Q}}/\mathbb{Q})$, in Motives 2, Proc. Symp. Pure Math. 55, AMS, 1994, 639-676.

[S] J-P. Serre, *Formes modulaires et fonctions zêta p-adiques*, SLNM vol. 350 (1973), 191-268.

School of Mathematics,

TIFR, Homi Bhabha Road,

Bombay 400 005, INDIA.

e-mail: shekhar@math.tifr.res.in

ON EXCEPTIONS OF INTEGRAL QUADRATIC FORMS

A. K. LAL AND B. RAMAKRISHNAN

Dedicated to Srinivasa Ramanujan

ABSTRACT. A short account of the works (including our recent observations) related to the determination of the possible exceptions of integral quadratic forms (even, unimodular and positive definite) of $8k$ variables is presented.

1. INTRODUCTION

Let f_n denote a positive definite integral quadratic form of n variables. A famous theorem of Tartakowsky states that when $n \geq 5$, a natural number a has an integral p-adic representation by f_n at each prime provided a is sufficiently large [18]. Tartakowsky's proof uses the circle method of Hardy and Littlewood. An arithmetic proof of this theorem was given by M. Kneser [7] (See [6].) Another approach to the question of representability by quadratic forms is by using the theory of modular forms. Explicit computation of the bounds for even unimodular positive definite quadratic forms f_n (where 8 divides n) is effectively done using the theory of modular forms starting with the work of M. Peters [15].

A natural number m is said to be an *exception for* n, if there exists an even, unimoudlar, positive-definite, integral quadratic form in n variables which does not represent $2m$. We denote by $\mathbf{T_n}$ the set of all such exceptions. Equivalently, m belongs to $\mathbf{T_n}$ if and only if there is a theta series whose m-th Fourier coefficient vanishes. Then one can ask the following question.

Question A.
  (i) Is $\mathbf{T_n}$ finite?
  (ii) Determine the set $\mathbf{T_n}$.

There is another related question. A modular form $g \in M_{n/2}(1)$ (the space of all modular forms of weight $n/2$ for the full modular group $SL_2(\mathbb{Z})$) is said to *behave like a theta series* if it satisfies the following conditions.

$$a_g(0) = 1, a_g(m) \in \mathbb{Z}; a_g(m) \geq 0 \text{ for all} m. \tag{1}$$

$$a_g(1) \leq 2n(n-1). \tag{2}$$

(In the above, $a_g(m)$ denotes the m-th Fourier coefficient of g.)
Note: The condition (2) is imposed because of (6). (See Remark 2.1 below.)
Define the set $\mathbf{S_n}$ as follows.

1991 *Mathematics Subject Classification.* Primary 11F11, 11F12, 11F27, 11F30.
Key words and phrases. quadratic forms, theta series, modular forms.
This is an expanded version of the talk presented by the second author at the International Conference on Discrete Mathematics and Number Theory held at Tiruchirapalli during January 3–6, 1996.

© 1998 American Mathematical Society

A natural number m belongs to $\mathbf{S_n}$ if and only if there exists a modular form in $M_{n/2}(1)$ which behaves like a theta series such that its m-th Fourier coefficient is zero. Since any theta series satisfies (1) and (2), the set $\mathbf{T_n}$ is contained in $\mathbf{S_n}$. It turns out to be more tractable to consider the following question instead of question A.

Question A'.

(i) Is $\mathbf{S_n}$ finite?
(ii) Determine the set $\mathbf{S_n}$.

Question A was first considered by Peters [15], who found an efficient and explicit bound for the set $\mathbf{T_n}$, where $24 \leq n \leq 64$, using the Ramanujan-Petersson estimate proved by Deligne. The results of Peters were improved and extended to further cases (up to the case $n = 72$) by A. M. Odlyzko and N. J. A. Sloane [13]. In fact, for the cases $48 \leq n \leq 72$, Odlyzko and Sloane considered question A'. In continuation to the work of the second author with M. Manickam [10], we, in collaboration with K. Chakraborty [1], have recenlty extended the results of Peters and Odlyzko–Sloane and determined the set $\mathbf{S_n}$ up to the case $n = 112$ using simpler method. A table of the results obtained so far is provided in section 2.

In this paper, we further extend our results to the cases $120 \leq n \leq 184$, while presenting a brief survey of the works done in this direction. The present article is a revised version of an earlier article written by the second author. The paper is organized as follows. In section 2, we give some preliminaries connecting quadratic forms and modular forms. Sections 3 and 4 contain a brief survey of the earlier works. In section 5, we present our recent work done for the cases $n = 120, 128, 136; 144, 152, 160; 168, 176$ and 184. In section 6, we review the observations made by the second author.

2. Preliminaries

Assume that 8 divides n. Let f_n denote an even, unimodular, positive definite integral quadratic form of n variables.

For $X = \begin{pmatrix} x_1 \\ \vdots \\ x_n \end{pmatrix}$, one may write

$$f_n(X) = A[X] := X^t \, A \, X \text{ (here } X^t \text{ means the transpose of } X), \tag{3}$$

where $A = (a_{ij})$ is a symmetric $n \times n$ matrix with integer coefficients. The form f_n is said to be *even* (resp. *unimodular*) (resp. *positive definite*) if $a_{ij} \in \mathbb{Z}$, $1 \leq i, j \leq n$; $a_{ii} \in 2\mathbb{Z}$, $1 \leq i \leq n$ or in other words $f_n(X)$ is even (resp. $\det A = 1$) (resp. $A[X] > 0$ if $X \neq 0$). Assume from now on that f_n is even, unimodular and positive definite.

Associated to f_n one defines a theta series as follows.

$$\theta_{f_n}(z) = \sum_{X \; integral} e^{\pi i f_n(X) z}, \quad z = x + iy, \quad y > 0. \tag{4}$$

It is known that $\theta_{f_n}(z)$ is a modular form of weight $k = n/2$ for the group $SL_2(\mathbb{Z})$. The proof of this fact uses the Poisson summation formula and the generalized Jacobi inversion formula. For details see ([5, Chapter VI, Theorem 1]).

Then
$$\theta_{f_n}(z) = \sum_{m=0}^{\infty} a_{n/2}(m) e^{2\pi i m z}, \tag{5}$$

where $a_{n/2}(m)$ is the number of times f_n represents $2m$.

Remark 2.1. There is a famous bound for the Fourier coefficient $a_{n/2}(1)$ due to Witt (cf. [12, p145 ff]), viz.,

$$a_{n/2}(1) \leq 2n(n-1) \text{ if } n \geq 16. \tag{6}$$

Also there is a unique quadratic form D_n for which equality holds in (6). (See [13].)

Remark 2.2. In each dimension n, there exists a unique theta series given by the following q expansion (here $q = e^{2\pi i z}$)

$$1 + 0 \cdot q + 0 \cdot q^2 + \cdots + 0 \cdot q^{[n/24]} + \cdots, \tag{7}$$

which is known as the *extremal* theta series. In the above $[x]$ denotes the greatest integer function of x. (See [13].)

List of earlier works.

Though it is known (see [21]) for a long time (due to Mordell and Witt - in 1940's) that the set $\mathbf{T_n}$ is empty for $n = 8$ and 16, question A was considered by M. Peters [15] in 1979 for $24 \leq n \leq 64$ and followed by A. M. Odlyzko and N. J. A. Sloane [13] in 1980. To date, $\mathbf{T_n}$ is determined exactly only for the cases up to $n = 40$ and for the remaining cases only the set $\mathbf{S_n}$ have been determined. The works of Peters and Odlyzko-Sloane have been extended by us in collaboration with K. Chakraborty up to the cases $n \leq 112$. As mentioned in the introduction, we study the cases $120 \leq n \leq 184$ in section 5.

The following are the results known so far.

$\mathbf{T}_8 = \mathbf{T}_{16} = \emptyset$	Mordell and Witt
$\mathbf{T}_{24} = \mathbf{T}_{32} = \mathbf{T}_{40} = \{1\}$; $\mathbf{S}_{48} = \{1,2,3\}$; $\mathbf{S}_{56} = \{1,2,4\}^\star$; $\mathbf{S}_{64} = \{1,2,5\}^\star$.	Peters and Odlyzko - Sloane
$\mathbf{S}_{72} = \{1,2,3,4,6\}$	Odlyzko - Sloane
$\mathbf{S}_n = \begin{cases} \{1,2\} & \text{if } n = 56, 64 \\ \{1,2,3,5\} & \text{if } n = 80 \\ \{1,2,3,6\} & \text{if } n = 88 \\ \{1,2,3,4,5,7\} & \text{if } n = 96 \\ \{1,2,3,4,6,11\} & \text{if } n = 104 \\ \{1,2,3,4\} & \text{if } n = 112. \end{cases}$	Chakraborty, Lal and Ramakrishnan

* Though we stated here (and also in the next section) the results of Odlyzko and Sloane in these cases as such, they have been corrected in our paper [1]. See Remark 3.5 and Theorem 4.1.

In the remaining part of this article we mostly discuss the set $\mathbf{S}_n$.

3. A short account of the works of Peters and Odlyzko–Sloane

The work of Peters [15].

By the theory of modular forms, we can write θ_{f_n} in terms of the Eisenstein series E_k (throughout this article $k = n/2$) and a cusp form of weight k. For the cases $24 \leq n \leq 40$, the space of cusp forms of weight k for the group $SL_2(\mathbb{Z})$, denoted by $S_k(1)$, is one dimensional. So,

$$\theta_{f_n}(z) = \sum_{m=0}^{\infty} a_{n/2}(m) q^m = E_k(z) + c_{f_n} g(z), \tag{8}$$

where $E_k(z) = 1 + A_k \sum_{m=0}^{\infty} \sigma_{k-1}(m) q^m$ (here $A_k = -\frac{2k}{B_k}$, B_k is the k-th Bernoulli number and $\sigma_r(m)$ denotes the sum of the r-th powers of the divisors of m), $g(z) = \sum_{m=1}^{\infty} a_g(m) q^m$, $a_g(1) = 1$ and $c_{f_n} \in \mathbb{C}$.

We have the Ramanujan's estimate, proved by Deligne [4], for $a_g(m)$ given by

$$|a_g(m)| \leq \sigma_0(m) m^{(n-2)/4}. \tag{9}$$

In terms of the Fourier coefficients, (8) is equivalent to

$$a_k(m) = A_k \sigma_{k-1}(m) + c_{f_n} a_g(m). \tag{10}$$

Putting $m = 1$ in the above equation, we get

$$c_{f_n} = a_{n/2}(1) + \frac{n}{B_{n/2}}. \tag{11}$$

Using the bound (6) in the above equation (11), and noticing that $B_{n/2}$ is negative, we get

$$|c_{f_n}| \leq max\left(2n(n-1), \left|\frac{n}{B_{n/2}}\right|\right) = 2n(n-1). \tag{12}$$

By equation (10), a is not an exception if $|c_{f_n} a_g(a)| \leq A_k \sigma_{k-1}(a)$. Therefore, using (9) and (12) one finds that a is not an exception if

$$\frac{a^{(n-2)/4}}{\sigma_0(a)} \geq 2(n-1)|B_{n/2}|. \tag{13}$$

In other words, using the estimate

$$\sigma_0(m) \leq \sqrt{3\,m} \quad \forall m \geq 1, \tag{14}$$

it follows that a is not an exception if

$$a^{n/4-1} \geq 2\sqrt{3}(n-1)|B_{n/2}|. \tag{15}$$

In particular,

$$a \geq \begin{cases} 2 & \text{if } n = 24 \\ 3 & \text{if } n = 32 \\ 4 & \text{if } n = 40. \end{cases} \tag{16}$$

Theorem 3.1. *(Peters ([15, Theorem]))*
a is integrally represented by $\frac{1}{2} f_n$ provided

$$a \geq \begin{cases} 2 & \text{if } n = 24 \\ 3 & \text{if } n = 32 \\ 4 & \text{if } n = 40. \end{cases}$$

In other words,

$$\mathbf{T}_n \subseteq \begin{cases} \{1\} & \text{if } n = 24 \\ \{1,2\} & \text{if } n = 32 \\ \{1,2,3\} & \text{if } n = 40. \end{cases}$$

Remark 3.1. In the above discussion Peters used Ramanujan's estimate for $\sigma_0(a)$ [16, 200], namely,

$$\sigma_0(a) \leq 8\sqrt[3]{\frac{3a}{35}}, \tag{17}$$

instead of (14) and this gives $a \geq 3$ instead of $a \geq 2$ in (16) for $n = 24$. Though Ramanujan's estimate (17) is better than (14), for small values (i.e., for $n \leq 72$) (14) works better.

Remark 3.2. In the case $n = 24$, there is Leech's lattice (see [15, 19]) which does not represent 1 and hence we have in the above theorem $\mathbf{T}_{24} = \{1\}$.

Remark 3.3. In his paper [15], M. Peters obtained similar results for the cases $n = 48, 56$ and 64. Also he remarked that this method can be generalized, which implies that $\mathbf{T_n}$ is a finite set. However, we have found a gap in his arguments for the cases mentioned above. In these cases the spaces of cusp forms are generated by two elements and due to this one has to deal with two constants c_1 and c_2 in (8). In page 198 of [15], it was remarked that using the Fourier coefficients of q and q^2 in the expansion of the theta series, one can estimate for c_1 and c_2. This is crucial to get the required results as claimed in the paper. Note that these two conditions together with the theta bound are not sufficient to control the constants c_1 and c_2 and hence, it is not possible to get estimates using the two conditions as indicated in the paper. Even for the cases $24 \leq n \leq 40$, Peters used two conditions to derive the estimate. As our method [10, 1] (see section 4 below) suggests, one needs at least $\ell + 1$ (ℓ is the dimension of the space of cusp forms in the particular case) conditions to obtain the necessary results. So, the possibility of extending Peters' work to higher dimensions is questionable. Hence, in the general situation, question A'(i) (about the finiteness of the set $\mathbf{T_n}$) remains open.

The work of Odlyzko and Sloane [13].

Let $n = 32$. As we have already noted, θ_{f_n} can be written as

$$\theta_{f_n}(z) = E_{16}(z) + \lambda E_4(z) \Delta(z),$$

where $\Delta(z) \in S_{12}(1)$ is the well known discriminant function. ($S_k(1)$ is the subspace of $M_k(1)$ consisting of all cusp forms.) By Theorem 3.1, we have $\mathbf{T}_{32} \subseteq \{1,2\}$. Now, $a_{n/2}(1) = 0$ gives the extremal theta series. Examples of quadratic forms with exception 1 are known due to Chernyakov [2, 15] and Conway–Pless [3, 13]. On the other hand, if we assume that $a_{n/2}(2) = 0$, then $a_{n/2}(1) < 0$, which is a contradiction. Hence, $\mathbf{T}_{32} = \{1\}$.

A similar argument implies that $\mathbf{T}_{40} = \{1\}$. In this case, the existence of a quadratic form with exception 1 was shown by McKay [11, 13].

For the cases $48 \leq n \leq 64$, one proceeds as follows.

Write

$$\theta_{f_n}(z) = E_k(z) - (A_k - l)E_{k-12}\Delta(z) + BE_{k-24}\Delta^2(z), \tag{18}$$

where $l = a_{n/2}(1)$ and B is a rational number. Since $l = a_{n/2}(1)$, we get a bound for l as $0 \leq l \leq 2n(n-1)$. The fact that $a_{n/2}(m)$ is non-negative implies that the points (l, B) lie in or on a bounded region in $\mathbb{R}^2$. It also follows that the only coefficients $(a_{n/2}(m))$ that can vanish are $a_{n/2}(i_k) = 0$, which constitute three sides of the bounded region, where

$$i_k = \begin{cases} 1,2,3 & \text{if } k = 24 \\ 1,2,4 & \text{if } k = 28 \\ 1,2,5 & \text{if } k = 32. \end{cases}$$

In other words,

$$\mathbf{S}_n = \begin{cases} \{1,2,3i\} & \text{if } k = 24 \\ \{1,2,4\} & \text{if } k = 28 \\ \{1,2,5\} & \text{if } k = 32. \end{cases}$$

A similar method is applied for the case $n = 72$.

We summarize the results of Odlyzko and Sloane in the following.

Theorem 3.2. *(Odlyzko and Sloane [13])*

$$\mathbf{T}_{32} = \mathbf{T}_{40} = \{1\}.$$
$$\{1,2\} \subseteq \mathbf{T}_{48} \subseteq \mathbf{S}_{48} = \{1,2,3\}.$$
$$\mathbf{S}_n = \begin{cases} \{1,2,4\} & \text{if } n = 56 \\ \{1,2,5\} & \text{if } n = 64 \\ \{1,2,3,4,6\} & \text{if } n = 72. \end{cases}$$

Remark 3.4. Examples of quadratic forms with exceptions 1 and 2 for the case $n = 48$, which provide the example of the extremal theta series in this case, were given by Leech–Sloane [8] and Venkov [20] ([13]).

Remark 3.5. The work of Odlyzko and Sloane has been corrected in [1] for the cases $n = 56$ and 64. See Theorem 4.1 below.

4. Another related problem

In this section, we state another interesting problem which gives information about question A$'$.

Question B: Determine all normalized modular forms in $M_k(1)$ whose Fourier coefficients are non-negative rational integers.

This question was posed by M. Ozeki [14] in connection with the following (somewhat difficult) questions.

Question C:
  (i) Determine all quadratic forms (even, unimodular and positive definite) of n variables ($8|n$).
  (ii) Determine all modular forms which are theta series associated to the above type of quadratic forms.

Note that question B is a weakened form of question C(ii).

Remark 4.1. Question B was solved by Ozeki [14] for the cases $k = 12, 16$ and we have extended Ozeki's results for the further cases $20 \leq k \leq 56$, $4|k$ in collaboration with M. Manickam and K. Chakraborty [1, 10]. Our method is different from Ozeki's and it is simpler in the sense that one can extend this method to further cases.

Remark 4.2. Question C(i) has been completely determined for the case $n = 24$ by H.-V. Niemeier [12] in his thesis in 1968. He proved that there are 24 positive definite even integral quadratic forms of 24 variables. M. Ozeki's [14] construction gives 8191 modular forms $\phi_s(\tau)$ which behave like theta series. In the notation of [14], writing

$$\phi_s(\tau) = 1 + \sum_{n \geq 1}^{\infty} b(n) q^n,$$

where

$$b(n) = \frac{65520}{691}(\sigma_{11}(n) - \tau(n)) + (1104 - 384s)\tau(n),$$

$384s \in \mathbb{Z}$ and $0 \leq (1104 - 384s) \leq 8190$,

we see that Niemeier's construction gives theta series corresponding to the values of $b(1) = 1104 - 384s = 0, 48, 72, 96, 120, 144, 168, 192, 216, 240, 288, 312, 336, 384, 432, 528, 600, 720$ and 1104. Among them there are two quadratic forms giving rise to the same theta series given by the values $b(1) = 144, 240, 288, 432$ and 720. It seems from Ozeki's remark ([14, p. 203]) that he wanted to know what are the modular forms in his construction that are theta series associated to Niemeier's quadratic forms. Since in this case the dimension of the space $M_{12}(1)$ is two, any modular form (with constant term 1) is uniquely determined by its q coefficient in the Fourier expansion. This gives the answer to Ozeki's question, namely, the modular forms constructed by Ozeki are in fact theta series determined by Niemeier's quadratic forms.

In the following we shall demonstrate our method briefly for the case $k = 24$ and details are available in [1, 10].

Let $\phi(z)$ be a modular form of weight k for $SL_2(\mathbb{Z})$. The problem is to determine all modular forms ϕ which behave like theta series. i.e., determine all $\phi(z) = 1 + \sum_{m=1}^{\infty} a(m) q^m$, where

$$0 \leq a(m) \in \mathbb{Z} \text{ for all } m \text{ and } a(1) \leq 4k(2k-1).$$

Expressing ϕ as a linear combination of E_k and two cusp forms one can prove the following.

(I) If $a(0), a(1)$ and $a(2)$ are integers, then $a(m) \in \mathbb{Z}$ for all m.
(II) $a(n) \geq 0$ for all n if and only if $a(i) \geq 0$, $1 \leq i \leq 6$.

In the above, (I) is an easy consequence of Lemma A.1 of [10], which states that (i) a modular form in $M_k(1)$ is completely determined by its first r Fourier coefficients, where r is the dimension of $M_k(1)$, (ii) if the first r Fourier coefficients of a modular form in $M_k(1)$ are rational integers, then all the Fourier coefficients of the modular form are rational integers.

Let us briefly sketch the proof of (II). Assuming $a(m) \geq 0$, for all m, we get a bounded region, say R_{24}, in $\mathbb{R}^2$ (in the first quadrant) determined by the equations $a(i) \geq 0$, $i = 1, 2, 3, 5$ and 6, which are given below.

$$X = a(1) \geq 0$$
$$Y = a(2) \geq 0$$
$$4368000 + 16305X - 4Y = a(3) \geq 0$$
$$14061745152 + 95013X - 32Y = a(5) \geq 0$$
$$12161804928000 - 27291648X + 3995Y = a(6) \geq 0.$$

(19)

($a(1) \geq 0$ and $a(2) \geq 0$ implies automatically that $a(4) \geq 0$.)

From this bounded region, we get upper bounds for $a(1)$ and $a(2)$ (considering $a(1)$ and $a(2)$ as the coordinates of the region) given by

$$0 \leq X \leq 901973, \quad 0 \leq Y \leq 3117528477.$$

These upper bounds and the estimates of the Fourier coefficients of the two cusp forms (which can be obtained from the Ramanujan's estimate, proved by Deligne, for the Fourier coefficients of the Hecke eigenforms) imply that $a(n) \geq 0$ for $n \geq 7$.

As a consequence of (I) and (II), we get a one-to-one correspondence between the lattice points inside and on the boundary of R_{24} and the set of all normalized modular forms in $M_{24}(1)$, whose Fourier coefficients are non-negative integers. A sketch of the region R_{24} has been provided in [1, Figure A.].

Since we are interested in modular forms which behave like theta series, we impose the theta condition (2) that $a(1) \leq 4512$. This gives a subregion (let us denote this by $\mathbf{R}\theta_{24}$) of R_{24} and from the above discussion, it follows that the set of all modular forms in $M_{24}(1)$, which behave like theta series are obtained from the lattice points inside and on the boundary of $\mathbf{R}\theta_{24}$. We give a sketch of the region $\mathbf{R}\theta_{24}$ in Figure 1. (Compare the corresponding figure given in [13].) Since all the modular forms which are obtained from the lattice points in the interior of $\mathbf{R}\theta_{24}$ have positive Fourier coefficients, the only Fourier coefficients which can vanish comes from the boundary. Since the boundary of $\mathbf{R}\theta_{24}$ is given by the equations $a(1) = 0$, $a(2) = 0$, $a(3) = 0$ and $a(1) = 4512$, it follows that $\mathbf{S_{48}} \subseteq \{1, 2, 3\}$. The point $(0, 0)$ provides the example of the extremal theta series and we have already remarked (see Remark 3.4) that an example of a quadratic form which corresponds to this extremal theta series was given by Leech and Sloane. This implies that $\{1, 2\} \subseteq \mathbf{T}_{48}$. Further, the point $(0, 1092000)$, which turns out to be a vertex of $\mathbf{R}\theta_{24}$, corresponds to the modular form $1 + 1092000q^2 + 40186692000q^4 + \cdots$ which omits the coefficients 1 and 3. Hence one concludes the following.

$$\{1, 2\} \subseteq \mathbf{T}_{48} \subseteq \mathbf{S}_{48} = \{1, 2, 3\}.$$

Notice that the method of Odlyzko–Sloane also gives a bounded region, but the determination of the modular forms which behave like theta series using their method is not so explicit as in our method.

We state in the following our results obtained in [1].

Theorem 4.1. *(K. Chakraborty, A. K. Lal and B. Ramakrishnan [1])*

$$\mathbf{S}_n = \begin{cases} \{1, 2\} & \text{if } n = 56, 64 \\ \{1, 2, 3, 5\} & \text{if } n = 80 \\ \{1, 2, 3, 6\} & \text{if } n = 88 \\ \{1, 2, 3, 4, 5, 7\} & \text{if } n = 96 \\ \{1, 2, 3, 4, 6, 11\} & \text{if } n = 104 \\ \{1, 2, 3, 4\} & \text{if } n = 112. \end{cases}$$

Remark 4.3. Examples of modular forms with exceptions cited in the above theorem have been provided in [1].

5. Recent Results

We feel that our method can be generalized. To check whether we get any pattern for the set $\mathbf{S}_n$, we have carried out the computations for the cases $120 \leq n \leq 184$ and we present these recent works in this section.

Since the method is similar to the one done in our earlier paper [1], we make a brief report of our work and we use mostly the notations as in our earlier paper [1]. As we are interested in only the modular forms which behave like theta series, in these cases we have determined only the theta regions imposing the condition (2). Let n be as above. Let $\dim_{\mathbb{C}} S_{n/2}(1) = \ell$. Then

$$\ell = \begin{cases} 5 & \text{if } 120 \leq n \leq 136 \\ 6 & \text{if } 144 \leq n \leq 160 \\ 7 & \text{if } 168 \leq n \leq 184. \end{cases}$$

Let $\phi_k(z) = 1 + \sum_{m=1}^{\infty} a_k(m) q^m \in M_k(1)$ be such that

$$a_k(m) \geq 0; \ a_k(m) \in \mathbb{Z} \text{ for all } m \geq 1 \text{ and } a_k(1) \leq 4k(2k-1). \tag{20}$$

Then our problem is to find all $\phi_k(z)$ satisfying (20). Note that $k = n/2$. Writing $\phi_k(z)$ in terms of a basis $\{\theta_k^e(z), g_{1,k}(z), g_{2,k}(z), \cdots, g_{\ell,k}(z)\}$ of $M_k(1)$, as we have done in our earlier work, we can write $a_k(m)$ as

$$a_k(m) = A_k B_k(m) + \sum_{i=1}^{\ell} \alpha_{i,k}(m) X_{k,i}, \tag{21}$$

where $a_k(i) = X_{k,i} \in \mathbb{Z}$, $1 \leq i \leq \ell$, and

$$\theta_k^e(z) = 1 + \sum_{m \geq 1} A_k B_k(m) q^m; \ A_k B_k(m) \in \mathbb{Z}, \forall m$$

$$g_{i,k}(z) = \sum_{m \geq i} \alpha_{i,k}(m) q^m; \ \alpha_{i,k}(m) \in \mathbb{Z}, \forall m, 1 \leq i \leq \ell$$

Following our method in [1], we get the following theorem.

Theorem 5.1. Let $a_k(1) \leq 4k(2k-1)$. Then $a_k(m) \geq 0$ if and only if $a_k(i_k) \geq 0$, where

$$i_k = \begin{cases} 1,2,3,4,5,6,7,8,10,14,15 & \text{if } k = 60 \\ 1,2,3,4,5,7,8,9,11,16,17 & \text{if } k = 64 \\ 1,2,3,4,5,8,9,10,12,13,18 & \text{if } k = 68 \\ 1,2,3,4,5,6,7,8,9,10,11,13,14,20,21 & \text{if } k = 72 \\ 1,2,3,4,5,6,8,9,10,11,12,15,23 & \text{if } k = 76 \\ 1,2,3,4,5,6,9,10,13,16,17,25 & \text{if } k = 80 \\ 1,2,3,4,5,6,7,8,9,10,11,12,14,18,19,27 & \text{if } k = 84 \\ 1,2,3,4,5,6,7,9,10,11,12,13,15,20,30 & \text{if } k = 88 \\ 1,2,3,4,5,6,7,9,10,11,13,14,16,17,21,22 & \text{if } k = 92. \end{cases}$$

Remark 5.1. As done in the previous cases, one can explicitly write down the vertices of the theta region $\mathbf{R}\theta_{n/2}$ in $\mathbb{R}^\ell$. Corresponding to each lattice point inside or on the boundary of this theta region one gets a modular form satisfying (20).

For the cases $120 \leq n \leq 136$ (i.e., for the cases $60 \leq k \leq 68$), we have determined the set $\mathbf{S}_n$ completely. For the remaining cases, apart from the easy examples, we have not tried to get examples of modular forms having the exception. In the following, we consider each case separately.

The case $n = 120$.

By Theorem 5.1, for the determination of the set $\mathbf{S}_{120}$, the hyperplanes to be considered are $a_{60}(i) = 0$, where $i = 1, 2, 3, 4, 5, 6, 7, 8, 10, 14, 15$. We have checked that $a_{60}(10) = 0$ and $a_{60}(15) = 0$ do not have any lattice points within the theta region. In the following we give examples of modular forms having the exceptions $1, 2, 3, 4, 5, 6, 7, 8$ and 14. First, we shall give the extremal theta series corresponding to the point $(0, 0, 0, 0, 0)$, which gives the exceptions $1, 2, 3, 4, 5$.

$$1 + 45792819072000 \ q^6 + 406954241261568000 \ q^7 \\ + 1074181924443974346000 \ q^8 + \cdots . \qquad (22)$$

We notice that in this case, there are (only) four lattice vertices of the theta region. (We make an observation in this regard in the last section.) The vertices $(0, 0, 0, 0, 381606825600)$ and $(28560, 0, 0, 0, 183666299211360)$ give the exception 6. The modular forms corresponding to these vertices are given below.

$$1 + 381606825600 \ q^5 + 409633121177280000 \ q^7 \\ + 1074080203328542410000 \ q^8 + \cdots . \qquad (23)$$

$$1 + 28560q + 183666299211360 \ q^5 + 97365490240971993600 \ q^7 \\ + 1237231549615220024429200 \ q^8 + \cdots . \qquad (24)$$

The points $(0, 0, 0, 261458060058, 8192)$ and $(4096, 0, 0, 9076620836378, 0)$ give the exception 7. The corresponding modular forms are given below.

$$1 + 261458060058 \ q^4 + 8192 \ q^5 + 49593141031320192 \ q^6 \\ + 980728039244565940000 \ q^8 + 112373008392426153998745 \mathbf{6} \ q^9 + \cdots \quad (25)$$

$$1 + 4096q + 9076620836378 \ q^4 + 1723256092754082432 \ q^6 \\ + 15426907238006046900000 \ q^8 + 10079146359777292464033792 \ q^9 + \cdots (26)$$

The point $(0, 0, 236073145, 5584790874726, 0)$ gives the exception 8 and the corresponding modular form is

$$1 + 236073145 \ q^3 + 5584790874726 \ q^4 + 1074038558777524704 \ q^6 \\ + 1622527807208490 \ q^7 + 12205867076325468355 66272 \ q^9 + \cdots . \qquad (27)$$

(0, 213432461426, 12276946072211613, 325103155000, 1794100091981023299317) is a point inside the region, which gives the exception 14 and the corresponding modular form is

$$1 + 213432461426 \ q^2 + 12276946072211613 \ q^3 + 325103155000 \ q^4$$
$$+ 1794100091981023299317 \ q^5 + 60067910949332571736000 \ q^6$$
$$+ 4471313839178937323646006 22 \ q^7 + 492823306559379006758334 62928 \ q^8$$
$$+ 10506730640172692312767814 16750 \ q^9$$
$$+ 17848053817093885193987795 26652 \ q^{10}$$
$$+ 996240024685358449302287940 652375 \ q^{11} \tag{28}$$
$$+ 2542977717494263775508126142 6778400 \ q^{12}$$
$$+ 14856701688096667761111443424 4512517 \ q^{13}$$
$$+ 64251816880634793352181042871 09089034 \ q^{15} + \cdots .$$

Hence, we conclude the following.

$$\mathbf{S}_{120} = \{1, 2, 3, 4, 5, 6, 7, 8, 14\}.$$

The case $n = 128$.

In this case, by Theorem 5.1, the hyperplanes to be considered are $a_{64}(i) = 0$, where $i = 1, 2, 3, 4, 5, 7, 8, 9, 11, 16, 17$. The hyperplanes $a_{64}(9) = 0$, $a_{64}(11) = 0$, $a_{64}(16) = 0$ and $a_{64}(17) = 0$ are not having any lattice points which lie in the theta region. The extremal theta series corresponding to the origin $(0, 0, 0, 0, 0)$, which gives the exceptions $1, 2, 3, 4, 5$ is given as follows.

$$1 + 6445658419200 \ q^6 + 106526207036620800 \ q^7$$
$$+ 479671330783886841600 \ q^8 + \cdots . \tag{29}$$

The points $(657, 0, 0, 0, 46034812488590)$ and $(0, 0, 0, 797558046592, 14915340)$ give exceptions 7 and 8 respectively. The respective modular forms are given by

$$1 + 657q + 46034812488590 \ q^5 + 5623815395160960 \ q^6$$
$$+ 2372149612311194080000 \ q^8 + 23596688594906356350164 37 \ q^9 + \cdots \tag{30}$$

$$1 + 797558046592 \ q^4 + 14915340 \ q^5 + 129899941148419488 \ q^6$$
$$+ 38899749300632530000 \ q^7 + 75134212693008776790 2664 \ q^9 + \cdots . \tag{31}$$

From the above examples, we conclude that

$$\mathbf{S}_{128} = \{1, 2, 3, 4, 5, 7, 8\}.$$

The case $n = 136$.

In this case, the hyperplanes to be considered are $a_{68}(i) = 0$, where $i = 1, 2, 3, 4, 5, 8, 9, 10, 12, 13, 18$. The hyperplanes $a_{68}(9) = 0$, $a_{68}(13) = 0$ and $a_{68}(18) = 0$ are not having any lattice points which lie in the theta region. The extremal theta series corresponding to the origin $(0,0,0,0,0)$, which gives the exceptions $1,2,3,4,5$ is given as follows.

$$1 + 707360371200 \; q^6 + 21669552237772800 \; q^7 \\ + 1664605075760744 79600 \; q^8 + \cdots . \tag{32}$$

The point $(0, 0, 3685, 408058, 50817868878303)$ gives the exception 8; $(0, 6949968, 34376470, 333145691415927, 818557713691412585)$ gives the exception 10; and $(0, 3587670, 2654967090285, 5307342733378048, 14869277264120394999)$ gives the exception 12. The respective modular forms are given by

$$1 + 3685 \; q^3 + 408058 \; q^4 + 50817868878303 \; q^5 + 18295199226294072 \; q^6 \\ + 598072586122652470 \; q^7 + 4551505659873 19525110186 \; q^9 + \cdots , \tag{33}$$

$$1 + 6949968 \; q^2 + 34376470 \; q^3 + 333145691415927 \; q^4 + 818557713691412585 \; q^5 \\ + 320877338827199850888 \; q^6 + 3762990896978 8481773032 \; q^7 \\ + 143545200606690908 6129512 \; q^8 + 144400515090836323482218694 \; q^9 \\ + 17314021917305106377 00242688650 \; q^{11} + \cdots, \tag{34}$$

$$1 + 3587670 \; q^2 + 2654967090285 \; q^3 + 5307342733378048 \; q^4 \\ + 14869277264120394999 \; q^5 + 57896691916 05633832992 \; q^6 \\ + 68733761419982669937 0790 \; q^7 + 6708666135 0549891213299760 \; q^8 \\ + 133902101258002 35733475477946 \; q^9 \\ + 6679771804836276 23563318894004 \; q^{10} \\ + 230748622124262993 37779186198615 \; q^{11} \\ + 2012179979993965244 5790188390881415 \; q^{13} + \cdots . \tag{35}$$

Therefore, we conclude that

$$\mathbf{S}_{136} = \{1, 2, 3, 4, 5, 8, 10, 12\} \, .$$

In the cases when $144 \leq n \leq 184$, we have not tried to exclude some of the possibilities of $a_{n/2}(i) = 0$ having lattice points inside the theta regions as we have done in the previous cases. However, we give below the extremal theta series in all the three cases and in the case when $n = 144, 168$, we give some more examples.

The case $n = 144$.
   The extremal theta series is
$$1 + 3486157968384000 \; q^7 + 45569082381053868000 \; q^8 \\ + 195211487788636999680000 \; q^9 + \cdots . \tag{36}$$

The points (which are vertices in the theta region) $(0, 0, 0, 0, 0, 24209430336000)$ and $(41184, 0, 0, 0, 0, 19321006011832720)$ give exception 7. The modular forms corresponding to these points are:

$$1 + 24209430336000 \; q^6 + 45814856517824940000 \; q^8 \\ + 195200187988608811008000 \; q^9 + \cdots . \tag{37}$$

$$1 + 41184\,q + 19321006011832720\,q^6 + 20473052501730201890400\,q^8$$
$$+ 482990025781711274716 44384\,q^9 + \cdots. \tag{38}$$

The case $n = 152$.
The extremal theta series in this case is
$$1 + 445669657804800\,q^7 + 9972668981775099600\,q^8$$
$$+ 68429983333723275264000\,q^9 + \cdots. \tag{39}$$

The case $n = 160$.
The extremal theta series in this case is given by
$$1 + 46211033088000\,q^7 + 1765470763309176000\,q^8$$
$$+ 19404836920555806720000\,q^9 + \cdots. \tag{40}$$

The case $n = 168$.
The extremal theta series is
$$1 + 256206274225902000\,q^8 + 4499117081888292864000\,q^9$$
$$+ 28249079519300191531499520\,q^{10} + \cdots. \tag{41}$$

The points (which are vertices in the theta region) $(0, 0, 0, 0, 0, 0, 1525037346582750)$ and $(56112, 0, 0, 0, 0, 0, 1828569594426496350)$ give exception 8. The modular forms corresponding to these points are:
$$1 + 1525037346582750\,q^7 + 4520254099511929779000\,q^9$$
$$+ 28247938547358972181259520\,q^{10} + \cdots, \tag{42}$$

$$1 + 56112\,q + 1828569594426496350\,q^7 + 3312083059253906502903912\,q^9$$
$$+ 12556635537393804951091979520\,q^{10} + \cdots. \tag{43}$$

The case $n = 176$.
The extremal theta series in this case is
$$1 + 30517846810308000\,q^8 + 86121086096611 7376000\,q^9$$
$$+ 824152881844587049058304 0\,q^{10} + \cdots. \tag{44}$$

The case $n = 184$.
The extremal theta series in this case is given by
$$1 + 3034192667130000\,q^8 + 137290127714549760000\,q^9$$
$$+ 2002491671933024672808960\,q^{10} + \cdots. \tag{45}$$

Since we have not ruled out some possibilities, we can conclude from Theorem 5.1 only the following.

$$\mathbf{S}_n \subseteq \begin{cases} \{1,2,3,4,5,6,7,8,9,10,11,13,14,20,21\} & \text{if } n = 144 \\ \{1,2,3,4,5,6,8,9,10,11,12,15,23\} & \text{if } n = 152 \\ \{1,2,3,4,5,6,9,10,13,16,17,25\} & \text{if } n = 160 \\ \{1,2,3,4,5,6,7,8,9,10,11,12,14,18,19,27\} & \text{if } n = 168 \\ \{1,2,3,4,5,6,7,9,10,11,12,13,15,20,30\} & \text{if } n = 176 \\ \{1,2,3,4,5,6,7,9,10,11,13,14,16,17,21,22\} & \text{if } n = 184. \end{cases}$$

Remark 5.2. From the examples provided for the cases $144 \leq n \leq 184$, one can also conclude the following.

$$\{1,2,3,4,5,6,7\} \subseteq \mathbf{S}_{144}$$
$$\{1,2,3,4,5,6\} \subseteq \mathbf{S}_n, \ n = 152, 160$$
$$\{1,2,3,4,5,6,7,8\} \subseteq \mathbf{S}_{168}$$
$$\{1,2,3,4,5,6,7\} \subseteq \mathbf{S}_n, \ n = 176, 184.$$

6. SOME OBSERVATIONS

Let $\dim_{\mathbb{C}} S_{n/2}(1) = \ell$. In the earlier version of this paper, a pattern for the set $\mathbf{S}_n$ was observed considering the results up to the case $\ell \leq 4$. But, our recent calculations reveal that no pattern seems to be possible for the set $\mathbf{S}_n$.

Regarding the set $\mathbf{T_n}$, we pose the following question.

Is

$$\mathbf{T_n} = \begin{cases} \{1,2,\cdots,\ell,\ell+1\} & \text{if } n \equiv 0 \pmod{24} \\ \{1,2,\cdots,\ell\} & \text{if } n \equiv 8, 16 \pmod{24} \end{cases} ? \qquad (46)$$

We shall analyze the reasons to pose such a question.

In our paper [1], we have observed that one gets a bounded region $\mathbf{R}\theta_{n/2}$ in $\mathbb{R}^\ell$ (possibly in general) which determines all the theta series. It follows from our results that the exceptions are determined only from the lattice points on the boundary of $\mathbf{R}\theta_{n/2}$.

We have found an interesting fact about the vertices of the region $\mathbf{R}\theta_{n/2}$. We list below those vertices which are lattice points (here we consider the case $n \geq 48$).

n	Lattice vertices
48	$(0,0), (4512,0), (4512, 19484040), (0, 1092000)$
56	$(0,0), (6160,0)$
64	$(0,0), (8064,0)$
72	$(0,0,0), (10224,0,0), (10224,0,8193582046), (0,0,86363550)$
80	$(0,0,0), (12640,0,0)$
88	$(0,0,0), (15312,0,0)$
96	$(0,0,0,0), (18240,0,0,0), (18240,0,0,1459597604100),$ $(0,0,0,5894441280)$
104	$(0,0,0,0), (21424,0,0,0)$
112	$(0,0,0,0), (24864,0,0,0)$
120	$(0,0,0,0,0), (28560,0,0,0,0), (28560,0,0,0,183666299211360),$ $(0,0,0,0,381606825600)$
128	$(0,0,0,0,0), (32512,0,0,0,0)$
136	$(0,0,0,0,0), (36720,0,0,0,0)$
144	$(0,0,0,0,0,0), (41184,0,0,0,0,0), (41184,0,0,0,0,19321006011832720),$ $(0,0,0,0,0,24209430336000)$
152	$(0,0,0,0,0,0), (45904,0,0,0,0,0)$
160	$(0,0,0,0,0,0), (50880,0,0,0,0,0)$
168	$(0,0,0,0,0,0,0), (0,0,0,0,0,0,1525037346582750)$ $(56112,0,0,0,0,0,0), (56112,0,0,0,0,0,1828569594426496350)$
176	$(0,0,0,0,0,0,0), (61600,0,0,0,0,0,0)$
184	$(0,0,0,0,0,0,0), (67344,0,0,0,0,0,0)$

In all the cases considered so far, the bounded region $\mathbf{R}\theta_{n/2}$ has only 4 lattice vertices when $n \equiv 0 \pmod{24}$ and 2 lattice vertices when $n \equiv 8, 16 \pmod{24}$. Let

$$v_1 = \underbrace{(0, 0, \cdots, 0)}_{\ell \text{ elements}}$$
$$v_2 = \underbrace{(2n(n-1), 0, \cdots, 0)}_{\ell \text{ elements}}$$
$$v_3 = \underbrace{(2n(n-1), 0, \cdots, 0, v_{3,\ell})}_{\ell \text{ elements}}$$
$$v_4 = \underbrace{(0, 0, \cdots, 0, v_{4,\ell})}_{\ell \text{ elements}},$$

where $v_{3,\ell}$ and $v_{4,\ell}$ are described below. Then if $n \equiv 0 \pmod{24}$, the four vertices are v_1, v_2, v_3, v_4 and if $n \equiv 8, 16 \pmod{24}$, the two vertices are v_1, v_2.

Among these, v_1 (provided the extremal theta series has non-negative Fourier coefficients) and v_2 are trivial vertices (trivial in the sense that these are certainly lattice vertices of the region) and v_3 and v_4 are non-trivial vertices. With regard to the vertex v_1, we shall make a remark (see Remark 6.1) at the end of this section. $v_{3,\ell}$ and $v_{4,\ell}$ are integers for the cases considered so far and it seems likely that this fact will be true in general. $v_{3,\ell}$ and $v_{4,\ell}$ are given by the following.

$$v_{3,\ell} = \frac{-(A_k B_k(\ell+1) + 4k(2k-1)\alpha_{1,k}(\ell+1))}{\alpha_{\ell,k}(\ell+1)}$$
$$v_{4,\ell} = \frac{-A_k B_k(\ell+1)}{\alpha_{\ell,k}(\ell+1)}.$$
(47)

These are obtained from the equation $a_k(\ell+1) = 0$ by putting $X_{k,1} = 2n(n-1)$, $X_{k,i} = 0, 2 \leq i \leq \ell-1$ for $v_{3,\ell}$ and by putting $X_{k,i} = 0, 1 \leq i \leq \ell-1$ for $v_{4,\ell}$. We remark here that the Fourier coefficients of $\theta_k^e(z)$ and $g_{i,k}(z)$, $1 \leq i \leq \ell$ (i.e., $A_k B_k(m)$ and $\alpha_{i,k}(m)$) can be explicitly given by nice formulas involving the Fourier coefficients of the Eisenstein series $E_k(z)$ and the Ramanujan τ-function. We also notice that $\theta_k^e(z)$ is the extremal theta series. Now by (47), we have the following.

$\alpha_{\ell,k}(\ell+1)$ divides $A_k B_k(\ell+1)$; and $\alpha_{\ell,k}(\ell+1)$ divides $4k(2k-1)\alpha_{1,k}(\ell+1)$

i.e., the $(\ell+1)$-th Fourier coefficient of the cusp form $g_{\ell,k}(z)$ divides the $(\ell+1)$-th Fourier coefficient of the extremal theta series and it also divides $4k(2k-1)$ times the $(\ell+1)$-th Fourier coefficient of the cusp form $g_{1,k}(z)$.

The reason why we are investigating the lattice vertices of $\mathbf{R}\theta_{n/2}$ is the fact that so far, examples of quadratic forms are known only for $n \leq 48$ and all of them correspond to the vertex v_1. This gives us the feeling that quadratic forms may exist only corresponding to these lattice vertices. Noticing that the vertices v_1 and v_2 give exceptions $1, 2, \cdots, \ell$ and the vertices v_3 and v_4 (when $n \equiv 0 \pmod{24}$) give exceptions $2, 3, \cdots, \ell-1, \ell+1$ and $1, 2, \cdots, \ell-1, \ell+1$ respectively, we think that (46) is clarified now.

By Remark 2.1, there is a unique quadratic form D_n for which the corresponding theta series has the property that $a_{n/2}(1) = 2n(n-1)$. That means that there is a unique lattice point on the hyperplane $a_{n/2}(1) = 2n(n-1)$ which gives the theta series (we shall refer this to Witt-theta series) corresponding to D_n. From the above discussion, it is interesting to know whether the Witt-theta series (corresponding to D_n) corresponds to the vertex v_2 if $n \equiv 8, 16 \pmod{24}$ or to one of the vertices v_2, v_3 if $n \equiv 0 \pmod{24}$. Here we want to make one comment on the example given by Odlyzko-Sloane. In their paper [13], for the case $n = 48$, they have given the example (for the exception 3) corresponding to the vertex $v_4 = (0, 1092000)$ and in fact the interesting example, in view of the above, could be the one corresponding to the vertex (vertex S of the diagram in [13]) $v_3 = (4512, 19484040)$ which is given below.

$$1 + 4512\, q + 19484040\, q^2 + 114555632736\, q^4 + 6517468628160\, q^5 + \cdots.$$
(48)

In the following, we list the interesting examples which are not given in the earlier works.

i) $n = 48$; $v_2 = (4512, 0)$.
$$1 + 4512\, q + 935233920\, q^3 + 93512869536\, q^4 + 6810508589760\, q^5 + \cdots.$$
(49)

ii) $n = 56$; $v_2 = (6160, 0)$.
$$1 + 6160\, q + 950308800\, q^3 + 412917166480\, q^4 + 35804396083680\, q^5 + \cdots.$$
(50)

iii) $n = 64$; $v_2 = (8064, 0)$.

$$1 + 8064\, q + 407585280\, q^3 + 728079149952\, q^4 + 170086527924480\, q^5 + \cdots. \tag{51}$$

iv) $n = 72$; $v_2 = (10224, 0, 0)$.

$$1 + 10224\, q + 589937907312\, q^4 + 403050965986080\, q^5 + \cdots. \tag{52}$$

v) $n = 72$; $v_3 = (10224, 0, 8193582046)$.

$$1 + 10224\, q + 8193582046\, q^3 + 423403823788344\, q^5 \\ + 66666031042644480\, q^6 + \cdots. \tag{53}$$

vi) $n = 80$; $v_2 = (12640, 0)$.

$$1 + 12640\, q + 245090535520\, q^4 + 487309907246400\, q^5 + \cdots. \tag{54}$$

vii) $n = 88$; $v_2 = (15312, 0)$.

$$1 + 15312\, q + 60700631376\, q^4 + 332478144931680\, q^5 + \cdots. \tag{55}$$

viii) $n = 96$; $v_2 = (18240, 0, 0, 0)$.

$$1 + 18240\, q + 140121369993600\, q^5 + 295009795187020800\, q^6 + \cdots. \tag{56}$$

ix) $n = 96$; $v_3 = (18240, 0, 0, 1459597604100)$.

$$1 + 18240\, q + 1459597604100\, q^4 + 301525438891723200\, q^6 + \cdots. \tag{57}$$

x) $n = 104$; $v_2 = (21424, 0, 0, 0)$.

$$1 + 21424\, q + 37888238692512\, q^5 + 181572485682931200\, q^6 + \cdots. \tag{58}$$

xi) $n = 112$; $v_2 = (24864, 0, 0, 0)$.

$$1 + 24864\, q + 7076775609792\, q^5 + 75310188757009920\, q^6 + \cdots. \tag{59}$$

The other interesting examples (apart from (48) to (59) and the extremal theta series) which are given already are:

$n = 48$; $v_4 = (0, 1092000)$ [13, p. 215],
$n = 72$; $v_4 = (0, 0, 86363550)$ [13, p. 216],
$n = 96$; $v_4 = (0, 0, 0, 5894441280)$ [1, eqn.(24)].

For the cases $120 \leq n \leq 184$, one can refer section 5 for the examples corresponding to the vertices v_1, v_3, and v_4. In the following we give examples of modular forms corresponding to the vertex v_2 for the cases $120 \leq n \leq 184$ in the same order.

$$1 + 28560\, q + 22039955905363200\, q^6 \\ + 96076152820508246400\, q^7 + 123772113050239782550800\, q^8 + \cdots, \tag{60}$$

$$1 + 32512\, q + 4618113916354560\, q^6 \\ + 39530508819558860800\, q^7 + 91474313271091383801600\, q^8 + \cdots, \tag{61}$$

$$1 + 36720\, q + 725294463264000\, q^6 \\ + 12024972400490640000\, q^7 + 49639052886027169878000\, q^8 + \cdots, \tag{62}$$

$$1 + 41184\, q + 2782224865703911680\, q^7$$
$$+ 20276905648698076116960\, q^8 + 48308020696369162417369824\, q^9 + \cdots \quad (63)$$

$$1 + 45904\, q + 4920070592521388 80\, q^7$$
$$+ 63685133754923157 35760\, q^8 + 251608059707613612 74070864\, q^9 + \cdots \quad (64)$$

$$1 + 50880\, q + 68695315762137600\, q^7$$
$$+ 156544865112655761 1200\, q^8 + 1021657090457720778 3756480\, q^9 + \cdots \quad (65)$$

$$1 + 56112\, q + 307199691863651386800\, q^8 + 3286739084675155263492912\, q^9$$
$$+ 12558003600021571078601195520\, q^{10} + \cdots, \quad (66)$$

$$1 + 61600\, q + 48122408265197508000\, q^8$$
$$+ 849707561434115641725600\, q^9 + 5072962916073882876765143040\, q^{10} + \cdots \quad (67)$$

$$1 + 67344\, q + 6170292895987270800\, q^8$$
$$+ 178700117613214653171984\, q^9 + 1663101624261219182270300160\, q^{10} + \cdots \quad (68)$$

In view of the observations explained above, it seems that only examples of quadratic forms corresponding to these examples together with the extremal theta series are of interest.

Remark 6.1. In their paper [9], C. L. Mallows, A. M. Odlyzko and N. J. A. Sloane proved that the extremal theta series has negative Fourier coefficients for sufficiently large weights. In fact, they proved that $A_k B_k(\ell + 1)$ is positive and $A_k B_k(\ell + 2)$ is negative for sufficiently large k. (We recall again that $A_k B_k(m)$ is the m-th Fourier coefficient of the extremal theta series.) This shows that for large weights the theta regions (assuming that in general one gets a bounded theta region) do not contain the origin, because the extremal theta series corresponds to the origin. For the cases considered so far, origin is the only vertex which gives the exception 1, whenever $24 \nmid n$. Since 1 is always an exception (see the remark below), in order to cope with our question (46), we feel that whenever the extremal theta series has negative Fourier coefficients (in fact, the $\ell + 2$-th Fourier coefficient is negative as mentioned above), the intersection of $a_k(\ell + 2) = 0$ with $a_k(1) = 0$ is a lattice vertex, at least for the case when $24 \nmid n$. So, in such a situation, $\ell + 2$ is also an exception. This shows that, in these cases the set $\mathbf{T_n}$ may contain $\ell + 2$ apart from the elements given in (46).

One final remark. With the help of the existing examples of quadratic forms of 24, 32, 40 and 48 variables it is possible to observe a fact about the set $\mathbf{T_n}$, which we feel is worth mentioning. It is presented in the following remark.

Remark 6.2. Using the examples of quadratic forms of $24, 32, 40$ variables (which are mentioned in section 3), which do not represent 2 (i.e., with exception 1), one can construct a quadratic form of n variables, $n \geq 24$, which does not represent 2. In other words, $\{1\} \subset \mathbf{T_n}$ whenever $n \geq 24$. Due to similar reasoning, using the examples of quadratic forms of 48 variables (refer Remark 3.4), one can conclude that $\{1, 2\} \subset \mathbf{T_n}$ whenever $48|n$.

Acknowledgements: We are grateful to the referee for making some comments which improved substancially the presentation of this article. The first author acknowledges the warm hospitality at the Mehta Research Institute during his recent visit in which the joint work was carried out. The second author thanks Professor R. Balasubramanian and Dr. M. Manickam for some fruitful discussions which detected a gap in the paper of Peters.

References

[1] K. Chakraborty, A. K. Lal and B. Ramakrishnan, *Modular forms which behave like theta series*, To appear in Math. Comp. (1997).

[2] A. G. Chernyakov, *An example of a 32-dimensional even unimodular lattice (in Russian)*, Zap. Naučn. Sem. Leningrad, Otdel. Mat. Inst. Steklov (LOMI) **86**, 170–179 (1979).

[3] J. H. Conway and V. Pless, *On the enumeration of self-dual codes*, J. Combinatorial Theory **28 A**, 26–53, (1980).

[4] P. Deligne, *La Cojecture de Weil I*, Publ. IHES **43**, 273–307 (1974).

[5] R. C. Gunning, *Lectures on modular forms*, Princeton University Press, Princeton 1962.

[6] J. S. Hsia, Y. Kitaoka and M. Kneser, *Representations of positive definite quadratic forms*, J. reine angew Math. **301**, 132–141 (1978).

[7] M. Kneser, *Quadratische Formen*, Vorlesungs-Ausarbeitung, Göttingen 1973/74.

[8] J. Leech and N. J. A. Sloane, *Sphere packings and error-correcting codes*, Canad. J. Math. **23**, 718–745 (1971).

[9] C. L. Mallows, A. M. Odlyzko and N. J. A. Sloane, *Upper Bounds for Modular Forms, Lattices, and Codes*, J. Algebra **36**, 68–76 (1975).

[10] M. Manickam and B. Ramakrishnan, *On normalized modular forms of weights $20, 24$ and 28 with non-negative integral Fourier coefficients*, Appendix to [1] above.

[11] J. McKay, *A setting for the Leech lattice*, in Finite Groups '72 (ed. by T. Gagen, M. P. Hale, Jr., and E. E. Shult), Amsterdam, 117–118 (1973).

[12] H. -V. Niemeier, *Definite quadratische Formen der Dimension 24 und Diskriminant 1*, J. Number Theory **5**, 142–178 (1973).

[13] A. M. Odlyzko and N. J. A. Sloane, *On exceptions of integral quadratic forms*, J. reine angew Math. **321**, 212–216 (1981).

[14] M. Ozeki, *On modular forms whose Fourier coefficients are non-negative integers with the constant term unity*, Math. Ann. **206**, 187–203 (1973).

[15] M. Peters, *Exceptions of integral quadratic forms*, J. reine angew Math. **314**, 196–199 (1980).

[16] S. Ramanujan, *Highly composite numbers*, Collected papers, 78–128, Cambridge 1927, Chelsea Reprint 1962.

[17] R. A. Rankin, *Modular Forms and Functions*, Cambridge, 1977.

[18] V. A. Tartakowsky, *Die Gesamt beit der Zahlen, die durch eine positive quadratische Form $F(x_1, \cdots, x_s)$ $(s \geq 4)$ darstellbar sind*, Izv. Akad. Nauk. SSSR **7**, 111–122, 165–195 (1929).

[19] B. B. Venkov, *On the classification of integral even unimodular 24-dimensional quadratic forms (in Russian)*, Trudy Mat. Inst. Steklov. **148**, 65–76, (1978).

[20] B. B. Venkov, *On odd unimodular lattices (in Russian)*, Zap. Naučn. Sem. Leningrad, Otdel. Mat. Inst. Steklov (LOMI) **86**, 40–48 (1979).

[21] E. Witt, *Eine Identität zuischen Modulformen zweiten Grades*, Abh. Math. Sem. Hansischen Univ. **14**, 323–337 (1941).

Department of Mathematics, Indian Institute of Technology, Kanpur 208 016, India.
E-mail address: arlal@iitk.ernet.in

Mehta Research Institute of Mathematics and Mathematical Physics, Chhatnag Road, Jhusi, Allahabad 221 506, India.
E-mail address: ramki@mri.ernet.in

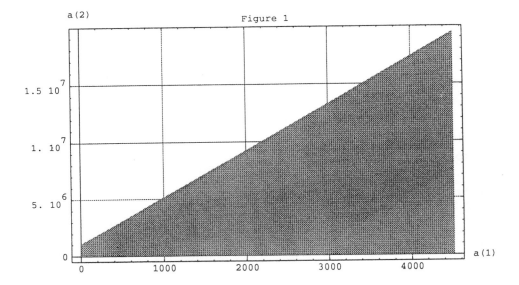

Figure 1

A brief survey on the theta correspondence

Dipendra Prasad

The following is an expanded version of the notes for a survey lecture on the theta correspondence given at Trichy in January, 1996. The aim of this lecture was to give an idea of this theory and some of the directions in which the theory has been developing. Perhaps some of the material contained in sections 2 and 3 is new.

The theory of theta correspondence gives one of the few general methods of constructing automorphic forms of groups over number fields, or admissible representations of groups over local fields. The method has its origin in the classical construction of theta functions which are modular forms–perhaps of half integral weight–on the upper half plane. Here is a special case of this:

Let q be a positive definite quadratic form on a lattice $L \subseteq \mathbb{R}^n$, $q : L \to \mathbb{Z}$. Let p be a homogeneous polynomial on $\mathbb{R}^n$ which is harmonic with respect to q, i.e. $\Delta_q p = 0$ where Δ_q denotes the Laplacian with respect to q (Δ_q is the unique homogeneous differential operator of order 2 invariant under the orthogonal group of q). Then,

$$\theta(z) = \sum_{v \in L} p(v) e^{\pi i q(v) z}$$

is a modular form on the upper half plane of weight $\frac{n}{2} + \deg p$. It is a cusp form whenever the polynomial p is of positive degree.

We refer the reader to the original papers of Howe [Ho1], [Ho2], [Ho3], of Weil [We1], [We2], the book of Gelbart [Ge1] for the one-dimensional case, the book of Moeglin-Vigneras-Waldspurger [MVW] for a comprehensive account of the local theory, and the exposition of Gelbart [Ge2] as well as an earlier exposition of the author [P1] for more details on theta correspondence.

---

[0]1991 *Mathematics Subject Classification.* Primary 11F27, 11R39, 22E50

The plan of this paper is as follows :

(i) Construction of the Weil Representation.
(ii) K-type of the Weil representation.
(iii) Character of the Weil representation.
(iv) Dual reductive pairs and the local theta correspondence.
(v) Global theta correspondence.
(vi) Towers of theta lifts.
(vii) An example of global theta lift: A theorem of Waldspurger.
(viii) Functoriality of theta correspondence.
(ix) The Siegel-Weil formula.
(x) Application to cohomology of Shimura Variety.
(xi) Recent generalisations of theta correspondence.

Acknowledgement : The work on this paper was begun during a trip of the author to MSRI, Berkeley in 1995, and was completed during a stay at Max-Planck-Institute, Bonn in 1996. The author would like to thank these Institutions for the support, and excellent working conditions which made this work possible. The author thanks B.H. Gross for posing the question on K-type of the Weil representation answered in section 2.

1 Construction of the Weil Representation

Let k be a field which is not of characteristic 2, and is either a local field or a finite field, and let W be a finite dimensional vector space over k together with a non-degenerate symplectic form:

$$W \times W \to k$$

$$\omega_1, \omega_2 \to <\omega_1, \omega_2>.$$

Define the Heisenberg group $H(W) = \{(w,t) | w \in W, t \in k\}$ with the law of multiplication :

$$(w_1, t_1) \cdot (w_2, t_2) = (w_1 + w_2, \ t_1 + t_2 + \frac{1}{2} <w_1, w_2>).$$

The Heisenberg group is a central extension of W by k :

$$0 \to k \to H(W) \to W \to 0.$$

The most important property about the representation theory of the Heisenberg group is the following uniqueness theorem due to Stone and von Neumann.

Theorem 1 *For any non-trivial character $\psi : k \to \mathbb{C}^*$, there exists a unique irreducible representation ρ_ψ of $H(W)$ on which $k \subseteq H(W)$ acts by ψ.*

Now we observe that the symplectic group $\mathrm{Sp}(W)$ operates on $H(W)$ by $g(w,t) = (gw,t)$. By the uniqueness of ρ_ψ, $\mathrm{Sp}(W)$ acts by intertwining operators on ρ_ψ. Namely, there exists $\omega_\psi(g)$, unique up to scalars, such that

$$\rho_\psi(gw,t) = \omega_\psi(g)\rho_\psi(w,t)\omega_\psi(g)^{-1}.$$

The map $g \to \omega_\psi(g)$ is a projective representation of the symplectic group $\mathrm{Sp}(W)$, and gives rise to an ordinary representation of a two fold covering of $\mathrm{Sp}(W)$. We will denote this two fold cover by $\overline{\mathrm{Sp}}(W)$, and the representation of it so obtained again by $\omega_\psi(g)$. The group $\overline{\mathrm{Sp}}(W)$ is called the metaplectic group, and the representation $\omega_\psi(g)$ the Weil or metaplectic representation. In this paper we will often abuse terminology to call the Weil representation of the metaplectic group $\overline{\mathrm{Sp}}(W)$ as simply the Weil representation of the symplectic group $\mathrm{Sp}(W)$.

Remark 1: If W is a symplectic vector space over a separable field extension K of a local field k, then $\mathrm{tr} <,>$ gives a symplectic structure on W with its k vector space structure, to be denoted by $R_{K/k}W$. If ψ is a character on k, then $\psi_K(x) = \psi(\mathrm{tr}\, x)$ gives a character on K. We have the inclusion of $\mathrm{Sp}(W)$ in $\mathrm{Sp}(R_{K/k}W)$, and it is easy to see that the restriction of the Weil representation of $\mathrm{Sp}(R_{K/k}W)$ for the character ψ of k to $\mathrm{Sp}(W)$ gives the Weil representation of $\mathrm{Sp}(W)$ for the character ψ_K of K. Moreover, if W_1 and W_2 are symplectic vector spaces, then $\mathrm{Sp}(W_1) \times \mathrm{Sp}(W_2)$ is contained in $\mathrm{Sp}(W_1 \oplus W_2)$, and the restriction of the Weil representation of $\mathrm{Sp}(W_1 \oplus W_2)$ to $\mathrm{Sp}(W_1) \times \mathrm{Sp}(W_2)$ is the tensor product of the Weil representations of $\mathrm{Sp}(W_1)$ and $\mathrm{Sp}(W_2)$. Obvious as these properties are, the reader will of course notice how special these are to the Weil representation: that their restriction to such small subgroups remains of finite length.

1.1 Explicit realisation of the Weil representation

Let $W = X \oplus Y$ where X and Y are subspaces of W on which the symplectic form is identically zero. The Weil representation of $\text{Sp}(W)$ can be realised on the Schwarz space $\mathcal{S}(X)$ which is the space of locally constant, compactly supported functions on X if k is non-Archimedean, and has the usual definition if k is Archimedean; the action of $\text{Sp}(W)$ on $\mathcal{S}(X)$ is as follows:

$$\begin{pmatrix} A & 0 \\ 0 & {}^tA^{-1} \end{pmatrix} f(X) = |\det A|^{\frac{1}{2}} f({}^tAX)$$

$$\begin{pmatrix} I & B \\ 0 & I \end{pmatrix} f(X) = \psi(\tfrac{{}^tXBX}{2}) f(X)$$

$$\begin{pmatrix} 0 & I \\ -I & 0 \end{pmatrix} f(X) = \gamma \hat{f}(X)$$

where γ is an 8-th root of unity, and $\hat{f}$ denotes the Fourier transform of f, and ψ is a non-trivial additive character of k.

Remark 2: It will be interesting to construct a model of the Weil representation which is defined over a number field. Since all the known models require the additive character ψ in an essential way, it does not seem obvious if it can be done at all. We note that the Weil representation of $\text{SL}(2)$ can be defined over a number field because it is sum of its even and odd pieces, both of which are defined over number fields: the even piece because it occurs in an explicit principal series, and the odd piece because it is induced from compact open subgroup.

2 K-type of the Weil representation

The decomposition of a representation of a p-adic group when restricted to a maximal compact subgroup is an important information used in the representation theory of p-adic groups. In this section we carry out this decomposition for the Weil representation of $\text{Sp}(W)$ where W is a symplectic vector space of dimension $2n$ over a non-Archimedean local field k. We let $<.,.>$ denote the symplectic form on W, and let ψ be the character of k of conductor 0 used in the definition of the Weil representation.

We will assume that the cardinality of the residue field of k is q

which is odd so that the metaplectic covering splits over the maximal compact subgroup $\mathrm{Sp}(\mathcal{L})$ of $\mathrm{Sp}(W)$ stabilising a lattice $\mathcal{L}$ on which the symplectic form $<.,.>$ is non-degenerate. The Weil representation of $\mathrm{Sp}(W)$ becomes a representation of $\mathrm{Sp}(\mathcal{L})$. Let $\Gamma(i)$ be the standard filtration on $\Gamma = \mathrm{Sp}(\mathcal{L})$, so that $\Gamma(0) = \mathrm{Sp}(\mathcal{L})$, and $\Gamma(i) = \ker[\mathrm{Sp}(\mathcal{L}) \to \mathrm{Sp}(\mathcal{L}/\pi^i \mathcal{L})]$.

We will prove the following theorem in this section.

Theorem 2 *The Weil representation ω of $\mathrm{Sp}(W)$ when restricted to the compact open subgroup $\mathrm{Sp}(\mathcal{L})$ decomposes as the sum of irreducible representations as follows:*

$$\omega = \omega_0 \oplus \sum_{m \geq 1} (\omega_{2m}^+ \oplus \omega_{2m}^-)$$

where ω_0 is the trivial representation of $\mathrm{Sp}(\mathcal{L})$, and $\omega_{2m}^+, \omega_{2m}^-$ are irreducible representations of $\mathrm{Sp}(\mathcal{L}/\pi^{2m}\mathcal{L})$ of dimension $\frac{1}{2}[q^{2mn} - q^{2(m-1)n}]$ for $m \geq 1$.

Proof : The proof of this theorem will be carried out in the lattice model of the Weil representation, cf. [MVW], which is most suitable for the study of K-types. We recall that the Weil representation of $\mathrm{Sp}(W)$ in this model is realised on the space $\mathcal{S}_{\mathcal{L},\psi}(W)$ of locally constant compactly supported functions on W such that

$$f(x + a) = \psi(<x, a>)f(x)$$

for all $x \in W$, and $a \in \mathcal{L}$. The action of an element g of $\mathrm{Sp}(\mathcal{L})$ on a function f in $\mathcal{S}_{\mathcal{L},\psi}(W)$ is given by

$$(g \cdot f)(x) = f(gx).$$

Let $\mathcal{S}_m \subset \mathcal{S}_{\mathcal{L},\psi}(W)$ be the subspace of functions on W which are supported on $\pi^{-m}\mathcal{L}$. From the functional equation $f(x + a) = \psi(<x, a>)f(x)$, we find that the functions in $\mathcal{S}_m$ are invariant under translation by $\pi^m \mathcal{L}$, and therefore $\mathcal{S}_m$ can be thought of as a space of functions on $\pi^{-m}\mathcal{L}/\pi^m \mathcal{L}$ with a specified property under translation by $\mathcal{L}/\pi^m \mathcal{L}$. It follows that

$$\dim \mathcal{S}_m = \sharp(\pi^{-m}\mathcal{L}/\mathcal{L}) = q^{2mn}.$$

Define $\mathcal{S}_m^+$ (resp. $\mathcal{S}_m^-$) to be the subspace of $\mathcal{S}_m$ consisting of even (resp. odd) functions. Define ω_{2m}^+ (resp. ω_{2m}^-) to be $\mathcal{S}_m^+/\mathcal{S}_{m-1}^+$ (resp. $\mathcal{S}_m^-/\mathcal{S}_{m-1}^-$). We find that

$$\dim \omega_{2m}^+ = \dim \omega_{2m}^- = \frac{1}{2}[q^{2mn} - q^{2(m-1)n}].$$

Clearly, Γ leaves $\mathcal{S}_m, \mathcal{S}_m^+, \mathcal{S}_m^-$ invariant, and it is easy to see that $\Gamma(2m)$ acts trivially on $\mathcal{S}_m$. Since $\Gamma(m)/\Gamma(2m)$ is an abelian group, it acts by characters which we now find. Let $g = 1 + \pi^m \gamma \in \Gamma(m), f \in \mathcal{S}_m$, and $x \in \pi^{-m}\mathcal{L}$, then

$$\begin{aligned}(g \cdot f)(x) &= f(gx) \\ &= f(x + \gamma \pi^m x) \\ &= \psi(<x, \gamma \pi^m x>)f(x) \\ &= \psi(<x, gx>)f(x).\end{aligned}$$

It follows that the action of $\Gamma(m)/\Gamma(2m)$ on $\mathcal{S}_m$ is via the characters $g \to \psi(<x, gx>)f(x)$ where $x \in \pi^{-m}\mathcal{L}$. We denote this character of $\Gamma(m)/\Gamma(2m)$ by ψ_x, so $\psi_x(g) = \psi(<x, gx>)$. It is easy to see that $\psi_x = \psi_y$ if $x \equiv y \mod \mathcal{L}$, or if $x \equiv -y \mod \mathcal{L}$. Conversely, if $\psi_x = \psi_y$ then either $x \equiv y \mod \mathcal{L}$, or $x \equiv -y \mod \mathcal{L}$. Moreover, $\psi_{Ax}(g) = \psi_x(A^{-1}gA)$, for $A \in \text{Sp}(\mathcal{L})$. It is a well-known fact that $\text{Sp}(\mathcal{L})$ acts transitively on vectors in $\pi^{-m}\mathcal{L}$ which do not belong to $\pi^{-m+1}\mathcal{L}$. Therefore, if ψ_x appears in an irreducible representation of $\text{Sp}(\mathcal{L})$ for one value of x in $\pi^{-m}\mathcal{L}$ not belonging to $\pi^{-m+1}\mathcal{L}$, ψ_y for all values of y in $\pi^{-m}\mathcal{L}$ not in $\pi^{-m+1}\mathcal{L}$ appears. Since $\psi_x = \psi_y$ if and only if $x \equiv y \mod \mathcal{L}$, or $x \equiv -y \mod \mathcal{L}$, we find that the dimension of an irreducible representation of $\text{Sp}(\mathcal{L})$ containing ψ_x is divisible by $\frac{1}{2}[q^{2mn} - q^{2(m-1)n}]$. Since $\omega_{2m}^+, \omega_{2m}^-$ are representations of $\text{Sp}(\mathcal{L})$ of dimension $\frac{1}{2}[q^{2mn} - q^{2(m-1)n}]$, these must therefore be irreducible, proving the decomposition

$$\omega = \omega_0 \oplus \sum_{m \geq 1}(\omega_{2m}^+ \oplus \omega_{2m}^-).$$

3 Character of the Weil representation for SL(2)

In this section we calculate the character of the Weil representation of $SL(2, k)$ for a non-Archimedean local field k. Because of the relation of the character formula obtained in this case to transfer factors, it would be very interesting to calculate the character of the Weil representation of the general symplectic group. For some work along this direction in the Archimedean case, see [Ad2].

The calculation of the character will be done in 2 steps. The first step consists in identifying those characters of K^1 (the subgroup of norm 1 elements of a quadratic extension K of k) which appear in the Weil representation. The answer is in terms of an epsilon factor. This is a result due to Rogawski [Ro1], based on an earlier work of Moen. Next we need to add all the characters of K^1 which appear in the Weil representation to find the character of the Weil representation restricted to K^1. This kind of summation of characters was done by the author (for a different purpose!) in [P2]. We set up some notation now.

Let K be a quadratic field extension of the local field k with $\omega_{K/k}$, the quadratic character of k^* associated to K by local class field theory. The group K^1 of norm 1 elements of K is contained in $SL(2,k)$, and the metaplectic cover of $SL(2,k)$ splits over K^1. However, there is no natural choice for this splitting. For each character χ of K^* whose restriction to k^* is $\omega_{K/k}$, one can construct such a splitting. Let ψ be an additive character of k and δ be an element of K^* with $\text{tr}(\delta) = 0$. This choice of δ gives rise to a symplectic structure on the 2-dimensional k vector space $K : x, y \to \text{tr}(\delta x \bar{y})$. Let $\omega(\psi, \delta)$ denote the Weil representation of $SL(2,k)$ associated to the additive character ψ and this symplectic structure on K. Let $\omega(\chi, \psi, \delta)$ denote the restriction of $\omega(\psi, \delta)$ to K^1 via the splitting given by χ. The following theorem is due to Rogawski [Ro1, Prop. 3.4].

Theorem 3 *Let ϕ be a character of K^1. Then ϕ appears in $\omega(\chi, \psi, \delta)$ if and only if*

$$\epsilon(\phi_K \chi^{-1}, \psi_K) = \phi(-1)\chi(2\delta)$$

where $\phi_K(x) = \phi(x/\bar{x}), \psi_K(x) = \psi(\text{tr } x)$.

The following theorem is Lemma 3.1 of [P2] proved there only for odd residue characteristic.

Theorem 4 *For the character ψ_0 of K defined by $\psi_0(x) = \psi(tr[-\frac{\delta x}{2}])$,* we have

$$\sum_{\substack{\epsilon(\nu,\psi_0)=1 \\ \nu|_{k^*}=\omega_{K/k}}} \nu(x) = \epsilon(\omega_{K/k},\psi)\frac{\omega_{K/k}\left(\frac{x-\bar{x}}{\delta-\bar{\delta}}\right)}{\left|\frac{(x-\bar{x})^2}{x\bar{x}}\right|_k^{\frac{1}{2}}}$$

where, as is usual, the summation on the left is by partial sums over all characters of K^* of conductor $\leq n$.

Since $\epsilon(\phi_K\chi^{-1},\psi_K) = \phi(-1)\chi(2\delta)\epsilon(\phi_K\chi^{-1},\psi_0)$, it is easy to see that the above two theorems imply the following.

Theorem 5 *The character of the Weil representation $\omega(\chi,\psi,\delta)$ at an element $y = x/\bar{x} \in K^1$ is equal to*

$$\epsilon(\omega_{K/k},\psi)\chi(-x)\frac{\omega_{K/k}\left(\frac{x-\bar{x}}{\delta-\bar{\delta}}\right)}{\left|\frac{(x-\bar{x})^2}{x\bar{x}}\right|_k^{\frac{1}{2}}}.$$

Remark 1: We note that any 2 of the theorems of this section implies the third. Theorem 3 is valid for any residue characteristic, but Theorem 4 is known only for odd residue characteristic, and therefore Theorem 5 is available only in odd residue characteristic.

Remark 2 : One can use Remark 2 of section 1 together with the character formula for the Weil representation of $SL(2)$ to calculate the character of the Weil representation at many elements of $\mathrm{Sp}(n)$ for general n. This presumes of course that, in the notation of that remark, the restriction of the character of $\mathrm{Sp}(R_{K/k}W)$ to $\mathrm{Sp}(W)$ is the character of the Weil representation of $\mathrm{Sp}(W)$, which can be justified.

4 Dual reductive pairs and the local theta correspondence

Definition: A pair of subgroups (G_1, G_2) in $\mathrm{Sp}(W)$ is called a dual reductive pair if

1. $Z(G_1) = G_2$, and $Z(G_2) = G_1$ where $Z(G_1)$ (resp. $Z(G_2)$) denotes the centraliser of G_1 (resp. G_2) in $\mathrm{Sp}(W)$.

2. G_1 and G_2 are reductive groups, i.e., any G_1 invariant subspace of W has a G_1-invariant complement; similarly for G_2.

We refer to [MVW] and [P1] for a detailed discussion on dual reductive pairs. Here we only note the following examples.

(i) Let V be an orthogonal space (i.e. a finite dimensional vector space over k with a non-degenerate quadratic form), and let W be a symplectic space. Then $V \otimes W$ is a symplectic space in natural way, and we have a map

$$O(V) \times \text{Sp}(W) \to \text{Sp}(V \otimes W).$$

The pair $(O(V), \text{Sp}(W))$ is a dual reductive pair.

(ii) Let K be a quadratic extension of k, V a Hermitian and W a skew-Hermitian space over K. Then the k-vector space $V \otimes_K W$ is naturally a symplectic space under the pairing

$$< v_1 \otimes w_1, v_2 \otimes w_2 > = \text{tr}_{K/k}(< v_1, v_2 >< w_1, w_2 >)$$

and gives rise to a dual reductive pair $(U(V), U(W))$ in $\text{Sp}(V \otimes_K W)$.

For a dual reductive pair (G_1, G_2) in $\text{Sp}(W)$, let $\bar{G}_1$ be the inverse image of G_1 in $\overline{\text{Sp}}(W)$, and $\bar{G}_2$ the inverse image of G_2. The groups $\bar{G}_1$ and $\bar{G}_2$ are known to commute. The Weil representation ω_ψ of $\overline{\text{Sp}}(W)$ can therefore be restricted to $\bar{G}_1 \times \bar{G}_2$. Let π be an irreducible representation of $\bar{G}_1$. Define

$$A(\pi) = \omega_\psi / \cap \text{Ker}\phi : \phi \in \text{Hom}_{\bar{G}_1}(\omega_\psi, \pi).$$

$A(\pi)$ is a smooth representation of $\bar{G}_1 \times \bar{G}_2$, and can be written as $A(\pi) = \pi \otimes \theta_0(\pi)$ for a smooth representation $\theta_0(\pi)$ of $\bar{G}_2$.

The following theorem is due to Waldspurger [Wa2] building on the earlier work of Howe.

Theorem 6 *The representation $\theta_0(\pi)$ is of finite length, and, if the residue characteristic of k is not 2, it has a unique irreducible quotient.*

Notation : The unique irreducible quotient of $\theta_0(\pi)$ is called the local theta lift of π, and is denoted by $\theta(\pi)$. We warn the reader that the representation ω_ψ and hence $\theta(\pi)$ depends on the choice of the additive character $\psi : k \to \mathbb{C}^*$.

Examples:

(i) For $V = K$, a quadratic extension of k thought of as a 2-dimensional vector space over k with the norm form as the quadratic form, $SO(V) = K^1$, and the theta lift from $O(V)$ to $SL(2)$ can be used to construct representations of $GL(2)$ from characters of K^*. This construction is due to Shalika and Tannaka.

(ii) For $V = D$, the quaternion division algebra over k, together with the reduced norm as the quadratic form, $SO(V)$ is essentially $D^1 \times D^1$, and the theta lift from $SL(2)$ to $D^1 \times D^1$ is related to the Jacquet-Langlands correspondence between discrete series representations of $GL(2)$ and representations of D^*.

Remark : The question about the field of definition of the Weil representation (cf. Remark 2 at the end of section 1) is interesting also because one could use it to give a field of definition to $\theta(\pi)$ in terms of a field of definition for π.

5 Global Theta Correspondence

Let $\mathbf{A}$ be the adele ring of a number field F. Let $\psi \colon \mathbf{A}/F \to \mathbb{C}$ be a non-trivial character. Let $W = X \oplus Y$ be a symplectic vector space over F with X and Y maximal isotropic subspaces. We have $\mathcal{S}(X(\mathbf{A})) = \otimes_v \mathcal{S}(X(F_v))$, and there is a pojective representation of $\mathrm{Sp}(W)(\mathbf{A})$ on it by taking the tensor product of local representations. This projective representation of $\mathrm{Sp}(W)(\mathbf{A})$ becomes an ordinary representation of a 2-fold cover $\overline{\mathrm{Sp}}(W)(\mathbf{A})$ of $\mathrm{Sp}(W)(\mathbf{A})$.

The following theorem is due to Weil [We1].

Theorem 7 *The covering* $\overline{\mathrm{Sp}}(W)(\mathbf{A}) \to \mathrm{Sp}(W)(\mathbf{A})$ *splits over* $\mathrm{Sp}(W)(F)$.

Because of this theorem, $\mathrm{Sp}(W)(F)$ operates on $\mathcal{S}(X(\mathbf{A}))$. Define a distribution θ on $\mathcal{S}(X(\mathbf{A}))$ by $\theta(\phi) = \sum_{x \in X(F)} \phi(x)$. This distribution is $\mathrm{Sp}(W)(F)$-invariant. Therefore, the function $g \to \theta(g \cdot \phi) = \theta_\phi(g)$ defines a function $\theta_\phi : \mathrm{Sp}(W)(F) \backslash \overline{\mathrm{Sp}}(W)(\mathbf{A}) \to \mathbb{C}$ These are called theta functions. They are slowly increasing and therefore automorphic. For an appropriate choice of ϕ on $\mathbf{A}$, this gives the adelic analogue of the classical theta function of weight $= \frac{1}{2}$ on the upper half plane given by
$$\theta(z) = \sum_{n \in \mathbb{Z}} e^{\pi i n^2 z}.$$

Now let (G_1, G_2) be a dual reductive pair in $\mathrm{Sp}(W)$ which is defined over F. Let π_1 be a cuspidal representation of G_1, realised on a space of cusp forms on $G_1(F)\backslash G_1(\mathbf{A})$. For a function $\phi \in \mathcal{S}(X(\mathbf{A}))$ and $f \in \pi_1$, define

$$\theta_\phi(f)(g_2) = \int_{G_1(F)\backslash G_1(\mathbf{A})} \theta_\phi(g_1,g_2) f(g_1) dg_1.$$

$\theta_\phi(f)$ is an automorphic form on G_2, and is called the theta lift of f.

Example: The theta function $\theta(z) = \sum_{v \in L} p(v) e^{\pi i q(v)}$ defined in the introduction is such a lift from $O(V)$ to $SL(2)$ where V is the quadratic space $\mathbb{R}^n$ with q as the qudratic form.

Basic questions for the global theta lift are

(i) When is the space generated by $\theta_\phi(f)$ for $\phi \in \mathcal{S}(X(\mathbf{A}))$ and f in the representation space π_1 not identically zero?

(ii) What is the relation of the space spanned by $\theta_\phi(f)$ to local theta correspondence?

As regards (ii) we have the following due to Rallis [Ra1].

Theorem 8 *If $\theta_\phi(f)$ consists of cusp forms, then the space generated by them is irreducible and is $\otimes \theta_v(\pi_v^\vee)$ where $\pi_v^\vee$ denotes the contragredient of π_v.*

Rallis has a theory about when $\theta_\phi(f)$ is cuspidal in terms of "towers of theta lifts", which we review next; this theory also partially answers part (i) of the above question (see Theorem 5(ii) below).

6 Towers of theta lifts

Let V be an even dimensional quadratic space over a number field which does not represent any zero. Let $\mathbb{H}$ be the two dimensional hyperbolic quadratic space (with quadratic form XY). The "towers of theta lifts" consists in looking at theta lifts from $\mathrm{Sp}(W)$ to various $O(V + n\mathbb{H})$. The following theorem is due to S. Rallis [Ra1].

Theorem 9 *(i) Let π be a cuspidal automorphic representation of $\mathrm{Sp}(W)$, and $\theta_n(\pi)$ its theta lift to $O(V + n\mathbb{H})$. Let n_0 be the smallest integer ≥ 0 such that $\theta_{n_0}(\pi) \neq 0$. Then $\theta_{n_0}(\pi)$ is cuspidal, and $\theta_n(\pi)$, $n \geq n_0$, are non-zero, and never cuspidal for $n > n_0$.*

(ii) $\theta_n(\pi) \neq 0$ for $n \geq \dim W$.

There is an analogous statement in which W is varying in the tower $\mathbb{H}, 2\mathbb{H}, \ldots$ ($\mathbb{H}$ now hyperbolic symplectic), or for dual reductive pairs consisting of unitary groups.

Remark: Let I_j be the space of cusp forms on $\text{Sp}(W)$ whose theta lifts to $O(V + j\mathbb{H})$ is non-zero but is zero to $O(V + i\mathbb{H})$, $i < j$. Then the theorem above gives an interesting decomposition of the space S of cusp forms on $\text{Sp}(W)$ as

$$S = I_0 \oplus I_1 \oplus \ldots I_{2n}$$

where $\dim W = 2n$. The cusp forms in different I_j's behave quite differently.

The theory of towers of theta lifts works in the local case also when one replaces the word cuspidal by supercuspidal, cf. the work of Kudla in [Ku]. This work of Kudla in [Ku] determines the principal series in which $\theta_n(\pi)$ lies as a subquotient in terms of $\theta_{n_0}(\pi)$, and is based on the calculation of the Jacquet functor of the Weil representation of $\text{Sp}(V \otimes W)$ with respect to the unipotent radical of maximal parabolics of $O(V)$ and $\text{Sp}(W)$.

Recently Harris, Kudla and Sweet have used these towers of theta lifts to construct a very interesting infinite family of supercuspidals on Unitary groups of dimension tending to infinity from one supercuspidal. We review this work briefly.

Let V and W be two Hermitian spaces over a quadratic extension K of a local field k. Let δ be a non-zero element of K whose trace to k is 0. Multiplication by δ turns a Hermitian space into a skew-Hermitian space, and therefore $(U(V), U(W))$ form a dual reductive pair in $\text{Sp}(V \otimes_K W)$. By [HKS] the metaplectic cover of $\text{Sp}(V \otimes_K W)$ splits over $U(V) \times U(W)$, the splitting depending on choice of characters χ_1, χ_2 of K^* such that $\chi_1|_{k^*} = \omega_{K/k}^{\dim W}$, and $\chi_2|_{k^*} = \omega_{K/k}^{\dim V}$. Here $\omega_{K/k}$ is the quadratic character of k^* associated to the quadratic extension K of k by local class field theory.

Note that there are two isomorphism classes of Hermitian or skew-Hermitian spaces of a given dimension depending on their discriminant over a non-archimedean local field.

The following theorem is due to Harris, Kudla and Sweet [HKS]. Here we fix the splitting of the metaplectic cover of $Sp(V \otimes_K W)$ over $(U(V), U(W))$ for the choice of the characters χ_1, χ_2 of K^* with $\chi_1 = \chi_2 = \chi$.

Theorem 10 *(i) Suppose that π is a supercuspidal representation of $U(V)$. Then the theta lift of π to $U(W)$ is non-zero for exactly one Hermitian space W with $\dim(W) = \dim(V)$.*

(ii) Given a Hermitian space W with $\dim(W) = \dim(V)$, the theta lift of π to $U(W)$ is non-zero if and only if

$$\omega_{K/k}(\mathrm{disc}V) = \epsilon(BC(\pi) \otimes \chi, \psi_K)\omega_\pi(-1)\chi(\delta^{-n})\omega_{K/k}(\mathrm{disc}W),$$

where $\epsilon(BC(\pi)\otimes\chi, \psi_K)$ is defined and studied via the "doubling-method" in [HKS].

This theorem implies that starting with a supercuspidal representation π on $U(V)$, one can construct a supercuspidal representation on $U(W \oplus n\mathbb{H})$ for $n > 0$, $\dim W = \dim V$, and the space W has the property that the theta lift of π to $U(W)$ is zero. Repeating the process one gets an infinite family of supercuspidal representations on unitary groups of increasing dimensions. This theorem of Harris, Kudla and Sweet also generalises Theorem 3 due to Rogawski.

7 An example of global theta lift: A theorem of Waldspurger

The theorem of Waldspurger [Wa1] completely describes the global theta correspondence between $PGL(2)$ and $\overline{SL}(2)$, and is proto-type of theorems expected in general.

Theorem 11 *(i) For a cuspidal automorphic representation π of $PGL(2)$, $\theta_\pi \neq 0$ on $\overline{SL}(2)$ if and only if*

$$L(\pi, \frac{1}{2}) \neq 0.$$

(ii) For an automorphic representation π of $\overline{SL}(2)$, $\theta_\pi \neq 0$ if and only if π has a ψ-Whittaker model.

(iii) For an automorphic representation π on $PGL(2)$, $\otimes_v \theta_v(\pi_v)$ is automorphic if and only if the sign in the functional equation for the L-function of π, $\varepsilon(\pi, \frac{1}{2}) = 1$.

Remark : For generalisations of Waldspurger's theorem above, see the book of Rallis [Ra3], and also a recent paper of Furusawa [Fu].

8 Functoriality of theta correspondence

The basic question about local theta correspondence is to classify those representations π_1 of G_1 for which $\theta(\pi_1) \neq 0$, and then to understand $\theta(\pi_1)$ which is a representation of G_2 in terms of the representation π_1 of G_1. Here, for simplicity, we assume that the metaplectic cover of the symplectic group splits over $G_1 \times G_2$, and that we have fixed such a splitting to regard the Weil representation of the metaplectic group as a representation of $G_1 \times G_2$, to define the theta correspondence between representations of G_1 and G_2.

We recall that according to conjectures of Langlands, representations of a reductive group G over a local field k are parametrised by certain homomorphisms of the Weil-Deligne group W_k' of k into a complex group associated to G, called the L-group of G and denoted by ^LG :

$$\phi : W_k' \to {}^LG.$$

This conjecture is known in the Archimedean case and in some other cases. The conjecture implies in particular that if there is a map between L-groups $^LG_1 \to {}^LG_2$, then there is a way of associating representations of G_2 to representations of G_1. Correspondence between representations of two groups arising in this way is said to be "functorial".

We refer to the work of Rallis [Ra2] in this regard which proves the functoriality of the theta correspondence for spherical representations. (Actually, the theta correspondence for spherical representations is not given by a mapping of the L-groups, but a simple modification of a mapping of L-groups.) When the groups G_1 and G_2 are of "similar" size then again the theta lifting seems functorial. There are some conjectures and evidences on this by Adams in the Archimedean case [Ad1], which are refined and extended by this author [P3], [P4]. However theta correspondence in general does not respect functoriality; see [Ad1], and [HKS].

We also refer to the recent work of Gelbart, Rogawski and Soudry [GRS] for rather complete information about theta lifting from $U(2)$ to $U(3)$ both locally and globally and its relation to functoriality.

9 The Siegel-Weil Formula

The aim of this section is to describe the Siegel-Weil formula which is at the heart of many applications of theta correspondence. It relates the integral of a theta function to an Eisenstein series. The simplest case of the Siegel-Weil formula is the identity

$$\sum \frac{\theta_Q(z)}{\sharp(\mathrm{Aut}Q)} = E_n(z)$$

where Q is the set of integral positive definite even unimodular quadratic forms in $2n$ variables, and $E_n(Z)$ is the Eisenstein series of weight n for $SL_2(\mathbf{Z})$.

We will be working with a number field F in this section. Let $G = \mathrm{Sp}(W)$ be the symplectic group of rank n, and $H = O(V)$ be the orthogonal group of a quadratic space V over F which we assume has even dimension $= 2m$. Then G and H form a dual reductive pair inside $\mathrm{Sp}(V \otimes W)$. Since the dimension of V is assumed to be even, the metaplectic cover of $\mathrm{Sp}(V \otimes W)$ splits over $G \times H$ both locally and globally. Weil representation of $\mathrm{Sp}(V \otimes W)$ therefore gives rise to a representation ω of $G(\mathbf{A}) \times H(\mathbf{A})$ on the Schwartz space $\mathcal{S}(V(\mathbf{A})^n)$. For $\phi \in \mathcal{S}(V(\mathbf{A})^n)$, define the theta function as usual by

$$\theta(g, h; \phi) = \sum_{x \in V(F)^n} \omega(g)\phi(h^{-1}x).$$

It is a theorem of Weil [We2] that the integral

$$I(g, \phi) = \int_{H(F)\backslash H(\mathbf{A})} \theta(g, h; \phi) dh,$$

converges if either V is an-isotropic (i.e., V does not represent any zero over F), or if r is the dimension of the maximal isotropic subspace of V, then $m > n + r + 1$.

For $s \in \mathbf{C}$, let $\Phi(g, s)$ be the function on $G(\mathbf{A})$ defined by

$$\Phi(g, s) = (\omega(g)\phi)(0)|a(g)|^{s-s_0}.$$

Here if $g = pk$ where $p \in P(\mathbf{A})$ for P the Siegel parabolic with $GL(n)$ as its Levi subgroup embedded as $\begin{pmatrix} X & 0 \\ 0 & {}^t X^{-1} \end{pmatrix}$ inside $\mathrm{Sp}(2n)$, and $k \in K = \prod_v K_v$ a standard maximal compact subgroup of $\mathrm{Sp}(n)(\mathbf{A})$, then $a(g) = \det X$.

Let $(.,.)$ denote the Hilbert symbol of the number field F. For the quadratic space V with discriminant $d(V)$, let χ_V denote the character of the idele class group of F defined by $\chi_V(x) = (x, (-1)^m d(V))$. Let $I_n(s, \chi_V)$ denote the principal series representation of $\text{Sp}(n)(\mathbf{A})$ induced from the character $g \to |a(g)|^s \chi_V(a(g))$ of the Siegel parabolic subgroup. It is easy to see that the map sending ϕ to $\Phi(g, s_0)$ defines a $G(\mathbf{A})$ intertwining map from $\mathcal{S}(V(\mathbf{A})^n)$ to $I_n(s_0, \chi_V)$. Define now the Eisenstein series by

$$E(g, s, \Phi) = \sum_{\gamma \in P(k) \backslash G(k)} \Phi(\gamma g, s).$$

The above series is absolutely convergent for $Re(s) > \frac{n+1}{2}$, and has analytic continuation to all of complex plane; it is known to be holomorphic at $s_0 = \frac{m-n-1}{2}$.

The identity expressed in the following theorem is called the Siegel-Weil formula.

Theorem 12 *Assume that either V is an-isotropic over F, or if r is the dimension of the maximal isotropic subspace of V, then $m > n + r + 1$ so that the integral defining $I(g, \phi)$ is absolutely convergent. Let $s_0 = \frac{m-n-1}{2}$, then with the notation as above, we have*

$$E(g, s_0, \Phi) = \kappa I(g, \phi)$$

where $\kappa = 2$ if $m \leq n+1$, and 1 otherwise.

This theorem was proved by Weil in [We2] when $m > 2n+2$, i.e., the case in which the Eisenstein series involved converged absolutely; the general case of the Siegel-Weil formula, including the holomorphicity of the Eisenstein series at s_0 is due to Kudla and Rallis [KR1], [KR2]. Weil proved it for general dual reductive pairs in [We2], and not just for the $(O(n), \text{Sp}(m))$ case considered here. The extension of this formula to other dual reductive pairs where the Eisenstein series does not converge absolutely is not yet complete.

Remark : The local case of the map $\phi \to \Phi$ defined by

$$\Phi(g, s_0) = (\omega(g)\phi)(0),$$

is also very important. By a theorem of Rallis [Ra1], this induces an injection from $\mathcal{S}(V^n)_{O(V)}$ into $I_n(s_0, \chi_V)$. (Where for a representation

X of a group E, X_E denotes the maximal quotient of X on which E acts trivially.) The method of Weil representation gives an important method for the study of the Jordan-Hölder series of $I_n(s_0, \chi_V)$. For instance, Kudla and Rallis [KR3] prove that if $m \leq n+1$ then the image $R_n(V)$ of $\mathcal{S}(V^n)$ in $I_n(s_0, \chi_V)$ is irreducible, and if V_1 and V_2 are the two distinct quadratic spaces with the same discriminant (if $m = 2$, and the discriminant $= -1$, then there is only one quadratic space), then $R_n(V_1)$ and $R_n(V_2)$ are distinct irreducible submodules of $I_n(s_0, \chi)$ ($\chi = \chi_{V_1} = \chi_{V_2}$) such that $I_n(s, \chi)/[R_n(V_1) + R_n(V_2)]$ is also irreducible, and non-zero if and only if $m < n+1$. If $n+1 < m \leq 2n$ (if $\chi = 1$, we exclude the value $m = 2n$; see [KR3] for this case), then $R_n(V_1)$ and $R_n(V_2)$ are maximal submodules of $I_n(s_0, \chi)$, and $R_n(V_1) \cap R_n(V_2)$ is irreducible. These results on Jordan-Holder series in turn find important application to theta liftings as in the work of [HKS] in the unitary case.

10 Application of theta correspondence to cohomology of Shimura variety

A Shimura variety is a topological space of the form $\Gamma \backslash G/K$ where G is a real Lie group, K a maximal compact subgroup of G, and Γ an arithmetic subgroup of G (An example : $SL_2(\mathbb{Z}) \backslash SL(2, \mathbb{R})/SO(2)$).

According to a theorem of Matsushima, if $\Gamma \backslash G/K$ is compact, then

$$H^i(\Gamma \backslash G/K, \mathbb{C}) = \oplus_\pi H^i(\mathcal{G}, K, \pi)^{m(\pi)}$$

where $L^2(G \backslash \Gamma) = \sum m(\pi) \pi$.

Therefore the calculation of the cohomology of a Shimura variety is equivalent to finding those automorphic representations whose component at infinity have non-trivial $(\mathcal{G}, K)$ cohomology.

The construction of cohomological automorphic representations of G depends on realising G as a member of a dual reductive pair (G, G') with G' small, and lifting automorphic representation of G' which are discrete series at infinity. Cuspidal automorphic representations of a group with prescribed discrete series component at infinity are known to exist in plenty according to a theorem of Savin, and then the problem splits into two parts:

(i) Realisation of cohomological representations as theta lifts at infinity.

(ii) Determination of the condition under which the global theta lift is non-zero.

For (i), one knows by Vogan and Zuckermann [V-W] an explicit description of cohomological representations. The theta lifts are calculated via the method of Jacquet functors, [Li1].

For (ii), observe that $\theta_\phi(f) \neq 0$ if and only if the inner product $<\theta_\phi(f), \theta_\phi(f)> \neq 0$. There is a formula for this inner product by Rallis [Ra3] involving special values of L-functions. Therefore we know it to be non-zero when the special value is at a point of absolute convergence. Actually, there are finitely many "bad" factors which might contribute a zero and which one needs to deal with also; see [Li2], and [KR4] in a greater generality. The formula of Rallis generalises Waldspurger's result mentioned before, and is proved by expressing an integral of theta functions by an Eisenstein series by the Siegel-Weil formula, and then unfolding the integral.

Non-vanishing theorems about cohomology of Shimura variety via the method of theta correspondence were first proved by Kazhdan for $SU(n,1)$, and the most recent results are due to J.-S. Li. Here is an example from the work of Li ([Li2], corollary 1.3).

Theorem 13 *Let $G = SO(p,q)$, Γ an arithmetic subgroup of G constructed by a skew-Hermitian form over a quaternion division algebra over a totally real field such that D is split at all but one real place. Then for $p \geq q$, $p + q = 2n \geq 8$, and for Γ deep enough, the qth Betti number of Γ is non-zero.*

11 Recent generalisations of theta correspondence

There has been much activity recently in constructing analogues of the Weil representation, and understanding the associated theta correspondence.

The characteristic property of the Weil representation taken for the purposes of this generalisation is the fact that the Weil representation is a rather small representation, in fact the smallest representation after the trivial representation. We will make this concept precise, but one can get some idea about it from the dimension formula for the Weil representation over finite field $\mathbf{F}_q$. The dimension of the Weil representation of $\mathrm{Sp}(2n, \mathbf{F}_q)$ is q^n, and it splits into two irreducible representation of dimension $\frac{q^n+1}{2}$ and $\frac{q^n-1}{2}$, whereas the dimension of a generic representation of $\mathrm{Sp}(2n, \mathbf{F}_q)$ is of the order of q^{n^2}.

We now make precise the concept of smallness of a representation. For an irreducible representation π of G, define a distribution Θ_π on the Lie algebra $\mathcal{G}$ of G as follows :

$$\Theta_\pi(f) = \mathrm{tr}\left(\int_{\mathcal{G}} f(X)\pi(expX)dX\right)$$

Theorem 14 (Harish-Chandra) *For each nilpotent orbit θ, there exists a complex number C_θ such that for all functions f with small support around the origin of $\mathcal{G}$,*

$$\Theta_\pi(f) = \sum C_\theta \int \hat{f}\mu_\theta,$$

the summation is over the (finite) set of nilpotent orbits, and where μ_θ is a G-invariant measure on the nilpotent orbit θ suitably normalised.

Definition:
(1) Wave front set $WF(\pi) = \cup_{C_\theta \neq 0} \bar{\theta}$.
(2) Gelfand-Kirillov dimension of π, denoted $\dim d(\pi)$ is $\max_{C_\theta \neq 0} \frac{1}{2} \dim \theta$.

Proposition 1 *Let $G(n)$ be the principal congruence subgroup of G of level n. Then for an irreducible admissible representation V of G,*

$$\lim_{n\to\infty} \frac{\dim V^{G(n)}}{q^{nd(\pi)}} = 1.$$

Definition : A representation π is called minimal if $WF(\pi) = \bar{\theta}_{min}$ where θ_{min} is a non-trivial minimal nilpotent orbit. (Such a nilpotent orbit is unique over the algebraic closure.)

Minimal representations have been constructed by many people starting with the work of Kazhdan, by Kazhdan, Savin, Gross-Wallach among others. Here is an example. Consider the Satake parameter corresponding to:

$$F^* \xrightarrow{\psi} SL(2, \mathbb{C}) \xrightarrow{\phi_0} {}^LG$$

where $\psi : x \to \begin{pmatrix} |x|^{\frac{1}{2}} & 0 \\ 0 & |x|^{-\frac{1}{2}} \end{pmatrix}$, and ϕ_0 corresponds to the subregular unipotent orbit. The following theorem is due to G. Savin [Sa].

Theorem 15 *Let G be a simple, simply connected, simply laced split group over a local field F. Then the spherical representation π with the above as Satake parameter is minimal.*

One can classify dual reductive pairs for general groups, cf. [Ru], and in each case when minimal representations are constructed, one would like to prove if the analogue of the Howe duality conjecture is true. Next one would like to prove that the local minimal representations are part of global automorphic representations, and if so, this gives a construction of global automorphic representation on one member of a dual reductive pair in terms of the other just as in the case of the Weil representation. This is a very active area of research at the moment. There are several recent papers of J-S. Li on these exceptional theta correspondence, cf. [Li3], [Li4]. Here is an example from the work of Gross and Savin[G-S].

The pair $(G_2, PG\mathrm{Sp}(6))$ is a dual reductive pair in E_7. There exists a form of G_2 over $\mathbf{Q}$ which is compact at infinity, and split at all the finite places. The pair $(G_2, PG\mathrm{Sp}(6))$ sits inside a form of E_7 which is of rank 3 at ∞ and split at all finite places. There is an automorphic form on E_7 constructed by H. Kim [Ki] which is minimal at all places. The theta lifts from automorphic form on G_2 to $PG\mathrm{Sp}(6)$ using this is expected to correspond to the mapping of L-groups given by $G_2(\mathbb{C}) \to \mathrm{Spin}7(\mathbb{C})$, as proved in [G-S] for many representations. The interest of this work of Gross and Savin is that they are able to lift automorphic forms from G_2 to $PG\mathrm{Sp}(6)$ using this "exceptional" theta correspondence, and the automorphic form they get on $PG\mathrm{Sp}(6)$ is of holomorphic kind, and so presumably has a motive associated to it, whose Galois group is G_2 (and not smaller as they take care to lift only those automorphic representations which have a Steinberg representation at a finite place, and for which the Satake parameter at some unramified finite place does not lie in the SL_2 corresponding to the Steinberg).

References

[Ad1] J.Adams,"L-functoriality for Dual Pairs", Astrisque 171-172, Orbites Unipotentes et Representations, vol.2(1989), pp. 85-129.

[Ad2]   J.Adams, "Lifting of characters on orthogonal and metaplectic groups," preprint.

[Fu]   M. Furusawa, "On the theta lift from $SO(2n+1)$ to $\tilde{Sp}(n)$," Crelle 466, 87-110 (1995).

[Ge1]   S.Gelbart, "Weil's Representation and the Spectrum of the Metaplectic group", LNM 530, Springer-Verlag (1975).

[Ge2]   S.Gelbart, "On Theta series liftings for Unitary groups", CRM proceedings and lecture notes, vol 1, edited by Ram Murty, (1993).

[GR]   S.Gelbart and J.Rogawski, "L-functions and Fourier-Jacobi coefficients for the unitary group $U(3)$," Inv. math 105, 445-472 (1991).

[GRS]   S.Gelbart, J.Rogawski and D.Soudry, "Endoscopy, theta-liftings, and period integrals for the unitary group in three variables", to appear in the Annals of Maths.

[GS]   B.H.Gross and G. Savin, "Motives with Galois groups of type G_2: An exceptional theta correspondence," preprint.

[H-K-S] M.Harris, S.Kudla and J.Sweet, "Theta dichotomy for unitary groups," to appear in the journal of the American Mathematical society.

[Ho1]   R.Howe, "Θ-series and invariant theory", In: Proc. Symp. Pure Math., vol 33, part 1, AMS, 1979, pp. 275-286.

[Ho2]   R.Howe, "Transcending classical invariant theory", J. of the Amer. Math. Soc. vol.2, no.3 (1989), pp. 535-552.

[Ho3]   R.Howe, "Remarks on classical invariant theory", Transactions of AMS 313, vol 2 (1989), pp. 539-570.

[Ki]   H. Kim, "Exceptional modular form of weight 4 on an exceptional domain contained in $\mathbf{C}^{27}$", Rev. Mat. Iberoamericana 9(1993) 139-200.

[Ku]   S.Kudla, "On the Local Theta Correspondence," Inv. Math. 83(1986), pp. 229-255.

[K-R1] S.S.Kudla and S. Rallis, "On the Weil-Siegel Formula", Crelle, 387 (1988) 1-68.

[K-R2] S.S. Kudla and S. Rallis, "On the Siegel-Weil Formula: the isotropic convergent case", Crelle, 391 (1988)65-84.

[K-R3] S.S.Kudla and S. Rallis, "Ramified principal series representations for Sp(n)", Israel J. of Math. 78 (1992) 209-256.

[K-R4] S.S. Kudla and S. Rallis, "A regularised Siegel-Weil formula: The first term identity," Annals of Mathematics, vol 140, 1-80 (1994).

[Li1] J.-S.Li, "Theta liftings for unitary representations with non-zero cohomology," Duke Math. J. 61,No. 3, 913-937 (1990).

[Li2] J.-S. Li, "Non-vanishing theorems for the cohomology for certain arithmetic quotients," Crelle 428, 177-217 (1992).

[Li3] J.-S.Li, "The correspondence of infinitesimal characters of reductive dual pairs in simple Lie groups," preprint.

[Li4] J.-S.Li, "Two dual reductive pairs in groups of type E," preprint.

[MVW] C.Moeglin, M.-F.Vigneras and J.L.Waldspurger, "Correspondence de Howe sur un corps p-adique", LNM 1291, Springer-Verlag(1987).

[P1] D.Prasad, "Weil Representation, theta correspondence, and the Howe duality," CRM proceedings and lecture notes, vol 1, edited by Ram Murty, 105-127(1993).

[P2] D.Prasad, "On an extension of a theorem of Tunnell", Compositio Math., vol. 94, 19-28 (1994).

[P3] D.Prasad, "On the local Howe duality correspondence," International mathematics research notices 11, 279-287(1993).

[P4] D.Prasad, "Theta correspondence for Unitary groups," preprint.

[Ra1] S.Rallis, "On the Howe duality conjecture", Compositio Math. 51(1984), pp. 333-399.

[Ra2] S.Rallis, "Langlands Functoriality and the Weil Representation", Amer. J. Math.104(1982), pp. 469-515.

[Ra3]   S. Rallis, "L-functions and the Oscillator representation", LNM 1245, Springer-Verlag, 1988.

[Ro1]   J. Rogawski, "The multiplicity formula for A-packets," in: The zeta functions of Picard Modular surface, edited by R. Langlands, and D. Ramakrishnan.

[Ru]   H. Rubenthaler, "Les paires duales dans les algebres de Lie reductives," Asterisque 219 (1994).

[Sa]   G. Savin, "Dual pair $G_\mathcal{J} \times PGL_2$, $G_\mathcal{J}$ is the automorphism group of the Jordan algebra $\mathcal{J}$," Inv. Math 118, 141-160(1994).

[V-Z]   D. Vogan and G. Zuckermann, "Unitary representations with non-zero cohomology," Compositio Math 53, 51-94(1984).

[Wa1]   J.-L. Waldspurger, "Correspondence de Shimura", J. Math. Pures Appl. 59, pp. 1-133(1980).

[Wa2]   J.-L. Waldspurger, "Demonstration d'une conjecture de dualité de Howe dans le cas p-adique, $p \neq 2$", In: Festschrift in honor of Piatetski-Shapiro, Israel Math. Conf. Proc., vol 2, pp. 267-324 (1990).

[We1]   A. Weil, "Sur certains groupes d'operateurs unitaires", Acta. Math. 111(1964), pp. 143-211.

[We2]   A. Weil, "Sur la formule de Siegel dans la theorie des groupes classiques", Acta Math. 113 (1965) 1-87.

Mehta Research Institute,
Allahabad-211 002, INDIA
Email: dprasad@mri.ernet.in

Deformations of Siegel-Hilbert Hecke eigensystems and their Galois representations

J. Tilouine*

1 Introduction

The goal of this paper is to put in perspective some results of a joint work with Eric Urban [44], [45] in the light of a non-abelian Main Conjecture. Let $G = GSp_{2n}$ ($n \geq 0$) be the group of symplectic similitudes viewed as a group scheme over $\mathbf{Z}$; let F be a totally real field and $\mathbf{G}_F = \text{Gal}(\bar{F}/F)$ its Galois group. Let $\mathbf{A}_F$ be the ring of adeles of F; let us consider a cuspidal representation π of $G(\mathbf{A}_F)$ which occurs in the cohomology of the Shimura variety of G of some level, say, N, with coefficients in the local system associated to a rational representation r of $G(F)$. Let us assume moreover that its infinity component π_∞ belongs to the holomorphic discrete series for $G(F \otimes \mathbf{R}))$ (for instance, for $n = 1$ and $F = \mathbf{Q}$, π is given by a classical cusp form of weight $k \geq 2$); its ramification set is denoted by $Ram(\pi)$. It is known that there exists a number field E_π containing the Hecke eigenvalues of π; for any prime $\mathbf{p}$ of E_π, let $\mathcal{O}$ be the valuation ring of the completion of E_π at $\mathbf{p}$. Let $\hat{G} = GSpin_{2n+1}$; note that for $n \leq 2$, $\hat{G}$ is canonically isomorphic to G. Let Σ_p be the set of places of F above p. Let us recall a well-known conjecture:

Conjecture 1 : *For π as above, there exists a Galois representation $\rho_\pi : \mathbf{G}_F \to \hat{G}(\mathcal{O})$, such that*

1) the ramification set of ρ_π is contained in $\text{Ram}(\pi) \cup \Sigma_p$,

2) (Eichler-Shimura relations) for each place v of F not in $\text{Ram}(\pi) \cup \Sigma_p$, *the characteristic polynomial of $\rho_\pi(Frob_v)$ is given by*

$$P_v(X) = \prod_{i=1}^{i=2^n} (X - t_{v,i})$$

---

*Université de Paris-Nord and Institut Universitaire de France

where the $t_{i,v}$ are the integral Satake parameters of π_v (that is, those given by the Satake transform not twisted by the $\delta^{-1/2}$-factor).

Comments: The explicit form of $P_v(X)$ for $n \leq 2$ is the following

- for $n = 0$:
$$\lambda_{\pi_v}(X - T_v)$$

- for $n = 1$:
$$\lambda_{\pi_v}(X^2 - T_v X + \mathbf{N}v S_v)$$

- for $n = 2$,
$$\lambda_\pi(X^4 - T_v X^3 + \mathbf{N}v(R_v + (\mathbf{N}v^2 + 1)S_v)X^2 - \mathbf{N}v^3 T_v S_v X + \mathbf{N}v^6 S_v^2)$$

where λ_{π_v} is the character of the local Hecke algebra for $G_{/F}$ at v associated to π_v, and T_v, resp. T_v and S_v, resp. T_v, R_v, S_v are the standard Hecke operators generating the Hecke algebra of $G(F_v)$. For $n = 2$, they correspond to the following matrices:

- $T_v \leftrightarrow \text{diag}(1, 1, \varpi_v, \varpi_v)$
- $R_v \leftrightarrow \text{diag}(1, \varpi_v, \varpi_v^2, \varpi_v)$
- $S_v \leftrightarrow \text{diag}(\varpi_v, \varpi_v, \varpi_v, \varpi_v)$

This conjecture has been proven

- for $n = 0$ (this is Class-Field Theory, see [47]),

- for $n = 1$ and $F = \mathbf{Q}$ by Eichler and Shimura for the weight 2 case ([12], [36] and [37]), by Deligne [10] for an arbitrary weight $k \geq 2$, its reinforcement providing an underlying Grothendieck motive has been proven by A. Scholl [33]; its analogue in $k = 1$ was proven by Deligne-Serre [11] (it corresponds to the limit of the holomorphic discrete series).

- For $n = 1$ and an arbitrary totally real field F, the conjecture has been proven as conjunction of works by Brylinski-Labesse [6], Ohta [30], Carayol [7], Wiles [49], R. Taylor [39] (these two works using congruences between representations), and a geometric proof has been finally given by Blasius-Rogawski [2] (for the limit of the holomorphic discrete series, a proof analogue to that by Deligne-Serre is due to Rogawski-Tunnell [32]).

- For $n = 2$, $F = \mathbf{Q}$, the existence of the Galois representation $\rho_\pi : Gal(\bar{F}/F) \to GL_4(\mathcal{O})$ is now proven by the works of Shimura [38], Chai-Faltings [9], R. Taylor [41], Laumon [25] and finally, Weissauer [48]; a congruence argument due to R. Taylor [40] extends the result to the limit of the holomorphic discrete series as well.

- For $n = 2$ and F arbitrary, several steps of Weissauer's work [48] go through (see [48]), but no definite result is available yet.

Remarks:

(1) The statement that ρ_π takes its values in $\hat{G}(\mathcal{O})$ is trivial for $n \leq 1$. For $n = 2$, one sees first that it takes values in $\hat{G}(\bar{\mathbf{Q}}_p)$ using [25] and [48], assuming that the following two assumptions are fulfilled

(i) $mult(\pi_f \otimes \pi_\infty^{Whitt}) = 1$ (recall that we assumed $\pi = \pi_f \otimes \pi_\infty^{hol}$ with π_∞^{hol} in the holomorphic discrete series),

(ii) $\rho_\pi \otimes \bar{\mathbf{Q}}_p$ is irreducible. (Recall that the absolute irreducibility of ρ_π is expected to hold if π is neither endoscopic nor CAP).

Indeed, the second assumption implies that the π_f-isotypic part of H^3 is a direct sum of copies of ρ_π (say m thereof). Then, one concludes by using assumption (i) together with an argument of R. Taylor (p.296 of [41]) showing that

$$1 = mult(\pi_f \otimes \pi_\infty^{hol}) = mult(\pi_f \otimes \pi_\infty^{Whitt}) = m$$

hence, the cup-product pairing on $\rho_\pi = H^3(\pi_f)$ does the job.

The question whether one can adjust the Pfaffian to be one is easily solved under the assumption that ρ_π is residually absolutely irreducible. In that case, assuming $mult(\pi) = 1$, one concludes that ρ_π takes values in $\hat{G}(\mathcal{O})$.

(2) By much more elementary arguments, one can show that if ρ_π is absolutely irreducible, the representation ρ_π with values *a priori* in $GL_4(\mathcal{O})$ takes its values in the subgroup of similitudes for a nondegenerate pairing which is *either* symmetric *or* symplectic. The argument is as follows: let $\chi : \mathbf{G}_F \to \mathcal{O}^\times$ be the character given for v prime to Np, by $\chi(Frob_v) = Nv^3 \lambda_\pi(S_v)$. It exists because up to a twist, it is the Galois avatar of the central character of π. Let ρ_π^* be the contragredient of ρ_π. Since the Langlands parameter of π_v belong to $GSp_4(\mathbf{C})$, we see by comparing the traces of Frobenii elements that

$$(0.1) \quad \rho_\pi \cong \rho_\pi^* \otimes \chi$$

Let $V \cong \mathcal{O}^4$ be the space of ρ_π; then, ρ_π acts on V by similitudes for a nondegenerate pairing $\psi : V \times V \to \mathcal{O}$ inducing the isomorphism ι_ψ of (0.1). This pairing is either orthogonal or symplectic. Indeed, let $\psi'(x,y) = \psi(y,x)$; ψ' yields another isomorphism $\iota_{\psi'} : V \to V^*$ and $\iota_\psi \circ \iota_{\psi'}^{-1}$ is an automorphism of V commuting to $\mathbf{G}_F$, hence by absolute irreducibility of V, it is a scalar. This scalar is ± 1 because $(\psi')' = \psi$. Therefore, the image of ρ_π is contained in $GO(V, \psi)$ or $GSp(V, \psi)$.

(3) For any totally real field F, under assumptions similar to those of Remark (1), a similar argument might apply. We plan to come back to this question (see [45]).

Actually, Langlands ([24], end of section 2) conjectured that conversely, "reasonable" (*i.e.* motivic) p-adic Galois representations ρ should come from automorphic forms in the sense that $\rho \cong \rho_\pi$ for some π. For $n = 0$, recall an even stronger question posed by A. Weil in [47] and answered positively by Serre, and finally, Henniart in [15]: for any number field F, for any continuous homomorphism $\rho : \mathbf{G}_F \to \mathbf{Q}^\times$ unramified outside a finite set of places Σ and such that there exists a number field E such that for $v \notin \Sigma$, $\rho(Frob_v) \in E$, there exists an A_0-type Hecke character π of F such that $\rho = \rho_\pi$ (see [47] for the construction of ρ_π). For $n > 0$, B. Mazur [26], [27], p.66, (Conjecture (*)) had the idea to formulate an infinitesimal version of this conjecture (at first, for $n = 1$ and $F = \mathbf{Q}$) in terms of deformations of a Galois representation, at least when one restricts the problem to representations with a prescribed local behaviour at p. A crucial difference with Langlands' conjecture however, is that one starts from a representation $\rho = \rho_\pi$ coming from a π; then only, one states that its deformations should come from deformations of π: nothing is said about whether ρ is "modular" (i.e. comes from a π; this obviously harder question is posed by Serre for $n = 1$ and $F = \mathbf{Q}$ in [35]; this is not however our present topic. Indeed, since the pioneering work of Hida in [16], [17] and [18], the idea has evolved that the universal ring representing a certain functor of deformations of a given absolutely irreducible "modular" Galois representation should be canonically isomorphic to a local component of the universal cuspidal Hecke algebra **h** of level p^∞. This has been proven to be true for ordinary or flat representations when $n = 1$, $F = \mathbf{Q}$, $p > 2$, by A. Wiles with the collaboration of R. Taylor [50] and [42] (assuming that the weight of the eigenform is 2 and $\bar\rho_\pi$ is absolutely irreducible as well as its symmetric square). For $n = 1$ and F totally real, K. Fujiwara in a recent preprint [13] proves a similar result under similar assumptions (actually, slightly stronger conditions; namely, $\bar\rho_\pi$ is absolutely irreducible even over $F(\sqrt{(-1)^{p-1/2}p})$, it is nearly ordinary or flat at each v above p, p does not divide h_F, does not ramify in F, F is linearly disjoint from $\mathbf{Q}(\zeta_p)$, the weight 2 Hilbert modular form has level N equal to the Artin conductor of $\bar\rho_\pi$, and at each $\mathbf{q}$ dividing N, $\bar\rho_\pi$ is of type A, B or C as in Wiles' paper [50]).

We want to discuss here evidences towards a generalization of Conjecture (*) of [27] to the case $n \leq 2$ and F totally real. At this stage of the discussion n can be arbitrary. Let us consider π as above and assume from now on that Conjecture 1 is true for π. Let $\mathcal{M}$ be the maximal ideal of $\mathcal{O}$ and $k = \mathcal{O}/\mathcal{M}$ its residue field; we consider the residual

representation
$$\bar{\rho}_\pi : \mathbf{G}_F \to \hat{G}(k)$$
defined as the reduction of ρ_π modulo $\mathcal{M}$. We assume that

(AI) $\bar{\rho}_\pi$ is absolutely irreducible.

(N.O.) It is nearly ordinary (section 2 below)

(Reg) and **(Reg*)** It is regular and dual regular at p (see sections 2 and 3)

In this paper, we discuss the Krull dimension of the universal ring of nearly ordinary deformations $R^{n.o}$ of $\bar{\rho}_\pi$ for $n \leq 2$: we find a lower bound for this dimension. Our results in [45] implies that this lower bound coincides with the dimension of the universal nearly ordinary Hecke algebra $S^{n.o}$ if and only if the Leopoldt conjecture is true for (F,p) (see Section 6 below). Moreover, assuming that a Katz-Messing theorem could be proven for the Galois representation ρ_π (it would be true if one knew that ρ_π is motivic in the sense of Grothendieck), one can define a ring homomorphism $\phi : R^{n.o} \to S^{n.o}$; moreover ϕ is linear for the action of the Hida-Iwasawa algebra Λ on these two rings (see Definition 4.2 and Section 5.2). Then, we risk a precise conjecture as follows: Assume that

(Unr) π *is unramified at all finite places outside* p,

then

Conjecture (*): (for $n = 2$) ϕ *is a surjective pseudo-isomorphism of* Λ-*algebras, and is an isomorphism for almost all* p'*s (the exceptional set depending only on* F)

Obviously, there must be a generalization of this conjecture for more general levels; however some care must be taken for formulating the deformation problem concerning the local behaviour of these deformations at places in the level. We hope to clarify these questions in another paper. Here, for simplicity, we consider only automorphic representations and Galois representations which are unramified outside the set $S_p \cup S_\infty$ of places of F above p or ∞. The author wishes to thank H. Hida, E. Urban and R. Weissauer for numerous discussions and clarifications.

2 Nearly ordinary Galois representations and their deformations

2.1 The deformation problem and its universal ring.

In the sequel all representations are assumed to be continuous. We assume that the reader is familiar with the ideas of [26]. The reference for the details on what follows are sections 6-10 of [43]. Let $n \geq 0$, G, $\mathcal{O}$, $k = \mathcal{O}/\mathcal{M}$ be as above; we put $H = \hat{G}$ and $\mathcal{H} = \text{Lie } H$. For each v in S_p, we fix a (connected) parabolic subgroup $P_v \subset H$ defined over $\mathcal{O}$ (i.e. a smooth subgroup scheme whose fibers are parabolic subgroups of H); let $\mathcal{P}_v = \text{Lie}(P_v \otimes k)$. For each $v \in S_p \cup S_\infty$, we denote by $\mathbf{G}_v$ the absolute Galois group of F_v.

Let $\mathbf{G}_{F,p}$ be the Galois group of the maximal extension F_p of F unramified outside p and ∞. For each $v \in S_p$, we fix a place $\bar{v}$ above v in F_p, hence a homomorphism of $\mathbf{G}_v \to \mathbf{G}_{F,p}$. It is an open question to decide whether this homomorphism is injective (see [28], Chapter 1). Hence, we must be careful to distinguish $\mathbf{G}_v$ from its image in $\mathbf{G}_{F,p}$. We give ourselves a "residual" representation $\bar{\rho}: \mathbf{G}_{F,p} \to H(k)$. Let $CNL_\mathcal{O}$ be the category whose objects are the complete noetherian $\mathcal{O}$-algebras with residue field k and the morphisms are the local $\mathcal{O}$-algebra homomorphisms. For any object A of $CNL_\mathcal{O}$, let $r: A \to k$ be the reduction map; it induces a reduction homomorphism $H(A) \to H(k)$ still denoted by r. We put $\hat{H}(A) = Ker\ r$. The functor $\hat{H}$ is represented by the formal completion of H along the unit section in $H \otimes k$.

Definition 1 *A deformation of $\bar{\rho}$ over A is a representation $\rho: \mathbf{G}_{F,p} \to H(A)$ such that $r \circ \rho = \bar{\rho}$.*

A deformation $\rho: \mathbf{G}_{F,p} \to H(A)$ of $\bar{\rho}$ is called nearly ordinary (N.O.) above p with respect to the near ordinarity data $(\mathbf{G}_v \to \mathbf{G}_{F,p}, P_v)_{v \in S_p}$, if it satisfies the following condition:

(N.O.) *for each v in S_p, there exists $h_v \in \hat{H}(A)$ such that $\rho(\mathbf{G}_v) \subset h_v P_v(A) h_v^{-1}$.*

Remarks: 1) A necessary condition for a deformation ρ of $\bar{\rho}$ to satisfy (N.O.) is that $\bar{\rho}$ satisfies:

(N.O.) for $\bar{\rho}$: *for each v in S_p, $\bar{\rho}(\mathbf{G}_v) \subset P_v(k)$.*

2) The set of nearly ordinary deformations over A of $\bar{\rho}$ is stable by conjugation of elements of $\hat{H}(A)$.

3) In the group $H = GSp_4$, there are 2 conjugacy classes of maximal parabolics, one conjugacy class of minimal parabolics (that is, of Borel subgroups), and the trivial parabolic subgroup H. In other words, P_v is either H or is conjugated to

1. the Siegel parabolic

$$P = \{\begin{pmatrix} A & B \\ 0 & x{}^tA^{-1} \end{pmatrix}; x \in \mathbf{G}_m, A \in GL_2, B \in M_2, A^tB = B^tA\}$$

2. the Klingen parabolic

$$P^* = \{\begin{pmatrix} a & * & b & * \\ 0 & u & * & * \\ c & * & d & 0 \\ 0 & 0 & * & v \end{pmatrix} \in GL_4; \begin{pmatrix} a & b \\ c & d \end{pmatrix} \in GL_2, ad - bc = uv\}$$

3. the standard Borel subgroup

$$B = \{\begin{pmatrix} * & * & * & * \\ 0 & * & * & * \\ 0 & 0 & * & 0 \\ 0 & 0 & * & * \end{pmatrix} \in GSp_4\}$$

Let
$$\mathcal{F}^{n.o} : CNL_\mathcal{O} \to Sets$$
be the following covariant functor

$$\mathcal{F}^{n.o}(A) = \{\Gamma \xrightarrow{\rho} H(A) \rho \otimes k = \bar{\rho}; \rho \text{ is } p\text{-nearly ordinary}\}/\hat{H}(A)$$

Proposition 2.1 *Assume that $\bar{\rho}$ is nearly ordinary above p for $(\mathbf{G}_v \to \mathbf{G}_{F,p}, P_v)_{v \in S_p}$. Assume moreover that*

(Irr) *$\bar{\rho}$ is absolutely irreducible*

and the regularity assumption

(Reg) *for any $v \in S_p$, $H^0(\mathbf{G}_v, \mathcal{H}/\mathcal{P}_v) = 0$,*

then, $\mathcal{F}^{n.o}$ is representable by a universal pair $(R^{n.o}, \rho^{n.o})$.

See the proof in [43], Proposition 6.1 of Section 6.

Comments: 1) In [43] Section 6, it is explained why because of the freedom of conjugacy by some element h_v in the definition of near ordinarity, $\mathcal{F}^{n.o}$ is not *a priori* a closed subfunctor of the functor $\mathcal{F}$ of all deformations of $\bar{\rho}$; however, the assumption of regularity makes it closed, explaining its representability as a closed formal subscheme of the universal scheme representing $\mathcal{F}$ (which exists by Th.3.3 of [43], an easy generalization of Proposition 1, Section 1.2 of [26]).

2) For $P_v = B$, the regularity assumption is implied by the more classical regularity condition: The four characters χ_1, χ_2, $\nu\chi_1^{-1}$ and $\nu\chi_2^{-1}$ defined by

$$\bar{\rho}|_{\mathbf{G}_v} \sim \begin{pmatrix} \chi_1 & * & * & * \\ 0 & \chi_2 & * & * \\ 0 & 0 & \nu\chi_1^{-1} & 0 \\ 0 & 0 & * & \nu\chi_2^{-1} \end{pmatrix}$$

are *mutually distinct*.

2.2 Nearly ordinary cohomology groups.

We assume from now on that (Irr), (N.O) and (Reg) are fulfilled. In order to estimate the Krull dimension of $R^{n.o}$, we introduce nearly ordinary cohomology groups $\tilde{H}^i_{n.o}(\mathbf{G}_{F,p}, \mathcal{H})$ ($i = 0, 1, 2$); these are subgroups of $H^i(\mathbf{G}_{F,p}, \mathcal{H})$:

$$\tilde{H}^i_{n.o}(\mathbf{G}_{F,p}, \mathcal{H}) = \mathrm{Ker}(H^i(\mathbf{G}_{F,p}, \mathcal{H}) \to \oplus_{v|p} H^i(\mathbf{G}_v, \mathcal{H})/im(H^i(\mathbf{G}_v, \mathcal{P}_v)))$$

where the map above is the direct sum of the restriction maps to $\mathbf{G}_v$ followed by the quotient maps.

Let $\mathcal{M}^{n.o}$ be the maximal ideal of $R^{n.o}$. By the universal property of $R^{n.o}$, there is a canonical isomorphism from $\mathcal{M}^{n.o}/((\mathcal{M}^{n.o})^2 + \mathcal{M}R^{n.o})$ to the tangent space $t_{\mathcal{F}^{n.o}} = \mathcal{F}^{n.o}(k[\varepsilon])$ of the functor $\mathcal{F}^{n.o}$.

Lemma 1 *There is a canonical isomorphism* $t_{\mathcal{F}^{n.o}} \cong \tilde{H}^1_{n.o}(\mathbf{G}_{F,p}, \mathcal{H})$.

Proof: Observe first that under the regularity assumption, the map

$$\oplus_{v|p} H^1(\mathbf{G}_v, \mathcal{H}) \oplus \mathrm{incl}_v : \oplus_{v|p} H^1(\mathbf{G}_v, \mathcal{P}_v) \to$$

is injective; hence $\tilde{H}^1_{n.o}(\mathbf{G}_{F,p}, \mathcal{H})$ is the group of cohomology classes $[c]$ in

H^1 such that for each $v \in S_p$, there exists a 1-cocycle c_v representing $[c]$ such that $c_v(\mathbf{G}_v) \subset \mathcal{P}_v$.

Consider $\dot{\rho} \in t_{\mathcal{F}^{n.o.}}$. We can write for any $\sigma \in \mathbf{G}_{F,p}$: $\rho(\sigma) = \exp(\varepsilon\, X(\sigma))\, \bar{\rho}(\sigma)$ for some 1-cocyle $X : \mathbf{G}_{F,p} \to \mathcal{H}$; furthermore, for each $v|p$, there exists $h_v \in \hat{H}(k[\varepsilon])$ such that

$$\forall \sigma \in \mathbf{G}_v, \qquad h_v \rho(\sigma) h_v^{-1} \in P_v(k[\varepsilon]) \qquad (*)$$

Since $h_v = \exp(\varepsilon\, Y_v)$, the relation (*) becomes

$$h_v \exp(\varepsilon X(\sigma))\, \bar{\rho}(\sigma)\, h_v^{-1} = \exp(\varepsilon(X(\sigma) + Y_v - \operatorname{Ad} \bar{\rho}(\sigma)\, Y_v))\, \bar{\rho}(\sigma) \in P_v(k[\varepsilon])$$

hence, for $[c] = cl(X)$, we have found a representative : c_v given by $c_v(\sigma) = X(\sigma) + (Y_v - \operatorname{Ad} \bar{\rho}(\sigma)\, Y_v)$ taking values in P_v. The converse is straightforward; the desired isomorphism follows. □

However, the obstruction for deforming a nearly ordinary representation into a nearly ordinary representation will be living in a slightly different $H^2_{n.o}$ than the one we have defined. Namely, let us consider the complex of cochains :

$$C^{\cdot}_{n.o}(\mathbf{G}_{F,p}, \mathcal{H}) = \operatorname{Ker}\left(C^{\cdot}(\mathbf{G}_{F,p}, \mathcal{H}) \to C^{\cdot}(\mathbf{G}_v, \mathcal{H})/C^{\cdot}(\mathbf{G}_v, \mathcal{P}_v)\right)$$

endowed with the usual differentials. Define the "cohomological" nearly ordinary cohomology groups $H^i_{n.o}(\mathbf{G}_{F,p}, \mathcal{H})$ by:

$$H^i_{n.o}(\mathbf{G}_{F,p}, \mathcal{H}) = H^i(C^{\cdot}_{n.o}(\mathbf{G}_{F,p}, \mathcal{H}))$$

See section 6, Proposition 2.2 of [43] for the comparison between these two kinds of nearly ordinary cohomology groups. We show there that the natural homomorphism:

$$H^2_{n.o}(\mathbf{G}_{F,p}, \mathcal{H})) \longrightarrow \tilde{H}^2_{n.o}(\mathbf{G}_{F,p}, \mathcal{H})$$

is surjective if $S_p = \{v\}$ and $H^0(\mathbf{G}_v, \mathcal{P}^*(1)) = 0$ (we shall need this assumption later, it will be called the dual regularity assumption).

2.3 The nearly ordinary obstruction map.

Let us consider an infinitesimal extension, that is, a morphism $\varphi : A_1 \to A_0$ in $CNL_\mathcal{O}$ whose kernel I is annihilated by the maximal ideal $\mathcal{M}_1$ of A_1. We define an obstruction map

$$\operatorname{obs}^{n.o}_\varphi : \mathcal{F}^{n.o}(A_0) \to H^2_{n.o}(\mathbf{G}_{F,p}, \mathcal{H}) \otimes I$$

as follows. Note that the complex $C_{n.o}^{\cdot}(\mathbf{G}_{F,p}, \mathcal{H})$ giving rise to the "cohomological" nearly ordinary cohomology groups can be also written as the fiber product complex:

$$C^{\cdot}(\mathbf{G}_{F,p}, \mathcal{H}) \times_{\prod_{v|p} C^{\cdot}(\mathbf{G}_v, \mathcal{H})} \prod_{v|p} C^{\cdot}(\mathbf{G}_v, \mathcal{P}_v)$$

where the projections are respectively: the sum of the restriction maps from $\mathbf{G}_{F,p}$ to $\mathbf{G}_v$ and the product of the map deduced from the inclusions $\mathcal{P}_v \subset \mathcal{H}$. We shall use this formulation, and view a cochain of $C_{n.o}^{\cdot}(\mathbf{G}_{F,p}, \mathcal{H})$ as a tuple $(c, (c_v)_{v|p})$.

Consider a class $\dot{\rho} \in \mathcal{F}^{n.o}(A_0)$. One can check that there exists a continuous section s of φ such that $s(\mathcal{P}_v(A_0)) \subset \mathcal{P}_v(A_1)$ and similarly for intersections $\bigcap_{v \in \Sigma} \mathcal{P}_v$ for any subset $\Sigma \subset S_p$. Then, $\tilde{\rho} = s \circ \rho : \mathbf{G}_{F,p} \to H(A_1)$ is a continuous map lifting ρ. Let us observe that for any $v \in S_p$, there exists $h_{v,1} \in \hat{H}(A_1)$ such that $^{h_{v,1}}\tilde{\rho}(\mathbf{G}_v) \subset \mathcal{P}_v(A_1)$ where we write $^h k = hkh^{-1}$. We define $c \in Z^2(\mathbf{G}_{F,p}, \mathcal{H}) \otimes I$ and $c_v \in Z^2(\mathbf{G}_v, \mathcal{P}_v) \otimes I$ by

(1) $\qquad \forall (\sigma, \tau) \in \mathbf{G}_{F,p} \times \mathbf{G}_{F,p} \qquad \tilde{\rho}(\sigma)\tilde{\rho}(\tau) = \exp(c_{\sigma,\tau})\tilde{\rho}(\sigma\tau)$

and

(2_v) $\qquad \forall (\sigma, \tau) \in \mathbf{G}_v \times \mathbf{G}_v \qquad h_{v,1}\tilde{\rho}(\sigma)\tilde{\rho}(\tau)h_{v,1}^{-1} = \exp(c_{v,\sigma,\tau}) h_{v,1}\tilde{\rho}(\sigma\tau)h_{v,1}^{-1}$

By comparing $(1)|_{\mathbf{G}_v}$ to (2_v); we get :

$$h_{v,1} \exp(c_{\sigma,\tau})\tilde{\rho}(\sigma\tau) h_{v,1}^{-1} = \exp(c_{v,\sigma,\tau}) h_{v,1} \tilde{\rho}(\sigma\tau) h_{v,1}^{-1}$$

Or : $h_{v,1} \exp(c_{\sigma,\tau}) h_{v,1}^{-1} = \exp(c_{v,\sigma,\tau})$.

Since $h_{v,1} \in \hat{H}(A_1)$, it amounts to :

$$c_{\sigma,\tau} = c_{v,\sigma,\tau} \quad \text{in} \quad Z^2(\mathbf{G}_v, \mathcal{H} \otimes I).$$

Hence $(c, (c_v)) \in Z^2(C_{n.o}^{\cdot}(\mathbf{G}_{F,p}, \mathcal{H})) \otimes I$. It defines a class in $H^2(C_{n.o}^{\cdot}(\mathbf{G}_{F,p}, \mathcal{H})) \otimes I$ denoted by $\mathrm{obs}_{\varphi,s}^{n.o}(\rho)$.

One checks that it does not depend on s and on the particular choice of $\rho \in \dot{\rho}$ that we have made. We may therefore denote this obstruction by $\mathrm{obs}_{\varphi}^{n.o}(\dot{\rho})$.

Fact : $\dot{\rho}$ lifts to $\dot{\rho}_1 \in \mathcal{F}^{n.o}(A_1) \Leftrightarrow \mathrm{obs}_{\varphi}^{n.o}(\dot{\rho}) = 0$.

Proof: It is well-known know that $\rho : \mathbf{G}_{F,p} \to H(A_1)$ lifts to a group homomorphism if and only if c is a coboundary $c_{\sigma,\tau} = (\delta X)_{\sigma,\tau}$; more precisely: $\hat{\rho}(\sigma) = \exp(X_\sigma)\tilde{\rho}(\sigma)$ is a homomorphism: $\hat{\rho} : \mathbf{G}_{F,p} \to H(A_1)$.

Let us put $^h k = hkh^{-1}$; then similarly, if $c_{v,\sigma,\tau} = (\delta X_v)_{\sigma,\tau}$, the map given by $\hat{\rho}_v(\sigma) = \exp(X_{v,\sigma})\,{}^{h_{v,1}}\tilde{\rho}(\sigma)$ is a homomorphism from $\mathbf{G}_v$ to $P_v(A_1)$.

Moreover, the relation $(c, c_v) = \delta(X, X_v)$ for $(X, X_v) \in C^1_{n.o}(\mathbf{G}_{F,p}, \mathcal{H}) \otimes I$. means that
$$c = \delta X$$
$$c_v = \delta X_v \quad \text{and} \quad X|_{\Gamma_v} = X_v.$$

Therefore for $\hat{\rho}(\sigma) = \exp(X_\sigma)\,\tilde{\rho}(\sigma)$, we have for $v|p$, and $h_{v,1} \in \hat{H}(A_1)$ as above :
$$\forall \sigma \in \mathbf{G}_v, \quad {}^{h_{v,1}}\hat{\rho}(\sigma) = \exp(X_\sigma)^{h_{v,1}}\tilde{\rho}(\sigma)$$
$$= \exp(X_{v,\sigma})^{h_{v,1}}\tilde{\rho}(\sigma) \in P_v(A_1).$$

since the two terms in the last product are respectively in $\exp(\mathcal{P}_v)$ and $P_v(A_1)$.

We conclude that $\hat{\rho}$ is a nearly ordinary lifting of ρ. The converse is now clear. $\square$

Let us state the functoriality property of this obstruction map. Consider a morphism of infinitesiomal extensions in $CNL_\mathcal{O}$:

$$\begin{array}{ccccccccc}
0 & \longrightarrow & I' & \longrightarrow & A'_1 & \xrightarrow{\varphi'} & A_0 & \longrightarrow & 0, \quad \mathcal{M}'_1 I' = 0 \\
& & \downarrow \lambda & & \downarrow \alpha & & \parallel & & \\
0 & \longrightarrow & I & \longrightarrow & A_1 & \xrightarrow{\varphi} & A_0 & \longrightarrow & 0, \quad \mathcal{M}_1 I = 0
\end{array}$$

where $\alpha \in \mathrm{Hom}_{\mathrm{loc}}(A'_1, A_1)$, hence $\lambda \in \mathrm{Hom}_{k-\mathrm{lin}}(I', I)$.

Then $\forall \dot{\rho} \in \mathcal{F}^{n.o}(A_0)$
$$(\mathrm{Id} \otimes \lambda)\,\mathrm{obs}^{n.o}_{\varphi'}(\dot{\rho}) = \mathrm{obs}^{n.o}_\varphi(\dot{\rho}).$$

where $\mathrm{Id} \otimes \lambda : H^2_{n.o}(\mathbf{G}_{F,p}, \mathcal{H}) \otimes I' \to \tilde{H}^2_{n.o}(\mathbf{G}_{F,p}, \mathcal{H}) \otimes I$.

Using this functoriality, one deduces an estimate for the Krull dimension of for $\bar{R}^{n.o} = R^{n.o} \otimes k$ as follows:

Let $\Sigma_{n.o} = \widehat{\mathrm{Sym}}\, H^1_{n.o}(\mathbf{G}_{F,p}, \mathcal{H})^*$ be the completed symmetric k-algebra of $H^1_{n.o}(\mathbf{G}_{F,p}, \mathcal{H})^*$. It is a free object of $CNL_\mathcal{O}$, hence, there exists an exact sequence in $CNL_\mathcal{O}$
$$0 \to J \to \Sigma_{n.o} \xrightarrow{\varphi} \bar{R}^{n.o} \to 0$$

and the minimal number of generators of J
$$s = \dim_k J/\mathcal{M}_{\Sigma_{n.o}} J$$

satisfies :
$$s \leq \tilde{h}_{n.o}^2 = \dim_k \tilde{H}_{n.o}^2(\mathbf{G}_{F,p}, \mathcal{H})$$

Actually, there is a canonical k-linear injection:
$$\operatorname{Hom}_{k-\operatorname{lin}}(J/\mathcal{M}_{\Sigma_{n.o}}J, k) \to H_{no}^2(\mathbf{G}_{F,p}, \mathcal{H})$$

given by $\lambda \mapsto (\operatorname{Id} \otimes \lambda) \operatorname{obs}_{\varphi'}(\bar{\rho}^{n.o})$ where $\varphi' : \Sigma_{n.o}/\mathcal{M}_{\Sigma_{n.o}}J \to \bar{R}^{n.o}$ is the infinitesimal extension deduced from $\Sigma_{n.o} \overset{\varphi}{\to} \bar{R}^{n.o}$.

Therefore, we get

Proposition 2.2
$$\dim \bar{R}^{n.o} \geq \tilde{h}_{n.o}^1 - h_{n.o}^2$$

where
$$\tilde{h}_{n.o}^1 = \dim_k \tilde{H}_{n.o}^1(\mathbf{G}_{F,p}, \mathcal{H})$$
$$h_{n.o}^2 = \dim H_{n.o}^2(\mathbf{G}_{F,p}, \mathcal{H}).$$

The proof proceeds exactly as that of Proposition 2 of Section 1.6 of [26].

3 Estimate for the Krull dimension of $R^{n.o}$

In this section, we study the number $\tilde{h}_{n.o}^1 - \tilde{h}_{n.o}^2$ by using a Poitou-Tate long exact sequence with local conditions. We keep assumptions as in section 2 above. But we assume moreover

(Reg*) for any $v \in S_p$, $H^0(\mathbf{G}_v, \mathcal{P}_v^*(1)) = 0$.

Recall that under (Reg*), one has the inequality $\tilde{h}_{n.o}^2 \leq h_{n.o}^2$ (under the assumption that S_p has one element: see Proposition 6.3 of [43]). Hence we have in that case $dim\ \bar{R}^{n.o} \geq \tilde{h}_{n.o}^1 - h_{n.o}^2 \geq \tilde{h}_{n.o}^1 - \tilde{h}_{n.o}^2$.

Remark: For each $v \in S_p$, let $\omega_v : \mathbf{G}_v \to \mathcal{O}^\times$ be the Teichmüller lifting of the cyclotomic character on the group of pth roots of unity of $\bar{F}_p$. Then, if the semisimplification of $\bar{\rho}|_{\mathbf{G}_v}$ is given by $diag(\chi_1, \ldots, \chi_n, \chi_{n+1}, \ldots, \chi_{2n})$ (with $\chi_i \chi_{n+i} = \nu$ for $i = 1, \ldots, n$), then if for each $v \in S_p$, $\chi_i/\chi_j \neq \omega_v$ for any $i < j$, then **(Reg*)** holds.

3.1 Euler characteristics.

Let $S = S_\infty \cup S_p$. Let $d = [F:\mathbf{Q}]$ and for each $v \in S_p$, $d_v = [F_v : \mathbf{Q}_p]$. For $v \in S_\infty$, we write $\mathbf{G}_v$ for the decomposition group at v.

Proposition 3.1 *Assume $\bar{\rho}$ satisfies the four assumptions: (Irr), (N.O.), (Reg) and (Reg*), then*
$$\tilde{h}_{n.o}^1 - \tilde{h}_{n.o}^2 = \sum_{v|p} d_v \dim \mathcal{P}_v + 1 - \sum_{v|\infty} h^0(\mathbf{G}_v, \mathcal{H})$$

Proof: We compare two Poitou-Tate exact sequences. First the usual one: (1)

$$0 \longrightarrow H^0(\mathbf{G}_{F,p}, \mathcal{H}) \xrightarrow{\beta^0} \bigoplus_{v \in S_p} H^0(\mathbf{G}_v, \mathcal{H}) \xrightarrow{\gamma^0} H^2(\mathbf{G}_{F,p}, \mathcal{H}^*(1))^*$$
$$\downarrow v_1$$
$$H^1(\mathbf{G}_{F,p}, \mathcal{H}^*(1))^* \xleftarrow{\gamma^1} \bigoplus_{v \in S_p} H^1(\mathbf{G}_v, \mathcal{H}) \xleftarrow{\beta^1} H^1(\mathbf{G}_{F,p}, \mathcal{H})$$
$$v_2 \downarrow$$
$$H^2(\mathbf{G}_{F,p}, \mathcal{H}) \xrightarrow{\beta^2} \bigoplus_{v \in S_p} H^2(\mathbf{G}_v, \mathcal{G}) \xrightarrow{\gamma^2} H^0(\mathbf{G}_{F,p}, \mathcal{H}^*(1))^* \longrightarrow 0$$

and the "nearly ordinary one", deduced from (1): (2)

$$0 \longrightarrow \tilde{H}_{n.o}^0(\mathbf{G}_{F,p}, \mathcal{H}) \longrightarrow \bigoplus_{v \in S_p} H^0(\mathbf{G}_v, \mathcal{P}_v) \longrightarrow H^2(\mathbf{G}_{F,p}, \mathcal{H}^*(1))^*$$
$$\downarrow$$
$$H^1(\mathbf{G}_{F,p}, \mathcal{H}^*(1))^* \longleftarrow \bigoplus_{v \in S_p} H^1(\mathbf{G}_v, \mathcal{P}_v) \longleftarrow \tilde{H}_{n.o}^1(\mathbf{G}_{F,p}, \mathcal{H})$$
$$\downarrow$$
$$\tilde{H}_{n.o}^2(\mathbf{G}_{F,p}, \mathcal{H}) \longrightarrow \bigoplus_{v \in S_p} H^2(\mathbf{G}_v, \mathcal{P}_v) \longrightarrow H^0(\mathbf{G}_{F,p}, \mathcal{H}^*(1))^* \longrightarrow 0$$

To show the existence of the exact sequence (2), one uses the three simple facts : assuming both regularity and dual regularity we have : $\forall v | p$

1. $H^0(\mathbf{G}_v, \mathcal{H}) = H^0(\mathbf{G}_v, \mathcal{P}_v)$
2. $H^1(\mathbf{G}_v, \mathcal{P}_v) \subset H^1(\mathbf{G}_v, \mathcal{H})$
3. $H^2(\mathbf{G}_v, \mathcal{P}_v) = 0$

1. and 2. result from $H^0(\mathbf{G}_v, \mathcal{H}/\mathcal{P}_v) = 0$ (regularity assumption).
By duality, 3. amounts to $H^0(\mathbf{G}_v, \mathcal{P}_v^*(1)) = 0$; this is nothing but the dual regularity assumption.

For any finite $k[\mathbf{G}_{F,p}]$-module M, we consider for each $v \in S$, the local Euler characteristic:

$$\chi(\mathbf{G}_v, M) = \sum_{i=0}^{2} (-1)^i \dim_k H^i(\mathbf{G}_v, M)$$

We introduce global Euler characteristics as well:

$$\chi(\mathbf{G}_{F,p}, \mathcal{H}) = \sum_{i=0}^{2} (-1)^i \dim_k H^i(\mathbf{G}_{F,p}, \mathcal{H}) = \sum_{i=0}^{2} (-1)^i h^i$$

$$\tilde{\chi}_{n.o}(\mathbf{G}_{F,p}, \mathcal{H}) = \sum_{i=0}^{2} (-1)^i \dim_k \tilde{H}_{n.o}^i(\mathbf{G}_{F,p}, \mathcal{H}) = \sum_{i=0}^{2} (-1)^i \tilde{h}_{n.o}^i$$

We deduce from (1) that: $\chi(\mathbf{G}_{F,p}, \mathcal{H}) + \chi(\mathbf{G}_{F,p}, \mathcal{H}^*(1)) = \sum_{v \in S_p} \chi(\mathbf{G}_v, \mathcal{H})$ and from (2) that:

$\tilde{\chi}_{n.o}(\mathbf{G}_{F,p}, \mathcal{H}) + \tilde{\chi}_{n.o}(\mathbf{G}_{F,p}, \mathcal{H}^*(1)) = \sum_{v \in S_p} \chi(\mathbf{G}_v, \mathcal{P}_v)$ Hence, by substraction:

$$\begin{aligned}
\tilde{\chi}_{n.o}(\mathbf{G}_{F,p}, \mathcal{H}) &= \chi(\mathbf{G}_{F,p}, \mathcal{H}) + \sum_{v|p}(\chi(\mathbf{G}_v, \mathcal{P}_v) - \chi(\mathbf{G}_v, \mathcal{H})) \\
&= \chi(\mathbf{G}_{F,p}, \mathcal{H}) - \sum_{v|p} \chi(\mathbf{G}_v, \mathcal{H}/\mathcal{P}_v).
\end{aligned}$$

The local Euler characterictics are computed by Tate local duality: for any finite-dimensional k-vector space, $\chi(\mathbf{G}_v, M) = -d_v \cdot \dim_k M$.

Now, $\tilde{h}_{n.o}^0 = h^0 = 1$ by (Irr); so, we obtain

$$\begin{aligned}
\tilde{h}_{n.o}^1 - \tilde{h}_{n.o}^2 &= h^1 - h^2 - \sum_{v|p} d_v \dim_k(\mathcal{G}/\mathcal{P}_v) \\
&= d \dim \mathcal{H} + 1 - \sum_{v|\infty} \dim H^0(\mathbf{G}_v, \mathcal{H}) - \sum_{v|p} d_v \dim_k(\mathcal{H}/\mathcal{P}_v) \\
&= \sum_{v|p} d_v \dim \mathcal{P}_v + 1 - \sum_{v|\infty} h^0(\mathbf{G}_v, \mathcal{H}).
\end{aligned}$$

□

Remark

One has $h^1 - h^2 \geq 1$ because

$$\sum_{v|\infty} h^0(\mathbf{G}_v, \mathcal{H}) \leq \#S_\infty \cdot \dim \mathcal{G} \leq d \dim \mathcal{H},$$

however $\tilde{h}_{n.o}^1 - \tilde{h}_{n.o}^2$ might well be negative.

Notation: *From now on, we put* $\tilde{\Delta}_{n.o} = \tilde{h}_{n.o}^1 - \tilde{h}_{n.o}^2$.

3.2 Examples.

Let us compute $\tilde{\Delta}_{n.o}$ for classical Hilbert and Siegel-Hilbert cusp eigenforms.

Consider first the case where $n = 1$, hence $H = GL_2$. If f is a classical cusp Hilbert eigenform of weight $k \geq 2$, and level dividing p^∞, let E_f be the number field generated by the Hecke eigenvalues; for any prime $\mathbf{p}$ of E_f above p, let $\rho_f : \mathbf{G}_{F,p} \to GL_2(E_{f,\mathbf{p}})$ be the Galois representation associated to f. We assume it is absolutely irreducible and is nearly ordinary above p with P_v equal to the standard Borel subgroup B for each $v \in S_p$. In Section 6 below, we recall a theorem of Wiles and Hida that if for any $v \in S_p$, the eigenvalue of T_v on f is a $\mathbf{p}$-adic unit, then ρ_f is nearly ordinary above p for $P_v = B$ for each $v \in S_p$. Furthermore we assume that for each $v \in S_p$, the characters χ_1 and χ_2 giving the semisimplification of $\bar{\rho}|\mathbf{G}_v$ are distinct and that $\chi_1 \neq \omega\chi_2$; this implies that (Reg) and (Reg*) are fulfilled. It is well known that ρ_f is totally odd: for any $v \in S_\infty$, let c_v be the generator of Gal $(\bar{F}_v/F_v)$, then

$$\rho_f(c_v) \sim \begin{pmatrix} 1 & 0 \\ 0 & -1 \end{pmatrix}$$

Therefore $h^0(\mathbf{G}_v, \mathcal{H}) = 2$; so, we find

$$\tilde{\Delta}_{n.o} = 3d + 1 - d = 2d + 1$$

Now, let $n = 2$ so $H = GSpin_5 = GSp_4$. Let $I = Hom_{\mathbf{Q}-alg}(F, \mathbf{R})$; let $\mathcal{Z}_2$ be the genus 2 Siegel half-space. Let f be a classical cusp eigenform on $\mathcal{Z}_2^I$ with level dividing p^∞, E_f the number field generated by the Hecke eigenvalues, $\mathbf{p}$ a prime of E_f above p and $\rho_f : \mathbf{G}_{F,p} \to H(E_{f,\mathbf{p}})$ be the Galois representation attached to $(f, \mathbf{p})$. Let us assume it satisfies (Irr) and (N.O) above p with $P_v = B$ the standard Borel, for each $v \in S_p$ (see also results below for a sufficient condition on f for ρ_f to satisfy (N.O)). One can show that ρ_f is totally odd of type II in the following sense: for each $v \in S_\infty$ and the v-complex conjugation c_v as above, then

$$\rho_f(c_v) \sim \begin{pmatrix} 1 & 0 & 0 & 0 \\ 0 & 1 & 0 & 0 \\ 0 & 0 & -1 & 0 \\ 0 & 0 & 0 & -1 \end{pmatrix}$$

[the proof is easy once one knows that the factor of similitudes of ρ_f is given for any v prime to p by:

$$\nu(\rho_f(Frob_v)) = (\mathbf{N}v)^3 \cdot \lambda_f(S_v)$$

(with the notations of the introduction); this implies $\nu(\rho_f(c_v)) = -1$ hence the result.]

From this, one computes $H^0(\mathbf{G}_v, \mathcal{H}) = \{\begin{pmatrix} A & 0 \\ 0 & \lambda - {}^t A \end{pmatrix}; \lambda \in \mathbf{G}_a, A \in M_2\}$, so $h^0(\mathbf{G}_v, \mathcal{H}) = 5$. On the other hand, $\dim_k \mathcal{P}_v = 7$; so,

$$\tilde{\Delta}_{n.o} = 7d + 1 - 5d = 2d + 1$$

4 The Hida-Iwasawa algebra

At this stage of the discussion, it becomes reasonable to simplify the treatment by assuming

(B-N.O) for all $v \in S_p$, P_v is the standard Borel subgroup of H

4.1 Amalgamation of local and global data.

Let us consider the $\mathcal{O}$-group scheme morphism given by the similitude factor:

$$\nu : H \to \mathbf{G}_m$$

(for $n = 1$ it is the determinant). It defines a morphism of functors from $\mathcal{F}^{n.o}$ to the functor of deformations of $\bar{\nu} \circ \bar{\rho}$ (where $\bar{\nu} = \nu \otimes k$). This functor is easy to represent (without assumption) and the universal ring is simply the completed group algebra over $\mathcal{O}$ of the largest abelian pro-p-quotient group Γ of $\mathbf{G}_{F,p}$. Hence, by universal property of this ring, we obtain a structural map, called the global Iwasawa algebra structure:

$$\Phi_{gl} : \mathcal{O}[[\Gamma]] \to R^{n.o}$$

Actually, the ring $R^{n.o}$ has additional local Iwasawa algebra structure coming from the near-ordinarity property: For each $v|p$, $\bar{\rho}_v = \bar{\rho}|_{\Gamma_v}$ takes values in $B(k)$. Let $B = TB^+$ be the Levi decomposition of B, with T the maximal torus and B^+ the unipotent radical. Consider the $\mathcal{O}$-group scheme morphism

$$B \xrightarrow{\varphi_v} T = B/B^+$$

Consider the local residual representation

$$\bar{\varphi}_v \circ \bar{\rho}_v : \mathbf{G}_v \to T(k)$$

The morphism φ_v induces a morphism of functors from $\mathcal{F}^{n.o}$ to the functor of deformations of $\bar{\varphi}_v \circ \bar{\rho}_v$, as follows. For $\dot{\rho} \in \mathcal{F}^{n.o}(A)$, for any $v \in S_p$, there exists $h_v \in \hat{H}(A)$ such that

$$^{h_v}\rho_v(\mathbf{G}_v) \subset B(A).$$

Moreover if h_v and h'_v have this property, then putting $\tilde{\rho}_v = {}^{h_v}\rho_v$ and $\tilde{h}_v = h'_v h_v^{-1}$, we have :

$$\begin{cases} \tilde{\rho}_v(\mathbf{G}_v) \subset B(A) \\ {}^{\tilde{h}_v}\tilde{\rho}_v(\mathbf{G}_v) \subset B(A) \end{cases}$$

Using the assumption (Reg), one sees that it implies $\tilde{h}_v \in \hat{B}(A)$, therefore the morphism of functors is well defined.

Let Γ_v be the maximal pro-p-quotient of $\mathbf{G}_v$ and I_v be the inertia subgroup of Γ_v; let $X^*(T)$ be the group of characters of T. The functor of deformations of $\bar{\varphi}_v \circ \bar{\rho}_v|I_v$ is representable by the Iwasawa algebra $\mathcal{O}[[X^*(T) \otimes I_v]]$; the morphism of functors provides a local ring homomorphism

$$\Phi_v : \mathcal{O}[[X^*(T) \otimes I_v]] \to R^{n.o}_{\bar{\rho}}.$$

Let us form

$$\Lambda_{\text{loc}} = \hat{\bigotimes}_{v|p} \mathcal{O}[[X^*(T) \otimes I_v]] = \mathcal{O}[[\bigoplus_{v|p} X^*(T) \otimes I_v]]$$

and a local ring homomorphism :

$$\Phi_{\text{loc}} = \hat{\bigotimes}_{v|p} \Phi_v : \Lambda_{\text{loc}} \to R^{n.o}.$$

Recall that we have also an algebra homomorphism $\Phi_{\text{gl}} : \Lambda_{\text{gl}} \to R^{n.o}$. Let us compare Ψ_{loc} and Φ_{gl}.

Consider, for that purpose, the natural maps :

(i)
$$\alpha_v : I_v \longrightarrow \Gamma$$

and induced by $\mathbf{G}_v \to \mathbf{G}_{F,p}$. It is injective by Class-Field Theory.

(ii) $\beta_v : \mathbf{Z} \to X^*(T)$ sending 1 to the similitude factor ν.

We then consider the following groups and homomorphisms:

$$H_{\mathrm{loc/gl}} = \bigoplus_{v|p} I_v$$
$$H_{\mathrm{loc}} = \bigoplus_{v|p} X^*(T) \otimes I_v$$
$$H_{\mathrm{gl}} = \Gamma$$

and

$$\alpha : H_{\mathrm{loc/gl}} \to H_{\mathrm{gl}}, \quad \alpha = \sum_{v|p} \alpha_v$$
$$\beta : H_{\mathrm{loc/gl}} \to H_{\mathrm{loc}}, \quad \beta = \bigoplus_{v|p} \beta_v \otimes \mathrm{Id}_v.$$

Let $\Lambda_{\mathrm{loc/gl}} = \mathcal{O}[[H_{\mathrm{loc/gl}}]]$ and $\Phi_\alpha : \Lambda_{\mathrm{loc/gl}} \to \Lambda_{\mathrm{gl}}$ resp. $\Phi_\beta : \Lambda_{\mathrm{loc/gl}} \to \Lambda_{\mathrm{loc}}$ be the ring homomorphisms defined by α resp. β by linearity and continuity.

Proposition 4.1 $\Phi_{\mathrm{gl}} \hat{\otimes} \Phi_{\mathrm{loc}}$ *factors through the local ring* $\Lambda = \Lambda_{\mathrm{gl}} \otimes_{\Lambda_{\mathrm{loc/gl}}} \Lambda_{\mathrm{loc}}$. *The resulting homomorphism* $\Phi_{\mathrm{gl}} \otimes_{\Lambda_{\mathrm{loc/gl}}} \Phi_{\mathrm{loc}}$ *endow therefore* $R^{n.o}$ *with a structure of Λ-algebra.*

Proof: See Proposition 8.1 of [43].

Definition 4.2 *(i) The Hida group is the amalgamated sum of H_{loc} and H_{gl} by α and β:*

$$H = \begin{array}{c} H_{\mathrm{gl}} \quad \oplus \quad H_{\mathrm{loc}} \\ {}_\alpha \nwarrow \qquad \nearrow {}_\beta \\ H_{\mathrm{loc/gl}} \end{array}$$

(ii) The Hida-Iwasawa algebra is defined as the completed group algebra of the Hida group over $\mathcal{O}$:

$$\Lambda = \mathcal{O}[[H]].$$

The terminology of the Hida group has been introduced by A. Panchishkin in [31], although the group he considers does not take the local conditions into account.

Comments: The Hida group is the largest quotient of $H_{\mathrm{loc}} \times H_{\mathrm{gl}}$ on which $\alpha = \beta$. Similarly, the Hida-Iwasawa algebra is the quotient of $\Lambda_{\mathrm{gl}} \hat{\otimes} \Lambda_{loc}$ by the closed ideal generated by $\{\alpha(h) \otimes \beta(h)^{-1} \; ; \; h \in H_{\mathrm{loc/gl}}\}$.

Since β is injective, we see that :

$$\text{rel.dim}_{\mathcal{O}}\Lambda = rk_{\mathbb{Z}_p} H = rk\, H_{\text{loc}} + rk\, H_{gl} - rk\, H_{\text{loc/gl}}.$$

Let us compute this rank :

$$\begin{aligned} rk\, H_{\text{loc/gl}} &= \sum_{v|p} d_v = d \\ rk\, H_{\text{loc}} &= \sum_{v|p} t \cdot d_v = td \\ rk\, H_{gl} &= 1 + \delta \end{aligned}$$

where $t = rk\, T$

So,

$$rk\, H = (t-1) \cdot d + 1 + \delta$$

where δ is the defect to Leopoldt conjecture for (F, p).

4.2 Examples.

(1) $G = GL_2$, F any totally real field, for each $v \in S_p$, $P_v =$ Borel, $t = 2$, hence:

$$rk\, H = d + 1 + \delta.$$

(2) $G = GSp_4$, F arbitrary totally real field, P_v is the standard Borel, $t = 3$, so

$$rk\, H = 2d + 1 + \delta.$$

4.3 Questions.

In the two examples above, we observe that

$$\text{reldim}_{\mathcal{O}} \Lambda \geq \tilde{\Delta}_{n.o} = \tilde{h}^1_{n.o} - \tilde{h}^2_{n.o}, \quad \text{for } \bar{\rho} \text{ odd, resp. odd II},$$

with equality if Leopoldt conjecture is true for (F, p). It is therefore tempting to ask:

Questions:

In examples 1 and 2,

- Do we have: $\operatorname{reldim}_{\mathcal{O}} R^{n.o} = \tilde{\Delta}_{n.o}$?

- Is $R^{n.o}$ finite torsion-free over Λ, or at least with pseudo-null torsion?

Actually, in Section 6 below, we shall formulate a precise conjecture for $n \leq 2$ relating $R^{n.o}$ to a universal local Hecke algebra associated to this situation. Then, our theorems of [44] and [45] about this Hecke algebra (see Section 7 below) can be considered as evidences towards a positive answer to these questions, [modulo the Leopoldt conjecture for (F, p)].

5 The universal nearly ordinary Hecke algebra

The definitions in this section are valid for $G = GSp_{2n}/F$ for any $n \geq 0$; the standard Borel of G is

$$B = \begin{pmatrix} A & * \\ 0 & x\,{}^t A^{-1} \end{pmatrix} \in G;\ A \in B_n,\ x \in \mathbf{G}_m\}$$

where B_n is the Borel of GL_n consisting in upper triangular matrices. Its unipotent radical is denoted by B^+. Let $G_\infty = G(F \otimes \mathbf{R})$, $G_f = G(F \otimes \hat{\mathbf{Z}})$ so that $G(\mathbf{A}_F) = G_f \times G_\infty$. Let $\mathbf{r}$ be the ring of integers of F and $\hat{\mathbf{r}} = \mathbf{r} \otimes \hat{\mathbf{Z}}$. For simplicity, we restrict ourselves to the level groups

$$U = G(\hat{\mathbf{r}}), \quad U_0(p^r) = \{g \in U; g \bmod.\ p^r \in B(\mathbf{r}/p^r\mathbf{r})\}$$

and

$$U_1(p^r) = \{g \in U; g \bmod.\ p^r \in B^+(\mathbf{r}/p^r\mathbf{r})\}.$$

for each $r \geq 0$.

Let U_∞ be the stabilizer in G_∞ of the map $h : (F \otimes \mathbf{C})^\times \longrightarrow G_\infty$ given by

$$h(x + iy) = \begin{pmatrix} x 1_n & y 1_n \\ -y 1_n & x 1_n \end{pmatrix}$$

We consider the Shimura varieties

$$S_r = S(U_1(p^r)) = G(F) \backslash G(\mathbf{A}_F) / U_1(p^r) U_\infty$$

Their connected components are Siegel-Hilbert modular varieties of dimension $\frac{n(n+1)}{2}d$. For $s \geq r$, there is a natural finite morphism $S_s \to S_r$; the varieties S_r together with these transition maps form an inverse system.

Let $\mathcal{O}$ be the valuation ring of a finite extension K of $\mathbf{Q}_p$. Let $\mathbf{r}_p = \mathbf{r} \otimes \mathbf{Z}_p$ and $F_p = F \otimes \mathbf{Q}_p$. We define local systems over S_r as follows.

Definition 2 *A character $\chi : T(\mathbf{r}_p) \to \mathcal{O}^\times$ is called p-arithmetic if there exists a character $\chi_0 \in X^*(T)$ and a finite order character $\varepsilon : \chi : T(\mathbf{r}_p) \to \mathcal{O}^\times$ such that $\chi = \chi_0 \varepsilon$. We say it is dominant if χ_0 is in the dominant cone of $X^*(T)$ for the order given by (B, T).*

Let $I \subset G(\mathbf{r}_p)$ be the Iwahori subgroup associated to B. For each p-arithmetic character χ, we consider the finite free $\mathcal{O}$-module:

$$L(\chi; \mathcal{O}) = \{f : I/B^+(\mathbf{r}_p) \to \mathcal{O}; f \text{ polynomial and } f(xt) = \chi^{-1}(t)f(x),\ t \in T(\mathbf{r}_p)\}$$

It is endowed with an action of I by left translation: $(g.f)(x) = f(g^{-1}x)$ for $g \in I$. Let $L(\chi) = L(\chi; \mathcal{O}) \otimes K/\mathcal{O}$. Let $\mathcal{L}(\chi)$ be the sheaf of locally constant sections of the covering

$$G(F) \backslash (G(\mathbf{A}_F) \times L(\chi)) / U_1(p^r) U_\infty$$
$$\downarrow$$
$$G(F) \backslash G(\mathbf{A}_F) / U_1(p^r) U_\infty = S_r$$

where the action defining the covering space is given by $\gamma(h, f) u u_\infty = (\gamma h u u_\infty, u_p^{-1}.f)$ for $\gamma \in G(F)$, $u u_\infty \in U_1(p^r) U_\infty$. For $r \geq 0$ and $q = 0, ..., n(n+1)d$, let $\mathcal{W}_{r,\chi}^q = H^q(S_r, \mathcal{L}(\chi))$ and $\mathcal{V}_r^q$ be the subgroup of $\mathcal{W}_{r,\chi}^q$ consisting in cohomology classes of compactly supported cocycles. $\mathcal{W}_{r,\chi}^q$ and $\mathcal{V}_r^q$ are Pontryagin dual to finitely generated $\mathcal{O}$-modules.

5.1 Hecke operators and action on the cohomology.

One defines representations of local Hecke algebras R_v at each finite place v of F on $\mathcal{W}_{r,\chi}^q$ and $\mathcal{V}_{r,\chi}^q$.

- For $v \notin S_p$ or if $r = 0$, R_v is the usual local Hecke $\mathcal{O}$-algebra.

- For $v \in S_p$ and $r \geq 1$, let $V = U_1(p^r)$ and V_v be the v-component of V. Consider the semigroup $\Delta_v = V_v D_v V_v$ in $G(F_v)$ where

$$D_v = \{\text{diag}\,(x_1, \ldots, x_n, \nu.x_1^{-1}, \ldots, \nu.x_n^{-1});$$
$$0 \leq ord(x_1),\ 0 \leq ord(x_{i+1}/x_i),\ 0 \leq ord(\nu \cdot x_n^{-2})\}$$

Then, one defines the commutative local Hecke algebra R_v as the convolution algebra of $\mathcal{O}$-valued locally constant functions with support in Δ_v for the Haar measure on $G(F_v)$ which gives the mass 1 to V_v.

Proposition 5.1 *For $v \in S_p$, once we have chosen a uniformizing parameter ϖ_v of $\mathbf{r}_v$, the algebra R_v is isomorphic to a polynomial algebra over the group algebra of $T(\mathbf{r}/p^r\mathbf{r})$: $\mathcal{O}[T(\mathbf{r}/p^r\mathbf{r})][T_{v,0}, \ldots, T_{v,n}]$ where for each $i < n$, $T_{v,i}$ is the V_v-double class of*

$$\mathrm{diag}\,(\underbrace{1,\ldots,1}_{i},\underbrace{\varpi_v,\ldots,\varpi_v}_{n-i},\underbrace{\varpi_v^2,\ldots,\varpi_v^2}_{i},\underbrace{\varpi_v,\ldots,\varpi_v}_{n-i})$$

and $T_{v,n}$ is the double class of

$$\mathrm{diag}\,(\underbrace{1,\ldots,1}_{n},\underbrace{\varpi_v,\ldots,\varpi_v}_{n})$$

Proof: See Sect.2 of [21] and Prop.1 of [45].

For each $v \in S_p$, one can define a natural left action of Δ_v on $I_v/B^+(\mathbf{r}_v)$. (see Section 3 of [21] and Section 3 of [45]). Let $\Delta_p = \prod_{v \in S_p} \Delta_v$; One obtains a left action of Δ_p^{-1} on $L(\chi; \mathcal{O})$ by $(\delta_p^{-1}.f)(x) = f(\delta_p.x)$. Let Δ be the restricted product of Δ_p and $G(\mathbf{A}_F^p)$. For any $\delta \in \Delta$, we have an action of the double class $[V\delta V]$ on $\mathcal{W}_{r,\chi}^q$ and $\mathcal{V}_{r,\chi}^q$: one defines first an algebraic correspondence on $(S_r, \mathcal{L}(\chi, \mathcal{O}))$ using the action of Δ on $\mathcal{L}(\chi, \mathcal{O}))$ through its quotient Δ_p. and one let it act on the cohomology by functoriality. These Hecke operators generate over $\mathcal{O}$ a subalgebra of $\mathrm{End}_{\mathcal{O}}\mathcal{V}_{r,\chi}^q$ denoted by $h_\chi^q(U_1(p^r); \mathcal{O})$. As already mentioned these Hecke algebras form a projective system when $r \geq 1$ varies and their limit is denoted by

$$h_\chi^q(U_1(p^\infty); \mathcal{O}).$$

It acts faithfully on
$$\mathcal{V}_\chi^q = \varinjlim \mathcal{V}_{r,\chi}^q$$

One can define in each Hecke algebra $h_\chi^q(U_1(p^r); \mathcal{O})$ an idempotent e_r called the nearly-ordinary idempotent as follows. Put $t = T_1 \ldots T_n$ and let

$$e_r = \lim_{k \to \infty} t^{k!}$$

These idempotents are compatible when r grows and their projective limit is an idempotent e of $h_\chi^q(U_1(p^\infty); \mathcal{O})$. Let us define the nearly ordinary part of the cohomology, resp. of the Hecke algebra by:

$$\mathcal{V}_\chi^{q\ n.o} = e\mathcal{V}_\chi^q$$

and
$$\text{for } r < \infty, \quad h_\chi^q(U_1(p^r); \mathcal{O})^{n.o} = e_r h_\chi^q(U_1(p^r); \mathcal{O})$$

and
$$\text{for } r = \infty, \quad h_\chi^{q\ n.o} = e h_\chi^q(U_1(p^\infty); \mathcal{O})$$

5.2 Dual construction of the Hida-Iwasawa algebra.

For any $q \in \{0, \ldots, n(n+1)d\}$, we endow $h_\chi^{q \ n.o}$ with a structure of algebra over the Hida-Iwasawa algebra Λ. However, here, this algebra is constructed in a way Langlands' dual to that of Section 4. Let

$$\iota : \mathbf{G}_m \to G$$

be the inclusion (viewed over $\mathbf{r}_p$) of the scalar matrices as central elements of G; the cocharacter ι is Langlands dual to the character $\nu : H = \hat{G} \to \mathbf{G}_m$. For each $r \geq 1$ and $U_r = \{x \in \hat{\mathbf{r}}^\times; x \equiv 1 \bmod. p^r\}$, ι followed by the central action induces

$$\phi_{gl,r} : \mathbf{G}_m(\mathbf{A}_F)/U_r\mathbf{G}_m(F)U_r\mathbf{G}_m(F_\infty)^+ \to h_\chi^q(U_1(p^r); \mathcal{O})^{n.o}, \quad z \mapsto <z> = [V\iota(z)V]$$

The group on the left hand side is the strict ray-class group $Cl_{p^r}^+$ of F of conductor p^r. These homomorphisms are compatible when $r > 0$ varies; hence taking inverse limits, we get:

$$\Phi_{gl} : \mathcal{O}[[Cl_{p^\infty}^+]] \to h_\chi^{q \ n.o}.$$

Consider also

$$\phi_{loc,r} : T(\mathbf{r}/p^r\mathbf{r}) \to h_\chi^q(U_1(p^r); \mathcal{O})^{n.o} \quad [x] \mapsto [VxV]$$

for $x \in T(\mathbf{r}_p)$, viewed in the right-hand side as an idele in Δ with trivial components outside S_p. Similarly, by taking the projective limit of these morphisms, one obtains taking the cartesian product of ϕ_{gl} and ϕ_{loc}, one obtains

$$\phi_{gl} \times \phi_{loc} : Cl_{p^\infty}^+ \times T(\mathbf{r}_p) \to h_\chi^{q \ n.o}$$

As in the Galois case, an amalgamation occurs: the homomorphism $\phi_{gl} \times \phi_{loc}$ factors through

$$H' = Cl_{p^\infty}^+ \underset{\mathbf{r}_p^\times}{\times} T(\mathbf{r}_p)$$

where the first map is

$$\mathbf{r}_p^\times \to \mathbf{r}_p^\times / closure(\mathbf{r}^\times) \subset Cl_{p^\infty}^+$$

and the second is the diagonal embedding

$$\mathbf{G}_m(\mathbf{r}_p) \to T(\mathbf{r}_p)$$

By definition of the Langlands dual of a torus, and by Class-Field Theory, the pro-p-Sylow of H' is isomorphic to the Hida group H of Definition 4.2. To see this, note that the standard maximal torus of $H = \hat{G}$ is canonically the Langlands dual $\hat{T}$ over $\mathcal{O}$ of the maximal torus T; however, since $G_{/\mathcal{O}}$ is split over F, there is a canonical isomorphism $\hat{T} = T$, thus, $X^*(T) = X_*(\hat{T}) = X_*(T)$ and $X^*(T) \times \mathbf{r}_p^\times$ becomes $\mathrm{Hom}(\mathbf{G}_m, T) \times \mathbf{r}_p^\times = T(\mathbf{r}_p)$. Hence the universal nearly ordinary Hecke algebra has a natural structure of algebra over the Hida-Iwasawa algebra $\Lambda = \mathcal{O}[[H]]$. This construction does not depend on the exceptional isomorphism $H \cong G$, hence it is valid for any $n \geq 1$.

5.3 Determination of the generic bottom degree

For $G = GSp_{2n}/F$, let $T = T_n$ be the standard maximal torus.

Definition 5.2 A dominant p-arithmetic character χ of $T = T_n$ is called sufficiently regular if for some constant C depending only on n and F, one has $C \leq a_1$, $C \leq a_2 - a_1, \ldots, C \leq a_n - a_{n-1}$, $C \leq b - 2a_n$.

Proposition 5.3 For $n \leq 2$, for such a character χ, the first degree $q = q(\chi)$ for which the cohomology group $H^q(S_r, \mathcal{L}(\chi; \mathbf{C}))$ does not vanish is the middle degree $\frac{n(n+1)}{2}d$.

Proof: For $n \leq 2$, we want for any $q < \frac{n(n+1)}{2}d$, $H^q(S_r, \mathcal{L}(\chi; \mathbf{C})) = 0$. For that purpose, we use

- the surjectivity of the restriction map $H^q \to H^q_\partial$ for $q > \frac{n(n+1)}{2}d$ which comes from the construction of Eisenstein classes (by Harder [14] for $n = 1$ and Schwermer [34] for $n = 2$, there, sufficient regularity of χ is needed), and

- the absence of ghost classes (see [14] for $n = 1$ and [34] for $n = 2$);

by Poincaré duality, this implies that the restriction maps are 0 for $q < \frac{n(n+1)}{2}d$. Hence, one is left with $H^q_!$. By a theorem of Borel, it embeds in $H^q_{L^2}$ so by a theorem of Knapp, the multiplicity of any representation occuring in H^q_{cusp} is the same as in $H^q_!$ and in $H^q_{L^2}$. Then by the classification of Vogan-Zuckerman of unitary cohomological representations π_∞, by direct inspection for $n = 1$, or by R. Taylor calculations (p.293-294 in [41]) for $n = 2$, one sees that the $(\mathcal{G}, \mathcal{K})$-cohomology of these representations vanishes.

□

6 Conjectures

Let $n \leq 2$ be an integer, F be any totally real field, p an odd rational prime; let us fix a p-arithmetic dominant character χ. Let $\pi = \pi_f \otimes \pi_\infty$ be a cuspidal representation unramified outside p with π_∞ in the holomorphic discrete series; let E be a field of definition for π_f; let $\mathbf{p}$ be a prime of E above p and $\mathcal{O}$ be the completed valuation ring of E at $\mathbf{p}$; the (Borel-) nearly ordinary idempotent e which is *a priori* defined over $\mathcal{O}$, can be defined over E; hence after fixing a complex embedding of E, one can define the **p**-nearly ordinary part, denoted $H^q_{n.o}(S_r, \mathcal{L}(\chi; \mathbf{C}))$ of $H^q(S_r, \mathcal{L}(\chi; \mathbf{C}))$. Let us assume that π occurs in the middle degree cohomology of $H^{\frac{n(n+1)}{2}d}_{n.o}(S_r, \mathcal{L}(\chi; \mathbf{C}))$ for an integer $r \geq 1$. We say that π is **p**-nearly ordinary cohomological. Note that in accordance with our simplifying assumption (B-N.O) made since the beginning of Section 4, we limit ourselves to the case of Borel-type near ordinarity at *each* place $v \in S_p$.

Note that in case π is unramified at all finite places, it also occurs in $H^{\frac{n(n+1)}{2}d}(S_0, \mathcal{L}_0(\chi; E))$, where $\mathcal{L}_0(\chi; E)$ is the local system of E-vector spaces on S_0 defined by the induction from B to G viewed as group schemes over F. It does not depend on p (contrary to the local system $\mathcal{L}(\chi; \mathcal{O})$ which is an induction from $B(\mathbf{r}_p)$ to the p-adic Iwahori subgroup I of $G(\mathbf{r}_p)$).

Let $\lambda_\pi : \mathbf{h}^{n.o}_\chi \to \mathcal{O}$ resp. $\rho_\pi : \mathbf{G}_{F,p} \to \hat{G}(\mathcal{O})$ be the character of the Hecke algebra, resp. the **p**-adic Galois representation, associated to π (in order to insure that ρ_π takes values in $\hat{G}$, recall that, due to our limited knowledge, we must assume that π has multiplicity one, as explained in the Introduction).

Conjecture:(Near Ordinarity Conjecture)

*Weak form: If π is **p**-nearly ordinary cohomological, then ρ_π is p-nearly ordinary;*

more precisely, one has a Strong Form specifying the diagonal characters:

for $n = 1$: for each $v \in S_p$, if one writes:

$$\rho_\pi|_{\mathbf{G}_v} \sim \begin{pmatrix} \nu\chi_1^{-1} & * \\ O & \chi_1 \end{pmatrix}$$

then, for $x \in F_v^\times$, one has

$$\chi_1((x, F_v)) = \lambda_\pi([V \cdot diag(x, 1) \cdot V])$$

where (x, F_v) is the local Artin reciprocity law of x and $V \subset G(\mathbf{r}_v)$ is the local component at v of the level group $U_1(p^r)$,

similarly, for $n = 2$:

$$\rho_\pi|_{\mathbf{G}_v} \sim \begin{pmatrix} \nu\chi_1^{-1} & * & * & * \\ 0 & \nu\chi_2^{-1} & * & * \\ 0 & 0 & \chi_1 & 0 \\ O & O & * & \chi_2 \end{pmatrix}$$

then, for $x \in r_v^\times$, one has

$$\chi_1((x, F_v)) = \mathbf{N}(x)^{-3}\lambda_\pi([V \cdot diag(x^{-1}, x^{-1}, 1, 1)V])$$

and
$$\chi_2((x, F_v)) = \mathbf{N}(x)^{-2}\lambda_\pi([V \cdot diag(x^{-1}, 1, x, 1)V])$$
$V \subset G(\mathbf{r}_v)$ being the local component at v of the level group $U_1(p^r)$ and $\mathbf{N}: F_v^\times \to \mathbf{Q}_p$ is the local norm.

Comments: For $n = 1$, this is a theorem of Hida [18] which relies on A. Wiles' Th.2.2.2 of [49]. If $n = 2$, it is only a conjecture even for $F = \mathbf{Q}$ because of our lack of knowledge of the geometry of the Siegel threefold; for $F = \mathbf{Q}$, the weak statement that the characters χ_1 and χ_2 exist (not their explicit form) *would* follow from [44] provided one had a Katz-Messing type theorem for $H^3(S_1(N), \mathcal{L}_0(\chi, \mathcal{O}))$ (N prime to p); see the comment after Th.7.1 below.

$\bar{\rho}_\pi$ is known to be odd (for $n = 1$) resp. odd II (for $n = 2$). Let us assume that $\bar{\rho}_\pi$ is absolutely irreducible and satisfies (N.O), (Reg) and (Reg*). Then, by Proposition 2.1, there exists a universal ring $R^{n.o}$ of deformations of $\bar{\rho}_\pi$. On the other hand, let us put:
$$\mathbf{h}_\chi^{n.o} = h_\chi^{n.o\ \frac{n(n+1)}{2}d}$$

Let $\lambda_\pi : \mathbf{h}_\chi^{n.o} \to \mathcal{O}$ be the $\mathcal{O}$-algebra homomorphism associated to π (sending a Hecke operator T to the eigenvalue of T on the π_f-isotypic component in $H_{n.o^2}^{\frac{n(n+1)}{2}d}(S_r, \mathcal{L}(\chi; Frac(\mathcal{O}))))$. Let $\bar{\lambda}_\pi$ the reduction of λ_π modulo $\mathcal{M} = \max(\mathcal{O})$. Let $S^{n.o}$ be the completion of $\mathbf{h}_\chi^{n.o}$ along the maximal ideal $Ker(\bar{\lambda}_\pi)$. We call $S^{n.o}$ the universal nearly ordinary Hecke algebra for $\bar{\lambda}_\pi$.

We propose then the following

Conjecture I: *Under the assumptions above, the universal nearly ordinary Hecke algebra $S^{n.o}$ for $\bar{\rho}_\pi$ is finite torsion-free over the Hida-Iwasawa algebra.*

Conjecture II: *There exists a nearly ordinary deformation of $\bar{\rho}_\pi$ to $S^{n.o}$.*

Comment: If Conjecture II holds, the corresponding ring-homomorphism $\phi : R^{n.o} \to S^{n.o}$ is Λ-linear (up to a twist by a power of the global cyclotomic character: for $n = 1$, see [27], Lemme 16) as one can easily check using the strong form of the Near Ordinarity Conjecture above and the formula for the similitude factor of $\rho_\pi(Frob_\ell)$.

Conjecture III: *The homomorphism ϕ is a surjective pseudo-isomorphism; if π is unramified at all finite places and occurs in $H^{\frac{n(n+1)}{2}d}(S_0, \mathcal{L}_0(\chi; E))$, for a $\chi \in X^*(T)$ regular dominant algebraic, then ϕ is an isomorphism for all primes p at which π is nearly ordinary, outside a finite set of primes depending only on χ*

Corollary: *The universal nearly ordinary deformation ring $R^{n.o}$ is finite over Λ, torsion-free in codimension one. Its Krull dimension over $\mathcal{O}$ is therefore equal to $nd + 1 + \delta$ ($n \leq 2$).*

7 Evidences.

For $G = GL_2$ and $F = \mathbb{Q}$, all these conjectures are proven, even with an auxiliary conductor (under some mild conditions, provided one modifies the deformation problem by taking into account the local properties of $\bar{\rho}_\pi$ at primes dividing the auxiliary conductor); see [50], [42]. For $n = 1$ and F totally real, Conjecture I is due to Hida [17] and Conjecture II is proven in [18], and [27] Lemma 16 and [22], Th.6.2.1 (ϕ is proven there to be surjective). Conjecture III is proven in a recent preprint by K. Fujiwara [13] assuming that p does not ramify in F and does not divide the class-number, provided also that π has minimal conductor, and under the same assumptions as in Wiles theorem.

For $n = 2$, one has the following evidence

Theorem 7.1 (*Urban-T.[44], [45]*) *Let $\chi \in X^*(T)$ be a regular dominant algebraic character, then there exists a finite set S_χ of rational primes, depending on F and χ such that for any $p \notin S_\chi$, the p-adic Hecke algebra $\mathbf{h}_\chi^{n.o}$ is finite torsion-free over Λ. In particular, the local component $S^{n.o}$ is finite torsion-free over the Hida-Iwasawa algebra; hence, it is $2d+1+\delta$-dimensional over $\mathcal{O}$. Moreover, assuming moreover that $F = \mathbf{Q}$ and that π has multiplicity one, there exists a Galois representation*

$$\rho_\pi^h : \mathbf{G}_\mathbf{Q} \to \hat{G}(S^{n.o})$$

deforming $\bar{\rho}_\pi$ and satisfying the (universal) Eichler-Shimura relations.

Comment: For $F = \mathbf{Q}$, it is known by a theorem of Faltings (J. A.M.S. 1989) that the module $H^{\cdot}(S_1(N), \mathcal{L}_0(\chi, \mathcal{O}))(\pi_f)$ (p prime to N) is crystalline at p; if one knew moreover that the crystalline Frobenius is annihilated by the automorphic Euler factor of π at p, then it would imply that the deformation ρ_π^h of the Theorem is nearly ordinary (see [44] and [45]).

The approach of [45] to establish the theorem above is cohomological: we use the natural faithful representation of $\mathbf{h}_\chi^{n.o}$ on the middle interior cohomology of the Siegel manifold of level p^∞. This is a method developed in a series of papers by Hida (see in particular [21]). We show the following properties of $\mathbf{h}_\chi^{n.o}$:

1. it does not depend on the weight χ of the local system used

2. it has the control property when localized at "arithmetic" primes in the Hida-Iwasawa algebra Λ,

3. it is torsion-free over Λ,

4. one glues the Galois representations associated to the eigensystems $\lambda_{\pi'}$ occuring in $S^{n.o.}$, assuming the residual absolute irreducibility, using a theorem of L. Nyssen on pseudo-representations. In this way we construct a big Galois representation $\rho^h : \mathbf{G}_{F,p} \to \hat{G}(S^{n.o.})$. Its near ordinarity could be proven using the control theorem provided it is proven for unramified representations π' occuring in $S^{n.o.}$.

Our final result deals with GSp_4 only because of lack of knowledge of the cohomology of arithmetic subgroups of Sp_{2n} in general. Nevertheless, a good deal of the intermediate results, like independence and control theorems for the full cohomology (that is, not interior or cuspidal), are very general; they seem to hold at least for reductive groups of type A,B,C or D. This remark has also been made by Hida [21] and independently by Ash-Stevens [1].

References

[1] A. Ash, G. Stevens: *p-adic deformations of cohomology classes of subgroups of $GL(n,\mathbf{Z})$*, to appear in Proc. Journées Arithmétiques 1995, Barcelone.

[2] D. Blasius, J. Rogawski: *Motives for Hilbert modular forms*, Inv. Math. 114 (1993), 55-87.

[3] D. Blasius, J. Rogawski: *Zeta functions of Shimura varieties* in Motives, PSPM 55, volume 2, AMS 1994.

[4] A.Borel J-P. Serre: *Corners and arithmetic groups*, Comment. Math. Helv. 48, 436-491 (1973).

[5] N. Bourbaki:*Commutative Algebra*, Chapters 1-7, Addison Wesley, 1972.

[6] J.-L. Brylinski, J.-P. Labesse: *Cohomologies d'intersection et fonctions L de certaines variétés de Shimura*, Ann. Sci. Ec. Norm. Sup. 4^{ieme} Série, 17, 361-412 (1984).

[7] H. Carayol: *Sur les représentations p-adiques associées aux formes modulaires de Hilbert*, Ann. Sci. Ecole Norm. Sup. 4^{ieme} Série, 19, 409-468 (1986).

[8] H. Carayol: *Formes modulaires et représentations galoisiennes dans un anneau local complet*, in p-adic Monodromy and the Birch and Swinnerton-Dyer Conjecture, G. Stevens and B. Mazur eds., Contemporary Math. 165, AMS 1994.

[9] C.-L. Chai, G. Faltings: *Degeneration of abelian varieties*, Erg. Math. $3^t e$ Folge, 22, Springer-Verlag 1992.

[10] P. Deligne: *Formes modulaires et représentations ℓ-adiques* Exp. 355, Séminaire Bourbaki 1968.

[11] P. Deligne, J.-P. Serre: *Formes modulaires de poids 1* , Ann. Sci. Ec Norm. Sup., 4^{ieme} Série, 7, 507-530, 1974.

[12] M. Eichler: *Quaternäre quadratische Formen und die Riemannsche Vermutung f'ur die Kongruenzzetafunktion*, Arch. Math. 5, 355-366 (1954).

[13] K. Fujiwara: *Deformation rings and Hecke algebras in the totally real case*, Preprint, Nagoya Univ. 1996.

[14] G. Harder: *Eisenstein cohomology of arithmetic groups: the case GL_2*, Inv. Math. 89, 37-118 (1987).

[15] G. Henniart: *Représentations ℓ-adiques abéliennes*, in Sém. Th. Nombres Paris 1980-81, p.107-126, ed. M.-J. Bertin, Birkhäuser Verlag 1982.

[16] H. Hida: *Galois representations into $GL_2(\mathbf{Z}_p[[X]])$ attached to ordinary cusp forms*, Invent. math., 545-613 (1986).

[17] H. Hida: *On p-adic Hecke algebras for GL_2 over totally real fields*, Ann. of Math., t.128, 1988, p.295-384.

[18] H. Hida: *Nearly ordinary Hecke algebras and Galois representations of several variables*, p.115-134,*in* Algebraic Analysis, Geometry, and Number Theory, Proc. of the JAMI inaugural conference, ed. J.-I. Igusa, Johns Hopkins Univ. Press, Baltimore, 1990.

[19] H. Hida: *p-Ordinary cohomology groups for $SL(2)$ for number fields*. Duke Math. J., Vol. 69, No. 2 (1993).

[20] H. Hida: *p-ordinary cohomology groups for $SL(2)$ over number fields*, Duke Math. J. 69 259-314, 1993.

[21] H. Hida: *Control theorems of p-nearly ordinary cohomology groups for $SL(n)$*, Bull. Soc. Math. France 123, 1995, p.425-475.

[22] H. Hida, J. Tilouine:*On the anticyclotomic main conjecture for CM fields*, Inv. Math.117, 89-147 (1994).

[23] H. Hida, J. Tilouine, E. Urban: *Adjoint modular Galois representations and their Selmer groups*, Proc. of Nat. Acad. Sci. Conf., Washington D.C., March 1996.

[24] R. P. Langlands: *Automorphic representations, Shimura varieties and motives, ein Märchen in* Automorphic forms, representations and L-functions, PSPM XXXIII, vol. 2, AMS 1979.

[25] G. Laumon: *Sur la cohomologie à supports compacts des variétés de Shimura pour $GSp(4)_{\mathbf{Q}}$*, to appear in Comp. Math. in 1997.

[26] B. Mazur: *Deforming Galois representations, in* Galois groups over $\mathbf{Q}$, Publ. MSRI 16, Springer 1990.

[27] B. Mazur, J. Tilouine: *Représentations galoisiennes, Différentielles de Kähler et conjectures principales* , Publ. Math. IHES 71, 1990, 65-103.

[28] J. S. Milne: *Arithmetic duality theorems*, Academic Press 1986.

[29] L. Nyssen: *Thèse de l'Université L. Pasteur*, Strasbourg, 8 Janvier 1996.

[30] M. Ohta: *On the zeta function of an abelian scheme over the Shimura curve*, Japan. J. Math. 9, 1-26 (1983).

[31] A. Panchishkin: *Admissible non-archimedean standard zeta functions associated with Siegel modular forms, in* Motives, PSPM 55, volume II, AMS 1994.

[32] J. Rogawski, J. Tunnell: *On Artin L functions associated to Hilbert modular forms of weight one*, Inv. Math. 74, 1-42 (1983).

[33] A. J. Scholl: *Motives for modular forms*, Inv. Math. 100, 419-430 (1990).

[34] J. Schwermer: *On arithmetic quotients of the Siegel upper half space of degree two* , Comp. Math. 58, 233-258, 1986.

[35] J.-P. Serre: *Sur les représentations modulaires de degré* 2 *de Gal* $(\bar{\mathbf{Q}}/\mathbf{Q})$, Duke Math. J. 54,179-230, 1987.

[36] G. Shimura: *Correspondances modulaires et les fonctions zeta des courbes algébriques*, J. Math. Soc. Japan, 10, 1-28 (1958).

[37] G. Shimura: *On the zeta functions of algebraic curves uniformized by certain automorphic functions*, J. Math. Soc. Japan, 13, 275-331 (1961).

[38] G. Shimura: *On modular correspondences for $Sp(N, \mathbf{Z})$ and their congruence relations*, Proc. NAS 49, 824-828 (1963).

[39] R. Taylor: *On Galois representations associated to Hilbert modular forms* , Inv. Math. 98, 265-280 (1989).

[40] R. Taylor: *Galois representations associated to Siegel modular forms of low weight*, Duke Math. J. 63, 281-332 (1992).

[41] R. Taylor: *On the ℓ-adic cohomology of Siegel threefold*, Inv. Math. 114 (1993), 289-310.

[42] R. Taylor, A. Wiles: *Ring-theoretic properties of certain Hecke algebras*, Ann. of Math. 141, 1995, 553-572.

[43] J. Tilouine: *Deformations of Galois representations and Hecke algebras*, Publ. Mehta Res. Inst., Narosa Publ., Delhi, 1996.

[44] J. Tilouine and E. Urban: *Familles p-adiques à trois variables de formes de Siegel et de représentations galoisiennes* C. R. A. S. Paris, t. 321,Série I,p. 5-10,1995.

[45] J. Tilouine and E. Urban: *Several variables p-adic families of Siegel-Hilbert cusp eigensystems and their Galois representations*, preprint.

[46] E. Urban: *Selmer groups and the Eisenstein-Klingen ideal*, preprint.

[47] A. Weil: *On a certain type of characters of the idèles class-group of an algebraic number field*, [1955c] *in* Collected Works vol. II, Springer Verlag 1983.

[48] R. Weissauer: *A special case of the fundamental lemma: the case GSp_4, I, II, III*, preprints.

[49] A. Wiles: *On p-adic representations for totally real fields*, Ann. of Math. 123, 407-456 (1986).

[50] A. Wiles *Modular elliptic curves and Fermat Last Theorem*, Ann. of Math.141, 1995, 443-551.

J. Tilouine, UA 742, Institut Galilée, Université de Paris-Nord, 93430 Villetaneuse, France
e-mail: tilouine@math.univ-paris13.fr

Part III

ELEMENTARY AND ANALYTIC NUMBER THEORY

A new key identity for Göllnitz' (Big) partition theorem

Krishnaswami Alladi and George E. Andrews

ABSTRACT. A new identity in three free parameters for partitions into distinct parts is proved using Jackson's q-analog of Dougall's summation. This identity is shown to be combinatorially equivalent to a reformulation of a deep partition theorem of Göllnitz. Applications to Schur's partition theorem and Jacobi's triple product identity are indicated.

1. Introduction

In a recent paper [2], by means of a new technique called *the method of weighted words*, substantial refinements and generalizations of the following (Big) theorem of Göllnitz [6] were obtained.

THEOREM G.

Let $C(n)$ denote the number of partitions of n into distinct parts $\equiv 2, 4$, or 5 $(\bmod\ 6)$.

Let $D(n)$ denote the number of partitions $h_1 + h_2 + \cdots + h_\nu$ of n such that $h_\nu \neq 1$ or 3 and $h_\mu - h_{\mu+1} \geq 6$ with strict inequality if $h_\mu \equiv 6, 7$ or $9 \,(\bmod\ 6)$.
Then $C(n) = D(n)$.

In Theorem G and throughout we have adopted the convention that $n \equiv \ell (\bmod\ M)$ means $n = tM + \ell$ with $n > 0$ and $t \geq 0$.

The main objective in [2] was to prove the following *key identity*

(1.1)
$$\sum_{i,j,k} a^i b^j c^k \sum_{\substack{i=\alpha+\delta+\varepsilon \\ j=\beta+\delta+\phi \\ k=\gamma+\varepsilon+\phi}} \frac{q^{T_s+T_\delta+T_\varepsilon+T_{\phi-1}}(1-q^\alpha(1-q^\phi))}{(q)_\alpha (q)_\beta (q)_\gamma (q)_\delta (q)_\varepsilon (q)_\phi} = (-aq)_\infty (-bq)_\infty (-cq)_\infty$$

and to view a three parameter refinement of Theorem G as emerging from (1.1) under the transformations

(1.2) $\begin{cases} \text{(dilation)}\ q \mapsto q^6, \\ \text{(translations)}\ a \mapsto aq^{-4},\ b \mapsto bq^{-2},\ c \mapsto cq^{-1}. \end{cases}$

1991 *Mathematics Subject Classification.* Primary: 05A15, 05A17, 11P83; Secondary: 05A19.
Key words and phrases. Göllnitz' theorem, partitions, new key identity.
Research of both authors were supported by grants from the National Science Foundation.

© 1998 American Mathematical Society

In (1.1) and elsewhere, $s = \alpha + \beta + \gamma + \delta + \varepsilon + \phi$ and $T_n = n(n+1)/2$ is the n-th triangular number. We shall refer to (1.2) as the *standard transformations* for Göllnitz' theorem. In (1.1) we have made use of standard notation

$$(a)_n = (a;q)_n = \prod_{j=0}^{n-1}(1-aq^j)$$

for any complex number a and positive integer n, and

$$(a)_\infty = \lim_{n\to\infty}(a)_n = \prod_{j=0}^{\infty}(1-aq^j), \text{ for } |q| < 1.$$

Our purpose here is to prove (see §4) the *new key identity*

(1.3) $$\sum_{i,j,k} \frac{a^i b^j c^k (-c)_i (-c)_j (-\frac{ab}{c}q)_k (-cq)_{i+j} q^{T_i+j+k}}{(q)_i (q)_j (q)_k (-c)_{i+j}} = (-aq)_\infty (-bq)_\infty (-cq)_\infty$$

and to show (see §3) that (1.3) is equivalent to a three parameter refinement of Theorem G.

While it is obvious that under the transformations (1.2) the product on the right in (1.1) and (1.3) is the generating function of a three parameter refinement of $C(n)$ in Theorem G, it is not at all clear (and indeed difficult to demonstrate) that the series on the left in (1.1) represents the generating function of a three parameter refinement of $D(n)$. In [2] we show that the series in (1.1) is the generating function for partitions into parts occuring in six possible colors and satisfying certain gap conditions which under the transformations (1.2) become the difference conditions governing $D(n)$. Identity (1.1) is very deep and its proof requires not only Watson's q-analogue of Whipple's transformation but the $_6\psi_6$ summation of Bailey as well (see [2]).

It is often the case that partition identities which can be refined can be proved combinatorially because the set up for the bijection is naturally suggested by the refinement. When Göllnitz [6] proved Theorem G in 1967, he established it in a refined form but no combinatorial proof of Theorem G was known. The multiparameter refinement given by (1.1) was a further compelling reason that a combinatorial proof or explanation ought to exist.

For combinatorial purposes, Alladi [1] recently noticed that it is preferable to consider the effect of the *cubic transformation*

(1.4) $$\begin{cases} \text{(dilation) } q \mapsto q^3, \\ \text{(translations) } a \mapsto aq^{-2},\ b \mapsto bq^{-1},\ c \mapsto c, \end{cases}$$

on (1.1). Then the product on the right in (1.1) becomes

(1.5) $$(-aq;q^3)_\infty (-bq^2;q^3)_\infty (-cq^3;q^3)_\infty$$

which is the refined generating function for partitions into distinct parts. Then the series on the left in (1.1) corresponds to the generating function for partitions with difference ≥ 3 between parts where each partition is counted with a certain weight. This leads to Theorem 2 of §2 which is a reformulation of Theorem G. The importance of Theorem 2 is that an explicit combinatorial bijection can be constructed to explain its validity (see [1]).

Utilizing extensions of certain ideas due to Bressoud [4] we show in §3 that under the cubic transformations (1.4), identity (1.3) is combinatorially equivalent to Theorem 2. Although simpler than (1.1), the new key identity (1.3) is still quite deep; its proof which makes use of Jackson's q-analog of Dougall's summation is given in §4.

It was already pointed out in [1] that Theorem 2 has several interesting applications. Some of these are mentioned briefly in §5 in the context of the new key identity.

2. A reformulation

First we briefly review the principal ideas in [2] involving the method of weighted words.

We consider the integer 1 as occuring in three primary colors a, b and c, and integers $n \geq 2$ as occuring in six colors, three of which a, b and c are primary, and the remaining three ab, ac and bc are secondary. We also use the letters d, e and f to signify ab, ac and bc respectively. An integer n occuring in color a will be denoted by the symbol a_n, while occuring in color b will be denoted by b_n, and so on. The usefulness of the method is because the alphabets a, b, c play a dual role; on the one hand they represent colors and on the other they are free parameters.

In order to define partitions we need an ordering of the symbols and the one we choose is

(2.1) $\qquad a_1 < b_1 < c_1 < d_2 < e_2 < a_2 < f_2 < b_2 < c_2 < d_3 < \ldots$

The effect of the standard transformations (1.2) is to convert the symbols to
(2.2)
$$\begin{cases} a_n \mapsto 6n-4, \ b_n \mapsto 6n-2, \ c_n \mapsto 6n-1, \text{ for } n \geq 1, \\ d_n = (ab)_n \mapsto 6n-6, \ e_n = (ac)_n \mapsto 6n-5, \ f_n = (bc)_n \mapsto 6n-3, \text{ for } n \geq 2. \end{cases}$$

and so (2.1) becomes

(2.3) $\qquad 2 < 4 < 5 < 6 < 7 < 8 < 9 < 10 < 11 < 12 < \ldots,$

the standard ordering among the integers. In Göllnitz' theorem the primary colors correspond to the residue classes 2,4 and 5(mod 6) and so the secondary colors are $2 + 4 \equiv 6$, $2 + 5 \equiv 7$ and $4 + 5 \equiv 9 \pmod 6$. Note that integers of secondary color occur only when $n \geq 2$ and so e_1 and f_1 are missing in (2.1). This is why integers 1 and 3 do not appear in (2.3) and this also explains the absence of 1 and 3 among the parts enumerated by $D(n)$ in Theorem G.

By a partition (word) π we mean a collection of symbols in non-increasing order as given by (2.1). By $\sigma(\pi)$ we mean the sum of the subscripts of the symbols in π. For example $d_3 c_2 c_2 a_2 c_1 a_1 a_1$ is a partition π of the integer $\sigma(\pi) = 12$. Also the gap between any two symbols is the absolute value of the difference between their subscripts. By a vector partition $\pi = (\pi_1; \pi_2; \pi_3)$ of n we mean that the π_i are partitions and that $\sigma(\pi) = \sigma(\pi_1) + \sigma(\pi_2) + \sigma(\pi_3) = n$.

The partitions of principal interest to us are Type 1 partitions π which are of the form $x_1 + x_2 + \ldots$, where the x_i are symbols from (2.1) with the condition that the gap between x_i and x_{i+1} is ≥ 1 with strict inequality if x_i is of secondary color. The principal result of [2] which is a substantial extension of Theorem G is:

THEOREM 1.
Let $C(n; i, j, k)$ denote the number of vector partitions $(\pi_1; \pi_2; \pi_3)$ of n such that π_1 has i distinct a-parts, π_2 has j distinct b-parts, and π_3 has k distinct c-parts.

Let $D(n; \alpha, \beta, \gamma, \delta, \varepsilon, \phi)$ denote the number of Type 1 partitions of n having α a-parts, β b-parts, ..., and ϕ f-parts.

Then
$$C(n; i, j, k) = \sum_{\substack{i=\alpha+\delta+\varepsilon \\ j=\beta+\delta+\phi \\ k=\gamma+\varepsilon+\phi}} D(n; \alpha, \beta, \gamma, \delta, \varepsilon, \phi).$$

It is clear that

(2.4) $$\sum_{i,j,k,n} C(n; i, j, k) a^i b^j c^k q^n = (-aq)_\infty (-bq)_\infty (-cq)_\infty.$$

It turns out that

(2.5) $$\sum_n D(n; \alpha, \beta, \gamma, \delta, \varepsilon, \phi) q^n = \frac{q^{T_\delta + T_\delta + T_\varepsilon + T_{\phi-1}}(1 - q^\alpha(1 - q^\phi))}{(q)_\alpha (q)_\beta (q)_\gamma (q)_\delta (q)_\varepsilon (q)_\phi}.$$

The proof of (2.5) is quite involved and goes by induction on $s = \alpha+\beta+\gamma+\delta+\varepsilon+\phi$, the length of the word, and by appeal to minimal partitions (see [2]).

We now describe the main ideas in [1].

Consider the effect of the cubic transformation (1.4) on the key identity (1.1). The product on the right in (1.1) becomes the product in (1.5). Under the transformation (1.4), the symbols in (2.1) become

(2.6) $$\begin{cases} a_n \mapsto 3n-2, \ b_n \mapsto 3n-1, \ c_n \mapsto 3n, \text{ for } n \geq 1, \\ d_n \mapsto 3n-3, \ e_n \mapsto 3n-2, \ f_n \mapsto 3n-1, \text{ for } n \geq 2. \end{cases}$$

Notice that in (2.6) each integer $n \geq 3$ occurs in two colors, one primary and one secondary. More precisely, if $n \equiv 3 \pmod 3$, then n occurs in colors c and ab, if $n \equiv 4 \pmod 3$, then n occurs in colors a and ac, and if $n \equiv 5 \pmod 3$, then n occurs in colors b and bc. The integer 1 occurs only in color a and the integer 2 only in color b. Thus under the cubic transformation, the function $C(n; i, j, k)$ in Theorem 1 enumerates the number of partitions of n into distinct parts of which i are $\equiv 1 \pmod 3$, j are $\equiv 2 \pmod 3$ and k are $\equiv 3 \pmod 3$. On the other hand the function D enumerates partitions of n into parts differing by ≥ 3 with each partition being counted with a weight as described below.

Let $\tilde{\pi}: h_1 + h_2 + \cdots + h_\nu$ be a partition with $h_\mu - h_{\mu+1} \geq 3$ for $1 \leq \mu < \nu$. We adopt the convention $h_{\nu+1} = -1$. Define the weight of each part h_μ as follows:

Congruence	Gap	Weight
$h_\mu \equiv 1 \pmod 3$	$h_\mu - h_{\mu+1} \leq 3$	a
$h_\mu \equiv 2 \pmod 3$	$h_\mu - h_{\mu+1} \leq 3$	b
$h_\mu \equiv 3 \pmod 3$	$h_\mu - h_{\mu+1} \leq 3$	c
$h_\mu \equiv 1 \pmod 3$	$h_\mu - h_{\mu+1} > 3$	$a + ac$
$h_\mu \equiv 2 \pmod 3$	$h_\mu - h_{\mu+1} > 3$	$b + bc$
$h_\mu \equiv 3 \pmod 3$	$h_\mu - h_{\mu+1} > 3$	$c + ab$

Finally, the weight $w(\tilde{\pi})$ of the partition $\tilde{\pi}$ is defined multiplicatively as

$$(2.7) \qquad w(\tilde{\pi}) = \prod_{\mu=1}^{\nu} w(h_\mu).$$

In [1] Alladi showed that under the effect of the cubic transformation, Theorem 1 becomes

THEOREM 2.
Let $\mathcal{D}$ denote the set of all partitions into distinct parts. For $\pi \in \mathcal{D}$, let $\nu_\ell(\pi')$ denote the number of parts of π' which are $\equiv \ell \pmod{3}$, $\ell = 1, 2, 3$.
Let $\mathcal{D}_3$ denote the set of all partitions into parts differing by ≥ 3. For $\tilde{\pi} \in \mathcal{D}_3$ let $w(\tilde{\pi})$ be as in (2.7).
As usual let $\sigma(\pi)$ denote the sum of the parts of a partition π. Then

$$\sum_{\substack{\pi' \in \mathcal{D} \\ \sigma(\pi')=n}} a^{\nu_1(\pi')} b^{\nu_2(\pi')} c^{\nu_3(\pi')} = \sum_{\substack{\tilde{\pi} \in \mathcal{D}_3 \\ \sigma(\tilde{\pi})=n}} w(\tilde{\pi}).$$

It is to be noted that while Theorem 2 is equivalent to a three parameter refinement of Theorem G, setting $a = b = c = 1$ in Theorem 2 does not yield a result equivalent to Göllnitz' theorem in the unrefined form.

In the next section, by means of a combinatorial study of partitions $\tilde{\pi} \in \mathcal{D}_3$, we will show that Theorem 2 is equivalent to the new key identity (1.3) under the cubic transformations (1.4).

3. Combinatorial equivalence

Let $\tilde{\pi}: h_1 + h_2 + \cdots + h_\nu$ satisfy $h_\mu - h_{\mu+1} \geq 3$ for $1 \leq \mu < \nu$. Put $h_{\nu+1} = -1$. Also let $\nu_1(\tilde{\pi}) = i$, $\nu_2(\tilde{\pi}) = j$ and $\nu_3(\tilde{\pi}) = k$ with ν_ℓ as in Theorem 2. So $i+j+k = \nu$. Subtract 0 from the smallest part h_ν, 3 from $h_{\nu-1}$, 6 from $h_{\nu-2}, \ldots, 3\nu - 3$ from h_1, and define

$$h^*_\mu = h_\mu - 3(\nu - \mu), \quad 1 \leq \mu \leq \nu.$$

Separate the integers h^*_μ into three groups G_1, G_2, G_3, where G_ℓ consists of the parts h^*_μ which are $\equiv \ell \pmod{3}$, for $\ell = 1, 2, 3$. Observe that $|G_1| = i$, $|G_2| = j$ and $|G_3| = k$. Note that it is possible for parts in each of the groups G_i to repeat.

Case 1: $h_\nu = 1$.
Redistribute the multiples of 3 as follows. Add 0 to the smallest part of G_1, 3 to the second smallest part of $G_1, \ldots, 3i - 3$ to the largest part of G_1 to get a partition π_1 into i distinct parts $\equiv 1 \pmod{3}$. Next add $3i$ to the smallest part of G_2, $3i + 3$ to the next smallest part of $G_2, \ldots, 3(i+j) - 3$ to the largest part of G_2 to get a partition π_2 into j distinct parts $\equiv 2 \pmod{3}$. Finally add $3(i+j)$ to the smallest part of G_3, $3(i+j)+3$ to the next smallest part of $G_3, \ldots, 3(i+j+k) - 3$ to the largest part of G_3 to get a partition π_3 into distinct parts $\equiv 3 \pmod{3}$.

Case 2: $h_\nu = 2$.
Redistribute the multiples of 3 as follows. Add 0 to the smallest part of G_2, 3 to the next smallest part of $G_2, \ldots, 3j - 3$ to the largest part of G_2. Then add $3j$ to the smallest part of G_1, $3j + 3$ to the next smallest part of $G_1, \ldots, 3(j+i) - 3$ to the largest part of G_1. Finally, distribute $3(j+i), 3(j+i)+3, \ldots, 3(j+i+k) - 3$ to the parts of G_3 exactly as in Case 1. Here also we have created partitions π_2, π_1, π_3 into distinct parts $\equiv 2, 1$ and $3 \pmod{3}$ respectively.

Case 3: $h_\nu = 3$.
Redistribute the multiples of 3 exactly as in Case 1.

Observe that after the redistribution, the partition $\tilde{\pi} \in \mathcal{D}_3$ is converted to a vector partition (π_1, π_2, π_3) with the following properties remaining invariant.
(i) $\pi_\ell \in \mathcal{D}_3$, for $\ell = 1, 2, 3$.
(ii) $\nu(\pi_\ell) = \nu_\ell(\tilde{\pi})$, for $\ell = 1, 2, 3$
(iii) The number of gaps between the parts of π_ℓ and $h_{\nu+1} = -1$ which are > 3 equals the number of parts h_μ of $\tilde{\pi}$ which satisfy $h_\mu \equiv \ell \pmod{3}$ and $h_\mu - h_{\mu+1} > 3$.

In view of invariants (i), (ii) and (iii), it follows that

$$(3.1) \qquad w(\tilde{\pi}) = w(\pi_1) w(\pi_2) w(\pi_3)$$

with w defined as in (2.7). Our purpose in replacing $\tilde{\pi}$ by (π_1, π_2, π_3) is because the generating function of this vector partition is more easy to calculate once we observe the following inequalities satisfied by their parts.

Case 1: $\lambda(\pi_1) = 1$, $\lambda(\pi_2) \geq 3i + 2$, $\lambda(\pi_3) \geq 3(i+j) + 3$
Case 2: $\lambda(\pi_2) = 2$, $\lambda(\pi_1) \geq 3j + 4$, $\lambda(\pi_3) \geq 3(i+j) + 3$
Case 3: All parts ≥ 3, $\lambda(\pi_2) \geq 3i + 5$, $\lambda(\pi_3) \geq 3(i+j) + 3$.

In the above inequalities, $\lambda(\pi_\ell)$ denotes the least part of π_ℓ.

We will now obtain the generating function of partitions π_1 in Case 1.

The partitions π_1 in Case 1 have i distinct parts $h_1 + h_2 + \cdots + h_i$ each $\equiv 1 \pmod{3}$ with $h_i = 1$. Each part h_μ is counted with weight a if $h_\mu - h_{\mu+1} \leq 3$, and with weight $a + ac$ if $h_\mu - h_{\mu+1} > 3$. For this purpose we take $h_{i+1} = -1$. Let $h_{t_1}, h_{t_2}, \ldots, h_{t_r}$ be the parts which are counted with weights $a(1+c)$. Now subtract 3 from each of the parts $h_1, h_2, \ldots, h_{t_1}$, to form new parts $h_1^{(1)}, h_2^{(1)}, \ldots, h_{t_1}^{(1)}$. The total amount subtracted is $3t_1$. Note that we have $h_1^{(1)} > h_2^{(1)} > \ldots h_{t_1}^{(1)} > h_{t_1+1} \cdots > h_i$. Now subtract 3 from $h_1^{(1)}, h_2^{(1)}, \ldots, h_{t_1}^{(1)}, \ldots, h_{t_2}$ to get new parts $h_1^{(2)} > h_2^{(2)}, \ldots, > h_{t_2}^{(2)}$. The total amount subtracted is $3t_2$. Observe that $h_1^{(2)} > h_2^{(2)} \ldots h_{t_2}^{(2)} > \cdots > h_i$. Continue this procedure until $3t_r$ is subtracted off at the rth stage. Thus the partition π_1 has been decomposed into a pair of partitions (π'_1, π''_1) where

$$(3.2) \quad \begin{aligned} \pi'_1 &= h_1^{(r)} + h_2^{(r)} + \cdots + h_{t_r}^{(r)} + h_{t_r+1} + \cdots + h_i \\ \pi''_1 &= 3t_r + 3t_{r-1} + \cdots + 3t_1. \end{aligned}$$

Note that all parts of π'_1 are distinct and $\equiv 1 \pmod{3}$. Also all parts of π''_1 are distinct with $1 \leq t_1 < t_2 < \cdots < t_r < i$. The partition π'_1 is counted with weight a^i and the partition π''_1 is counted with weight c^r. Thus the generating function of partitions (π'_1, π''_1) is

$$(3.3) \qquad \frac{a^i q^{(3i^2-i)/2}(-cq^3; q^3)_{i-1}}{(q^3; q^3)_{i-1}}.$$

The generating functions for the partitions π_2, π_3 in Case 1 as well as for all partitions in Cases 2 and 3 can be computed by utilizing the decomposition process underlying (3.2). We do not give the cumbersome details because they are similar to (3.2). So we now simply state the generating function for each of the three cases.

The generating function for Case 1:
(3.4)
$$\frac{a^i q^{(3i^2-i)/2}(-cq^3;q^3)_{i-1}}{(q^3;q^3)_{i-1}} \cdot \frac{b^j q^{3ij} q^{(3j^2+j)/2}(-c;q^3)_j}{(q^3;q^3)_j} \cdot \frac{c^k q^{3(i+j)k} q^{3(k^2+k)/2}(\frac{-ab}{c};q^3)_k}{(q^3;q^3)_k}$$

The generating function for Case 2:
(3.5)
$$\frac{b^j q^{(3j^2+j)/2}(-cq^3;q^3)_{j-1}}{(q^3;q^3)_{j-1}} \cdot \frac{a^i q^{3ij} q^{(3i^2+5i)/2}(-c;q^3)_i}{(q^3;q^3)_i} \cdot \frac{c^k q^{3(i+j)k} q^{3(k^2+k)/2}(\frac{-ab}{c};q^3)_k}{(q^3;q^3)_k}$$

The generating function for Case 3:
(3.6)
$$\frac{a^i q^{(3i^2+5i)/2}(-c;q^3)_i}{(q^3;q^3)_i} \cdot \frac{b^j q^{3ij} q^{(3j^2+7j)/2}(-c;q^3)_j}{(q^3;q^3)_j} \cdot \frac{c^k q^{3(i+j)k} q^{3(k^2+k)/2}(\frac{-ab}{c};q^3)_k}{(q^3;q^3)_k}$$

So by (3.4), (3.5) and (3.6), Theorem 2 is equivalent to the identity
(3.7)
$$\sum_{i,j,k} \frac{a^i q^{(3i^2-i)/2}(-cq^3;q^3)_{i-1}}{(q^3;q^3)_{i-1}} \frac{b^j q^{3ij} q^{(3j^2+j)/2}(-c;q^3)_j}{(q^3;q^3)_j} \frac{c^k q^{3(i+j)k} q^{3(k^2+k)/2}(\frac{-ab}{c};q^3)_k}{(q^3;q^3)_k}$$
$$+ \sum_{j,i,k} \frac{b^j q^{(3j^2+j)/2}(-cq^3;q^3)_{j-1}}{(q^3;q^3)_{j-1}} \frac{a^i q^{3ij} q^{(3i^2+5i)/2}(-c;q^3)_i}{(q^3;q^3)_i} \frac{c^k q^{3(i+j)k} q^{3(k^2+k)/2}(\frac{-ab}{c};q^3)_k}{(q^3;q^3)_k}$$
$$+ \sum_{i,j,k} \frac{a^i q^{(3i^2+5i)/2}(-c;q^3)_i}{(q^3;q^3)_i} \frac{b^j q^{3ij} q^{(3j^2+7j)/2}(-c;q^3)_j}{(q^3;q^3)_j} \frac{c^k q^{3(i+j)k} q^{3(k^2+k)/2}(\frac{-ab}{c};q^3)_k}{(q^3;q^3)_k}$$
$$= (-aq;q^3)_\infty (-bq^2;q^3)_\infty (-cq^3;q^3)_\infty.$$

The three triple summations on the left in (3.7) can be amalgamated into a single triple summation as we show next.

First consider the sum of all terms for which $i = 0$. This gives

$$\sum_{j\geq 1}\sum_{k\geq 0} \frac{b^j q^{(3j^2+j)/2}(-cq^3;q^3)_{j-1}}{(q^3;q^3)_{j-1}} \cdot \frac{c^k q^{3jk} q^{3(k^2+k)/2}(\frac{-ab}{c};q^3)_k}{(q^3;q^3)_k}$$
$$+ \sum_{j\geq 0}\sum_{k\geq 0} \frac{b^j q^{(3j^2+7j)/2}(-c;q^3)_j}{(q^3;q^3)_j} \cdot \frac{c^k q^{3jk} q^{3(k^2+k)/2}(\frac{-ab}{c};q^3)_k}{(q^3;q^3)_k}$$
(3.8)
$$= \sum_{k\geq 0} \frac{c^k q^{3(k^2+k)/2}(\frac{-ab}{c};q^3)_k}{(q^3;q^3)_k}$$
$$+ \sum_{j\geq 1}\sum_{k\geq 0} \frac{b^j q^{(3j^2+j)/2} q^{3j}(1+c)(-cq^3;q^3)_{j-1}}{(q^3;q^3)_j} \frac{c^k q^{3jk} q^{3(k^2+k)/2}(\frac{-ab}{c};q^3)_k}{(q^3;q^3)_k}$$
$$+ \sum_{j\geq 1}\sum_{k\geq 0} \frac{b^j q^{(3j^2+j)/2}(1-q^{3j})(-cq^3;q^3)_{j-1}}{(q^3;q^3)_j} \frac{c^k q^{3jk} q^{3(k^2+k)/2}(\frac{-ab}{c};q^3)_k}{(q^3;q^3)_k}.$$

Observe that
$$q^{3j}(1+c) + (1-q^{3j}) = 1 + cq^{3j}$$
and so the expression in (3.8) becomes

(3.9)
$$\sum_{j \geq 0} \sum_{k \geq 0} \frac{b^j q^{(3j^2+j)/2}(-cq^3;q^3)_j}{(q^3;q^3)_j} \cdot \frac{c^k q^{3jk} q^{3(k^2+k)/2}(\frac{-ab}{c};q^3)_k}{(q^3;q^3)_k}.$$

Similarly the sum of all terms having $j = 0$ in (3.7) will yield

(3.10)
$$\sum_{i \geq 0} \sum_{k \geq 0} \frac{a^i q^{(3i^2-i)/2}(-cq^3;q^3)_i}{(q^3;q^3)_i} \cdot \frac{c^k q^{3ik} q^{3(k^2+k)/2}(\frac{-ab}{c};q^3)_k}{(q^3;q^3)_k}.$$

Finally, the sum over all terms on the left in (3.7) with $i \geq 1$, $j \geq 1$ is
(3.11)
$$\sum_{i \geq 1} \sum_{j \geq 1} \sum_{k \geq 0} \frac{a^i(-cq^3;q^3)_{i-1}}{(q^3;q^3)_i} \cdot \frac{b^j q^{3ij}(-cq^3;q^3)_{j-1}}{(q^3;q^3)_j} \cdot \frac{c^k q^{3(i+j)k} q^{3(k^2+k)/2}(\frac{-ab}{c};q^3)_k \cdot f}{(q^3;q^3)_k}$$

where
(3.12)
$$f = q^{(3i^2+5i)/2}(1+c)q^{(3j^2+7j)/2}(1+c) + q^{(3i^2-i)/2}(1-q^{3i})(1+c)q^{(3j^2+j)/2}$$
$$+ q^{(3j^2+j)/2}(1-q^{3j})(1+c)q^{(3i^2+5i)/2}$$
$$= (1+c)q^{(3i^2-i)/2}q^{(3j^2+j)/2}(1+cq^{3(i+j)}).$$

In view of (3.12), the expression in (3.11) can be rewritten as

(3.13)
$$\sum_{i \geq 1} \sum_{j \geq 1} \sum_{k \geq 0} \frac{a^i q^{(3i^2-i)/2}(-c;q^3)_i}{(q^3;q^3)_i} \cdot \frac{b^j q^{3ij} q^{(3j^2+j)/2}(-c;q^3)_j}{(q^3;q^3)_j} \left(\frac{1+cq^{3(i+j)}}{1+c}\right)$$
$$\times \frac{c^k q^{3(i+j)k} q^{3(k^2+k)/2}(\frac{-ab}{c};q^3)_k}{(q^3;q^3)_k}.$$

Observe that if we formally put $i = 0$ in (3.13) we get (3.9), whereas if we formally set $j = 0$ in (3.13) we get (3.10). Thus from (3.8) thru (3.13) we see that (3.7) can be recast in the form

(3.14)
$$\sum_{i \geq 0} \sum_{j \geq 0} \sum_{k \geq 0} \frac{a^i q^{(3i^2-i)/2}(-c;q^3)_i}{(q^3;q^3)_i} \cdot \frac{b^j q^{3ij} q^{(3j^2+j)/2}(-c;q^3)_j}{(q^3;q^3)_j} \left(\frac{1+cq^{3(i+j)}}{1+c}\right)$$
$$\times \frac{c^k q^{3(i+j)k} q^{3(k^2+k)/2}(\frac{-ab}{c};q^3)_k}{(q^3;q^3)_k}$$
$$= (-aq;q^3)_\infty (-bq^2;q^3)_\infty (-cq^3;q^3)_\infty.$$

Thus Theorem 2 is the combinatorial version of (3.14). It is easy to see that under the cubic transformations (1.4), equation (1.3) becomes (3.14).

In the next section, we give a purely q-series proof of (1.3).

4. Proof of the new key identity.

The symbol $(a)_n$ introduced in §1 satisfies the relation

(4.1) $$(a)_n = \frac{(a)_\infty}{(aq^n)_\infty}.$$

The advantage in (4.1) is that it can be used to define $(a)_n$ for all real values of n. In particular, from (4.1) it follows that

(4.2) $$\frac{1}{(q)_n} = 0, \text{ for } n = -1, -2, -3, \ldots.$$

We begin by rewriting (1.3) in the form

(4.3) $$\sum_{i,j,k} \frac{a^i b^j c^k q^{T_{i+j+k}} (-c)_i (-c)_j (-\frac{abq}{c})_k (1 + cq^{i+j})}{(q)_i (q)_j (q)_k} = (-aq)_\infty (-bq)_\infty (-c)_\infty.$$

At this point we use two classical expansions, namely,

(4.4) $$(-zq)_\infty = \sum_{m=0}^{\infty} \frac{z^m q^{T_m}}{(q)_m}$$

with $z = a$, and $z = b$ in (4.3), and

(4.5) $$(-zq)_n = \sum_{m=0}^{n} z^m q^{T_m} \frac{(q)_n}{(q)_m (q)_{n-m}}$$

with $z = ab/c$ to rewrite (4.3) as

(4.6) $$\sum_{i,j,k,\ell} \frac{a^i b^j c^k q^{T_{i+j+k}} (-c)_i (-c)_j a^\ell b^\ell q^{T_\ell} c^{-\ell} (1 + cq^{i+j})}{(q)_i (q)_j (q)_\ell (q)_{k-\ell}}$$
$$= \sum_{i,j} \frac{a^i b^j q^{T_i + T_j} (-c)_\infty}{(q)_i (q)_j}.$$

Thus (4.3) will be proved if we show that the coefficient of $a^i b^j$ on both sides of (4.6) is the same. That is replacing $i + \ell$ by i and $j + \ell$ by j on the left side of (4.6), we wish to show that

(4.7) $$\sum_{k,\ell \geq 0} \frac{c^k q^{T_{i+j+k-2\ell}} (-c)_{i-\ell} (-c)_{j-\ell} q^{T_\ell} c^{-\ell} (1 + cq^{i+j-2\ell})}{(q)_{i-\ell} (q)_{j-\ell} (q)_\ell (q)_{k-\ell}} = \frac{q^{T_i + T_j} (-c)_\infty}{(q)_i (q)_j}.$$

Note that the term $1/(q)_{k-\ell}$ on the left becomes zero when $\ell > k$. Thus the sum on the left in (4.7) is over $k \geq \ell$. So, replacing $k - \ell$ by k in (4.7) gives

(4.8) $$\sum_{k,\ell \geq 0} \frac{c^k q^{T_{i+j+k-\ell}} (-c)_{i-\ell} (-c)_{j-\ell} q^{T_\ell} (1 + cq^{i+j-2\ell})}{(q)_{i-\ell} (q)_{j-\ell} (q)_\ell (q)_k} = \frac{q^{T_i + T_j} (-c)_\infty}{(q)_i (q)_j}.$$

On the left, for fixed ℓ, the sum over k is

(4.9) $$\sum_{k=0}^{\infty} \frac{c^k q^{T_k + k(i+j-\ell) + T_{i+j-\ell}}}{(q)_k} = q^{T_{i+j-\ell}} (-cq^{i+j-\ell+1})_\infty$$

by (4.4). So we must prove that
(4.10)
$$\sum_{\ell \geq 0} \frac{q^{T_{i+j-\ell}+T_\ell}(-cq^{i+j-\ell+1})_\infty (-c)_{i-\ell}(-c)_{j-\ell}(1+cq^{i+j-2\ell})}{(q)_{i-\ell}(q)_{j-\ell}(q)_\ell} = \frac{q^{T_i+T_j}(-c)_\infty}{(q)_i(q)_j}.$$

At this point we observe the relations

(4.11)
$$\frac{(A)_{i-\ell}}{(B)_{i-\ell}} = \frac{(A)_i}{(B)_i} \frac{(1-Bq^{i-\ell})\cdots(1-Bq^{i-1})}{(1-Aq^{i-\ell})\cdots(1-Aq^{i-1})}$$
$$= \frac{(A)_i}{(B)_i} \left(\frac{B}{A}\right)^\ell \frac{(q^{1-i}/B)_\ell}{(q^{1-i}/A)_\ell},$$

and

(4.12)
$$(Aq^{-\ell})_\infty = (1-Aq^{-\ell})(1-Aq^{-\ell+1})\cdots(1-Aq^{-1})(A)_\infty$$
$$= (-1)^\ell q^{-T_\ell} A^\ell \left(\frac{q}{A}\right)_\ell (A)_\infty.$$

Using (4.11) and (4.12) and the relation
$$T_{i+j-\ell} = T_{i+j} + T_{-\ell} - \ell(i+j) = T_i + T_j + T_{-\ell} + ij - i\ell - j\ell,$$

(4.10) can be recast in the form
(4.13)
$$\frac{(-c)_i(-c)_j}{(q)_i(q)_j}(-cq^{i+j+1})_\infty \sum_{\ell \geq 0} q^{T_{i+j-\ell}}(-1)^\ell (-cq^{i+j+1})^\ell \cdot \left(\frac{-1}{cq^{i+j}}\right)_\ell \left(\frac{c}{q}\right)^{-\ell} \left(\frac{c}{q}\right)^{-\ell}$$
$$\times \frac{(q^{-i})_\ell(q^{-j})_\ell}{(-\frac{q^{1-i}}{c})_\ell(-\frac{q^{1-j}}{c})_\ell} \cdot \frac{(1+cq^{i+j-2\ell})}{(q)_\ell} = \frac{q^{T_i+T_j}(-c)_\infty}{(q)_i(q)_j}.$$

Cancelling common factors in (4.13) reduces it to
(4.14)
$$\sum_{\ell \geq 0} \frac{q^{T_{-\ell}+ij-i\ell-j\ell} c^\ell q^{i\ell+j\ell+\ell} \left(\frac{-1}{cq^{i+j}}\right)_\ell c^{-2\ell} q^{2\ell}(q^{-i})_\ell(q^{-j})_\ell(1+cq^{i+j-2\ell})}{(-q^{1-j}/c)_\ell(-q^{1-i}/c)_\ell(q)_\ell} = \frac{(-c)_{i+j+1}}{(-c)_i(-c)_j}.$$

This may be further reduced to

(4.15) $$\sum_{\ell \geq 0} \frac{(\frac{1}{-cq^{i+j}})_\ell cq^{i+j}(1+c^{-1}q^{-i-j-2\ell})(q^{-i})_\ell(q^{-j})_\ell c^{-\ell} q^{T_\ell+ij}}{(q)_\ell(-\frac{q^{1-i}}{c})_\ell(-\frac{q^{1-j}}{c})_\ell} = \frac{(-c)_{i+j+1}}{(-c)_i(-c)_j}.$$

The final step is to use Jackson's q-analog of Dougall's summation (see (II.22) in Gasper and Rahman [5])

(4.16) $$\sum_{n=0}^\infty \frac{(A)_n(1-q^{2n})}{(q)_n} \cdot \frac{(B)_n(D)_n(-1)^n}{(\frac{Aq}{B})_n(\frac{Aq}{D})_n} \cdot \frac{q^{n(n-1)/2}A^n q^n}{B^n D^n} = \frac{(A)_\infty(\frac{Aq}{BD})_\infty}{(\frac{Aq}{B})_\infty(\frac{Aq}{D})_\infty}$$

with

(4.17) $$A = \frac{-1}{cq^{i+j}}, B = q^{-i}, D = q^{-j}, \frac{A}{BD} = \frac{-1}{c},$$

to get

$$\sum_{\ell \geq 0} \frac{(\frac{-1}{cq^{i+j}})_\ell (1+c^{-1}q^{-i-j+2\ell})}{(q)_\ell} \frac{(q^{-i})_\ell (q^{-j})_\ell (-1)^\ell q^{T_\ell} (\frac{-1}{cq^{i+j}})^\ell q^{i\ell+j\ell}}{(-q^{1-i}/c)_\ell (-q^{1-j}/c)_\ell}$$

$$= \frac{(\frac{-1}{cq^{i+j}})_\infty (-\frac{q}{c})_\infty}{(\frac{-q}{cq^j})_\infty (\frac{-q}{cq^i})_\infty}$$

(4.18)
$$= \frac{(1+\frac{1}{cq^{i+j}})(1+\frac{1}{cq^{i+j-1}})\cdots(1+\frac{1}{cq})}{(1+\frac{1}{cq^{j-1}})\cdots(1+\frac{1}{cq})(1+\frac{1}{cq^{i-1}})\cdots(1+\frac{1}{cq})(1+\frac{1}{c})}$$

$$= \frac{(-cq)_{i+j}}{(-cq)_{i-1}(-cq)_{j-1}} \frac{q^{T_{i-1}+T_{j-1}-T_{i+j}}}{c^2(1+\frac{1}{c})}$$

$$= \frac{c^{-2}q^{-ij-i-j}(-cq)_{i+j}}{(1+\frac{1}{c})(-cq)_{i-1}(-cq)_{j-1}}$$

$$= \frac{c^{-1}q^{-ij-i-j}(-c)_{i+j+1}}{(-c)_i(-c)_j}.$$

If we multiply both sides of (4.18) by cq^{i+j+ij} we get (4.15) and that completes the proof of the new key identity.

5. Applications

First take $c = 0$ in Theorem 2. This means on the left we are summing over partitions π' of n into distinct parts $\equiv 1$ or $2 \pmod{3}$. Since $c = 0$, we have $a + ac = a$ and $b + bc = b$. Thus all parts $\equiv 1 \pmod 3$ in $\tilde{\pi}$ have weight a and all parts $\equiv 2 \pmod 3$ in $\tilde{\pi}$ have weight b, regardless of the gap. Also, since $c = 0$, we see that consecutive multiples of 3 cannot occur as parts. Thus with $c = 0$, Theorem 2 reduces to Schur's famous partition theorem in the following refined form as observed in [1]:

THEOREM S.

Let $S(n; i, j)$ denote the number of partitions π' of n into distinct parts such that $\nu_1(\pi') = i$, $\nu_2(\pi') = j$ and $\nu_3(\pi') = 0$.

Let $S_1(n; i, j)$ denote the number of partitions $\tilde{\pi}$ of n into parts differing by ≥ 3 and no consecutive multiples of 3 such that $\nu_1(\tilde{\pi}) + \nu_3(\tilde{\pi}) = i$ and $\nu_2(\tilde{\pi}) + \nu_3(\tilde{\pi}) = j$.
Then
$$S(n; i, j) = S_1(n; i, j).$$

We now show how this relates to the new key identity. Letting $c \to 0$ in (1.3) yields

(5.1)
$$\sum_{i,j,k \geq 0} \frac{a^{i+k} b^{j+k} q^{T_{i+j+k}+T_k}}{(q)_i (q)_j (q)_k} = (-aq)_\infty (-bq)_\infty.$$

Now (5.1) may be rewritten in the form

(5.2)
$$\sum_{i,j} a^i b^j \sum_{m=0}^{\min(i,j)} \frac{q^{T_{i+j-m}+T_m}}{(q)_{i-m}(q)_{j-m}(q)_m} = (-aq)_\infty (-bq)_\infty.$$

In [3], Alladi and Gordon show that Theorem S emerges from (5.2) under the substitutions $q \to q^3$ and $a \mapsto aq^{-2}$, $b \mapsto bq^{-1}$.

Next consider the special case $c = -1$ and $ab = 1$ in Theorem 2. This means

(5.3) $$a + ac = b + bc = c + ab = 0.$$

Now (5.3) implies that among the parts $h_1, h_2, \ldots, h_\nu$ of $\tilde{\pi}$ and $h_{\nu+1} = -1$, there are no gaps > 3. Hence there are only two cases for $\tilde{\pi}$.

Case 1: $h_\nu = 1$, $h_\mu - h_{\mu+1} = 3$, for $1 \leq \mu < \nu$.
Case 2: $h_\nu = 2$, $h_\mu - h_{\mu+1} = 3$, for $1 \leq \mu < \nu$.

In Case 1, $\tilde{\pi}$ is the partition

(5.4) $$1 + 4 + 7 + \cdots + 3\nu - 2 \text{ of } n = \frac{3\nu^2 - \nu}{2}.$$

In Case 2, $\tilde{\pi}$ is the partition

(5.5) $$2 + 5 + 8 + \cdots + 3\nu - 1 \text{ of } n = \frac{3\nu^2 + \nu}{2}.$$

Also, observe that $c = -1$, $ab = 1$ yields

(5.6) $$\sum_n \left(\sum_{\substack{\pi' \in \mathcal{D} \\ \sigma(\pi') = n}} a^{\nu_1(\pi')} b^{\nu_2(\pi')} c^{\nu_3(\pi')} \right) q^n = (-aq; q^3)_\infty (-a^{-1} q^2; q^3)_\infty (q^3; q^3)_\infty.$$

Thus (5.4), (5.5) and (5.6) yield

(5.7) $$\sum_{\nu = -\infty}^{\infty} a^\nu q^{(3\nu^2 - \nu)/2} = (-aq; q^3)_\infty (-a^{-1} q^2; q^3)_\infty (q^3; q^3)_\infty$$

which is equivalent to Jacobi's triple product identity.

On the other hand, in (1.3), first replace a by aq^{-1}. Then take $c = -1$ and $ab = 1$. This yields

(5.8) $$1 + \sum_{i=1}^{\infty} a^i q^{i(i-1)/2} + \sum_{j=1}^{\infty} a^{-j} q^{j(j+1)/2} = \sum_{\nu = -\infty}^{\infty} a^\nu q^{\nu(\nu-1)/2} = (-a)_\infty (-a^{-1} q)_\infty (q)_\infty$$

which is Jacobi's triple product identity. Equation (5.7) is equivalent to (5.8) as can be seen by using the substitutions $q \mapsto q^3$ and $a \mapsto aq$ in (5.8).

References

1. K. Alladi, *A combinatorial correspondence related to Göllnitz' (Big) partition theorem and applications*, Trans. Amer. Math. Soc. (to appear).
2. K. Alladi, G. E. Andrews and B. Gordon, *Generalizations and refinements of a partition theorem of Göllnitz*, J. Reine Angew. Math. **460** (1995), 165-188.
3. K. Alladi and B. Gordon, *Generalizations of Schur's partition theorem*, Manus. Math. **79** (1993), 113-126.
4. D. M. Bressoud, *A new family of partition identities*, Pacific J. Math. **77** (1978), 71-74.
5. G. Gasper and M. Rahman, *Basic hyper-geometric series*, Encyclopedia of Mathematics and its Applications, Vol. 35, Cambridge (1990).
6. H. Göllnitz, *Partitionen mit Differenzenbedingungen*, J. Reine Angew. Math. **225** (1967), 154-190.

University of Florida, Gainesville, FL 32611, USA
E-mail address: alladi@math.ufl.edu

Pennsylvania State University, University Park, PA 16802, USA
E-mail address: andrews@math.psu.edu

SOME LOCAL-CONVEXITY THEOREMS FOR THE ZETA-FUNCTION-LIKE ANALYTIC FUNCTIONS-III

BY

R. BALASUBRAMANIAN and K. RAMACHANDRA

(TO PROFESSOR M.S. RANGACHARI ON HIS SIXTIETH BIRTHDAY)

§ 1. INTRODUCTION. In our paper [2] we proved Theorem 6-C and applied it to claim a result namely Theorem 6-D stated in the post-script to that paper. However we could not prove Theorem 6-D. In the introduction to the present paper we state the result of Theorem 6-D as a conjecture. Also we state some further conjectures. In the next few sections we prove some theorems which form a slight progress in the direction of these two conjectures.

CONJECTURE 1. *Given any positive constant δ, there exists a constant $\varepsilon(>0)$ depending on δ, such that*

$$max \mid \zeta(\tfrac{1}{2} + iu) \mid > t^{-\varepsilon}$$

where the maximum is taken over $t - \delta \leq u \leq t + \delta$ and $t \geq t_0(\delta, \varepsilon)$ and $t_0(\delta, \varepsilon)(>0)$ depends only on δ and ε. (Note that ε can be a large constant). However on the assumption of LINDELÖF hypothesis, $\delta > 0$ and $\varepsilon > 0$ can be taken to be arbitrarily small constants independent of each other.

CONJECTURE 2. *Assume RIEMANN hypothesis. Then given any positive constant δ (with $0 < \delta \leq 1$), there exists a positive constant $C(\delta)$ depending only on δ, such that*

$$max \mid \zeta(\tfrac{3}{4} + iu) - 1 \mid > (log\ t)^{-C(\delta)},$$

where $t \geq 10$ and the maximum is taken over $t - \delta \leq u \leq t + \delta$.

REMARK 1. We will make some remarks on results involving L-functions in the end. We feel that the results corresponding to Conjectures 1 and 2 cannot be proved at present. In passing we may note that (on page 319 of his note book [4]) S. RAMANUJAN has given explicitly all the terms of the series $\sum_{n=1}^{400} n\mu(n)$. This shows that he was interested in RIEMANN hypothesis which is equivalent to the fact that $\sum_{n=1}^{N} n\mu(n) = O_\varepsilon\left(N^{\frac{3}{2}+\varepsilon}\right)$, for every $\varepsilon > 0$.

REMARK 2. Conjecture 1 deals with the $\frac{1}{2}$-line and Conjecture 2 deals with the $\frac{3}{4}$-line. We may state Conjecture 1 for any σ-line with $\frac{1}{2} \leq \sigma < 1$ and Conjecture 2 for any σ-line with $\frac{1}{2} < \sigma < 1$.

In § 2 we state Theorem 6-C of [2] and in later sections apply it to prove results on generalised Dirichlet series and the Riemann zeta-function.

§ 2. A LOCAL CONVEXITY THEOREM. Let $s = \sigma + it$ and let $f(s)$ be an analytic function of s, defined in the rectangle $R : \{a \leq \sigma \leq b, t_0 - H \leq t \leq t_0 + H\}$ where a, b, H are constants satisfying $a < b$ and $H > 0$. Let the maximum of $|f(s)|$ taken over R be $\leq M (\geq 3)$. Let $\sigma_0, \sigma_1, \sigma_2$ be real numbers subject to $0 \leq a \leq \sigma_0 < \sigma_1 < \sigma_2 \leq b$. Let r be any positive integer, $0 < D \leq H$ and $s_1 = \sigma_1 + it_0$ (t_0 should not be confused with that in § 1; t_0 is now a variable). Finally we define

$$I_0 = \int_{|v| \leq D} |f(\sigma_0 + it_0 + iv)| \frac{dv}{|\sigma_0 - \sigma_1 + iv|}$$

and

$$I_2 = \int_{|v| \leq D} |f(\sigma_2 + it_0 + iv)| \frac{dv}{|\sigma_2 - \sigma_1 + iv|}.$$

We now state Theorem 6-C of [2] as Theorem 1.

THEOREM 1. (R. BALASUBRAMANIAN and K. RAMACHANDRA). *We have*

$$2\pi |f(s_1)| \leq \left\{ 2(I_0 + M^{-A})^{\frac{\sigma_2-\sigma_1}{\sigma_2-\sigma_0}} (I_2 + M^{-A})^{\frac{\sigma_1-\sigma_0}{\sigma_2-\sigma_0}} \right\} J^r$$

$$+\frac{2M(\sigma_2 - \sigma_0)}{D} \left\{ \left(\frac{I_0 + M^{-A}}{I_2 + M^{-A}}\right)^{\frac{\sigma_2-\sigma_1}{\sigma_2-\sigma_0}} + \left(\frac{I_2 + M^{-A}}{I_0 + M^{-A}}\right)^{\frac{\sigma_1-\sigma_0}{\sigma_2-\sigma_0}} \right\} J_1^r$$

where $J = Exp(min(C(\sigma_1 - \sigma_0), C(\sigma_2 - \sigma_1)))$ and $J_1 = \frac{2J}{CD}$. Here A and C are any two positive constants.

REMARK. On page 3 of [2] at the end of line 5 from the bottom there should be a comma.

We now make a deduction. Let

(i) $b = 2\sigma_2$, (ii) $|f(\sigma_1 + it_0)| \geq L > 0$ (where L is a constant)

and

(iii) $\max_{|v| \leq D} |f(\sigma_2 + it_0 + iv)| \leq U$ (where U is a constant).

Then we have

$$2\pi L \leq \left\{ 2(I_0 + M^{-A})^{\frac{\sigma_2-\sigma_1}{\sigma_2-\sigma_0}} \left(\frac{DU}{\sigma_2 - \sigma_1} + M^{-A}\right)^{\frac{\sigma_1-\sigma_0}{\sigma_2-\sigma_0}} \right\} J^r +$$

$$+\frac{2M(\sigma_2 - \sigma_0)}{D} \left\{ \left(1 + M^{A+1}\frac{DM}{\sigma_1 - \sigma_0}\right)^{\frac{\sigma_2-\sigma_1}{\sigma_2-\sigma_0}} + \left(1 + \frac{M^A DU}{\sigma_2 - \sigma_1}\right)^{\frac{\sigma_1-\sigma_0}{\sigma_2-\sigma_0}} \right\} J_1^r.$$

Hence we state Theorem 2.

THEOREM 2. *We have*

$$\pi L \leq (I_0 + M^{-A})^{\frac{\sigma_2-\sigma_1}{\sigma_2-\sigma_0}} \left(\frac{DU}{\sigma_2 - \sigma_1} + M^{-A}\right)^{\frac{\sigma_1-\sigma_0}{\sigma_2-\sigma_0}} J^r$$

$$+\frac{M^{A+3}(\sigma_2 - \sigma_0)}{D} \left\{ \left(1 + \frac{D}{\sigma_1 - \sigma_0}\right)^{\frac{\sigma_2-\sigma_1}{\sigma_2-\sigma_0}} + \left(1 + \frac{DU}{\sigma_2 - \sigma_1}\right)^{\frac{\sigma_1-\sigma_0}{\sigma_2-\sigma_0}} \right\} J_1^r.$$

Here A and C are any two positive constants,

$$J = Exp(min(C(\sigma_1 - \sigma_0), C(\sigma_2 - \sigma_1))) \text{ and } J_1 = \frac{2J}{CD}.$$

We now pass on to a further special result.

THEOREM 3. *Let* $0 < \varepsilon < 1, M \geq 10^{\varepsilon^{-1}}, A > 0, 2\sigma_1 = \sigma_0 + \sigma_2,$

$D = 6(\sigma_2 - \sigma_1)Exp(\frac{10}{\varepsilon}(A+4)), U \geq \max_{|v| \leq D} | f(\sigma_2 + it_0 + iv) |,$

$L \leq | f(\sigma_1 + it_0) |$. Then

$L \leq (I_0 + M^{-A})^{\frac{1}{2}}(U+1)^{\frac{1}{2}} Exp(\frac{5}{\varepsilon}(A+4))M^{\varepsilon} +$

$+ (U+1)Exp(\frac{10}{\varepsilon}(A+4))M^{-9A-37}.$

Note that

$$\begin{aligned} I_0 &= \int_{|v| \leq D} | f(\sigma_0 + it_0 + iv) | \frac{dv}{|\sigma_2 - \sigma_1 + iv|} \\ &< \frac{1}{\sigma_2 - \sigma_1} \int_{|v| \leq D} | f(\sigma_0 + it_0 + iv) | dv \\ &< 6 \, Exp(\frac{10(A+4)}{\varepsilon}) \max_{|v| \leq D} | f(\sigma_0 + it_0 + iv) |. \end{aligned}$$

PROOF. We take $C = (\sigma_2 - \sigma_1)^{-1}$ and note that

$$\left(\frac{DU}{\sigma_2 - \sigma_1} + M^{-A} \right)^{\frac{1}{2}} \leq \left(\frac{D}{\sigma_2 - \sigma_1} \right)^{\frac{1}{2}} (U+1)^{\frac{1}{2}} \leq 3 \left(Exp\left(\frac{5}{\varepsilon}(A+4)\right) \right) (U+1)^{\frac{1}{2}}.$$

Next we have

$$\left(\left(1 + \frac{D}{\sigma_2 - \sigma_1}\right)^{\frac{1}{2}} + \left(1 + \frac{DU}{\sigma_2 - \sigma_1}\right)^{\frac{1}{2}} \right) \left(\frac{\sigma_2 - \sigma_0}{D} \right)$$

$$\leq \frac{2(\sigma_2 - \sigma_1)}{D} \left(\left(1 + \frac{D}{\sigma_2 - \sigma_1}\right)^{\frac{1}{2}} + \left(1 + \frac{D}{\sigma_2 - \sigma_1}\right)^{\frac{1}{2}} (U+1)^{\frac{1}{2}} \right)$$

$$\leq 2 \left(\frac{\sigma_2 - \sigma_1}{D} \right)^{\frac{1}{2}} \left(\left(\frac{\sigma_2 - \sigma_1}{D} + 1 \right)^{\frac{1}{2}} (U+2) \right)$$

$$\leq (U+2)(2\sqrt{2} Exp(-\frac{5}{\varepsilon}(A+4))) \leq 3(U+1).$$

Since $\pi > 3$ this completes the proof of Theorem 3.

§3. SOME LEMMAS ON GENERALISED DIRICHLET SERIES.

Let $a_1 = \lambda_1 = 1, \lambda_1 < \lambda_2 < \cdots, \frac{1}{B} \leq \lambda_{n+1} - \lambda_n \leq B$, where $n = 1, 2, \cdots$ and $B \geq 1$ is a constant. Let $a_1 = 1, a_2, a_3, \cdots$ be a sequence of complex numbers with $| a_n | \leq E\lambda_n^d$ (where $E \geq 2$ and $d \geq 0$ are constants) and

$n = 2, 3, \cdots$. Let m be the least integer ≥ 2 for which $a_m \neq 0$ and let $|a_m|^{-1} \leq E\lambda_m^d$. Then the series

$$F(s) = \sum_{n=1}^{\infty} a_n \lambda_n^{-s}$$

defines an analytic function of $s = \sigma + it$ in $\sigma \geq d + 2$. Suppose that $F(s)$ can be continued analytically in $(\sigma \geq \frac{1}{2}, t_0 - H_0 \leq t \leq t_0 + H_0)$ where t_o is any positive real number and $0 < 2H_0 \leq t_0$ and H_0 exceeds a constant and there $\max |F(s)| \leq M(\geq 3)$. Put

$$F(s) = 1 + G(s).$$

Then we have

LEMMA 1. *Let $C \geq 10$ be any constant. Then for $\sigma \geq d + 1 + \lambda_m + 2B \log(4BCE)$, we have*

$$|G(s)| \leq \frac{1}{C}.$$

PROOF. We have

$$|G(s)| \leq E \left(\lambda_m^{d-\sigma} + \sum_{n \geq m+1} \left(\lambda_m + \frac{n-m}{B} \right)^{d-\sigma} \right).$$

Here the sum on the RHS does not exceed

$$\int_0^\infty \frac{du}{(\lambda_m + \frac{u}{B})^{\sigma-d}} = \frac{B\lambda_m^{d-\sigma+1}}{d - \sigma + 1}.$$

Since $\sigma \geq d + 1 + \lambda_m$ we have

$$|G(s)| \leq E\lambda_m^{d-\sigma}(1 + B) \leq 2BE\lambda_m^{d-\sigma}$$

$$\leq 2BE\lambda_m^{-1-\lambda_m - 2B \log(4BCE)}.$$

Noting that $\lambda_m^{-2B} \leq (1 + \frac{1}{B})^{-2B} = Exp(-2B \log(1 + \frac{1}{B}))$

$$\leq Exp\left(-\frac{2B}{B+1}\right) \leq e^{-1} \text{ (since } \log(1 + \frac{1}{B}) = -\log(1 - \frac{1}{B+1}) > \frac{1}{B+1}),$$

we have
$$\lambda_m^{-2B \log(4BCE)} \leq (4\,BCE)^{-1}$$

and this proves the lemma.

LEMMA 2. *Let $C \geq 10$ and $\sigma \geq 400B^2 CE\lambda_m \log(20CE^2 \lambda_m^{2d})$. Then $\log F(s)$ is regular and different from zero. Moreover we have,*
$$\mid a_m^{-1} \lambda_m^s \log F(s) - 1 \mid \leq \frac{1}{2}.$$

PROOF. We have (writing G for $G(s)$, for simplicity)
$$\log F(s) = G - \frac{G^2}{2} + \frac{G^3}{3} - \cdots$$

and so
$$a_m^{-1} \lambda_m^s \log F(s) - 1 = S_1 + S_2,$$

where
$$S_1 = a_m^{-1} \lambda_m^s G - 1 \text{ and } S_2 = a_m^{-1} \lambda_m^s \left(-\frac{G^2}{2} + \frac{G^3}{3} - \cdots\right).$$

It is not hard to check the inequality for σ (in the present lemma) implies the lower bound for σ required by Lemma 1 and so
$$\mid S_2 \mid \leq \mid a_m^{-1} \lambda_m^s G \mid \left(\frac{1}{C} + \frac{1}{C^2} + \cdots\right) \leq \frac{2}{C} \mid a_m^{-1} \lambda_m^s G \mid.$$

It suffices to prove that $\mid S_1 \mid \leq \frac{1}{C}$. For, in that case we have $\mid a_m^{-1} \lambda_m^s G \mid \leq 1 + \frac{1}{C}$ and so $\mid S_2 \mid \leq \left(1 + \frac{1}{C}\right)\left(\frac{1}{C} + \frac{1}{C^2} + \cdots\right) \leq \frac{2}{C}(1 + \frac{1}{C}) \leq \frac{4}{C}$ and so
$$\mid S_1 + S_2 \mid \leq \frac{1}{C} + \frac{4}{C} = \frac{5}{C} \leq \frac{1}{2}.$$

We now prove that $\mid S_1 \mid \leq \frac{1}{C}$. We have
$$\mid S_1 \mid \leq E\lambda_m^{d-\sigma} \sum_{n=m+1}^{\infty} E\lambda_m^{d-\sigma}$$
$$\leq E^2 \lambda_m^{d-\sigma} \left(\lambda_{m+1}^{d-\sigma} + \frac{B}{\sigma-d-1}\lambda_{m+1}^{d-\sigma+1}\right)$$

(by an argument similar to the one used in the proof of Lemma 1)

$$\leq E^2 \lambda_m^{d-\sigma} \left((\lambda_m + \tfrac{1}{B})^{d-\sigma} + \tfrac{B}{\sigma-d-1}(\lambda_m + \tfrac{1}{B})^{d-\sigma+1} \right)$$

$$\leq E^2 \lambda_m^{d-\sigma}(\lambda_m + \tfrac{1}{B})^{d-\sigma} \left(1 + \tfrac{B}{\sigma-d-1}(\lambda_m + \tfrac{1}{B})\right)$$

$$\leq 2E^2 \lambda_m^{2d} \left(1 + \tfrac{1}{B\lambda_m}\right)^{-\sigma}$$

(on observing that $B\lambda_m \log \lambda_m \geq B(1 + \tfrac{1}{B})(B+1)^{-1} = 1$)

$$\leq 2E^2 \lambda_m^{2d} Exp\left(-\tfrac{\sigma}{B\lambda_m}\right) \leq \tfrac{1}{C}.$$

This proves Lemma 2 completely.

We next state a theorem (see page 26 of [3]) in the notation therein as a lemma.

LEMMA 3 (BOREL-CARATHEODORY THEOREM). *Suppose $f(z)$ is analytic in $|z - z_0| \leq R$, and on the circle $z = z_o + Re^{i\theta}$ we have $\text{Re } f(z) \leq U$. Then in $|z - z_o| = r < R$ we have*

$$|f(z) - f(z_o)| \leq \tfrac{2r}{R-r}(U - \text{Re } f(z_0)).$$

LEMMA 4. *Let $\sigma_2 = 400B^2 E\lambda_m \log(200E^2 \lambda_m^{2d})$, $\sigma_0 = \tfrac{3}{4}$, and $2\sigma_1 = \sigma_0 + \sigma_2$. Then for $(\sigma_0 \leq \sigma \leq 2\sigma_2, t_0 - \tfrac{1}{10}H_0 \leq t \leq t_0 + \tfrac{1}{10}H_0)$ we have*

$$|\log F(s)| \leq 64\sigma_2 \log M$$

provided $40000\sigma_2 \leq H_0 \leq \tfrac{1}{2}t_0$ and $F(s) \neq 0$ in $(\sigma > \tfrac{1}{2}, t_0 - H_0 \leq t \leq t_0 + H_0)$.

PROOF. We apply Lemma 3 with $f(z) = \log F(z), z_0 = \sigma_2 + it_0, R$ such that the circle $|z - z_0| = R$ touches the line $\sigma = \tfrac{1}{2} + 10^{-5}, r$ such that the circle $|z - z_0| = r$ touches the line $\tfrac{5}{8} + 10^{-5}$. We have

$$|\log F(z)| \leq \tfrac{2r}{R-r}\log M + \tfrac{R+r}{R-r}|\log F(z_0)|$$
$$\leq 32\sigma_2(\log M + 2E\lambda_m^{d-\sigma_2})$$
$$\leq 64\sigma_2 \log M \text{ (since } 2E\lambda_m^{d-\sigma_2} \leq 1 \text{ and } M \geq 3).$$

§3. APPLICATIONS OF LOCAL CONVEXITY THEOREM.

THEOREM 4. *Let* $0 < \varepsilon \leq \frac{1}{2}, D = 6(d + \frac{3}{2} + \lambda_m + 2B \, log(40BE))Exp(\frac{60}{\varepsilon})$ *and* $M > Exp(\frac{100}{\varepsilon})$. *Then*

$$\int_{|v|\leq D} | F(\frac{1}{2}+it_0+iv)| \, dv > (d+1+\lambda_m+2B \, log(40BE))\left(Exp\left(-\frac{100}{\varepsilon}\right)\right)M^{-2\varepsilon}.$$

PROOF. In Lemma 1 take $C = 10$ and so in Theorem 3 we can take $L = \frac{1}{2}, U = 2, \sigma_0 = \frac{1}{2}, \sigma_1 = d + \frac{3}{2} + \lambda_m + 2B \, log(40BE), \sigma_0 + \sigma_2 = 2\sigma_1, 0 < \varepsilon \leq \frac{1}{2}, A = 2$ and we obtain

$$\frac{1}{2} \leq (I_0 + M^{-2})^{\frac{1}{2}}3^{\frac{1}{2}}(Exp(\frac{30}{\varepsilon}))M^{\varepsilon} + 3(Exp(\frac{60}{\varepsilon}))M^{-55}.$$

Let $3(Exp(\frac{60}{\varepsilon}))M^{-55} \leq \frac{1}{4}$ i.e. $M^{55} > 12 \, Exp(\frac{60}{\varepsilon})$ which is so if $M > 2 \, Exp(\frac{2}{\varepsilon})$ i.e. if $M > Exp(\frac{3}{\varepsilon})$. Thus we get

$$I_0 + M^{-2} \geq \frac{1}{48}(Exp(-\frac{60}{\varepsilon}))M^{-2\varepsilon}.$$

We next impose $M^{-2} \leq \frac{1}{96}(Exp(-\frac{60}{\varepsilon}))M^{-2\varepsilon}$ which is so if $M > 96 \, Exp(\frac{60}{\varepsilon})$ i.e. if $M > Exp(\frac{100}{\varepsilon})$. Thus we obtain Theorem 4 since $| \sigma_2 - \sigma_1 + iv |^{-1} \leq | \sigma_2 - \sigma_1 |^{-1} \leq | \sigma_1 - \frac{1}{2} |^{-1}$.

THEOREM 5. *Let* $0 < \varepsilon \leq \frac{1}{6}$ *and* $D = Exp(\frac{200}{\varepsilon})$. *Then*

$$\int_{|v|\leq D} | \zeta(\frac{1}{2} + it_0 + iv) | \, dv > t_0^{-\varepsilon}$$

provided only that $t_0 > Exp(\frac{3000}{\varepsilon^2})$.

REMARK 1. Theorem 5 (which is a corollary to Theorem 4) is only a slight progress towards CONJECTURE 1.

REMARK 2. The corresponding result for $L(\frac{1}{2} + it, \chi)(\chi \, mod \, q)$ for instance q being the product of all primes $\leq x$ will result in λ_m being $> x$. This increases D. But the lower bound for $\max_{|t-t_0|\leq D} | L(\frac{1}{2} + it, \chi) |$ will be $C_1(qt_0)^{-\varepsilon}$ where $C_1 > 0$ is independent of q and t_0. We have also to satisfy the condition $t_0 \gg_\varepsilon \lambda_m$ and in the special case in question reads $t_0 \gg_\varepsilon \log q$.

REMARK 3. Integrals over intervals of length $\gg \log\log t_0$ in place of D have been dealt with by K. RAMACHANDRA (see the references at the end of [1]).

REMARK 4. If in Theorem 4 we fix an $\varepsilon = \varepsilon_0$ and then use LINDELÖF hypothesis we obtain

$$\lim_{t_0 \to \infty} t_0^\varepsilon \int_{|v| \leq C_0} |\zeta(\frac{1}{2} + it + iv)| \, dv = \infty$$

for every fixed $\varepsilon > 0$ and a certain $C_0(> 0)$ independent of ε.

PROOF OF THEOREM 5. We apply Theorem 4 to $F(s) = \zeta(s)$. Note that we can take $d = 0, \lambda_m = 2, B = 1, E = 2$. Hence $d + \frac{3}{2} + \lambda_m + 2B \log(40BE) < 18$. Also in $(\sigma \geq \frac{1}{2}, t_0 - D \leq t \leq t_0 + D)$ we have $|\zeta(\sigma + it)| \leq 3t_0$, provided $t_0 \geq 2D$. Thus we obtain (with $0 < \varepsilon \leq \frac{1}{2}$)

$$\int_{|v| \leq D_1} |\zeta(\frac{1}{2} + it_0 + iv)| \, dv > 3 \left(Exp\left(-\frac{100}{\varepsilon}\right) \right) (3t_0)^{-2\varepsilon} \left(> t_0^{-3\varepsilon} \right)$$

provided $D_1 = 108 \, Exp(\frac{60}{\varepsilon})$ and $t_0 \geq Exp(\frac{300}{\varepsilon^2})$. Replacing ε by $\frac{\varepsilon}{3}$ we obtain Theorem 5.

We now apply Theorem 3 to $f(s) = \log F(s)$, using Lemmas 2 and 4. We use $C_2, C_3, \cdots$ to denote constants ≥ 2. Accordingly we assume that $F(s) \neq 0$ in the rectangle of Lemma 4. Let $0 < \varepsilon \leq \frac{1}{2}, C = A = 10$, and

$$D = H_0 = C_2 \lambda_m (\log(C_3 \lambda_m^{2d})) Exp(\frac{140}{\varepsilon})$$

where $C_2 = 40000 \times 400 B^2 E$ and $C_3 = 200 E^2$. We can take $L = \frac{1}{2} |a_m| \lambda_m^{-\sigma_1}$ and $U = 2 |a_m| \lambda_m^{-\sigma_2}$. By Theorem 3 we obtain

$$L \leq (I_0 + (C_4 \sigma_2 \log M)^{-10})^{\frac{1}{2}} (U+1)(Exp(\frac{70}{\varepsilon}))(C_4 \sigma_2 \log M)^\varepsilon +$$

$$+ (U+1)(Exp\frac{140}{\varepsilon})(C_4 \sigma_2 \log M)^{-127}$$

where $C_4 \geq 64$.

Note that
$$\frac{L}{U+1} = \frac{\mid a_m \mid \lambda_m^{-\sigma_1}}{2+4\mid a_m \mid \lambda_m^{-\sigma_2}} \geq \frac{\lambda_m^{-\sigma_1}}{2(E\lambda_m^d + 2\lambda_m^{-\sigma_2})} \geq \frac{1}{4E\lambda_m^{d+\sigma_2}}.$$

Hence
$$(4E\lambda_m^{d+\sigma_2})^{-1} \leq (I_0 + (C_4\sigma_2 \log M)^{-10})^{\frac{1}{2}}(Exp(\frac{70}{\varepsilon}))(C_4\sigma_2 \log M)^\varepsilon +$$
$$+(Exp(\frac{140}{\varepsilon}))(C_4\sigma_2 \log M)^{-127}.$$

Let $(Exp(\frac{140}{\varepsilon}))(C_4\sigma_2 \log M)^{-127} \leq (8E\lambda_m^{d+\sigma_2})^{-1}$. This is so if $C_4\sigma_2 \log M \geq (8E\lambda_m^{d+\sigma_2}Exp(\frac{140}{\varepsilon}))^{\frac{1}{127}}$. This is satisfied if we set $C_4 = E\lambda_m^{d+\sigma}Exp(\frac{16}{\varepsilon^2})$. Then
$$I_0 + (C_4\sigma_2 \log M)^{-10} \geq (8E\lambda_m^{d+\sigma_2})^2(C_4\sigma_2 \log M)^{-2\varepsilon}(Exp(-\frac{140}{\varepsilon})).$$

Let $(C_4\sigma_2 \log M)^{-10} \leq \frac{1}{2}(8E\lambda_m^{d+\sigma_2})^2(C_4\sigma_2 \log M)^{-2\varepsilon}(Exp(-\frac{140}{\varepsilon}))$. This is so if
$$(C_4\sigma_2 \log M)^9 \geq 128E^2\lambda_m^{2d+2\sigma_2}Exp(\frac{140}{\varepsilon}),$$

This is satisfied by our choice of C_4. Thus
$$I_0 > (128E^2\lambda_m^{2d+2\sigma_2})^{-1}(C_4\sigma_2 \log M)^{-2\varepsilon}Exp(-\frac{140}{\varepsilon})$$
$$> (128E^2\lambda_m^{2d+2\sigma_2}\sigma_2)^{-1}(Exp(-\frac{172}{\varepsilon}))(E\lambda_m^{d+\sigma_2})^{-2\varepsilon}(\log M)^{-2\varepsilon}$$
$$> (\log M)^{-3\varepsilon}$$

if $(\log M)^\varepsilon > (128E^3\lambda_m^{3d+3\sigma_2}\sigma_2)Exp(\frac{172}{\varepsilon})$ i.e. if $(\log M)^\varepsilon > (\sigma_2\lambda_m^{4\sigma_2})(Exp(\frac{172}{\varepsilon}))$ (since $B\lambda_m \log \lambda_m \geq 1$ and so $\lambda_m^{\sigma_2} = Exp(400BE(B\lambda_m \log \lambda_m)\log(200E^2\lambda_m^{2d})) \geq Exp(400BE \log(200E^2\lambda_m^{2d})) \geq 128E^3\lambda_m^{3d}$). Collecting we have the following result. Put $\sigma_0 = \frac{3}{4}, 0 < \varepsilon \leq \frac{1}{4}, \sigma_2 = 400B^2E\lambda_m \log(200E^2\lambda_m^{2d}), 2\sigma_1 = \sigma_0 + \sigma_2, D = 40000\sigma_2 Exp(\frac{140}{\varepsilon})$. Then
$$\int_{|v|\leq D} \mid \log F(\frac{3}{4} + it_0 + iv) \mid dv > \frac{1}{4}\sigma_2(\log M)^{-3\varepsilon}$$

provided $F(s) \neq 0$ in $(\sigma > \frac{1}{2}, t_0 - D \leq t \leq t_0 + D), (t_0 > 2D)$ and there $max \mid F(s) \mid \leq M (\geq ExpExp\{\frac{172}{\varepsilon^2} + \varepsilon^{-1}log(\sigma_2 \lambda_m^{4\sigma_2})\})$. Replacing ε by $\frac{\varepsilon}{3}$ we state

THEOREM 6. Let $0 < \varepsilon \leq \frac{1}{12}, \sigma_2 = 400B^2 E\lambda_m log(200E^2 \lambda_m^{2d}), \sigma_0 = \frac{3}{4}, 2\sigma_1 = \sigma_0 + \sigma_2, D = \sigma_2 Exp(\frac{440}{\varepsilon})$. Then

$$\int_{|v| \leq D} \mid log \, F(\frac{3}{4} + it_0 + iv) \mid dv > \frac{1}{4} \sigma_2 (log \, M)^{-\varepsilon},$$

provided $F(s) \neq 0$ in $(\sigma > \frac{1}{2}, t_0 - D \leq t \leq t_0 + D), (t_0 > 2D)$ and there $max \mid F(s) \mid \leq M \left(\geq ExpExp\{\frac{1548}{\varepsilon^2} + 3\varepsilon^{-1}log(\sigma_2 \lambda_m^{4\sigma_2})\}\right)$.

We now specialise this to $F(s) = \zeta(s)$. We can take $d = 0, E = 2, B = 1, \lambda_m = 2, 0 < \varepsilon \leq \frac{1}{12}, \sigma_2 = 1600 log(800), \frac{3}{4} + \sigma_2 = 2\sigma_1,$

$$D = 64000000(log(800))Exp(\frac{140}{\varepsilon})(< Exp(\frac{500}{\varepsilon})).$$

Note that in $(\sigma \geq \frac{1}{2}, t \geq 10)$ we have $\mid \zeta(s) \mid < 10 t^{\frac{1}{2}}$. Hence $t_0 > ExpExp(\frac{10000}{\varepsilon^2})$ will do. Thus we have

THEOREM 7. Let $0 < \varepsilon \leq \frac{1}{12}, D_0 = Exp(\frac{500}{\varepsilon})$ and $t_0 > ExpExp(\frac{10000}{\varepsilon^2})$. Then

$$\int_{|v| \leq D_0} \mid log\zeta(\frac{3}{4} + it_0 + iv) \mid dv > 1600(log \, t_0)^{-\varepsilon}$$

provided $\zeta(s) \neq 0$ in $(\sigma > \frac{1}{2}, t_0 - D_0 < t \leq t_0 + D_0)$.

COROLLARY. Put $\theta = \frac{(2 \, log \, t_0)^{-\varepsilon}}{100 D_0}$. Then under the conditions of the theorem we have the inequality

$$\max_{|v| \leq D_0} \mid \zeta(\frac{3}{4} + it_0 + iv) - 1 \mid > \theta.$$

PROOF OF THE COROLLARY. We just observe that if z is any complex number with $\mid log \, z \mid > \delta$ and $z = 1 + \theta$ with $\mid \theta \mid < \frac{\delta}{100}$ we obtain a contradiction

$$\delta < \mid log \, z \mid = \mid log(1 + \theta) \mid < \delta.$$

REMARK 1. The analogue to L-functions involves λ_m both in σ_2 and D. It is not hard to work out the analogue (to L functions) of Theorem 7.

REMARK 2. The corollary just stated is a slight progress towards CONJECTURE 2.

REMARK 3. We can obtain lower bounds for $max \mid (\zeta(\frac{3}{4}+it_0-iv))^\phi - 1 \mid$ where ϕ is any non-zero complex constant. Trivially ϕ can be any constant (or any function of t_0 and v) for which $\mid \phi \mid$ is bounded below by a positive constant.

REFERENCES

[1] R. BALASUBRAMANIAN and K. RAMACHANDRA, *Some local-convexity theorems for the zeta-function-like analytic functions*-I, Hardy-Ramanujan J., Vol. 11 (1988), 1-12.

[2] R. BALASUBRAMANIAN and K. RAMACHANDRA, *Some local-convexity theorems for the zeta-function-like analytic functions*-II, Hardy-Ramanujan J., vol. 20 (1997), 2-11.

[3] K. RAMACHANDRA, On the mean-value and Omega-theorems for the Riemann zeta-function, LN85, Published for TIFR by Springer-Verlag (1995).

[4] S. RAMANUJAN, Note book-II, Tata Institute of Fundamental Research (1957).

ADDRESS OF THE AUTHORS

1. R. BALASUBRAMANIAN
   SENIOR PROFESSOR
   MATSCIENCE
   THARAMANI P.O.
   MADRAS 600 113, INDIA
   e-mail : BALU@IMSC.IMSC.ERNET.IN

2. K. RAMACHANDRA
   SENIOR PROFESSOR
   SCHOOL OF MATHEMATICS
   TATA INSTITUTE OF FUNDAMENTAL RESEARCH
   HOMI BHABHA ROAD
   MUMBAI 400 005, INDIA

e-mail : KRAM@TIFRVAX.TIFR.RES.IN

MANUSCRIPT COMPLETED ON 29TH MARCH 1996.
CURRENT ADDRESS OF PROFESSOR K.RAMACHANDRA

1. PROFESSOR K. RAMACHANDRA
   N.I.A.S., I.I.Sc., CAMPUS
   BANGALORE -560 012, INDIA
   e-mail : KRAM@MATH.TIFRBNG.RES.IN

Perfect powers in products of terms in an arithmetical progression IV

R. Balasubramanian and T.N. Shorey

§ 1. **Introduction.** For an integer $x > 1$, we denote by $P(x)$ the greatest prime factor of x. Let $d \geq 1, k \geq 2$ and $m > 0$ be integers such that $(m, d) = 1$. Let $d_1, \cdots, d_t$ with $t \geq 2$ be distinct integers in the interval $[0, k)$. If $t = k > 2$ and $d > 1$, Shorey and Tijdeman [4] proved that

$$P((m + d_1 d) \cdots (m + d_t d)) > k$$

unless $m = 2, d = 7$ and $k = 3$. Let $\varepsilon > 0$. In [5, Theorem 2], the following equation was considered :

$$(m + d_1 d) \cdots (m + d_t d) = by^\ell$$

where $d > 1, \ell$ prime, $P(b) \leq k$ and

$$t \geq k - (1 - \varepsilon)k \frac{h(k)}{\log k}$$

with $h(k) = \log \log \log k$ if $\ell = 2, 3$ and $h(k) = \log \log k$ if $\ell \geq 5$. The assertion of [5, Theorem 2] is under the assumption that the left hand side of the equation is divisible by a prime exceeding k. In this paper, we discuss this assumption. We prove

Theorem 1. *Let $\epsilon > 0$. There exists an effectively computable number C_1 depending only on ε such that for $k \geq C_1$, the inequality*

(1) $$P((m + d_1 d) \cdots (m + d_t d)) \leq k$$

with

(2) $$t \geq k - k\frac{\log \log k}{\log k} + 2\varepsilon k\frac{\log \log \log k}{\log k}$$

implies that

(3) $$\frac{m + (k-1)d}{k} \leq (\log k)(\log \log k)^{-\varepsilon}.$$

On the other hand, we have

Theorem 2. *There exists an effectively computable absolute constant C_2 such that for every triple d, k, m with $k \geq C_2$ and (3), we can find distinct integers $d_1, \cdots, d_t$ in $[0, k)$ satisfying (1) and*

$$t \geq k - k\frac{\log \log k}{\log k} + O(k\frac{\log \log \log k}{\log k}).$$

By [1, Lemma 2], we obtain a subset S of $\{d_1, \cdots, d_t\}$ obtained in Theorem 2 by omitting at most $\pi(k)$ elements such that the corresponding product

$$\prod_{d_i \varepsilon S} (m + d_i d)$$

is a square. Compare this assertion with [1, Theorem 2(a)].

§ 2. Proof of Theorem 1. For every prime $p \leq k$, we take an $f(p) \in \{d_1, \cdots, d_t\}$ such that

$$ord_p(m + f(p)d) = \max_{1 \leq i \leq t} ord_p(m + d_i d).$$

We write S_1 for the set obtained by deleting all $f(p)$ with $p \leq k$ from $\{d_1, \cdots d_t\}$. We observe that $|S_1| \geq t - \pi(k)$ and we apply a well-known argument of Erdös for deriving that

$$\prod_{s \varepsilon S_1}(m+sd) \leq \prod_{p \leq k} p^{[\frac{k}{p}]+[\frac{k}{p^2}]+\cdots} = k! \leq k^k$$

which implies that

(4) $\qquad m^{t-\pi(k)} \leq k^k \text{ and } (t-\pi(k))! d^{t-\pi(k)} \leq k!$.

Finally we derive (3) from (4) and (2). □

§ 3. Lemmas for the proof of Theorem 2.

For $x \geq 2$ and positive integers D and M with $(D, M) = 1$, let $\pi(x; D, M)$ denote the number of primes p not exceeding x such that $p \equiv M \pmod{D}$. Then we derive from Prime number theorem for arithmetic progressions with error term (see Estermann[2, Chapter 2]) the following result.

Lemma 1. *Let $A > 0$ and $B > 0$. For $x \geq 3$, $D \leq (\log x)^A$ and $h \geq x(\log x)^{-B}$, we have*

$$\pi(x+h; D, M) - \pi(x; D, M) = \frac{h}{(\log x)\phi(D)}\left(1 + 0\left(\frac{\log \log x}{\log x}\right)\right)$$

where the constant implied by 0 symbol is an effectively computable number depending only on A and B.

Lemma 2. *For $x \geq 2$ and a positive integer D with $D < x$, we have*

(5) $\qquad \sum_{\substack{n \leq x \\ (n,D)=1}} n^{-1} \leq \frac{\phi(D)}{D}(\log x + C_3 \log \log D)$

where C_3 is an effectively computable absolute constant.

Proof. We have

$$\sum_{\substack{n\leq x \\ (n,D)=1}} n^{-1} = \sum_{n\leq x}\frac{1}{n}\sum_{\substack{d|D \\ d|n}}\mu(d)$$

$$= \sum_{d|D}\mu(d)\sum_{n=md\leq x}\frac{1}{md}$$

$$= \sum_{d|D}\frac{\mu(d)}{d}\sum_{m\leq x/d}\frac{1}{m}$$

$$= \sum_{d|D}\frac{\mu(d)}{d}(\log(x/d)+\gamma+O(d/x))$$

$$= (\log x+\gamma)\sum_{d|D}\frac{\mu(d)}{d}-\sum_{d|D}\frac{\mu(d)\log d}{d}+O(\sum_{d|D}1/x)$$

Since

$$\sum_{d|D}\frac{\mu(d)}{d} = \phi(D)/D \gg 1/\log\log D$$

and

$$|\sum_{d|D}\frac{\mu(d)\log d}{d}| \ll (\log\log D)\phi(D)/D,$$

the inequality (5) follows.

§ 4. Proof of Theorem 2. We may assume that d, k and m are positive integers satisfying (3). Further we may suppose that $k \geq k_o$ where k_o is a sufficiently large effectively computable absolute constant. We put $J = [k(\log k)^{-10}]$ and $S_2 = \{(m+Jd), \cdots, m+(k-1)d\}$. We write S_3 for the set of all elements of S_2 whose greatest prime factor exceeds k and S_4 for the complement of S_3 in S_2. Any element of S_3 is of the form $p\lambda$ where $p \equiv m_\lambda$

(mod d) with $m_\lambda = m\overline{\lambda}$ and $\lambda\overline{\lambda} \equiv 1 \pmod{d}$ is a prime exceeding k and λ is a positive integer satisfying

(6) $$\lambda < \frac{m+(k-1)d}{k} < (\log k)(\log \log k)^{-\varepsilon}$$

by (3). We have

(7) $$|S_3| \leq \sum_{\substack{(\lambda,d)=1 \\ \lambda \leq (m+(k-1)d)/k}} (\pi(\frac{m+(k-1)d}{\lambda}; d, m_\lambda) - \pi(\frac{m+Jd}{\lambda}; d, m_\lambda)).$$

By Lemma 1 with $A = B = 1$ which is admissible by (3) and (6), each summand is at most

(8) $$\frac{1}{\lambda} \frac{k}{\log k} \frac{d}{\phi(d)} (1 + C_4 \frac{\log \log k}{\log k})$$

for an effectively computable absolute constant C_4 and we derive from (5) and (3) that

(9) $$\sum_{\substack{(\lambda,d)=1 \\ \lambda \leq (m+(k-1)d)/k}} \lambda^{-1} \leq \frac{\phi(d)}{d}(\log(\frac{m+(k-1)d}{k}) + C_3 \log \log d)$$

$$\leq \frac{\phi(d)}{d}(\log \log k - \varepsilon \log \log \log k + C_3 \log \log d)$$

Finally we combine (7), (8), (9) and (3) to conclude that

$$|S_3| \leq k \frac{\log \log k}{\log k} + O(k \frac{\log \log \log k}{\log k}).$$

Consequently

$$|S_4| = k - J - |S_3| \geq k - k\frac{\log \log k}{\log k} + O(k\frac{\log \log \log k}{\log k}) \qquad \square$$

Remark. The assertion of Theorem 2 continues to be valid with (3) relaxed to
$$\frac{m + (k-1)d}{k} \leq (\log k)(\log \log k)^{C_5}$$
where C_5 is an arbitrary absolute positive constant.

References

[1] R. Balasubramanian and T.N. Shorey, *Squares in products from a block of consecutive integers*, Acta Arith. 65 (1994), 213-220.

[2] T. Estermann, *Introduction to modern Prime Number Theory*, Cambridge tracts in Mathematics and Mathematical Physics 41 (1969), Cambridge University Press.

[3] G.H. Hardy and E.M. Wright, *An introduction to the Theory of Numbers*, Clarendon Press, Oxford (1960), fourth edition.

[4] T.N. Shorey and R. Tijdeman, *On the greatest prime factor of an arithmetical progression*, A Tribute to Paul Erdös, edited by A. Baker, B. Bollobás and A. Hajnal, Cambridge University Press (1990), 385-389.

[5] T.N. Shorey and R. Tijdeman, *Perfect powers in products of terms in an arithmetical progression* III, Acta Arith. 61 (1992), 391-398.

R. Balasubramanian
The Institute of Mathematical Sciences
Madras 600 113, India

and

T.N. Shorey
School of Mathematics
Tata Institute of Fundamental Research
Homi Bhabha Road
Bombay 400 005, India

On some Connections between Probability and Number Theory

Jean-Marc Deshouillers

To Gregory Abelevich Freiman, on the occasion of his 70th birthday, with my warmest greetings

ABSTRACT. The talk aims at emphasizing the connection between probability and number theory by providing a few examples, the last of those being a new application, due to G.A. Freiman, A.A. Yudin and the author, of additive number theory to the study of the concentration of the sum of independant discrete random variables.

1. Introduction

Probability theory and number theory have a long common history, rich of striking breakthroughs, which is still waiting for its scribe(s). This lecture has the modest ambition to claim that time has come to realize this project and to illustrate how connected are those topics, with a special emphasis on the less known ties with additive number theory.

Not only is this topic close to my heart, but it is specially adequate for this meeting, due to the key rôle that Ramanujan played in the building of "probabilistic number theory", connected to multiplicative number theory (we'll speak about his study of the prime factors of an integer) as well as to the additive side of the field (partitions and circle method).

In this talk, we aim at stressing the following points :
A. Probability leads to models for the integers;
B. Probabilistic techniques permit to prove number theoretic results;
C. Number theoretic questions lead to probability questions;
D. Number theoretic methods permit to prove probability results.

We shall explore these ideas in the following way :
I - Multiplicative number theory

1991 *Mathematics Subject Classification*. Primary 11-02, 60-02 ; Secondary 11P82, 11B13, 60E10.

Text of the lecture delivered at the invitation of the Ramanujan Mathematical Society meeting on Discrete Mathematics and Number Theory, Tiruchirapalli, January 1996.

Member of the laboratory "Algorithmique Arithmétique Expérimentale" UMR 9936, supported by CNRS and Universités Bordeaux 1 et 2.

© 1998 American Mathematical Society

Prime factors $[A, B]$.
II - Additive number theory
  Sidon problem $[B]$;
  Partitions $[A, B]$;
  Sums of powers $[A, C]$.
III - Probability
  Concentration function $[D]$.

At the end of this paper we add to the list of specific references a short list of general ones.

2. Multiplicative number theory

We quote one of the first results in probabilistic number theory, due to Hardy and Ramanujan [13].

THEOREM 1. *Let $(\Psi(n))$ be an increasing sequence that tends to infinity with n. The set of integers n such that*
$$|\mathrm{Card}\{p; p|n\} - \log \log n| \leq \Psi(n)\sqrt{\log \log n}$$
has density 1.

Proof (sketch)

(1) $$\text{Let } \omega(n) := \mathrm{Card}\{p; p|n\} = \sum_{p|n} 1,$$

we have

(2) $$N^{-1}\sum_{n\leq N} \omega(n) = N^{-1}\sum_{p\leq N}[N/p] = \sum_{n\leq N}\frac{1}{p} + O(1)$$
$$= \log \log N + O(1),$$

as well as

(3) $$N^{-1}\sum_{n\leq N}\omega^2(n) = (\log \log N)^2 + O(\log \log N).$$

The number of exceptional n's up to N is thus at most

(4) $$\sum_{n\leq N}\frac{(\omega(n) - \log\log N)^2}{\Psi^2(n)\log\log n} = O\left(\frac{N}{\Psi^2(N)}\right).$$

We now rephrase this proof in a probabilistic way, according to Turán [19]:

(1') Let (X_p) (for prime p) be a family of independent random variables such that $Pr\{X_p = 1\} = 1/p$ and $Pr\{X_p = 0\} = 1$ and let $\omega := \sum_{p\leq N} X_p$.

(2') We have $E(\omega) = \sum_{n\leq N} E(X_p) = \sum_{n\leq N}\frac{1}{p} = \log\log N + O(1)$,

as well as
(3') $\mathrm{Var}(\omega) = \sum_{n\leq N}\mathrm{Var}(X_p) = \sum_{n\leq N}\frac{1}{p}(1-\frac{1}{p}) = \log\log N + O(1)$

By Chebyshev's inequality, we have for any Ψ:
(4') $Pr\{|\omega - E(\omega)| \geq \Psi\sqrt{\mathrm{Var}(\omega)}\} \leq \frac{1}{\Psi^2}$.

The probabilistic interpretation connects the Hardy-Ramanujan result with the so-called "weak law of large numbers". This suggests to understand what is the counterpart of the central limit theorem in this context, and this was achieved by Erdős and Kac [7], who proved in 1940 that we have

THEOREM 2. *For any real u, one has*

$$\frac{1}{N}|\{n \leq N; \omega(n) - \log\log n \leq u\sqrt{\log\log n}\}| \longrightarrow_{n\to\infty} \frac{1}{\sqrt{2\pi}} \int_{-\infty}^{u} e^{-t^2/2} dt$$

In the late 30's, Erdős played a key rôle in extending this result to additive functions (in the present context, it will be sufficient to restrict ourselves to strongly additive functions, i.e. functions satisfying, as ω does, the relation $f(n) = \sum_{p/n} f(p)$). For example, the Erdős-Wintner Theorem [9] (1939) states the following.

THEOREM 3. *Let f be a strongly additive arithmetic function. There exists a distribution function F on $\mathbb{R}$ such that the sequence $(\frac{1}{N}|\{n \leq N; f(x) \leq x\}|)_N$ converges towards $F(x)$ whenever F is continuous at x, if and only if the following three series converge :*

$$\sum_{|f(p)|>1} \frac{1}{p}, \sum_{|f(p)|\leq 1} \frac{f(p)}{p} \text{ and } \sum_{|f(p)|\leq 1} \frac{f^2(p)}{p}.$$

This statement bears an obvious connection with the "three series theorem", namely : the sum $\sum X_n$ of a sequence of independent random variables converges P-almost surely if and only if there exists a such that the three following series converge :

$$\sum_n Pr\{|X_n| > a\}, \sum_n E(X_n^{(a)}) \text{ and } \sum_n \text{Var}(X_n^{(a)}),$$

where $X_n^{(a)}(\omega) = X_n(\omega)$ when $|X_n(\omega)| \leq a$ and 0 otherwise.

More generally one may ask how far one can compare the quantities

$$\frac{1}{N}|\{n \leq N; f(n) \leq x\}| \text{ and } Pr\{\sum_{p\leq n} X_p \leq x\},$$

where the X_p are independent random variables such that $Pr\{X_p = f(p)\} = \frac{1}{p}$ and $Pr\{X_p = 0\} = 1 - \frac{1}{p}$. By sieve arguments, one can show that

$$\frac{1}{N}|\{n \leq N; \sum_{p|n, p\leq r} f(p) \leq x\}| - Pr\{\sum_{p\leq r} X_p \leq x\}$$

$$= O(N^{-1/15} + \exp(-\frac{1}{8}\frac{\log N}{\log r} \log\left(\frac{\log N}{\log r}\right))),$$

which implies that this difference is $o(1)$ when $r = N^{\epsilon(N)}$ tends to infinity in such a way that $\epsilon(N)$ tends to zero. This is called the "Kubilius model" ; Tenenbaum [18] has recently given a more precise error term.

3. Additive number theory

3.1. A question by Sidon. Sidon raised in 1932 the following question : can one find a sequence $\mathcal{A}$ of non-negative integers such that

(i) $\forall n \geq 0, \exists a, b \in \mathcal{A} : n = a + b$,

(ii) $|\{a \in \mathcal{A}; n - a \in \mathcal{A}\}| = o(n^\epsilon)$ for any $\epsilon > 0$?

Erdős [**6**] gave in 1954 the much more precise answer :
there exist $\mathcal{A}$ and $0 < c_1 < c_2$ such that

(iii) $c_1 \log n < |\{a; n - a \in \mathcal{A}\}| < c_2 \log n$, for all $n \geq 2$.

The proof is startling and, up to my knowledge, basically the only one given to Sidon's question :

Let $X_2, \cdots, X_n, \cdots$ be a sequence of independent random Bernoulli variables on a probability space $(\Omega, \mathcal{T}, P)$ such that :

$$(5) \qquad Pr\{X_n = 1\} = c\sqrt{\frac{\log n}{n}}$$

for some suitable c ($c = (\frac{1}{2}(1 + \frac{2}{\pi}))^{1/2}$ is convenient). Then, for almost all ω in Ω, the set $\mathcal{A}(\omega) = \{0, 1\} \cup \{n \geq 2; X_n(\omega) = 1\}$ satisfies (iii). The interested reader will find in the third chapter of the above-mentionned book of Halberstam and Roth a very detailed proof of this result, as well as other occurrences of similar ideas.

3.2. Partitions. This example underlies the probabilistic interpretation of the Ramanujan-Hardy-Littlewood circle method.

For a given N, let $q(N)$ denote the number of ways to write

$$N = n_1 + \cdots + n_r, \text{ with } 0 < n_1 < n_2 < \cdots < n_r (\leq N).$$

We can equivalently define $q(N)$ in the following way ; let $\mathcal{E}_N := \{0, 1\} \times \cdots \times \{0, N\}$, then

$$q(N) = \frac{|\{\boldsymbol{x} \in \mathcal{E}_N; \sum x_n = N\}|}{|\{\boldsymbol{x} \in \mathcal{E}_N\}|} \cdot 2^N,$$

which can be also seen in the following way : let $X_1, \cdots, X_N$ be independent random variables such that X_n takes the values 0 and n which probability $1/2$. We have

$$q(N) = 2^N Pr\{X_1 + \cdots + X_N = N\}.$$

If we assume that a local limit theorem holds for the random variable $S_N = X_1 + \cdots + X_N$, we would have

$$q(N) = 2^N \frac{1}{\sqrt{2\pi \text{ Var } (S_N)}} \left(\exp\left(-\frac{(N - E(S_N))^2}{2 \text{ Var } (S_N)} \right) + o(1) \right).$$

However, an easy computation leads to $E(S_N) \sim \frac{N^2}{4}$ and $\text{Var}(S_N) \sim \frac{N^3}{12}$, and we immediately notice that N is too far apart from the expectation of S_N to hope getting any success by this approach. We can however modify the definition of the X_n's to repair the situation.

Let $Y_1, \cdots, Y_N$ be independent random variables such that Y_n takes the values 0 and n with $Pr\{Y_n = n\} = p_n$. We clearly have

$$q(N) = \sum_{\substack{\boldsymbol{x} \in \mathcal{E}_N \\ x_1 + \cdots + x_N = N}} Pr\{(Y_1, \cdots, Y_N) = \boldsymbol{x}\} / \Pi_{x_n = n} p_n \Pi_{x_n = 0} (1 - p_n).$$

The advantage of the previous choice ($p_n = \frac{1}{2}$) was to give a constant weight in the right hand side : fortunately, the choice $p_n = e^{-\sigma n}/(1 + e^{-\sigma n})$ has the same property, and we get

$$q(N) = \frac{e^{-\sigma N}}{\Pi_{n=1}^{N}(1+e^{-\sigma n})} \sum_{\substack{x \in \mathcal{E}_N \\ x_1 + \cdots + x_N = N}} Pr\{(Y_1, \cdots, Y_N) = x\}$$

$$= \frac{e^{-\sigma N}}{\Pi_{n=1}^{N}(1+e^{-\sigma n})} Pr\{Y_1 + \cdots + Y_N = N\}.$$

We now choose σ in such a way that $E(Y_1 + \cdots + Y_N) = N$ (this can be achieved for some $\sigma = \frac{\pi}{2\sqrt{3N}}(1 - \frac{1}{8N} + O(\frac{1}{N^2})))$ and "the" local limit theorem (one can prove such a result, but this is of course the heart of the matter) leads to

$$q(N) \sim \frac{1}{4(3N)^{1/4}} \exp\{\frac{\pi}{\sqrt{3}}N^{1/2}\}.$$

Let us say one word about the proof : let φ_n be the characteristic function (i.e. the Fourier transform of the image measure) of Y_n; we have $\varphi_n(t) = (1 - p_n) + p_n \exp(2\pi i n t)$. Then (by the Fourier inverse transformation, or by the orthogonality of the characters $e_n(t) = \exp(2\pi i n t)$, we have the integral representation

(6) $$Pr\{Y_1 + \cdots + Y_N = N\} = \int_{\mathbb{R}/\mathbb{Z}} (\prod_{n \leq N} \varphi_n(t)) \exp(-2\pi i N t) dt.$$

The main term comes from the contribution of an interval around 0, the so-called major arc. The central point of the proof lies in getting a suitable upper bound for $|\prod_{n \leq N} \varphi_n(t)|$ outside the major arc.

This argument has recently been developed by Freiman and Pitman [**10**] in a more general context of certain "restricted" partitions. We must also stress here that the idea of using the integral representation (6) has been developped by Hardy, Littlewood and many other writters into a powerful method for attacking many additive problems, the "circle method".

3.3. Sum of powers. We shall be dealing with the following question : let $s \geq 2$ be an integer. Does the set of sums s integral s^{th} powers (e.g. sums of 3 cubes, sums of 4 biquadrates) have a positive (lower) density ?

For $s = 2$, the answer is known : the set of sums of 2 squares has a zero density; this is a direct consequence of the Dirichlet theorem for primes congruent to 1 modulo 4.

For no other value of s is the answer known, and this seems to be a very difficult question, which justifies heuristic approaches.

In 1960 Erdős and Rényi [**8**] considered "pseudo-s^{th} powers", i.e. random sequences $\mathcal{A}^{(s)}$ defined as for the Sidon problem, with (5) replaced by the following

(7) $$Pr\{X_n = 1\} = 1/(sn^{1-1/s}),$$

and outlined a proof that the number of representations $r_s(N)$ of an integer N as a sum of s elements from $\mathcal{A}^{(s)}$ should almost surely follow a Poisson law. The difficulty in this approach is the proof of some quasi-independence property for the sets which are involved; this question led Landreau [**16**] to give in 1995 a correlation inequality which reduces general quasi-independence to pairwise ones :

THEOREM 4. *Let $E_1, \cdots, E_N$ be independent events in a given probability space, and $A_1, \cdots, A_T$ be such that each A_t is an intersection of some of the E_n's. Then*

$$0 \leq P(\cap \overline{A}_t) - \Pi P(\overline{A}_t) \leq \sum_{1 \leq t < t' \leq T} (P(A_t \cap A_{t'}) - P(A_t)P(A_{t'})).$$

The Erdős-Rényi construction has the drawback of erasing the arithmetical structure of powers; it gives a positive density for sums of two pseudo-squares, and it seems therefore hazardous to derive from it any suggestion for an answer to our initial question, although the arithmetical structure of sums of s integral s^{th} powers is much weaker for $s \geq 3$ than for $s = 2$.

On the basis of some computation, Barrucand [1] conjectured in 1968 that the answer should be "no" at least for $s = 3$ and $s = 4$. In 1986, Hooley [14] provided heuristic supports to an affirmative answer for $s \geq 3$. Hennecart, Landreau and myself [3] have recently built a probabilistic model in the spirit of Erdős and Rényi, which also involves the arithmetical structure of sth powers and gives a positive density for sums of pseudo-s^{th} powers if and only if $s \geq 3$; moreover, our model is in accordance with some of Hooley's heuristics.

4. Probability

In this last section, we shall illustrate the possibility of applying number theoretic ideas to probability theory by outlining the proof of the following recent result due to Freiman, Yudin and myself [2]. In the restricted context of integral valued random variables, it shows that when the tail of the distribution of independent identically distributed random variables is large, the concentration function of their sum is small, modulo a necessary arithmetic condition.

THEOREM 5. *Let $\frac{\log 4}{\log 3} < \sigma < 2$, let $\epsilon > 0$ and let $X_1, \cdots, X_n$ be i.i.d. integral valued random variables such that*

(i) $\forall L \geq 2 : Pr\{|X_1| > L\} \geq L^{-\sigma}$,

(ii) $\max_{q \geq 2} \max_{0 \leq s < q} Pr\{X_1 \equiv s \bmod q\} \leq 1 - \epsilon$.

Then one has

$$\max_k Pr\{X_1 + \cdots + X_n = k\} \leq cn^{-1/\sigma},$$

where c depends on ϵ, σ and $\max_k P\{X_1 = k\}$ only.

Sketch of the proof :
Let φ be the characteristic function of X_1, i.e. $\varphi(t) = \sum_{l \in \mathbb{Z}} p_l \exp(2\pi i l t)$, where $p_l = Pr\{X_1 = l\}$; if we denote by S_n the sum $X_1 + \cdots + X_n$, we have

$$Pr\{S_n = k\} = \int_0^1 \varphi^n(t) \exp(-2\pi i k t) dt,$$

whence $\max_k Pr\{S_n = k\} \leq \int_0^1 |\varphi(t)|^n dt$.

We are thus interested in the large values of $|\varphi(t)|$. It follows from Bochner theorem (or directly form the Cauchy inequality) that the sets

$$E(\theta) := \{t \in \mathbb{R}/\mathbb{Z} : |\varphi(t)| \geq \cos \theta\} \quad (\text{for } 0 \leq \theta \leq \pi/2)$$

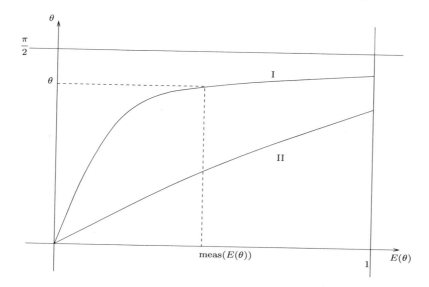

FIGURE 1. Two examples of graphs for g.

have the following property :
Whenever $\theta_1 \geq 0, \theta_2 \geq 0$ and $\theta_1 + \theta_2 \leq \frac{\pi}{2}$, then
$$(8) \qquad E(\theta_1) + E(\theta_2) \subset E(\theta_1 + \theta_2),$$
where $A + B$ denotes the set $\{a + b / a \in A, b \in B\}$.

Relation (8) is a quantitative formulation of the fact that $|\varphi(t_1 + t_2)|$ is large as soon as $|\varphi(t_1)|$ and $|\varphi(t_2)|$ are large. Let us consider the function g (a kind of a rearrangement function of $|\varphi|$) from $[0,1]$ to $[0, \frac{\pi}{2}]$ such that $g(u)$ is the infinum of the θ's for which meas $(E(\theta)) \geq u$. Figure 1 shows two types of behaviour for g.

The case I, where g is quickly increasing around the origin is the "good" case : the set of large values for $|\varphi|$ has a small measure, and so the integral $\int_0^1 |\varphi(t)|^n dt$ is small and our result is proved.

The case II, where g has a more or less linear increase rate around the origin is the "bad" case : we have to show that our hypotheses exclude this case. If the function g is more or less linear, then meas $E(2\theta)$ is close to 2 meas $E(\theta)$ and relation (8) enters the game by implying that, in this case, we have :

$$(9) \qquad \text{meas } (E(\theta) + E(\theta)) \text{ is not much larger than 2 meas } (E(\theta)).$$

This is a crucial information : a little digression will help us to understand it .

Structure theory of set addition. Let us first consider a finite set of integers, A. It is readily seen that one has :

$$2|A| - 1 \leq |A + A| \leq \frac{1}{2}|A|(|A| + 1).$$

In the late 50's an early 60's, Freiman has developed a study of the *structure* of the set A when $|A + A|$ is close to its lower bound. It is an easy exercise to show that the lower bound can be attained only when A is an arithmetic progresssion; one of the first results of Freiman's consists in showing that A is included in a short arithmetic progression when $|A + A|$ is small compared to $|A|$, namely

THEOREM 6. *Let A be a finite subset of $\mathbb{Z}$ such that*
$$|A + A| \leq 2|A| - 1 + b \quad \text{and} \quad b \leq |A| - 3;$$
then there exists an arithmetic progression L with (at most) $|A| + b$ elements such that $A \subset L$.

Results of this type have been obtained by different authors. Freiman himself, in collaboration with Moskvin and Yudin, showed that a measurable subset A of $\mathbb{R}/\mathbb{Z}$ such that meas $(A + A) \leq (2 + c)$ meas (A) for some $c < 1$ is concentrated around the vertices of a regular polygon. We can close our digression on this qualitative statement and go back to our main stream.

Relation (9) can now be transformed into an information concerning the structure of $E(\theta)$ for some small θ : $E(\theta)$ has to be located close to the vertices of a regular polygon. However, if this polygon has one vertex, this means that $|\varphi|$ has large values close to the origin, that is to say that the measure of X_1 is close to a Dirac measure, which is ruled out by the fact that the distribution of X_1 has a large tail. What if $E(\theta)$ is located close to the vertices of a regular polygon with more than one side ? Then the support of X_1 is essentially located on some arithmetic progression, but this is ruled out by our hypotheses on X_1. If follows that case II is not possible, and our theorem is proven.

References

[1] P.-A. Barrucand, *Sur la distribution empirique des sommes de 3 cubes ou de 4 bicarrés*, CRAS **267** (1968), 409–411.

[2] J-M. Deshouillers, G.A. Freiman and A.A. Yudin, *On bounds for the concentration function, 1*. Prépublications IHES **M/95/37**, 15p.

[3] J-M. Deshouillers , F. Hennecart and B. Landreau, *Sums of powers : an arithmetic refinement to the probabilistic model of Erdős and Rényi*, Prépublication **96-12** de l'UMR 9936.

[4] J-M. Deshouillers et G. Tenenbaum *Survol de la théorie probabiliste des nombres et de quelques progrès récents*, Sem. Th. Nb. Paris 1983-84, PM 59, Birkhäuser, 1985, pp. 49–69.

[5] P.D.T.A. Elliott, *Probabilistic Number Theory (2 vol.)*, Springer, 1979.

[6] P. Erdős, *On a problem of Sigon in additive number theory*, Acta Sci. Math. (Szeged), **15** (1954), 255–259.

[7] P. Erdős and M. Kac, *The Gaussian law of errors in the theory of additive number theory functions*, Amer. J. Math **62** (1940), 738–742.

[8] P. Erdős and A. Rényi, *Additive properties of random sequences of positive integers*, Acta Arith. **6** (1960), 83–110.

[9] P. Erdős and A. Wintner, *Additive arithmetical functions and statistical independance*, Amer. J. Math **61** (1939), 713–721.

[10] G.A. Freiman and J. Pitman, *Partitions onto distinct large parts*, J. Austr. Math. Soc. **57** (1994), 386–416.

[11] H.Halberstam and K.F. Roth, *Sequences*, Oxford, at the Clarendon Press, 1966.

[12] R.R. Hall and G. Tenenbaum, *Divisors*, Cambridge tracts in Mathematics 90, Cambridge University Press, 1988.

[13] G.H. Hardy and S. Ramanujan, *The normal number of prime factors of a number n*, Quart. J. Math. **48** (1920), 76–92.

[14] Ch. Hooley, *On some topics connected with Waring's problem*, J. Reine Angew. Math. **369** (1986), 110–153.

[15] J. Kubilius, *Probabilistic methods in the theory of numbers*, Trans. Math. Monographs, vol 11, AMS, 1964.

[16] B. Landreau, *Étude probabiliste des sommes de s puissances s-ièmes*, Compositio Math. **99** (1995), 1–31.

[17] G. Tenenbaum, *Introduction à la théorie analytique et probabiliste des nombres*, cours spécialisé 1, SMF, 1995, (An English version has appeared at Cambridge University Press).

[18] G. Tenenbaum, *Crible d'Erathosthène et modèle de Kubilius*, preprint, 1996, 4p.
[19] P. Turán, *On a Theorem of Hardy and Ramanujan*, J. London Math. Soc. **9** (1934), 274–276.

MATHÉMATIQUES STOCHASTIQUES, UNIVERSITÉ VICTOR SEGALEN, F-33076 BORDEAUX
E-mail address: `J-M.Deshouillers@u-bordeaux2.fr`

Inhomogeneous Minima of a Class of Quaternary Quadratic Forms of Type (2,2).

Madhu Raka and Urmila Rani[1]

ABSTRACT. It is proved that the first five minima for positive values of zero quaternary quadratic forms of type (2, 2) are 16, 16/3, 81/16, $7^4/8^3$ and 4 respectively. All the critical forms are also determined.

1. Introduction

Let $Q(x_1,...,x_n)$ be a real indefinite quadratic form in n variables of determinant $D \neq 0$ and of type (r,s), $r + s = n$. Let $\Gamma_{r,s}$ denote the infimum of all real numbers Γ for which the inequality

(1.1) $\qquad 0 < Q(x_1 + c_1,...,x_n + c_n) \leq (\Gamma|D|)^{1/n}$

is solvable in integers $x_1,...,x_n$, for any real numbers $c_1,...,c_n$. In other words, if we define

$$M_I^+(Q) = \sup_{c_1,...,c_n \in R^n} \inf_{\substack{x_i \in Z \\ Q(x_1+c_1,...,x_n+c_n) > 0}} Q(x_1 + c_1,...,x_n + c_n)/|D|^{1/n}$$

then $\Gamma_{r,s}^{1/n} = \sup M_I^+(Q)$ where the supremum is taken over all real indefinite quadratic forms Q in n variables of type (r,s) and determinant $D \neq 0$.

The values of $\Gamma_{r,s}$ are known for all r,s except for $\Gamma_{1,4}$; for references see Dumir and Sehmi [12]. Davenport and Heilbronn [4] showed that $\Gamma_{1,1}$ is not isolated. For incommensurable forms (forms which are not a multiple of rational forms) in $n \geq 3$ variables it follows from the results of Watson [21] and Margulis [19] that $M_I^+(Q) = 0$. For rational forms in $n \geq 3$ variables, it follows

---

1991 *Mathematics subject classification* 11E20

  1 Research supported by CSIR is gratefully acknowledged.

from Vulakh [20] (see the remarks added by the referee) that $M_1^+(Q)$ is discrete i. e. for any $\varepsilon > 0$, there exist finitely many inequivalent forms Q_i of determinant D such that $M_1^+(Q) < \varepsilon$ for Q not equivalent to ρQ_i, $\rho > 0$.

Let $n \geq 3$ and $\Gamma_{r,s}^{(k)}$ denote the kth successive inhomogeneous minima. It is known that

$\Gamma_{2,1}^{(2)} = 8/3$, Madhu Raka and Urmila Rani [18].

$\Gamma_{1,2}^{(2)} = 27/4$, Dumir and Hans Gill [9].

$\Gamma_{3,1}^{(2)} = 4$, $\Gamma_{r+2,r}^{(2)} = 2^{2r}$, for all $r \geq 2$, Madhu Raka and Urmila Rani [16], [17]

$\Gamma_{2,3}^{(2)} = 16$, Bambah, Dumir, and Hans-Gill [1].

$\Gamma_{3,2}^{(2)} = 8$, $\Gamma_{r+1,r}^{(2)} = 2^{2r-1}$, for all $r \geq 2$, Dumir and Sehmi [10],[11].

For the class of zero forms, the first four minima for positive values of ternary quadratic forms of type (2,1) have been obtained by Madhu Raka [15].

In this paper we shall obtain the first five minima (namely 16, 16/3, 81/16, $7^4/8^3$ and 4) for positive values of zero quaternary quadratic forms of type (2,2). The first minimum has already been obtained by Dumir [7].

For all the known values of $\Gamma_{r,n-r}^{(k)}$, $k \geq 1$, the real numbers $c_i, 1 \leq i \leq n$, for which the critical form attains its minimum, have been either congruent to 0 or 1/2 (mod 1). It is for the first time in the history of $\Gamma_{r,n-r}$ that we get a critical form Q_{13} (see the table below) which attains its minimum for $(c_1, c_2, c_3, c_4) = (0,0,1/4,0)$. Also for the first time, we get critical forms (Q_9, Q_{10}, Q_{11} and Q_{12} in the table below) which attain their minima for two incongruent values of c_j's. More precisely we prove:

THEOREM. Let $Q(x_1, x_2, x_3, x_4)$ be a real, zero quaternary quadratic form of type (2,2) and determinant $D \neq 0$. Then given any real numbers c_j, $1 \leq j \leq 4$, there exist $x_j \equiv c_j$ (mod 1), satisfying

(1.2) $\qquad 0 < Q(x_1, x_2, x_3, x_4) < (4|D|)^{1/4}$

except when $Q \sim \rho Q_i$, $1 \leq i \leq 14$, $\rho \neq 0$ and (c_1, c_2, c_3, c_4) is equivalent to $(c_1^{(i)}, c_2^{(i)}, c_3^{(i)}, c_4^{(i)})$, where Q_i and $c_j^{(i)}$ are given in the following table

i	Q_i	$(c_1^{(i)}, c_2^{(i)}, c_3^{(i)}, c_4^{(i)})$	$\Gamma_{2,2}(Q_i)$
1	$x_1 x_2 + x_3 x_4$	(0,0,0,0)	16
2	$(x_1 + x_2/2)x_2 + x_3 x_4$	(1/2,0,0,0)	16
3	$(x_1 + x_2/2 + x_3/2 + x_4/2)x_2 - x_3^2/2 + 3x_4^2/2$	(1/2,0,0,0)	16/3

4	$x_1 x_2 + (x_3 + x_4/4)x_4$	(1/2,0,0,0)	81/16
5	$x_1 x_2 + (x_3 + 3x_4/4)x_4$	(0,0,0,0)	81/16
6	$(x_1 + x_2/2)x_2 + (x_3 + x_4/4)x_4$	(0,0,1/2,0)	81/16
7	$(x_1 + x_2/2)x_2 + (x_3 + 3x_4/4)x_4$	(1/2,0,1/2,0)	81/16
8	$x_1 x_2 + x_3^2/2 - x_4^2$	(0,0,1/2,1/2)	$7^4/8^3$
9	$x_1 x_2 + 2x_3 x_4$	(0,0,0,0) or (0,0,1/2,0)	4
10	$(x_1 + x_2/2)x_2 + 2x_3 x_4$	(1/2,0,0,0) or (1/2,0,1/2,0)	4
11	$x_1 x_2 + 2(x_3 + x_4/2)x_4$	(0,0,0,0) or (0,0,1/2,0)	4
12	$(x_1 + x_2/2)x_2 + 2(x_3 + x_4/2)x_4$	(1/2,0,0,0) or (1/2,0,1/2,0)	4
13	$x_1 x_2 + 2(x_3 + x_4/4)x_4$	(0,0,1/4,0)	4
14	$(x_1 + x_2/2 + x_4/2)x_2 + 2(x_3 + x_4/4)x_4$	(1/2,0,1/2,0)	4

For the exceptional form Q_i, the inhomogeneous minimum $\Gamma_{2,2}(Q_i)$ is as shown in the table.

The first minimum 16 is attained for two inequivalent forms Q_1 and Q_2 and not for three forms as stated in Dumir [7]. In fact the form $x^2 - y^2 - z^2 + t^2$ stated in [7] is equivalent to $x^2 - y^2 - 2zt$ by means of the transformation $x \to y + z - t, y \to y + z, z \to y - t$ and $t \to x$. Also note that $2Q_2 \sim x^2 - y^2 - 2zt$.

For non zero forms one has to use different methods and the problem becomes harder there.

2. Some Lemmas

Let α, β ($\alpha < \beta$) be any real numbers and $f(x_1, \ldots, x_n)$ be any polynomial. We say that the inequality

(2.1) $\qquad \alpha < f(x_1, \ldots, x_n) < \beta$

is solvable, if for any real numbers $c_1, \ldots, c_n$ there exist $(x_1, \ldots, x_n) \equiv (c_1, \ldots, c_n)$ (mod 1) satisfying (2.1).

LEMMA 1. Let α, β, γ be real numbers with $\gamma > 1$. Let m be the integer defined by $m < \gamma \leq m + 1$. Then

(2.2) $\qquad 0 < (x_1 + \alpha)^2 + \beta < \gamma$,

(2.3) or $\qquad 0 < -(x_1 + \alpha)^2 + \beta < \gamma$

are solvable provided

$\qquad -m^2/4 < \beta < \gamma - 1/4$,

or $\qquad 1/4 < \beta < \gamma + m^2/4$, respectively.

This follows from Dumir [6] and [5].

LEMMA 2. *Let $0 \le t \le 15/9$ be a real number. Let $\phi(x_2, x_3, x_4)$ be an indefinite ternary quadratic form of type (2,1) and determinant $D < 0$. Then*

$$-t\{f(t)|D|\}^{1/3} < \phi(x_2, x_3, x_4) \le \{f(t)|D|\}^{1/3}$$

is solvable with

(2.4) $f(t) = \begin{cases} 4/(1+t)^2(1+5t) & \text{if } 0 \le t \le 1/7 \\ 27/t(t+9)^2 & \text{if } 1/7 < t \le 15/9 \end{cases}$

This follows from a result of Dumir [8] since one finds that

$$32/(1+t)(7-t)(1+9t) \le 27/t(t+9)^2$$

if and only if $(5t-3)^2(11t-21) \le 0$;

and

$$27/(1+4t)(11-t)(1+t) \le 27/t(t+9)^2$$

if and only if $(t-1)^2(5t-11) \le 0$.

LEMMA 3. *Let $0 \le t \le 1/9$ be a real number and $\phi(x_2, x_3, x_4)$ be as in Lemma 2. Then*

$$t\{f(t)|D|\}^{1/3} < \phi(x_2, x_3, x_4) \le \{f(t)|D|\}^{1/3}$$

is solvable with $f(t) = 4/(1-t)^2(1-5t)$.

This is a result of Hans-Gill and Raka [13].

Let d be equal to $(4|D|)^{1/4}$ throughout this paper.

LEMMA 4. *Let Q be as given in the Theorem. If Q represents μ at a primitive point such that $0 < \mu < d/3$, then (1.2) is solvable.*

Proof: Applying a suitable unimodular transformation we can suppose that

$$Q = \mu\{(x_1 + a_2 x_2 + a_3 x_3 + a_4 x_4)^2 - \phi(x_2, x_3, x_4)\}$$

where $\phi(x_2, x_3, x_4)$ is an indefinite ternary quadratic form of type (2,1) and determinant $-D/\mu^4$. By homogeneity we can assume that $\mu = 1$ so that $d > 3$. Let $n(\ge 3)$ be an integer such that $n < d \le n+1$. Then by Lemma 1, it is enough to solve

(2.5) $-(d - 1/4) < \phi(x_2, x_3, x_4) < n^2/4$.

Let $t = (4d-1)/n^2$, so that $t \le 15/9$. Now (2.5) follows from Lemma 2, if we have

(2.6) $\{f(t)|D|\}^{1/3} < n^2/4$

where $f(t)$ is as given in (2.4). Substituting for t and D and simplifying this is so if

(2.7) $g_1(d) = (n^2 + 4d - 1)(n^2 + 20d - 5)d^{-4} > 64$, for $t \le 1/7$;

(2.8) $g_2(d) = (4d - 1)(9n^2 + 4d - 1)^2 d^{-4} > 432$, for $t > 1/7$.

For $d \le n + 1$ and $n \ge 3$, one can easily check that (2.7) and (2.8) are true, proving thereby (2.6) and hence the Lemma.

LEMMA 5. *Let Q be as given in the Theorem. If Q represents $-v$ such that $0 < v < d/5$, then (1.2) is solvable.*

Proof: Working as in Lemma 4, we can suppose that

$$Q = -(x_1 + a_2 x_2 + a_3 x_3 + a_4 x_4)^2 + \phi(x_2, x_3, x_4)$$

where $\phi(x_2, x_3, x_4)$ is an indefinite ternary quadratic form of type (2,1) and determinant $-D$, $d > 5$. Let n (≥ 5) be an integer such that $n < d \le n + 1$. Then by Lemma 1, it is enough to solve

(2.9) $1/4 < \phi(x_2, x_3, x_4) < d + n^2/4$.

Let $t = 1/(4d + n^2)$, so that $0 \le t \le 1/9$. Now (2.9) follows from Lemma 3, if we have

$$\{4|D|/(1-t)^2 (1-5t)\}^{1/3} < d + n^2/4.$$

Substituting for t and D and simplifying we see that this is so if

(2.10) $f(d) = (n^2 + 4d - 1)^2 (n^2 + 4d - 5)d^{-4} > 64$.

For $d \le n + 1$ and $n \ge 5$, one can easily check that (2.10) is true and hence the Lemma.

LEMMA 6: *Let α, β, γ be real numbers with $\alpha > 0$, $\gamma > 0$. Let 2h and k be integers, such that*

(2.11) $|h - k^2 \alpha| + 1/2 < \gamma$.

Suppose that either $\alpha \ne h/k^2$ or $\beta \equiv h/k \pmod{1/k, 2\alpha}$. Then for any real number v, there exist integers x and y satisfying

(2.12) $0 < \pm x + \beta y \pm \alpha y^2 + v < \gamma$.

This follows from Lemma 6 of Macbeath [14] (See Lemma 11 of Dumir and Sehmi [10]).

LEMMA 7: *Let $Q(x_1, x_2, x_3, x_4) = (x_1 + a_2 x_2)x_2 + x_3^2/2 - x_4^2$, $c_3 = c_4 = 1/2$. If the section of Q namely $Q_s = Q(x_1, x_2, 0, 0)$ represents a real number μ primitively, such that*

    (i) $0 < \mu < d/2$,

or   (ii) $\mu = -v$, $d/5 \le v < d/3$,

then (1.2) is solvable.

Proof: Here $d = (1/2)^{1/4}$.

If (i) holds we can assume, replacing Q by an equivalent form, that

$$Q = \mu(x_1 + b_2 x_2)^2 - x_2^2/4\mu + x_3^2/2 - x_4^2.$$

Let $m (\geq 2)$ be an integer such that $m < d/\mu \leq m + 1$. Applying Lemma 1, we find that (1.2) is solvable if

$$-m^2/4 < -x_2^2/4\mu^2 + x_3^2/2\mu - x_4^2/\mu < d/\mu - 1/4$$

(2.13) i.e. $\qquad -m^2\mu^2 < -x_2^2 + 2\mu x_3^2 - 4\mu x_4^2 < 4d\mu - \mu^2$

is solvable.

If $m \geq 3$, take $x_3 = x_4 = 1/2$ and $x_2 \equiv c_2 \pmod 1$ such that $0 \leq |x_2| \leq 1/2$. For $m = 2$, take $x_3 = 3/2$, $x_4 = 1/2$ and $x_2 \equiv c_2 \pmod 1$ such that $1/2 \leq |x_2| \leq 1$. Now one can easily see that (2.13) is satisfied for $\mu \geq d/(m+1)$.

In (ii), we can assume that

$$Q = -\upsilon(x_1 + b_2 x_2)^2 + x_2^2/4\upsilon + x_3^2/2 - x_4^2.$$

Let m be an integer such that $m < d/\upsilon \leq m + 1$, so that $m = 3$ or 4. Applying Lemma 1, we see that (1.2) is solvable if

$$1/4 < x_2^2/4\upsilon^2 + x_3^2/2\upsilon - x_4^2/\upsilon < d/\upsilon + m^2/4$$

(2.14) i.e. if $\qquad \upsilon^2 < x_2^2 + 2\upsilon x_3^2 - 4\upsilon x_4^2 < 4d\upsilon + m^2\upsilon^2$

is solvable.

On taking $x_3 = x_4 = 1/2$ and $x_2 \equiv c_2 \pmod 1$ such that $1/2 \leq |x_2| \leq 1$, one can easily see that (2.14) is satisfied for $m = 3$ and 4 separately.

LEMMA 8. *Let* $Q = (x_1 + a_2 x_2 + x_3/2 + x_4/2)x_2 - x_3^2/2 + 3x_4^2/2$, $c_3 = c_4 = 0$. *If the section of Q namely $Q_s = Q(x_1, x_2, 0, 0)$ represents a real number μ primitively, such that*

(i) $0 < \mu < d/2$,

or $\qquad$ (ii) $\mu = -\upsilon$, $d/11 < \upsilon < d/2$,

then (1.2) is solvable.

Proof: Here $d = (3/4)^{1/4}$. In (i) we can assume that

$$Q = \mu(x_1 + b_2 x_2)^2 - x_2^2/4\mu - x_3^2/2 + 3x_4^2/2.$$

In (ii) we can assume that

$$Q = -\upsilon(x_1 + b_2 x_2)^2 + x_2^2/4\upsilon - x_3^2/2 + 3x_4^2/2.$$

We apply Lemma 1 and work as in Lemma 7. On taking $x_3 = x_4 = 0$; $x_2 \equiv c_2$ (mod 1) such that $0 \leq |x_2| \leq 1/2$ in case (i) and $1/2 \leq |x_2| \leq 1$ in case (ii), one can easily find that (1.2) is solvable.

3. Proof of The Theorem

If Q is a zero incommensurable form, then following the proof of Theorem 2 of Birch [2] we find that $M_1^+(Q) = 0$. Let, therefore, Q be a zero rational form of type (2,2). Applying a suitable unimodular transformation we can assume that

$$Q = (x_1 + a_2 x_2 + a_3 x_3 + a_4 x_4)x_2 + \phi(x_3, x_4)$$

where $\phi(x_3, x_4)$ is an indefinite binary quadratic form of discriminant $\Delta^2 = 4d^4$ and a_i, $2 \leq i \leq 4$, are some real numbers satisfying

(3.1) $\qquad 0 \leq a_i < 1, 2 \leq i \leq 4.$

Applying a suitable unimodular transformation, we can assume that

(3.2) $\qquad -1/2 < c_j \leq 1/2, 1 \leq j \leq 4.$

LEMMA 9. *Let $\alpha, \beta, \gamma, \delta$ be any real numbers with $\gamma > 0$. Then*

$$0 < (x_1 + \delta x_2 + \alpha)x_2 + \beta < \gamma$$

is solvable provided

(i) $c_2 \neq 0$ and $\gamma > 1/2$.
(ii) $c_2 = 0$ and $\gamma > 1$.

Proof is trivial.

We distinguish the cases when ϕ is a zero or non zero form.

3.1. $\phi(x_3, x_4)$ **is a zero form.** Following Birch reduction [2, Lemma 12] and Watson [22, page 21] we can assume that

$$Q = (x_1 + a_2 x_2 + a_4 x_4)x_2 + m(x_3 + b_4 x_4)x_4$$

where m is a positive integer, given by

(3.3) $\qquad d^4/4 = |D| = m^2/16$ i.e. $m = 2d^2$;

$a_3 = 0$ and b_4 is a real number satisfying

(3.4) $\qquad 0 \leq b_4 < 1, b_4 = 0$ implies $a_4 = 0.$

Changing x_1 to $x_1 + x_4$, x_3 to $-x_3$ and x_4 to $-x_4$, if necessary, we can assume (from (3.1) and (3.2)) that

(3.5) $\qquad 0 \leq a_4 \leq 1/2$ and $0 \leq c_3 \leq 1/2.$

Since $m \geq 1$, so (3.3) implies that $d \geq 1/\sqrt{2} > 1/2$, therefore by Lemma 9, we can assume that

(3.6) $\qquad c_2 = 0$ and $d \leq 1.$

Thus from (3.3), we have

(3.7) $\qquad m = 1$ and $d = 1/\sqrt{2}$ or $m = 2$ and $d = 1$.

LEMMA 10. *If $b_4 = 0$, then (1.2) is solvable unless $Q \sim Q_i$, $i = 1,2,4,5,9,10$ and $c_j \sim c_j^{(i)}$, $1 \le j \le 4$, as listed in the Theorem.*

Proof: Here $Q = (x_1 + a_2 x_2)x_2 + m x_3 x_4$, by (3.4).

If $c_4 \ne 0$, choose $x_4 \equiv c_4$ (mod 1) such that $0 < |x_4| \le 1/2$, take $x_2 = 0$ and choose $x_3 \equiv c_3$ (mod 1) such that

$$0 < Q = m x_3 x_4 \le m|x_4| \le m/2 \le d.$$

Strict inequality holds unless $m = 2$, $d = 1$ and $c_4 = 1/2$. Therefore, there being symmetry in x_3 and x_4, (1.2) is solvable unless

(i) $m = 2$, $c_3, c_4 = 0$ or $1/2$,
(ii) $m = 1$, $c_3 = c_4 = 0$.

Case (I) $m = 2$, $d = 1$.

If $c_3 = c_4 = 1/2$, then

$$0 < Q(x_1, 0, 1/2, 1/2) = 1/2 < d = 1.$$

So we need consider $(c_3, c_4) = (0,0)$ or $(1/2, 0)$ only. Now

$$Q = (x_1 + a_2 x_2)x_2 + 2x_3 x_4.$$

Take $x_2 = \pm 1$, $x_1 = c_1$ and $x_3 x_4 = 0$, then (1.2) is solvable unless

$$\pm c_1 + a_2 \equiv 0 (\text{mod } 1).$$

This gives $(c_1, a_2) = (0,0)$ or $(1/2, 1/2)$. Thus (1.2) is solvable unless

$$Q = x_1 x_2 + 2x_3 x_4 = Q_9, \quad c_j = c_j^{(9)}, \quad 1 \le j \le 4,$$

or

$$Q = (x_1 + x_2/2)x_2 + 2x_3 x_4 = Q_{10}, \quad c_j = c_j^{(10)}, \quad 1 \le j \le 4.$$

Case (II) $m = 1$, $d = 1/\sqrt{2}$.

Here $(c_2, c_3, c_4) = (0,0,0)$, $Q = (x_1 + a_2 x_2)x_2 + x_3 x_4$.

If $a_2 = 0$, choose $x_1 \equiv c_1$ (mod 1) such that $0 \le |x_1| \le 1/2$, so that

$$0 \le Q(x_1, \pm 1, 0, 0) = |x_1| \le 1/2 < d.$$

Thus (1.2) is solvable unless $c_1 = 0$ that is $Q = Q_1$ and $c_j = c_j^{(1)}$, $1 \le j \le 4$.

Let $0 < a_2 < 1$. Since Q represents a_2, by Lemma 4, we can assume that $a_2 \ge d/3$. Now, for integers x and y, we have

$$Q(c_1, y, x, 1) = x + c_1 y + a_2 y^2.$$

By Lemma 6, it is enough to find integers 2h and k such that

(3.8) $$|h - a_2 k^2| + 1/2 < d.$$

On taking

$h = 1, k = 2$ for $d/3 \leq a_2 < (d + 1/2)/4$,
$h = 1/2, k = 1$ for $(d + 1/2)/4 \leq a_2 < d$,
$h = 3, k = 2$ for $d \leq a_2 < (d + 5/2)/4$,
$h = 1, k = 1$ for $(d + 5/2)/4 \leq a_2 < 1$,

we find that (3.8) is satisfied. Therefore by Lemma 6, (1.2) is solvable unless

(i) $a_2 = 1/4$, $c_1 \equiv 1/2 \pmod{1/2, 1/2}$ i.e. $c_1 = 0$ or $1/2$,
(ii) $a_2 = 1/2$, $c_1 \equiv 1/2 \pmod 1$ i.e. $c_1 = 1/2$,
(iii) $a_2 = 3/4$, $c_1 \equiv 3/2 \pmod{1/2, 3/2}$ i.e. $c_1 = 0$ or $1/2$.

When $a_2 = 1/4$ and $c_1 = 0$, then $0 < Q(0,1,0,0) = 1/4 < d$.
When $a_2 = 3/4$ and $c_1 = 1/2$, then $0 < Q(-1/2,1,0,0) = 1/4 < d$.
Thus (1.2) is solvable unless

$$Q = (x_1 + x_2/4)x_2 + x_3 x_4 \sim Q_4, \quad c_j \sim c_j^{(4)}, \quad 1 \leq j \leq 4;$$
or
$$Q = (x_1 + x_2/2)x_2 + x_3 x_4 = Q_2, \quad c_j = c_j^{(2)}, \quad 1 \leq j \leq 4;$$
or
$$Q = (x_1 + 3x_2/4)x_2 + x_3 x_4 \sim Q_5, \quad c_j \sim c_j^{(5)}, \quad 1 \leq j \leq 4.$$

LEMMA 11. *If $0 < b_4 < 1$, then (1.2) is solvable unless $Q \sim Q_i$, $i = 2, 4, 6, 7, 11, 12, 13, 14$, and $c_j \sim c_j^{(i)}$, $1 \leq j \leq 4$, as listed in the Theorem.*

Proof: If $c_4 \neq 0$, choose $x_4 \equiv c_4 \pmod 1$ such that $0 < |x_4| \leq 1/2$, take $x_2 = 0$, and then choose $x_3 \equiv c_3 \pmod 1$ satisfying

$$0 < Q = m(x_3 + b_4 x_4)x_4 \leq m|x_4| \leq m/2 \leq d,$$

because of (3.6) and (3.7). Further strict inequality holds unless $m = 2$, $d = 1$ and $c_4 = 1/2$. But in that case, take $x_4 = \pm 1/2$, $x_2 = 0$ and $x_3 \equiv c_3 \pmod 1$ such that $0 \leq |x_3| \leq 1/2$, so that

$$0 < Q = 2(x_3 + b_4 x_4)x_4 = |x_3| + b_4/2 \leq 1/2 + b_4/2 < d.$$

Thus we must have

(3.9) $$c_4 = 0.$$

Since Q represents mb_4, so by Lemma 4, we can assume that $mb_4 \geq d/3$. Now, for integers x and y, we have

$$Q(x+c_1, \pm 1, c_3, y) = \pm x + \beta y + mb_4 y^2 + \nu$$

where ν is some real number and $\beta = \pm a_4 + mc_3$.

Take

$h = 1, k = 2$ for $d/3 \leq b_4 < (d + 1/2)/4,$
$h = 1/2, k = 1$ for $(d + 1/2)/4 \leq b_4 < d,$
$h = 3, k = 2$ for $d \leq b_4 < (d + 5/2)/4,$ } for $m = 1, d = 1/\sqrt{2}$;
$h = 1, k = 1$ for $(d + 5/2)/4 \leq b_4 < 1,$

and

$h = 1/2, k = 1$ for $1/6 \leq b_4 < 1/2,$
$h = 1, k = 1$ for $1/2 \leq b_4 < 3/4,$ } for $m = 2, d = 1$.
$h = 3/2, k = 1$ for $3/4 \leq b_4 < 1,$

With h and k as above, we find that $|h - mb_4 k^2| + 1/2 < d$.
Therefore by Lemma 6, (1.2) is solvable unless

(3.10) $\quad b_4 = 1/4, \pm a_4 + mc_3 \equiv 1/2 \pmod{1/2}$,
(3.11) $\quad b_4 = 1/2, \pm a_4 + mc_3 \equiv 1/2 \pmod 1$,
(3.12) $\quad b_4 = 3/4, \pm a_4 + mc_3 \equiv 3/2 \pmod{1/2, 3/2}$.

For $m = 2$, we must have

(3.13) $\quad (a_2, c_1) = (0,0)$ or $(1/2, 1/2)$

because otherwise $0 < Q(c_1, \pm 1, c_3, 0) = \pm c_1 + a_2 < 1 = d$.

For $m = 1$, we have

$$Q(c_1, y, x + c_3, \pm 1) = \pm x + (c_1 \pm a_4)y + a_2 y^2 + \nu$$

where ν is some real number. Applying Lemma 6 and working as in Lemma 10, Case (ii), we see that (1.2) is solvable unless

(3.14) $\quad \begin{cases} a_2 = 0 \\ a_2 = 1/4, c_1 \pm a_4 \equiv 1/2 \pmod{1/2} \\ a_2 = 1/2, c_1 \pm a_4 \equiv 1/2 \pmod 1 \\ a_2 = 3/4, c_1 \pm a_4 \equiv 3/2 \pmod{1/2, 3/2} \end{cases}$

Further for $a_2 = 0$, if $c_1 \neq 0$, we have $0 < Q(c_1, \pm 1, c_3, 0) = |c_1| \leq 1/2 < d$
Therefore we must have

(3.15) $\quad c_1 = 0$ for $a_2 = 0$

We discuss these cases one by one.

Case (I) $b_4 = 1/4, m = 2$.

Here,

$$Q = (x_1 + a_2 x_2 + a_4 x_4)x_2 + 2(x_3 + x_4/4)x_4.$$

From (3.10), we get $\pm 2a_4 + 4c_3 \equiv 0 \pmod 1$. This, together with (3.5) gives $c_3 = 0, 1/8, 1/4, 3/8$ or $1/2$. Further for $c_3 = 1/4$ we get $a_4 = 0$ or $1/2$; for

$c_3 = 3/8$, $a_4 = 1/4$ and for $c_3 = 1/2$, $a_4 = 0$ or $1/2$. Also (3.9) and (3.13) hold here. The following table gives the solution of (1.2) in different cases, '-' means it can take any arbitrary value.

c_3	a_4	(a_2, c_1)	$Q(x_1, x_2, x_3, x_4)$
0	-	-	$Q(c_1, 0, 0, 1) = 1/2$
1/8	-	-	$Q(c_1, 0, 1/8, 1) = 3/4$
1/4	1/2	(0,0)	$Q(1, 1, 1/4, -1) = 1/2$
1/4	1/2	(1/2, 1/2)	$Q(1/2, 1, 1/4, -1) = 1/2$
3/8	1/4	(0,0)	$Q(1, 1, 3/8, -1) = 1/2$
3/8	1/4	(1/2, 1/2)	$Q(1/2, 1, 3/8, -1) = 1/2$
1/2	0	(0,0)	$Q(1, -1, 1/2, 1) = 1/2$
1/2	0	(1/2, 1/2)	$Q(3/2, -1, 1/2, 1) = 1/2$
1/2	1/2	(0,0)	$Q(0, -2, 1/2, -1) = 1/2$

Thus, we see that (1.2) is solvable unless (c_3, a_4, a_2, c_1) is equal to $(1/4, 0, 0, 0)$ or $(1/4, 0, 1/2, 1/2)$ or $(1/2, 1/2, 1/2, 1/2)$. This gives

$$Q = x_1 x_2 + 2(x_3 + x_4/4)x_4 = Q_{13}, \quad c_j = c_j^{(13)}, \quad 1 \leq j \leq 4.$$

or

$$Q = (x_1 + x_2/2)x_2 + 2(x_3 + x_4/4)x_4 \sim Q_{13}, \quad \text{by the transformation}$$

$$x_1 \to x_1 - x_2 - 2x_3 - x_4, \quad x_2 \to x_2, \quad x_3 \to x_3, \quad x_4 \to x_4 + x_2;$$

or

$$Q = (x_1 + x_2/2 + x_4/2)x_2 + 2(x_3 + x_4/4)x_4 = Q_{14}, \quad c_j = c_j^{(14)}, \quad 1 \leq j \leq 4..$$

Case (II) $b_4 = 1/2$, $m = 2$.

Here $Q = (x_1 + a_2 x_2 + a_4 x_4)x_2 + 2(x_3 + x_4/2)x_4$.

From (3.11), we get $\pm a_4 + 2c_3 \equiv 0 \pmod 1$. This, together with (3.5), gives $c_3 = 0, 1/4$ or $1/2$. Further for $c_3 = 0$ or $1/2$, we get $a_4 = 0$; when $c_3 = 1/4$, we have

$$0 < Q(x_1, 0, 1/4, 1) = 1/2 < d = 1.$$

Thus we see, using (3.9) and (3.13) that (1.2) is solvable unless

$$Q = x_1 x_2 + 2(x_3 + x_4/2)x_4 = Q_{11}, \quad c_j = c_j^{(11)}, \quad 1 \leq j \leq 4.$$

or

$$Q = (x_1 + x_2/2)x_2 + 2(x_3 + x_4/2)x_4 = Q_{12}, \quad c_j = c_j^{(12)}, \quad 1 \leq j \leq 4.$$

Case (III) $b_4 = 3/4$, $m = 2$.

Here $Q = (x_1 + a_2 x_2 + a_4 x_4)x_2 + 2(x_3 + 3x_4/4)x_4$.

From (3.12), we get $\pm 2a_4 + 4c_3 \equiv 0 \pmod 1$. This, together with (3.5), gives $c_3 = 0, 1/8, 1/4, 3/8$ or $1/2$. Further for $c_3 = 0$, we get $a_4 = 0$ or $1/2$; for $c_3 = 1/8, a_4 = 1/4$; and for $c_3 = 1/4$ we get $a_4 = 0$ or $1/2$. Also (3.9) and (3.13) hold here.

The following table gives us the solutions of (1.2) in different cases:

c_3	a_4	(a_2, c_1)	$Q(x_1, x_2, x_3, x_4)$
0	0	(0,0)	$Q(1, -1, 0, 1) = 1/2$
0	0	(1/2, 1/2)	$Q(3/2, -1, 0, 1) = 1/2$
0	1/2	(0,0)	$Q(0, 2, -1, 1) = 1/2$
1/8	1/4	(0,0)	$Q(0, -6, 1/8, 1) = 1/4$
1/8	1/4	(1/2, 1/2)	$Q(3/2, -2, 1/8, 1) = 1/4$
1/4	1/2	(0,0)	$Q(0, 1, 1/4, -1) = 1/2$
1/4	1/2	(1/2, 1/2)	$Q(-1/2, 1, 1/4, -1) = 1/2$
3/8	-	-	$Q(x_1, 0, 3/8, -1) = 3/4$
1/2	-	-	$Q(x_1, 0, 1/2, -1) = 1/2$

Thus (1.2) is solvable unless $(c_3, a_4, a_2, c_1) = (0, 1/2, 1/2, 1/2)$ or $(1/4, 0, 0, 0)$ or $(1/4, 0, 1/2, 1/2)$. This gives

$$Q = (x_1 + x_2/2 + x_4/2)x_2 + 2(x_3 + 3x_4/4)x_4 \sim Q_{14}, \text{ by the transformation}$$

$$x_1 \to x_1 + 2x_4, \quad x_2 \to x_2 - 2x_4, \quad x_3 \to x_1 + x_3 + x_4, \quad x_4 \to x_4.$$

or

$$Q = x_1 x_2 + 2(x_3 + 3x_4/4)x_4 \sim -Q_{13},$$

or

$$Q = (x_1 + x_2/2)x_2 + 2(x_3 + 3x_4/4)x_4 \sim -Q_{13}, \text{ by the transformation}$$

$$x_1 \to -x_1 + 2x_3 - x_4 - 2x_2, \quad x_2 \to x_2, \quad x_3 \to x_3 + x_4, \quad x_4 \to -x_4 - x_2.$$

Case (Iv) $b_4 = 1/4$, $m=1$.

Here $Q = (x_1 + a_2 x_2 + a_4 x_4)x_2 + (x_3 + x_4/4)x_4$.

From (3.10), we get $\pm 2a_4 + 2c_3 \equiv 0 \pmod 1$. This, together with (3.5), gives $c_3 = 0, 1/4$ or $1/2$. For $c_3 = 1/2$, we have $a_4 = 0$ or $1/2$; but then from (3.14) and (3.15) we get

$c_1 = 0$ for $a_2 = 0$,

$c_1 = 0$ or $1/2$ for $a_2 = 1/4$ or $3/4$,

$(a_4, c_1) = (0, 1/2)$ or $(1/2, 0)$, for $a_2 = 1/2$.

The following table gives us the solutions of (1.2) in different cases:

c_3	a_4	a_2	c_1	$Q(x_1, x_2, x_3, x_4)$
0	-	-	-	$Q(c_1, 0, 0, 1) = 1/4$
1/4	-	-	-	$Q(c_1, 0, 1/4, 1) = 1/2$
1/2	0	1/4	0	$Q(0, 1, 1/2, 0) = 1/4$
1/2	0	1/4	1/2	$Q(1/2, -1, 1/2, 1) = 1/2$
1/2	0	3/4	0	$Q(-1, 1, 1/2, 1) = 1/2$
1/2	0	3/4	1/2	$Q(-1/2, 1, 1/2, 0) = 1/4$
1/2	1/2	0	0	$Q(0, -1, 1/2, 1) = 1/4$
1/2	1/2	1/4	0	$Q(0, 1, 1/2, 0) = 1/4$
1/2	1/2	1/2	0	$Q(-3, 1, 1/2, 2) = 1/2$
1/2	1/2	3/4	1/2	$Q(-1/2, 1, 1/2, 0) = 1/4$

For $(c_3, a_4, a_2, c_1) = (1/2, 0, 0, 0), (1/2, 0, 1/2, 1/2), (1/2, 1/2, 1/4, 1/2)$ and $(1/2, 1/2, 3/4, 0)$ we get critical forms namely

$$Q = x_1 x_2 + (x_3 + x_4/4)x_4 = Q_4, \quad c_j = c_j^{(4)}, \quad 1 \leq j \leq 4,$$

or

$$Q = (x_1 + x_2/2)x_2 + (x_3 + x_4/4)x_4 = Q_6, \quad c_j = c_j^{(6)}, \quad 1 \leq j \leq 4,$$

or

$$Q = (x_1 + x_2/4 + x_4/2)x_2 + (x_3 + x_4/4)x_4 \sim Q_4, \text{ by the transformation}$$

$$x_1 \to x_1 - x_3, \quad x_2 \to x_2, \quad x_3 \to -x_2 + x_3, \quad x_4 \to x_4 + x_2,$$

or

$$Q = (x_1 + 3x_2/4 + x_4/2)x_2 + (x_3 + x_4/4)x_4 \sim Q_6.$$

Case (v) $b_4 = 1/2$, $m=1$.

Here $Q = (x_1 + a_2 x_2 + a_4 x_4)x_2 + (x_3 + x_4/2)x_4$.

From (3.11), we get $\pm a_4 + c_3 \equiv 1/2 \pmod 1$. This gives $c_3 = 0$ or $1/2$. For $c_3 = 1/2$ we have $a_4 = 0$; but then from (3.14) and (3.15) we get

$c_1 = 0$ for $a_2 = 0$,
$c_1 = 0$ or $1/2$ for $a_2 = 1/4$ or $3/4$,
$c_1 = 1/2$ for $a_2 = 1/2$.

The following table gives us the solutions of (1.2) in different cases:

c_3	a_4	a_2	c_1	$Q(x_1, x_2, x_3, x_4)$
0	-	-	-	$Q(c_1, 0, 0, 1) = 1/2$
1/2	0	1/4	0	$Q(0, 1, 1/2, 0) = 1/4$
1/2	0	3/4	1/2	$Q(-1/2, 1, 1/2, 0) = 1/4$

For remaining values of $(c_3, a_4, a_2, c_1) = (1/2, 0, 0, 0), (1/2, 0, 1/4, 1/2), (1/2, 0, 1/2, 1/2)$ or $(1/2, 0, 3/4, 0)$ we get critical forms, namely

$Q = x_1 x_2 + (x_3 + x_4/2)x_4 \sim Q_2$,
or
$Q = (x_1 + x_2/4)x_2 + (x_3 + x_4/2)x_4 \sim Q_6$,
or
$Q = (x_1 + x_2/2)x_2 + (x_3 + x_4/2)x_4 \sim Q_2$, by the transformation
$x_1 \to -x_1 + x_3$, $x_2 \to x_4$, $x_3 \to x_1 - x_4$, $x_4 \to x_2 + x_4$,
or
$Q = (x_1 + 3x_2/4)x_2 + (x_3 + x_4/2)x_4 \sim Q_7$, $c_j = c_j^{(7)}$, $1 \le j \le 4$.

Case (vi) $b_4 = 3/4$, $m = 1$.

Here $Q = (x_1 + a_2 x_2 + a_4 x_4)x_2 + (x_3 + 3x_4/4)x_4$.

From (3.12), we get $\pm 2a_4 + 2c_3 \equiv 0 \pmod{1}$. This gives $c_3 = 0, 1/4$ or $1/2$.
Further for $c_3 = 0$ we have $a_4 = 0$ or $1/2$; and then from (3.14) and (3.15) we get

$c_1 = 0$ for $a_2 = 0$,
$c_1 = 0$ or $1/2$ for $a_2 = 1/4$ or $3/4$,
$(a_4, c_1) = (0, 1/2)$ or $(1/2, 0)$ for $a_2 = 1/2$.

The following table gives us the solutions of (1.2) in different cases:

c_3	a_4	a_2	c_1	$Q(x_1, x_2, x_3, x_4)$
1/4	-	-	-	$Q(c_1, 0, 1/4, -1) = 1/2$
1/2	-	-	-	$Q(c_1, 0, -1/2, 1) = 1/4$
0	0	1/4	0	$Q(0, 1, 0, 0) = 1/4$
0	0	1/4	1/2	$Q(1/2, 1, -1, 1) = 1/2$
0	0	3/4	0	$Q(-1, 1, 0, 1) = 1/2$
0	0	3/4	1/2	$Q(-1/2, 1, 1, 0) = 1/4$
0	1/2	0	0	$Q(0, -1, 0, 1) = 1/4$
0	1/2	1/4	0	$Q(0, 1, 1, 0) = 1/4$
0	1/2	1/2	0	$Q(0, 1, 1, 0) = 1/2$
0	1/2	3/4	1/2	$Q(-1/2, 1, 0, 0) = 1/4$

For remaining values of $(c_3, a_4, a_2, c_1) = (0, 0, 0, 0)$, $(0, 0, 1/2, 1/2)$, $(0, 1/2, 1/4, 1/2)$ or $(0, 1/2, 3/4, 0)$ we get critical forms, namely

$Q = x_1 x_2 + (x_3 + 3x_4/4)x_4 = Q_5$, $c_j = c_j^{(5)}$, $1 \le j \le 4$.
or
$Q = (x_1 + x_2/2)x_2 + (x_3 + 3x_4/4)x_4 = Q_7$, $c_j = c_j^{(7)}$, $1 \le j \le 4$.
or
$Q = (x_1 + x_2/4 + x_4/2)x_2 + (x_3 + 3x_4/4)x_4 \sim Q_7$, by the transformation
$x_1 \to x_1 - x_3 - x_4$, $x_2 \to x_2$, $x_3 \to -x_2 + x_3$, $x_4 \to x_2 + x_4$.
or
$Q = (x_1 + 3x_2/4 + x_4/2)x_2 + (x_3 + 3x_4/4)x_4 \sim Q_5$, by the transformation
$x_1 \to x_1 - x_2 - x_3 - x_4$, $x_2 \to x_2$, $x_3 \to -x_2 + x_3$, $x_4 \to x_2 + x_4$.

3.2. $\phi(x_3, x_4)$ Is a non-zero form.

By the Markoff Chain Theorem (see eg [3], Chapter II), if $\phi \sim \rho\phi_1 = \rho(x_3^2 + x_3 x_4 - x_4^2)$, there exist integers u, v not both zero with g.c.d.$(u,v)=1$, such that $a = \phi(u,v)$ and $|a| \leq \Delta/\sqrt{8}$; $\Delta^2 (=4d^4)$ being the discriminant of ϕ. Further if $|a| > \Delta/3$, then ϕ is a Markoff form and so represents both a and $-a$. In that case we can assume $a > 0$. Since ϕ and hence Q represents a, we have $a \geq d/3$ if $a > 0$, by Lemma 4; and if $a = -b < 0$ we have $b \geq d/5$ by Lemma 5. Replacing ϕ by an equivalent form we need to distinguish the following cases:

(3.16) $\quad \phi(x_3, x_4) = a(x_3 + \lambda x_4)^2 - t x_4^2$, where $t = \Delta^2/4a = d^4/a$,
$d/3 \leq a \leq \Delta/\sqrt{8} = d^2/\sqrt{2}$, $-1/2 < \lambda \leq 1/2$.

(3.17) $\quad \phi(x_3, x_4) = -b(x_3 + \lambda x_4)^2 + t x_4^2$, where $t = \Delta^2/4b = d^4/b$,
$d/5 \leq b \leq \Delta/3 = 2d^2/3$, $-1/2 < \lambda \leq 1/2$.

(3.18) $\quad \phi \sim -b\phi_1 = -b(x_3^2 + x_3 x_4 - x_4^2)$, $b > 0$.

3.2.1. ϕ is of the form (3.16). Here

$$Q = (x_1 + a_2 x_2 + a_3 x_3 + a_4 x_4)x_2 + a(x_3 + \lambda x_4)^2 - t x_4^2$$
$$= a(x_3 + \lambda x_4 + a_3 x_2 /2a)^2 + (x_1 + a_2' x_2 + a_4' x_4)x_2 - t x_4^2,$$

where a_2' and a_4' are some real numbers and

(3.19) $\quad d \geq \sqrt{2}/3$, $\sqrt{2}/d \leq d/a \leq 3$.

If $c_2 \neq 0$, we have $d \leq 1/2$ from Lemma 9. Then $2 < 2\sqrt{2} \leq \sqrt{2}/d \leq d/a \leq 3$. Now by Lemma 1, (1.2) is solvable if

$$-1 < (x_1 + a_2' x_2 + a_4' x_4)x_2/a - t x_4^2/a < d/a - 1/4,$$

i.e. if

(3.20) $\quad 0 < F(x_1, x_2, x_4) = (x_1 + a_2' x_2 + a_4' x_4)x_2 - t x_4^2 + a < d + 3a/4$

is solvable.

Since from (3.19), $d + 3a/4 \geq 5d/4 \geq 5\sqrt{2}/12 > 1/2$; (3.20) is solvable, by Lemma 9, on taking x_4 arbitrarily and hence (1.2) is solvable.

Let now $c_2 = 0$ and therefore $d \leq 1$ by Lemma 9.

Take $x_3 = y + c_3$, $x_1 = x + c_1$, $x_2 = \pm 1$ and x_4 arbitrarily, so that

(3.21) $\quad Q = \pm x + \beta y + a y^2 + \nu$, with $\beta = \pm a_3 + 2a(c_3 + \lambda x_4)$,

and ν is some real number. Suppose there exist integers $2h$ and k satisfying

(3.22) $\quad |h - ak^2| + 1/2 < d$

then, by Lemma 6, (1.2) is solvable, unless $a = h/k^2$ and $\beta \equiv h/k \pmod{1/k, 2a}$. Taking $h = 1/2$, $k = 1$ and using (3.19) we see that (3.22) is satisfied unless $a + d \leq 1$.

Thus, we are left with the cases:

(i) $a+d \leq 1$, $a < 1/2$,

(ii) $a = 1/2$, $\beta = \pm a_3 + (c_3 + \lambda x_4) \equiv 1/2 \pmod 1$.

On taking $x_4 = c_4$ and $1+c_4$ simultaneously in case (ii), we get $\lambda = 0$, Therefore $\pm a_3 + c_3 \equiv 1/2 \pmod 1$, which gives $(a_3, c_3) = (0, 1/2)$ or $(1/2, 0)$.

LEMMA 12. *If $c_2 = 0$, $a+d \leq 1$, then (1.2) is solvable.*

Proof: From (3.19), $d \leq 1-a \leq 1-d/3$, so we have

(3.23) $\qquad d \leq 3/4$.

Also

$$a \leq \min(1-d, d^2/\sqrt{2}) = \begin{cases} 1-d & \text{if } d > 0.6764... \\ d^2/\sqrt{2} & \text{if } d \leq 0.6764... \end{cases}$$

which gives that $(3 \geq) d/a > 2$.

Now working as before we find that (1.2) has a solution if (3.20) is solvable. If $a > d^2/2$, take $x_2 = 0$, $x_4 \equiv c_4 \pmod 1$ such that $0 \leq |x_4| \leq 1/2$ and x_1 arbitrarily to get

$$0 < -d^4/4a + a \leq F(x_1, x_2, x_4) = -tx_4^2 + a \leq a < d + 3a/4.$$

Let therefore

(3.24) $\qquad d/3 \leq a \leq d^2/2$, and so $d \geq 2/3$.

Let $x_1 = x+c_1$, $x_2 = \pm 1$, $x_4 = y+c_4$, so that $F = \pm x + \beta y - ty^2 + \nu$, for some real number β and ν. Using (3.23) and (3.24) one can easily check that $|h-tk^2| + 1/2 < d + 3a/4$ for $h = k = 1$. Therefore, by Lemma 6 applied to F, we find that (3.20) and hence (1.2) is solvable unless $t = 1$ i.e. $a = d^4$.

From (3.24), we get

(3.25) $\qquad (1/3)^{1/3} \leq d \leq (1/2)^{1/2}$, and so $a = d^4 \leq 1/4$.

Further taking $h = 1$, $k = 2$ and using (3.25) we find that (3.22) is satisfied, for Q as given in (3.21). Therefore (1.2) is solvable unless $a = 1/4$ and $\beta \equiv 1/2 \pmod{1/2, 1/2}$, i.e. $\pm 2a_3 + c_3 + \lambda x_4 \equiv 0 \pmod 1$. On taking $x_4 = c_4$ and $1+c_4$ simultaneously, we find that $\lambda = 0$. But then $\phi(x_3, x_4) = x_3^2/4 - x_4^2$, is a zero form and have already been discussed in § 3.1. This completes the proof of the Lemma.

LEMMA 13. *If $a = 1/2$, $c_2 = 0$, $d \leq 1$, $\lambda = 0$, $(a_3, c_3) = (1/2, 0)$ or $(0, 1/2)$ then (1.2) is solvable unless $Q \sim Q_8$ and $c_j \sim c_j^{(8)}$, $1 \leq j \leq 4$.*

Proof: Here
$$Q = (x_1 + a_2 x_2 + a_3 x_3 + a_4 x_4) x_2 + x_3^2/2 - tx_4^2, \quad t = d^4/a = 2d^4.$$

Also from (3.16),

(3.26) $\qquad 1/2 = a \leq d^2/\sqrt{2}$ i.e. $t = 2d^4 \geq 1$ i.e. $d \geq (1/\sqrt{2})^{1/2}$.

If $d = 1$ i.e. $t = 2$, then $\phi(x_3, x_4) = x_3^2/2 - 2x_4^2$ is a zero form, and is already discussed.

Let now $d<1$ i.e. $t<2$.

If $c_3=0$, take $x_2=0$, $x_3=1$ and $x_4 \equiv c_4$ (mod 1) such that $0 \leq |x_4| \leq 1/2$, so that
$$0 < 1/2 - t/4 < Q = 1/2 - t x_4^2 \leq 1/2 < d.$$
Let therefore $(a_3, c_3) = (0, 1/2)$.

Changing x_4 to $-x_4$ if necessary we can suppose that $c_4 \geq 0$. Let $x_1 = x+c_1$, $x_2 = \pm 1$, $x_3 = 1/2$, $x_4 = y+c_4$, so that $Q = \pm x + \beta' y - t y^2 + \nu$, where $\beta' = \pm a_4 - 2tc_4$ and ν is some real number. Taking

$$h = 1, \quad k = 1 \text{ if } 1 \leq t < d + 1/2,$$
$$h = 3/2, \quad k = 1 \text{ if } d + 1/2 \leq t < 2,$$

we see, using (3.26) that $|h - tk^2| + 1/2 < d$. Therefore by Lemma 6, (1.2) has a solution unless

(i) $t = 1$, $\pm a_4 - 2c_4 \equiv 0 \pmod{1}$ i.e. $c_4 = 0, 1/4$ or $1/2$,
(ii) $t = 3/2$, $\pm a_4 - 3c_4 \equiv 0 \pmod{1}$ i.e. $c_4 = 0, 1/6, 1/3$ or $1/2$.

The following table gives us the solutions of (1.2) in different cases:

t	c_4	$Q(x_1, x_2, x_3, x_4)$
1, 3/2	0	$Q(c_1, 0, 1/2, 0) = 1/8$
1	1/4	$Q(c_1, 0, 1/2, 1/4) = 1/16$
3/2	1/6	$Q(c_1, 0, 1/2, 1/6) = 1/12$
3/2	1/3	$Q(c_1, 0, 5/2, 4/3) = 11/24$
3/2	1/2	$Q(c_1, 0, 3/2, 1/2) = 3/4$

Therefore we are left with the case $t=1$, $c_4 = 1/2$, which further gives $a_4 = 0$. Hence
$$Q = (x_1 + a_2 x_2) x_2 + x_3^2/2 - x_4^2, \quad (c_2, c_3, c_4) = (0, 1/2, 1/2).$$

If $0 < a_2 < d/2$, then $Q_s = Q(x_1, x_2, 0, 0)$ represents a_2, so (1.2) is solvable by Lemma 7. If $d/2 \leq a_2 < 1/2$, then Q represents $-\nu = a_2 - 1/2$, so (1.2) is solvable by Lemma 5, since $0 < \nu = 1/2 - a_2 < d/5$. If $1/2 < a_2 < 1/2 + d/3$, then Q represents $\mu = a_2 - 1/2$, so again (1.2) is solvable by Lemma 4. If $1/2 + d/3 \leq a_2 \leq 1 - d/5$, then $Q_s = Q(x_1, x_2, 0, 0)$ represents $-\nu = a_2 - 1$, where $d/5 \leq \nu = 1-a_2 < d/3$. If $1 - d/5 < a_2 < 1$, then Q represents $-\nu = a_2 - 1$, $0 < \nu < d/5$, therefore (1.2) is solvable by Lemmas 7 and 5, except when $a_2 = 0$ or $1/2$.

For $a_2 = 0$, $Q = x_1 x_2 + x_3^2/2 - x_4^2 = Q_8$, $c_j = c_j^{(8)}$, $1 \leq j \leq 4$; because if $c_1 \neq 0$, choose $x_1 \equiv c_1$ (mod 1) such that $0 < |x_1| \leq 1/2$ and then choose x_2 integer such that
$$0 < Q = x_1 x_2 + x_3^2/2 - x_4^2 \leq |x_1| \leq 1/2 < d.$$
If $a_2 = 1/2$, $Q = (x_1 + x_2/2) x_2 + x_3^2/2 - x_4^2 \sim Q_8$, by the transformation
$$x_1 \to x_1 - x_2 - x_3, \quad x_2 \to x_2, \quad x_3 \to x_2 + x_3, \quad x_4 \to x_4.$$

3.2.2. ϕ is of the form (3.17). Here

$$Q = (x_1 + a_2 x_2 + a_3 x_3 + a_4 x_4)x_2 - b(x_3 + \lambda x_4)^2 + tx_4^2$$
$$= -b(x_3 + \lambda x_4 - a_3 x_2 /2b)^2 + (x_1 + a_2' x_2 + a_4' x_4)x_2 + tx_4^2, \text{ and}$$

(3.27) $\quad t = d^4/b,\ d/5 \le b \le \Delta/3 = 2d^2/3,\ d \ge 3/10.$

We can assume, by Lemma 9 that $d \le 1/2$, if $c_2 \ne 0$ and $d \le 1$ if $c_2 = 0$. Let N be an integer such that $N < d/b \le N+1$, then since $5 \ge d/b \ge 3/2d$, we get

(3.28) $\qquad N = 2, 3$ or 4, if $c_2 \ne 0$, also $d = 1/2$ for $N = 2$; and

(3.29) $\qquad N = 1, 2, 3$ or 4, if $c_2 = 0$.

Applying Lemma 1 we see that, (1.2) is solvable if

(3.30) $\quad 0 < F(x_1, x_2, x_4) = (x_1 + a_2' x_2 + a_4' x_4)x_2 + tx_4^2 - b/4 < d + (N^2 - 1)b/4$

is solvable.

Further, by Lemma 9, we can assume that

$\qquad (N+3)d/4 \le d + (N^2 - 1)b/4 \le 1/2, \qquad$ if $c_2 \ne 0,$
and $\quad (N+3)d/4 \le d + (N^2 - 1)b/4 \le 1, \qquad$ if $c_2 = 0.$

This gives

(3.31) $\qquad 3/10 \le d \le 2/(N+3), \qquad$ if $c_2 \ne 0,$
(3.32) and $\quad 3/10 \le d \le 4/(N+3), \qquad$ if $c_2 = 0.$

LEMMA 14. *If $c_2 \ne 0$, then (3.30) and hence (1.2) is solvable.*

Proof: We can rewrite F as
$$F = t(x_4 + a_4' x_2 /2t)^2 + (x_1 + a_2'' x_2)x_2 - b/4.$$

Let $p(\ge 1)$ be an integer such that $p < \{d + (N^2 - 1)b/4\}/t \le p+1$. Applying Lemma 1, we see that (3.30) is solvable if

(3.33) $\qquad 0 < (x_1 + a_2'' x_2)x_2 - b/4 + p^2 t/4 < d + (N^2 - 1)b/4 + (p^2 - 1)t/4$

is solvable.

From (3.28) and (3.31), we get that $N = 3$ and $d \le 1/3$. Further for $N = 3$, $\{d + (N^2 - 1)b/4\}/t = (d + 2b)b/d^4 > 3/8d^2 > 3$; so that $p \ge 3$. Now since

$$d + (N^2 - 1)b/4 + (p^2 - 1)t/4 \ge \{(N+3)/4\} \cdot \{(p+3)/4\} \cdot d \ge (3/2) \cdot (3/2) \cdot d > 1/2,$$

we see that (3.33) is solvable, by Lemma 9.

LEMMA 15. *If $c_2 = 0, N = 1$, then (1.2) is solvable unless $Q \sim Q_3$, and $c_j \sim c_j^{(3)},\ 1 \le j \le 4.$*

Proof: Here
(3.34) $\quad d/2 \leq b \leq 2d^2/3,\ d \leq 1$ and so $d \geq 3/4$.

Let $x_2 = \pm 1$, $x_1 = x+c_1$, $x_3 = y+c_3$, and x_4 arbitrary. So that
$$Q = \pm x + \beta_1 y - by^2 + \nu, \text{ where } \beta_1 = \pm a_3 - 2b(c_3 + \lambda x_4).$$
Suppose there exist integers $2h$ and k satisfying
(3.35) $\quad |h - bk^2| + 1/2 < d$,

then by Lemma 6, (1.2) is solvable unless $b = h/k^2$ and $\beta_1 \equiv h/k \pmod{1/k, 2b}$.

Using (3.34) one can easily verify that (3.35) is satisfied for $h = 1/2$ and $k = 1$. Therefore (1.2) is solvable unless $b = 1/2$ and $\beta_1 = \pm a_3 - (c_3 + \lambda x_4) \equiv 1/2 \pmod 1$. Taking $x_4 = c_4$ and $1+c_4$ simultaneously we get $\lambda = 0$ and then $(a_3, c_3) = (1/2, 0)$ or $(0, 1/2)$. Further $b = 1/2 \leq 2d^2/3$ implies $d \geq \sqrt{3/4}$, so that $9/8 \leq t = 2d^4 \leq 2$.

If $t = 2$, we have $\phi(x_3, x_4) = -x_3^2/2 + 2x_4^2$, which is a zero form.
So let therefore $t < 2$ i.e. $d < 1$.

Taking x_3 arbitrarily, we have
$$Q(x+c_1, \pm 1, x_3, y+c_4) = \pm x + \beta_2 y + ty^2 + \nu, \text{ where } \beta_2 = \pm a_4 + 2tc_4.$$
Suppose there exist integers $2h$ and k satisfying
(3.36) $\quad |h - tk^2| + 1/2 < d$,

then, by Lemma 6, (1.2) is solvable unless $t = h/k^2$ and $\beta_2 \equiv h/k \pmod{1/k, 2t}$.
We take
$$h=1,\ k=1 \quad \text{for } t = 2d^4 < d+1/2$$
$$h=3/2,\ k=1 \quad \text{for } d+1/2 \leq t < 2,$$
and find that (3.36) is satisfied, therefore (1.2) is solvable unless $t = 3/2$ and $\beta_2 = \pm a_4 + 3c_4 \equiv 1/2 \pmod 1$. Hence
$$Q = (x_1 + a_2 x_2 + a_3 x_3 + a_4 x_4)x_2 - x_3^2/2 + 3x_4^2/2,\ d = (3/4)^{1/4},$$
and $c_4 = 0, 1/6, 1/3$ or $1/2$. (We can assume, as before, that $c_4 \geq 0$.)
When $c_3 = 0$, take $x_2 = x_3 = 0$, $x_4 = c_4$ to find that (1.2) has a solution unless $c_4 = 0$.
When $c_3 = 1/2$, (1.2) has a solution on taking $x_2 = 0$, $x_4 = c_4$ and $x_3 = 3/2, 3/2, 1/2$ or $1/2$ for $c_4 = 0, 1/6, 1/3$ or $1/2$ respectively.

Thus (1.2) has a solution unless $c_3 = c_4 = 0$; but then $a_3 = a_4 = 1/2$.
Therefore $Q = (x_1 + a_2 x_2 + x_3/2 + x_4/2)x_2 - x_3^2/2 + 3x_4^2/2$.

If $0 < a_2 < d/2$, the section of Q namely $Q_s = Q(x_1, x_2, 0, 0)$ represents a_2, so (1.2) is solvable by Lemma 8. If $d/2 \leq a_2 < 1/2$, Q represents $-\nu = -2 + 4a_2$, so (1.2) is solvable by Lemma 5. If $1/2 < a_2 < 1/2 + d/8$, then section Q_s represents $\mu = -2 + 4a_2$, and if $1/2 + d/8 \leq a_2 < 1 - d/11$, section Q_s represents

$-\nu = -1 + a_2$, so (1.2) is solvable by Lemma 8. If $1 - d/11 \le a_2 < 1$, Q represents $-\nu = -1 + a_2$, so (1.2) is solvable by Lemma 5. Thus we are left with $a_2 = 0$ or $1/2$.

If $a_2 = 0$ but $c_1 \ne 0$, then $Q(c_1, \pm 1, 0, 0) = |c_1|$ and if $a_2 = 0$, $c_1 = 0$ then $Q(1, 2, -1, 0) = 1/2$, gives a solution of (1.2).

If $a_2 = 1/2$,
$$Q = (x_1 + x_2/2 + x_3/2 + x_4/2)x_2 - x_3^2/2 + 3x_4^2/2 = Q_3.$$

For Q_3 we must have $c_1 = 1/2$, because otherwise
$$0 < Q(c_1, \pm 1, 0, 0) = -|c_1| + 1/2 < d.$$

Hence $c_j = c_j^{(3)}$, $1 \le j \le 4$.

LEMMA 16. *If $c_2 = 0$, $N = 2$, then (1.2) is solvable.*

Proof: If $t < d$, take $x_2 = 0$, $x_4 \equiv c_4 \pmod{1}$ such that $1/2 \le |x_4| \le 1$ and $x_3 \equiv c_3 \pmod{1}$ such that $0 \le |x_3 + \lambda x_4| \le 1/2$, so that
$$0 < -b/4 + t/4 < Q = -b(x_3 + \lambda x_4)^2 + tx_4^2 \le t < d,$$
as $t = d^4/b \ge 3d^2/2 > 2d^2/3 \ge b$ (by (3.27)). So let therefore
(3.37) $\qquad\qquad t \ge d.$

Here, since $b \ge d/3$, we have $d \le t = d^4/b \le 3d^3$; therefore $d \ge 1/\sqrt{3}$. Also from (3.32) we have $d \le 4/5$. Now
$$F(x + c_1, \pm 1, y + c_4) = \pm x + \beta_3 y + ty^2 + \nu, \text{ where } \beta_3 = \pm a_4' + 2tc_4.$$

Working as usual, existence of integers $2h$ and k satisfying
(3.38) $\qquad |h - tk^2| + 1/2 < d + (N^2 - 1)b/4,$

implies that (3.30) is solvable unless $t = h/k^2$ and $\beta_3 \equiv h/k \pmod{1/k, 2t}$. We take

$\qquad h = 1/2, k = 1 \qquad$ for $d \le t < d + 3b/4$,
$\qquad h = 1, \quad k = 1 \qquad$ for $d + 3b/4 \le t < d + 3b/4 + 1/2$ and
$\qquad h = 3/2, k = 1 \qquad$ for $d + 3b/4 + 1/2 \le t \le 3d^3$.

(Note that $3d^3 \ge t \ge d + 3b/4 \ge 5/4 d$ implies $d > \sqrt{5/12} > 3/5$.)

For the above choice of h and k, one finds that (3.38) is satisfied. Therefore (3.30) and hence (1.2) is solvable unless $t = 1$ or $3/2$.

Case (I) $t = 1$.

Here, $d/3 \le b = d^4 < d/2$, $(1/3)^{(1/3)} \le d \le (1/2)^{(1/3)}$ and so $(1/3)^{(4/3)} \le b = d^4 \le (1/2)^{(4/3)}$.

One can check that (3.35) is satisfied on taking
$\qquad h = 1, \quad k = 2 \qquad$ for $b < (2d+1)/8$
$\qquad h = 1/2, k = 1 \qquad$ for $b = d^4 \ge (2d+1)/8$ (This gives $d > 7/10$.)

Thus (1.2) is solvable unless $b = 1/4$ and $2\beta_1 \equiv 0 \pmod 1$, i.e. $\pm 2a_3 - (c_3 + \lambda x_4) \equiv 0 \pmod 1$. On taking $x_4 = c_4$ and $1 + c_4$ simultaneously we get $\lambda = 0$. But then $\phi = -x_3^2/4 + x_4^2$ is a zero form.

Case (II) $t = 3/2$.

Here $d/3 \leq b = 2d^4/3 < d/2$, so $(1/2)^{1/3} \leq d \leq (3/4)^{1/3}$ and so $1/4 < b < 1/2$. Now (3.35) is satisfied on taking

$$h=1, \quad k=2 \quad \text{for } b < (2d+1)/8,$$
and
$$h=1/2, \quad k=1 \quad \text{for } b \geq (2d+1)/8.$$

Therefore (1.2) is solvable by Lemma 6.

LEMMA 17. *If $c_2 = 0, N=3$, then (1.2) is solvable.*

Proof: We can assume as in Lemma 16, that $t \geq d$.

Here since $b \geq d/4$, $d \leq t = d^4/b \leq 4d^3$, we have $d \geq 1/2$. Also from (3.32), $d \leq 2/3$. Taking

$$h=1/2, \quad k=1 \quad \text{for } 1/2 \leq t < d+2b,$$
$$h=1, \quad k=1 \quad \text{for } d+2b \leq t \leq 4d^3,$$

and noting that $4d^3 \geq t \geq d+2b \geq 3d/2$ implies $d \geq \sqrt{(3/8)}$, one finds that (3.38) is satisfied. Therefore (1.2) is solvable unless

(i) $t = d = 1/2$, $\beta_3 = \pm a'_4 + c_4 \equiv 1/2 \pmod 1$ i.e. $c_4 = 0$ or $1/2$,

or (ii) $t = 1$, $\pm a'_4 + 2c_4 \equiv 0 \pmod 1$, i.e. $c_4 = 0, \pm 1/4$, or $1/2$.

Case (I) $t = d = 1/2$.

Here $b = 1/8$ and therefore

$$Q = (x_1 + a_2 x_2 + a_3 x_3 + a_4 x_4) x_2 - (x_3 + \lambda x_4)^2/8 + x_4^2/2.$$

If $c_4 = 1/2$, take $x_2 = 0$, $x_4 = 1/2$, $x_3 \equiv c_3 \pmod 1$ such that $0 \leq |x_3 + \lambda/2| \leq 1/2$, so that

$$0 < -1/32 + 1/8 \leq Q = -(x_3 + \lambda/2)^2/8 + 1/8 \leq 1/8 < d.$$

If $c_4 = 0$, take $x_2 = 0$, $x_4 = \pm 1$, $x_3 \equiv c_3 \pmod 1$ such that $0 \leq |x_3 \pm \lambda| \leq 1/2$, so that

$$0 < -1/32 + 1/2 \leq Q = -(x_3 \pm \lambda)^2/8 + 1/2 \leq 1/2 = d.$$

Thus (1.2) is satisfied with strict inequality unless $c_3 \pm \lambda \equiv 0 \pmod 1$. This gives $(\lambda, c_3) = (0,0)$ or $(1/2, 1/2)$.

For $(\lambda, c_3) = (0,0)$, $Q(x, 0, 1, 1) = 3/8$, and for $(\lambda, c_3) = (1/2, 1/2)$, $Q(x_1, 0, 1/2, 1) = 3/8$. This gives a solution of (1.2).

Case (II) $t = 1$.

Here since $d/4 \leq b = d^4 < d/3$, we have

(3.39) $\quad (1/4)^{1/3} \leq d < (1/3)^{1/3} \quad \text{and } b < 1/4.$

If $c_4 = 1/2$, take $x_4 = 1/2$, $x_2 = 0$ and choose $x_3 \equiv c_3$ (mod 1) such that $1/2 \le |x_3 + \lambda/2| \le 1$, so that

$$0 < -b + 1/4 \le Q = -b(x_3 + \lambda/2)^2 + 1/4 \le -b/4 + 1/4 < d.$$

If $c_4 = \pm 1/4$, take $x_4 = \pm 1/4$, $x_2 = 0$ and choose $x_3 \equiv c_3$ (mod 1) such that $0 \le |x_3 \pm \lambda/4| \le 1/2$, so that

$$0 < -b/4 + 1/16 \le Q = -b(x_3 \pm \lambda/4)^2 + 1/16 \le 1/16 < d.$$

If $c_4 = 0$ and $9b/4 > (1-d)$, take $x_2 = 0$, $x_4 = 1$, and choose $x_3 \equiv c_3$ (mod 1) such that $3/2 \le |x_3 + \lambda| \le 2$, so that

$$0 < -4b + 1 \le Q = -b(x_3 + \lambda)^2 + 1 \le -9b/4 + 1 < d.$$

So let now $b \le 4(1-d)/9$ ($< 1/6$). Taking $h = 3/2$, $k = 3$, and using (3.39) we find that (3.35) is satisfied, so (1.2) is solvable by Lemma 6.

LEMMA 18. *If $c_2 = 0$, $N = 4$, then (1.2) is solvable.*

Proof: We can assume, as in Lemma 16, that $t \ge d$. Here since $b \ge d/5$, $d \le t = d^4/b \le 5d^3$, we get $d \ge 1/\sqrt{5}$. Also from (3.32), we have $d \le 4/7$. On taking

$h = 1/2$, $k = 1$ for $t < d + 15b/4$,

and $\qquad h = 1$, $k = 1$ for $t \ge d + 15b/4$,

we find that (3.38) is satisfied, so (3.30) and hence (1.2) is solvable unless $t = 1/2$ and $\beta_3 = \pm a_4' + c_4 \equiv 1/2$(mod 1), i.e. $c_4 = 0$ or $1/2$. Then $b = 2d^4 < d/4$ gives $d < 1/2$ and $b < 1/8$.

If $c_4 = 0$, take $x_2 = 0$, $x_4 = 1$, $x_3 \equiv c_3$ (mod 1) such that $1 \le |x_3 + \lambda| \le 3/2$, so that

$$0 < -9b/4 + 1/2 \le Q = -b(x_3 + \lambda)^2 + 1/2 \le -b + 1/2 < d.$$

If $c_4 = 1/2$, take $x_2 = 0$, $x_4 = 1/2$, $x_3 \equiv c_3$ (mod 1) such that $0 \le |x_3 + \lambda/2| \le 1/2$, so that

$$0 < -b/4 + 1/8 \le Q = -b(x_3 + \lambda/2)^2 + 1/8 \le 1/8 < d.$$

Thus (1.2) is solvable in all the cases.

3.2.3. ϕ *is of the form (3.18).* Here

$Q = (x_1 + a_2 x_2 + a_3 x_3 + a_4 x_4)x_2 - b(x_3 + x_4/2)^2 + 5bx_4^2/4$, $b = \sqrt{(4d^4/5)}$

Since Q represents $b > 0$, we can assume, by Lemma 4, that

(3.40) $\qquad b \ge d/3$, which implies $d \ge \sqrt{(5/36)}$.

If $c_2 \ne 0$, and so $d \le 1/2$, we have $3 \ge d/b = \sqrt{(5/4d^2)} \ge 2\sqrt{(5/4)} > 2$.

Working as in § 3.2.2 with $t = 5b/4$, we find that $N = 2$ here and $d \le 2/5$ from (3.31). Further for $N = 2$,

$$\{d + (N^2 - 1)b/4\}/t = (2 + 3d/\sqrt{5})/\sqrt{5}d$$

Thus if p is the integer as defined in Lemma 14, we have $p \geq 2$.
Now using (3.40), we have

$$d + (N^2 - 1)b/4 + (p^2 - 1)t/4 \geq (N+3)(p+3)d/16 \geq 25d/16 > 1/2.$$

Therefore (3.33) and hence (1.2) is solvable.

Let now $c_2 = 0$; $d \leq 1$.

We can assume, as in Lemma 16, that $t = 5b/4 \geq d$, this gives $d \geq \sqrt{(4/5)}$. Taking $h = 1/2$, $k = 1$, and observing that $1/2 < b < d$ here, we find that (3.35) is satisfied, therefore, (1.2) is solvable by Lemma 6.

This completes the proof of the theorem.

ACKNOWLEDGEMENT. The authors are grateful to Professors V.C.Dumir and R.J.Hans-Gill for many useful comments during the preparation of this paper.

References

1. R.P.Bambah, V.C.Dumir, R.J.Hans-Gill, *Positive values of non-homogeneous indefinite quadratic forms II*, J. Number Theory, **18** (1984) 313-341.
2. B.J.Birch, *The inhomogeneous minima of quadratic forms of signature zero*, Acta Arith., **4**(1958), 85-98.
3. J.W.S.Cassels, *An introduction to Diophantine Approximation*, Cambridge: University Press, (1957).
4. H.Davenport, *Non-homogeneous binary quadratic forms*, Proc. Kon. Nederl.Akad. Wetensch., **49** (1946) 815-821.
5. V.C.Dumir, *Asymmetric inequality for non-homogeneous ternary quadratic forms*, Proc. Camb Phil Soc, **63**(1967) 291-303.
6. V.C.Dumir, *Positive values of inhomogeneous quadratic forms I*, J. Austral Math Soc, **8**(1968) 87-101.
7. V.C.Dumir, *Positive values of inhomogeneous quadratic forms II*, J. Austral Math Soc, **8**(1968) 287-303.
8. V.C.Dumir, *Asymmetric inequality for non-homogeneous ternary quadratic forms*, J. Number Theory, **1**(1969) 326-345.
9. V.C.Dumir and R.J.HansGill, *The second minimum for non homogeneous ternary quadratic forms of type (1,2)*, unpublished.
10. V.C.Dumir and Ranjeet Sehmi, *Positive values of non homogeneous indefinite quadratic forms of type (3,2)*, Indian J.Pure Appl Math, **23**(7)(1992), 827-853.
11. V.C.Dumir and Ranjeet Sehmi, *Positive values of non homogeneous indefinite quadratic forms of signature 1*, Indian J.Pure Appl Math., **23**(1992) 855-864.
12. V.C.Dumir and Ranjeet Sehmi, *Positive values of non homogeneous indefinite quadratic forms of type (1,4)*, Proc. Indian Acad.Sci.(Math Sci.), **104**(1994) 557-579.
13. R.J.Hans-Gill and Madhu Raka, *An equality for indefinite ternary quadratic form of type (2,1)*, Indian J.Pure Appl. Math., **11**(8) (1980) 994-1006.
14. A.M.Macbeath, *A new sequence of minima in the geometry of numbers*, Proc. Camb Phil Soc, **47**(1951) 266-273.

15. Madhu Raka, *Inhomogeneous minima of a class of ternary quadratic forms*, J.Austral Math Soc (Series A) **55**(1993) 334-354.
16. Madhu Raka and Urmila Rani, *Positive values of non homogeneous indefinite quadratic forms of type (3,1)*, Osterreich Acad. Wiss Math Natur Kl Abt II, **203**(1994) 175-197.
17. Madhu Raka and Urmila Rani, *Positive values of non homogeneous indefinite quadratic forms of Signature 2*, Osterreich Acad. Wiss Math Natur Kl Abt II, **203**(1994) 199-213.
18. Madhu Raka and Urmila Rani, *Positive values of inhomogeneous indefinite ternary quadratic forms of type (2,1)*, Hokkaido Mathematical Journal, Vol **25**(1996) 215-230
19. G.A.Margulis, *Indefinite quadratic forms and unipotent flows on homogeneous spaces*, Conples Rendus Acad Sci Paris, Vol **304**, Series I, (1987) 249-253.
20. L.Ya. Vulakh, *On minima of rational indefinite quadratic forms*, J.Number Theory, **21** (1985) 275-285.
21. G.L.Watson, *Indefinite quadratic polynomials*, Mathematica, **7** (1960) 141-144.
22. G.L.Watson, *Integral quadratic forms*, Cambridge Univ Press, London/ New York (1960).

Department Of Mathematics, Panjab University, Chandigarh 160014, INDIA.

On an Arithmetical Inequality, II

S. Srinivasan

ABSTRACT.[1] Here we describe the work done towards a conjecture, from K. Alladi, P. Erdös and J.D. Vaaler, about multiplicative functions. In this context, we also observe that a conjecture in Combinatorics implies (qualitatively) more than the known results.

1. Statement

Let h be a non-negative submultiplicative function defined on natural numbers satisfying, $h(p) \leq c$, for primes p and some fixed $c > 0$. Let $0 < \alpha < 1$. With squarefree integers n, we denote by $\mathbf{K}(\mathbf{c}, \alpha)$ the following statement: The partial sum of values $h(d)$ over divisors d of n not exceeding n^α is more than a constant κ times the complete sum of values $h(d)$ over all divisors d of n, with a positive $\kappa = \kappa(c, \alpha)$.

In 1987, Alladi, Erdös and Vaaler conjectured in [1] that $\mathbf{K}(\mathbf{c}, \alpha)$ is true for $\alpha = c/(c+1)$ while c is the reciprocal of an integer. In fact, for their purpose there, it was sufficient to have the above statement under the stronger condition $\alpha > c/(c+1)$, and this case was proved by them and some others, (cf., [1], p.7) leaving the case $\alpha = c/(c+1)$ open. So, in what follows, this relation between α and c is assumed. (It is not difficult to show that $\mathbf{K}(\mathbf{c}, \alpha)$ is false, if α is smaller than $c/(c+1)$.) Then, they also conjectured in [2] that perhaps κ is an absolute constant (as $c \leq 1$).

For multiplicative functions h, $\mathbf{K}(\mathbf{c}, \alpha)$ was confirmed in [2], [3] and [5], but with $\kappa \to 0$ as $c \to 0$. The proof in [5], being different, could cover even the additional case of *rational* c. Moreover, these results *automatically* extend to submultiplicative functions h, in view of [6]. However, the questions of the nature of κ and the case of other values of c remain unsettled. (Here it should be mentioned that - as the author of [5] communicated privately as being implicit in [5] - if an *absolute* κ works for all integral c^{-1}, then for all real c (< 1), $\mathbf{K}(\mathbf{c}, \alpha)$ holds with an absolute constant κ.)

---

[1]1991 *Mathematics Subject Classification.* 11N64.

2. Conditional Proof of K(c, α), for small c

Let $x_1, \cdots, x_k$ be real numbers with sum $= 0$. For a subset A of $S := \{1, \cdots, k\}$, let x_A denote the sum of all x_i for which $i \in A$. Denote by $N(\ell)$ the number of A, with $|A| = \ell$, for which $x_A \leq 0$. It is conjectured by Manickam and Singhi, in [4] (cf. Conjecture 1.4, for a more specific form), that

$$(1) \qquad N(\ell) \geq \frac{\ell}{k} \binom{k}{\ell},$$

provided ℓ/k is small (i.e., not exceeding some fixed absolute constant). Indeed, they prove (1), when ℓ divides k.

Now we show that (1) implies $K(c, \alpha)$, when c is small. In view of Theorem 2 of [6], we can assume that h is multiplicative with $h(p) = c$, for all primes p. Let $n = p_1 \cdots p_k$ be a squarefree number with k prime factors. Set $x_i = \{((\log p_i)/(\log n)) - k^{-1}\}$, for $i = 1, \cdots, k$. So each divisor d of n corresponds to a subset A of S and *vice versa*, by means of the subscripts i in A coming from divisors p_i of d. We have thus that $\log d \leq \frac{\omega(d)}{k} \log n$, if $x_A \leq 0$, where $\omega(d)$ denotes the number of prime divisors of d. Hence the partial sum under consideration (in $K(c, \alpha)$) is, by (1),

$$\geq \frac{1}{2(c+1)} \sum_{\ell \leq \alpha k + 1} \frac{\ell}{k} \binom{k}{\ell} c^\ell \geq \frac{c}{2(c+1)} \sum_{(\ell-1) \leq \alpha(k-1)} \binom{k-1}{\ell-1} c^{\ell-1},$$

for *small c*. Since the ratio of the last sum here to the complete sum in $K(c, \alpha)$ is asymptotically $\frac{1}{2(c+1)}$ (cf., (3.4) of [2]), as $\alpha = c/(c+1)$, this proves $K(c, \alpha)$, for small c. (Here it is implicit that αk can be assumed to be sufficiently large, being used for inclusion of the last term in the first sum of the above formula.)

3. References

1. K. Alladi, P. Erdös, and J.D. Vaaler, *Multiplicative functions and small divisors*, Progress in Math. **70** (1987), 1-13.

2. ———, *Multiplicative functions and small divisors, II*, J. Number Theory, **31** (1989), 183-190.

3. B. Landreau, *Majorations de Fonctions Arithmétiques en moyenne sur des ensembles de faible densité* (Ph. D. Thesis), L' Université de Bordeaux **I** (1987).

4. N. Manickam and N.M. Singhi, *First Distribution Invariants and EKR Theorems*, J. Comb. Theory, Series A, **48** (1988), 91-103.

5. K. Soundararajan, *An Inequality for mutliplicative functions*, J. Number Theory, **41** (1992), 225-230.

6. S. Srinivasan, *On an Arithmetical Inequality*, Glasgow Math. J., **36** (1994), 81-86.

School of Mathematics, Tata Institute of Fundamental Research,
Homi Bhabha Road, Mumbai 400 005, India
E-mail address: ssvasan@tifrvax.tifr.res.in

Part IV

TRANSCENDENTAL NUMBER THEORY

Algebraic numbers close to 1: results and methods

Francesco Amoroso

§1 Introduction.

Let $\alpha \neq 1$ be an algebraic number and denote by $F(x) = a(x-\alpha_1)\cdots(x-\alpha_d)$ ($a > 0$) its minimal polynomial over $\mathbb{Z}$. We are interested in lower bounds for $|\alpha-1|$ depending on the degree d of α and on its Mahler measure

$$M(\alpha) = M(F) = a\prod_{h=1}^{d} \max(|\alpha_h|, 1).$$

Liouville's inequality gives

$$\log|\alpha - 1| \geq -(d-1)\log 2 - \log M(\alpha). \tag{1.1}$$

This bound is sharp if $M(\alpha)$ is large, say $\log M(\alpha) \geq \text{(constant)} \times d$. Otherwise, better results are known, obtained using two different methods. In 1979 M. Mignotte [M1] gave the lower bound:

$$\log|\alpha - 1| \geq -4\sqrt{d}\log(4d) \tag{1.2}$$

provided that $M(\alpha) \leq 2$. In this range, (1.2) improves (1.1) by a factor $\sqrt{d}(\log d)^{-1}$. To prove (1.2), Mignotte applied the Gel'fond method. He used Siegel's Lemma to construct a polynomial P with integer coefficients and low height vanishing at 1 with relatively high multiplicity. Then (1.2) follows by applying Liouville's inequality.

Recently M. Mignotte and M. Waldschmidt [MW] improved (1.2) by finding for any $\mu > \log M(\alpha)$ the lower bound

$$\log|\alpha - 1| \geq -\frac{3}{2}\sqrt{d\mu\log^+(d/\mu)} - 2\mu - \log^+(d/\mu), \tag{1.3}$$

where $\log^+ x = \max(\log x, 0)$ ($x > 0$). The approach of M. Mignotte and M. Waldschmidt differs from Mignotte's paper essentially by two arguments: firstly, they use the Schneider method; secondly they employ an interpolation determinant, which avoids the use of Siegel's Lemma.

The inequality (1.3) was slightly improved upon by Y. Bugeaud, M. Mignotte and F. Normandin [BMN] and by A. Dubickas [Du], who found better values for the constants involved, also giving simpler proofs.

1991 *Mathematics Subject Classification* Primary 11J68. Secondary 11J85.

The aim of this paper is to briefly describe the crucial steps in the proofs of these lower bounds. In section 2 we first describe Mignotte's approach (Gel'fond's method and Siegel's Lemma: Theorem 2.1). Then we study the relation between the behaviour of the height of a product of cyclotomic polynomials with prescribed vanishing at 1 and the lower bounds for $|\alpha - 1|$ (Theorem 2.2). This allows us to describe the limit of Gel'fond's method. We also show that the determinant which appears in Mignotte and Waldschmidt's approach (Schneider's method and interpolation determinant) is a product of cyclotomic polynomials with high vanishing at 1 and low height, hence a "good" auxiliary function for Gel'fond's method. This remark gives, by using Theorem 2.2, an alternative proof of the main result of [MW] (Theorem 2.3). In section 3 we describe the main result of [A2], an explicit construction of algebraic numbers close to 1, which shows that the inequality (1.3) is almost sharp. Finally, in section 4 we discuss a generalization in several variables of these results.

Acknowledgements. I am indebted to Michel Waldschmidt for his encouragement during a discussion we had at the airport of Madras. The main part of this paper was written during the International Conference on Discrete Mathematics and Number Theory which took place in Tiruchiripally, India, in January 3-6, 1996. I am grateful to the organizers, especially to R. Balakrishnan, K. Murty and M. Waldschmidt. I am also indebted to Yann Bugeaud for his useful remarks.

§2 1-dimensional results.

In this section we shall prove bounds of the shape $|\alpha-1| \geq f(\deg \alpha, M(\alpha))$. In order to prove such a bound, we assume $|\alpha| \leq 1$, since otherwise $\beta = 1/\alpha$ satisfies $|\beta| < 1$, $|\beta - 1| \leq |\alpha - 1|$, $\deg \beta = \deg \alpha$ and $M(\beta) = M(\alpha)$. Obviously, we also assume $\alpha \neq 0$.

We start with Mignotte's approach, which uses Gel'fond's method and Siegel's Lemma.

First step: construction of the auxiliary function.

Let $m < N$ be two positive integers; the Bombieri and Vaaler version of Siegel's Lemma (see [BV], Theorem 1) gives a non-zero polynomial $P \in \mathbb{Z}[x]$ of degree $< N$, vanishing at 1 with multiplicity $m(P) \geq m$ and such that its height $H(P)$ (i.e. the maximum modulus of its coefficients) satisfies the inequality [1]

$$\log H(P) \leq \frac{m^2}{2(N-m)} \log \frac{cN}{m}, \qquad c = \frac{1}{4} \exp \frac{3}{2}.$$

Since the maximum modulus $|P|$ of P on the unit circle is bounded by $N \cdot H(P)$, we deduce the inequality

$$\log |P| \leq \frac{m^2}{2(N-m)} \log \frac{cN}{m} + \log N. \qquad (2.1)$$

---

[1] By using the box-principle we easily obtain the weaker estimate

$$\log H(P) \leq \frac{m(m+1)}{2(N-m)} \log N.$$

The bound of [BV] allows us to save a factor $\sqrt{2}$ in Theorem 2.1.

Second step: multiplicity estimate.
  To ensure that $P(\alpha) \neq 0$ we must have $N \leq m + d$. It is convenient to choose N as large as possible: therefore we fix $N = m + d$.
Third step: maximum principle.
  Let P be any polynomial of degree $\leq N$, vanishing at 1 with multiplicity $\geq m$ such that $P(\alpha) \neq 0$. For any embedding $\sigma \colon \mathbb{Q}(\alpha) \to \mathbb{C}$ we have

$$|\sigma P(\alpha)| \leq \max\{1, |\sigma \alpha|\}^N |P|. \tag{2.2}$$

Moreover, by the maximum principle on the disk $|z| = N/(N - m)$,

$$|P(\alpha)| \leq |\alpha - 1|^m \frac{N^N}{m^m (N - m)^{N-m}} |P| \leq |\alpha - 1|^m \left(\frac{eN}{m}\right)^m |P|. \tag{2.3}$$

Since $P(\alpha) \neq 0$, the inequalities (2.2) and (2.3) together give

$$1 \leq a^N \prod_\sigma |\sigma P(\alpha)| \leq |\alpha - 1|^m \left(\frac{eN}{m}\right)^m M(\alpha)^N |P|^d,$$

whence we obtain

$$-\log|\alpha - 1| \leq 1 + \log \frac{N}{m} + \frac{N}{m} \log M(\alpha) + \frac{d}{m} \log |P|. \tag{2.4}$$

Last step: choice of the parameter m.
  Substituting (2.1) and the contraint $N = m + d$ into (2.4), we obtain

$$-\log |\alpha - 1| \leq \log(2eM(\alpha)) + \frac{d}{m} \log(2dM(\alpha)) + \frac{m}{2} \log \frac{2cd}{m} \tag{2.5}$$

for any positive integer $m \leq d$. We choose

$$m = 1 + \left\lceil 2\sqrt{\frac{d \log(2dM(\alpha))}{\log(2d)}} \right\rceil.$$

Assume first $2\sqrt{d \log(2dM(\alpha))} < d\sqrt{\log(2d)}$. Then $m \leq d$ and (2.5) yields, using the inequality $2cd/m \leq c\sqrt{d} < \sqrt{2d}$,

$$-\log |\alpha - 1| < \log(2eM(\alpha)) + \frac{1}{4} \log(2d) + \sqrt{d \log(2d) \log(2dM(\alpha))}. \tag{2.6}$$

Otherwise, if $2\sqrt{d \log(2dM(\alpha))} \geq d\sqrt{\log(2d)}$, the right hand side of (2.6) is

$$\geq \log M(\alpha) + \frac{1}{2} d \log(2d) > \log M(\alpha) + (d - 1) \log 2$$

and Liouville's inequality (1.1) implies (2.6). Therefore, (2.6) holds in any case. We have proved:

Theorem 2.1.
Let $\alpha \neq 1$ be an algebraic number of degree d. Then
$$|\alpha - 1| \geq 2^{-5/4} e^{-1} d^{-1/4} M(\alpha)^{-1} \exp\left\{-\sqrt{d \log(2d) \log(2dM(\alpha))}\right\}.$$

It can be shown that any polynomial with integer coefficients having small degree and high vanishing at 1 must also vanish at several roots of unity (see [BV], section 5, and [A1]).

Let $\alpha = \exp(2j\pi i/k)$ be a primitive k-th root of unity ($k > 1$). Then $d = \deg \alpha = \phi(k)$ and $|\alpha - 1| = 2|\sin(j\pi/k)| \geq 2\sin(\pi/k)$. Since $\sin x \sim x$ for $x \to 0$ and
$$\liminf_{k \to +\infty} \frac{\phi(k) \log \log k}{k} = e^{-\gamma}$$
where γ is Euler's constant (see [HW], Theorem 328), we have
$$|\alpha - 1| \geq (1 + o_d(1)) \frac{2\pi}{e^\gamma} (d \log \log d)^{-1} \tag{2.7}$$
where $o_d(1)$ is a function of d satisfying $\lim_{d \to +\infty} o_d(1) = 0$. Hence, in order to find lower bounds for $|\alpha - 1|$, we can assume that α is not a root of unity. [2]

Denote by $\Phi_k(x)$ the k-th cyclotomic polynomial and let
$$\Gamma = \{x^{e_0} \Phi_1^{e_1} \cdots \Phi_k^{e_k}, \text{ such that } k \in \mathbb{N} \text{ and } e_0, \ldots, e_k \in \mathbb{Z}\}.$$

Also let, for $t \geq 0$, $\Gamma(t)$ be the set of non-constant polynomials $P \in \Gamma$ satisfying the inequality $\log|P| \leq t \cdot \deg P$. Notice that $\Gamma(t) \neq \emptyset$, since $x \in \Gamma(t)$ for any $t \geq 0$. We define a function $r: [0, +\infty) \to [0, 1]$ by setting
$$r(t) = \sup_{P \in \Gamma(t)} \frac{m(P)}{\deg P},$$
where $m(P)$ denotes the multiplicity of P at $x = 1$. Let α be an algebraic number of degree d which is not a root of unity and let $h = d^{-1} \log M(\alpha) > 0$ be its logarithmic height. Let also $t > 0$; from the definition of $r(t)$ it follows that for any $\varepsilon > 0$ there exists a polynomial $P \in \Gamma$ of degree N, vanishing at 1 with multiplicity m such that $\log|P| \leq t \cdot N$ and $m/N \geq (1+\varepsilon)^{-1} r(t)$. Then (2.4) implies
$$-\log|\alpha - 1| \leq 1 + \log \frac{1+\varepsilon}{r(t)} + (1+\varepsilon) \frac{d(h+t)}{r(t)}.$$

Since this inequality holds for any $\varepsilon > 0$, we have:

---

[2] The lower bound (2.7) is stronger than the result available in the general case. This suggests that also for algebraic numbers of very small height (say $M(\alpha)$ bounded by an absolute constant) strong results can be proved.

Theorem 2.2.
For any algebraic number α of degree d and logarithmic height $h > 0$ and for any $t > 0$ we have

$$\log |\alpha - 1| \geq -1 - \log \frac{1}{r(t)} - \frac{d(h+t)}{r(t)}.$$

For small values of h a good choice of the parameter t is $t = h$:

Corollary 2.1.
For any algebraic number α of degree d and logarithmic height $h > 0$ we have

$$\log |\alpha - 1| \geq -1 - \log \frac{1}{r(h)} - \frac{2dh}{r(h)}.$$

It can be shown that $r(t) \leq c\sqrt{t}$ for some absolute constant $c > 0$ (see [M1], [A1] and [BV], section 5). Hence the limit of Gel'fond's method seems to be a lower bound of the shape

$$\log |\alpha - 1| \geq -1 - \log \frac{1}{c} - \frac{1}{2} \log \frac{d}{\log M(\alpha)} - \frac{2}{c} \sqrt{d \log M(\alpha)}.$$

We now consider Schneider's approach. The auxiliary function is a polynomial with coefficients in $\mathbb{Q}(\alpha)$ vanishing at several powers of α. The corresponding interpolation determinant is

$$\Delta = \mathrm{Det}\left(\alpha^{ij}\right)_{\substack{i=0,\ldots,k-1;\\ j=0,\ldots,k-1}} = \prod_{0 \leq i < j \leq k-1} (\alpha^j - \alpha^i) \neq 0$$

where $k \in \mathbb{N}$ is a parameter at our disposal. We consider this determinant as a polynomial $\Delta(\alpha) \in \mathbb{Z}[\alpha]$. [3] Then $\Delta \in \Gamma$ and an easy computation shows that Δ has degree $\frac{1}{6}(k-1)k(2k-1)$, vanishes at 0 with multiplicity $\frac{1}{6}(k-1)k(k-2)$ and vanishes at 1 with multiplicity $m := \frac{1}{2}k(k-1)$. Moreover, by Hadamard's inequality, $|\Delta| \leq k^{k/2}$. Consider the polynomial $P(x) = x^{-(k-1)k(k-2)/6} \Delta(x) \in \Gamma$ of degree $N := \frac{1}{6}(k-1)k(k+1)$. We have

$$\frac{\log |P|}{N} \leq (1 + o_k(1)) \frac{3 \log k}{k^2}, \qquad \frac{m}{N} = (1 + o_k(1)) \frac{3}{k},$$

where $o_k(1) \to 0$ for $k \to +\infty$. Let $t > 0$; choosing $k = 1 + \left[\sqrt{\frac{3}{2t} \log \frac{1}{t}}\right]$ we obtain

$$\frac{\log |P|}{N} \leq (1 + o_t(1))t, \qquad \frac{m}{N} = (1 + o_t(1)) \sqrt{\frac{6t}{\log 1/t}},$$

where $o_t(1) \to 0$ for $t \to 0$. This proves that

$$r(t) \geq (1 + o_t(1)) \sqrt{\frac{6t}{\log 1/t}}.$$

---

[3] This polynomial was firstly introduced by Dobrowolski (see [D]).

Let α be an algebraic number of degree d and logarithmic height $h > 0$. By applying Corollary 2.1 we find that for any $\varepsilon \in (0,1)$ there exists $C(\varepsilon) > 1$ such that

$$\log |\alpha - 1| \geq -2 - \frac{1}{2}\log\left(\frac{1}{6h}\log\frac{1}{h}\right) - (1+\varepsilon/2)d\sqrt{\frac{2}{3}h\log 1/h}, \qquad (2.8)$$

for any algebraic number α of degree d and logarithmic height $h \in (0, \log C(\varepsilon))$. Moreover

$$\log M(\alpha) \geq \frac{1}{4}\left(\frac{\log \log d}{\log d}\right)^3, \qquad (2.9)$$

by a recent result of [V] concerning the Lehmer problem. A fortiori, for any $\varepsilon > 0$ there exists $\delta(\varepsilon) > 0$ such that

$$2 + \frac{1}{2}\log\left(\frac{1}{3h}\log\frac{1}{h}\right) \leq \frac{\varepsilon}{2}d\sqrt{\frac{2}{3}h\log 1/h} \qquad (2.10)$$

provided that $d \geq \delta(\varepsilon)$. Collecting (2.8) and (2.10) together we find the following version of the main result of [MW]:

Theorem 2.3.
Let α be an algebraic number of degree d which is not a root of unity. Then for any $\varepsilon > 0$ there exist $C(\varepsilon) > 1$ and $\delta(\varepsilon) > 0$ such that

$$\log |\alpha - 1| \geq -(1+\varepsilon)\sqrt{\frac{2}{3}d\log M(\alpha)\log\frac{d}{\log M(\alpha)}}$$

provided that $M(\alpha) \leq C(\varepsilon)^d$ and $d \geq \delta(\varepsilon)$.

In the original paper [MW] the constant was slightly worse, 1 instead of $\sqrt{2/3}$. The constant $\sqrt{2/3}$ was firstly obtained in [BMN]. A further improvement has been recently proved by A. Dubickas, who replaces $\sqrt{2/3}$ by $\pi/4$ (see [D]). This improvement comes from the use of a more complicated determinant which leads, in our interpretation of the method, to a better asymptotic lower bound for $r(t)$ as $t \to +\infty$.

§3 Explicit construction of algebraic number close to 1.

In this section we recall the main result of [A2]:

Theorem 3.1.
Let r be a positive integer and consider the polynomial

$$G(z) = 1 + (z-1)\prod_{n=1}^{r}(z^{2n-1} + 1)$$

of degree $d = 1 + r^2$. Then there exists a root α of G such that $|\alpha - 1| \leq (r^2+1)2^{-r}$. Moreover, the Mahler measure of α is bounded by:

$$\exp\left\{\frac{1}{\pi^2}(\log r)^2 + 3\log r + 7\right\}. \qquad (3.1)$$

Notice that the bound (1.3) gives in this case

$$\log|\alpha - 1| \geq -(1+\varepsilon)\frac{\sqrt{2}}{\pi}r(\log r)^{3/2}$$

for any $\varepsilon > 0$ and for any r sufficiently large with respect to ε^{-1}. Hence the term $\sqrt{d}$ in (1.3) cannot be replaced by $d^{1/2-\delta}$ for any $\delta > 0$.

Proof of Theorem 3.1.

Let $\alpha = \alpha_1, \ldots, \alpha_d$ be the roots of G and assume that α is the root closest to 1. From

$$G'(z) = G(z) \sum_{i=1}^{d} \frac{1}{z - \alpha_i}, \qquad z \neq \alpha_1, \ldots, \alpha_d$$

we deduce, taking into account $G(1) = 1$ and $G'(1) = 2^r$,

$$|\alpha - 1| \leq d 2^{-r} = (r^2 + 1) 2^{-r}.$$

We now prove (3.1). Let

$$F(z) = \prod_{n=1}^{r} (z^{2n-1} - 1).$$

Since

$$|G(e^{it})| \leq 1 + 2|F(-e^{it})| \leq 3 \max\{|F(-e^{it})|, 1\}$$

and since, by Jensen's formula,

$$M(G) = \exp\left(\frac{1}{2\pi}\int_{-\pi}^{\pi} \log|G(e^{it})|dt\right),$$

we have

$$\log M(G) \leq \log 3 + \frac{1}{2\pi}\int_{-\pi}^{\pi} \log^+|F(e^{it})|\,dt. \tag{3.2}$$

Using standard Fourier analysis, it can be shown that

$$\frac{1}{2\pi}\int_{-\pi}^{\pi} \log^+|F(e^{it})|\,dt \leq \left(\frac{1}{2}\log r + 2\right)K_0 + 2\log r + 3$$

where

$$K_0 = \frac{1}{2\pi}\int_{-\pi}^{\pi}\left|\sum_{n=1}^{r}\cos(2n-1)t\right|dt$$

(see [A2], Theorem 1.2). An easy computation shows that $K_0 \leq \frac{2}{\pi^2}(\log r) + 1$. Hence

$$\frac{1}{2\pi}\int_{-\pi}^{\pi} \log^+|F(e^{it})|\,dt \leq \frac{1}{\pi^2}(\log r)^2 + 3\log r + 5$$

and, by (3.2),

$$\log M(G) \leq \frac{1}{\pi^2}(\log r)^2 + 3\log r + 7.$$

□

§4 Generalization in several variables.

The aim of this section is to generalize (1.3) for several algebraic numbers.

Theorem 4.1.
Let **K** be a number field of degree d and let $\alpha^{(1)}, \ldots, \alpha^{(n)}$ be multiplicatively independent elements of **K**. [4] Let $h = h(\alpha^{(1)}) \cdots h(\alpha^{(n)})$. Then

$$\log \max_{j=1,\ldots,n} |\alpha^{(j)} - 1| \geq -(n+1)d \left(h \log^+ \frac{1}{h} \right)^{1/(n+1)} \\ - 2d \sum_{j=1}^{n} h(\alpha^{(j)}) - \log(3nd^2). \tag{4.1}$$

Moreover, for any $\varepsilon > 0$ there exists $c = c(\varepsilon) > 0$ such that

$$\log \max_{j=1,\ldots,n} |\alpha^{(j)} - 1| \geq -(1+\varepsilon) \left(\frac{2}{3}(n+1) \right)^{n/(n+1)} d \left(h \log^+ \frac{1}{h} \right)^{1/(n+1)} \\ - \frac{d}{6} \sum_{j=1}^{n} h(\alpha^{(j)}) - \log(nd^2),$$

provided that $h \leq c^n$.

Although a slightly weaker form of this result already appeared in corollary 4 of [MW], but the proof we present here follows a different scheme. We shall see that an explicit auxiliary function is given by the natural generalisation of the determinant Δ of section 2. This auxiliary function is a polynomial in several variables with high vanishing at 1 and low height, hence, by using the method of §2, we easily obtain the proof of Theorem 4.1.

Proof of Theorem 4.1.
First step: choice of the determinant Δ.

Let $k_1, \ldots, k_n$ be positive integers such that $K := k_1 \cdots k_n > 1$. We denote by z the vector $(z_1, \ldots, z_n)$ and we use the standard convention $z^\lambda = z^{\lambda_1} \cdots z^{\lambda_n}$ for a multi-index λ. Let $\Lambda = \{(\lambda_1, \ldots, \lambda_n) \in \mathbb{Z}^n, 0 \leq \lambda_j \leq k_j - 1, j = 1, \ldots, n\}$. We fix an arbitrary total order $<$ on Λ and we consider the Vandermonde determinant

$$\Delta(z) = \text{Det}\left((z^\lambda)^j\right)_{\substack{\lambda \in \Lambda \\ j=0,\ldots K-1}} = \prod_{\substack{\lambda,\mu \in \Lambda \\ \mu < \lambda}} (z^\lambda - z^\mu).$$

Since $\alpha^{(1)}, \ldots, \alpha^{(n)}$ are multiplicatively independent, $\Delta(\alpha^{(1)}, \ldots, \alpha^{(n)}) \neq 0$.
Second step: upper bound for $|N_{\mathbb{Q}}^{\mathbf{K}} \Delta(\alpha^{(1)}, \ldots, \alpha^{(n)})|$ and main inequality.

An easy computation shows that Δ has partial degree

$$N_j := \frac{K(k_j - 1)(4Kk_j + K - 3k_j)}{12k_j}$$

---

[4] This assumption is in fact necessary. Otherwise, considering $\alpha^{(j)} = \alpha^j$, we could improve the main term $d^{1/2}$ in (1.3), which is impossible according to the result of section 3.

with respect to z_j and vanishes at 1 with multiplicity $m = K(K-1)/2$. Moreover, by Hadamard inequality,

$$\max_{|z_j|=1} |\Delta(z_1,\ldots,z_n)| \leq K^{K/2}.$$

Hence,

$$|\Delta(\sigma\alpha^{(1)},\ldots,\sigma\alpha^{(n)})| \leq K^{K/2} \prod_{j=1}^{n} \max\{1, |\sigma\alpha|\}^{N_j} \qquad (4.2)$$

for any embedding $\sigma: \mathbf{K} \to \mathbb{C}$. Assume $|\alpha^{(1)} - 1| \geq |\alpha^{(j)} - 1|$ $(j = 1, \ldots, n)$ and let $N = N_1 + \cdots + N_n$. If $|\alpha^{(1)} - 1| \geq m/(N-m)$, then, by (4.2),

$$|\Delta(\alpha^{(1)},\ldots,\alpha^{(n)})| \leq |\alpha^{(1)} - 1|^m \left(\frac{eN}{m}\right)^m K^{K/2} \prod_{j=1}^{n} \max\{|\alpha_j|, 1\}^{N_j}. \qquad (4.3)$$

Assume now $|\alpha^{(1)} - 1| < m/(N-m)$, let $\rho = m/(|\alpha^{(1)} - 1|(N-m))$ and consider the polynomial in one variable

$$Q(t) = \Delta\big((\alpha^{(1)} - 1)t + 1, \cdots, (\alpha^{(n)} - 1)t + 1\big).$$

By the maximum principle,

$$|\Delta(\alpha^{(1)},\ldots,\alpha^{(n)})| = |Q(1)| \leq \rho^{-m} \max_{|t|=\rho} |Q(t)|$$
$$\leq \frac{(|\alpha^{(1)} - 1|\rho + 1)^N}{\rho^m} K^{K/2}$$
$$\leq |\alpha^{(1)} - 1|^m \left(\frac{eN}{m}\right)^m K^{K/2}.$$

Hence (4.3) holds in any case. Let now $a_j > 0$ be the leading coefficient of the minimal equation of α_j over $\mathbb{Z}$, and let d_j its degree. By Lemma 4 of [MW],

$$A = a_1^{N_1 d/d_1} \cdots a_n^{N_n d/d_n} \prod_{l=1}^{d} \sigma_l \Delta(\alpha^{(1)},\ldots,\alpha^{(n)})$$

is an integer. By using (4.2) and (4.3) we have

$$|A| \leq |\alpha^{(1)} - 1|^m \left(\frac{eN}{m}\right)^m K^{dK/2} \prod_{j=1}^{n} \left(a_j^{d/d_j} \prod_{l=1}^{d} \max\{|\sigma_l \alpha^{(j)}|, 1\}\right)^{N_j}$$
$$\leq |\alpha^{(1)} - 1|^m \left(\frac{eN}{m}\right)^m K^{dK/2} \prod_{j=1}^{n} M(\alpha^{(j)})^{N_j d/d_j}$$
$$= |\alpha^{(1)} - 1|^m \left(\frac{eN}{m}\right)^m K^{dK/2} \exp\big\{d(N_1 h(\alpha^{(1)}) + \cdots + N_n h(\alpha^{(n)}))\big\}.$$

Since the left hand side of this inequality is ≥ 1, we have

$$\log \frac{1}{|\alpha^{(1)} - 1|} \leq 1 + \log \frac{N}{m} + \frac{d}{m}\Big(\frac{1}{2} K \log K + N_1 h(\alpha^{(1)}) + \cdots + N_n h(\alpha^{(n)})\Big). \qquad (4.4)$$

An easy computation shows that

$$\frac{N_j}{m} \leq \frac{2}{3}\frac{K}{K-1}(k_j - 1) + \frac{1}{6}. \tag{4.5}$$

From (4.4) and (4.5) we obtain the main inequality:

$$-\log \max_{j=1,\ldots,n} |\alpha^{(j)} - 1| \leq 1 + \log\left(\frac{2}{3}\frac{K}{K-1}\sum_{j=1}^{n}(k_j - 1) + \frac{n}{6}\right)$$
$$+ d\left(\frac{\log K}{K-1} + \frac{2}{3}\frac{K}{K-1}\sum_{j=1}^{n}(k_j - 1)h(\alpha^{(j)}) + \frac{1}{6}\sum_{j=1}^{n}h(\alpha^{(j)})\right). \tag{4.6}$$

Last step: choice of the parameter k.
Let $X > 1$ and choose

$$k_j = \left[(Xh)^{1/n}h(\alpha^{(j)})^{-1}\right] + 1, \qquad j = 1, \ldots, n.$$

Therefore $K = k_1 \cdots k_n \geq X > 1$. Since $K \mapsto (\log K)/(K-1)$ and $K \mapsto K/(K-1)$ are both decreasing for $K > 1$, we have, by (4.6),

$$-\log \max_{j=1,\ldots,n} |\alpha^{(j)} - 1| \leq d\frac{\log X + \frac{2}{3}nX^{(n+1)/n}h^{1/n}}{X-1} + \frac{d}{6}\sum_{j=1}^{n}h(\alpha^{(j)})$$
$$+ \log\left(\frac{\frac{2}{3}X^{(n+1)/n}h^{1/n}}{X-1}\sum_{j=1}^{n}h(\alpha^{(j)})^{-1} + \frac{n}{6}\right) + 1. \tag{4.7}$$

Let $\Omega = h^{-1/(n+1)}$. To prove (4.1) we distinguish three cases.
• First case: $\Omega < \frac{6}{5}$. The relation between the arithmetic and geometric means shows that

$$\sum_{j=1}^{n}h(\alpha^{(j)}) \geq nh^{1/n} = n\Omega^{-(n+1)/n} \geq \left(\frac{6}{5}\right)^{-2} > \log 2.$$

Hence

$$2\sum_{j=1}^{n}h(\alpha^{(j)}) > h(\alpha^{(1)}) + \log 2$$

and Liouville's inequality (1.1) implies (4.1).
• Second case: $\frac{6}{5} \leq \Omega < 4$. An easy computation shows that in this case

$$(n+1)\left(h\log\frac{1}{h}\right)^{1/(n+1)} = (n+1)\Omega^{-1}((n+1)\log\Omega)^{1/(n+1)} > \log 2.$$

Hence

$$(n+1)\left(h\log\frac{1}{h}\right)^{1/(n+1)} + 2\sum_{j=1}^{n}h(\alpha^{(j)}) > \log 2 + h(\alpha^{(1)})$$

and Liouville's inequality implies again (4.1).

- Third case: $\Omega \geq 4$. Choose

$$X = \left(\frac{\frac{3}{2}\log(1/h)}{n+1}\right)^{n/(n+1)} h^{-1/(n+1)}.$$

Then

$$\frac{\log X + \frac{2}{3}nX^{(n+1)/n}h^{1/n}}{X-1} = f_n(\Omega) \cdot \left(h \log \frac{1}{h}\right)^{1/(n+1)} \qquad (4.8)$$

and

$$\frac{\frac{2}{3}X^{(n+1)/n}h^{1/n}}{X-1} = g_n(\Omega), \qquad (4.9)$$

where

$$f_n(t) = \left(\frac{2}{3}(n+1)\right)^{n/(n+1)}\left(1 + \frac{n}{(n+1)^2}\frac{\log(\frac{3}{2}\log t)}{\log t}\right)\left(1 - \left(\frac{3}{2}\log t\right)^{-n/(n+1)}\frac{1}{t}\right)^{-1}$$

and

$$g_n(t) = \frac{\log t}{t(\frac{3}{2}\log t)^{n/(n+1)} - 1}.$$

Moreover, using for instance the inequality (2.9) of P. Voutier,

$$\sum_{j=1}^{n} h(\alpha^{(j)})^{-1} \leq 3nd^2. \qquad (4.10)$$

By (4.7), (4.8), (4.9) and (4.10) we have

$$-\log \max_{j=1,\ldots,n} |\alpha^{(j)} - 1| \leq f_n(\Omega) \cdot d \left(h \log \frac{1}{h}\right)^{1/(n+1)} + \frac{d}{6}\sum_{j=1}^{n} h(\alpha^{(j)}) \\ + \log\left(3d^2 n g_n(\Omega) + \frac{n}{6}\right) + 1. \qquad (4.11)$$

A computation shows that $f_n(t)$ and $g_n(t)$ both decrease for $t \geq 4$, $f_n(4) \leq n+1$ and $g_n(4) \leq \frac{3}{10}$. Therefore, by (4.11),

$$-\log \max_{j=1,\ldots,n} |\alpha^{(j)} - 1| \leq (n+1)d\left(h\log\frac{1}{h}\right)^{1/(n+1)} + \frac{d}{6}\sum_{j=1}^{n} h(\alpha^{(j)}) \\ + \log\left(\frac{9}{10}d^2n + \frac{n}{6}\right) + 1 \\ \leq (n+1)d\left(h\log\frac{1}{h}\right)^{1/(n+1)} + \frac{d}{6}\sum_{j=1}^{n} h(\alpha^{(j)}) + \log(3d^2n),$$

which implies (4.1).

To prove the second assertion of Theorem 4.1, we remark that

$$\lim_{t \to +\infty} f_n(t) = \left(\frac{2}{3}(n+1)\right)^{n/(n+1)} \qquad \text{and} \qquad \lim_{t \to +\infty} g_n(t) = 0.$$

Therefore, for any $\varepsilon > 0$ there exists a constant $C = C(\varepsilon) \geq 1$ such that if $\Omega \geq C$ we have $f_n(\Omega) \leq (1+\varepsilon)\left(\frac{2}{3}(n+1)\right)^{n/(n+1)}$ and $3g_n(\Omega) \leq 1/e - 1/6$. Let $c = C^{-2}$. If $h \leq c^n$ then $\Omega \geq C$ and, by (4.11),

$$-\log \max_{j=1,\ldots,n} |\alpha^{(j)} - 1| \leq (1+\varepsilon)\left(\frac{2}{3}(n+1)\right)^{n/(n+1)} d \left(h \log^+ \frac{1}{h}\right)^{1/(n+1)}$$
$$+ \frac{d}{6} \sum_{j=1}^{n} h(\alpha^{(j)}) + 2\log d + \log n.$$

The proof of Theorem 4.1 is now complete. □

REFERENCES.

[A1] F. Amoroso, "Polynomial with prescribed vanishing at roots of unity", Bollettino U.M.I.(7) 9-B, 1021-1045 (1995);

[A2] F. Amoroso, "Algebraic numbers close to 1 and variants of Mahler's measure", Journal of Number Theory 60, 80-96 (1996);

[BMN] Y. Bugeaud, M. Mignotte and F. Normandin, "Nombres algébriques de petite mesure et formes linéaires en un logarithme", C.R. Acad. Sci. Paris, t. 321, Serie I, p. 517-522 (1995);

[BV] E. Bombieri and J.D. Vaaler, "Polynomials with low height and prescribed vanishing", Analytic Number Theory and Diophantine Problems, Oklahoma 1984, Ed. A.C. Adolphson, J.B. Conrey, A. Ghosh and R.I. Yager, Birkhäuser PM 70 (1987), 53-73;

[D] E. Dobrowolski, "On a question of Lehmer and the number of irreducible factors of a polynomial", Acta Arith. 34, p. 391-401 (1979);

[Du] A. Dubickas, "On algebraic numbers of small measure", Liet. Mat. Rink. 35(4), 421-431(1995);

[HW] G.H. Hardy and E.M. Wright "An Introduction to The Theory of Numbers", Oxford University Press 1979;

[M1] M. Mignotte "Approximation des nombres algébriques par des nombres algébriques de grand degré", Ann. Fac. Sci. Toulouse 1, p.165-170 (1979);

[M2] M. Mignotte "Estimations élémentaires effectives sur les nombres algébriques", Journées Arithmétiques 1980, éd. J.V. Armitage, London Math. Soc. Lecture Notes Ser. 56, Cambridge U. Press, p.24-34 (1982);

[MW] M. Mignotte and M. Waldschmidt, "On Algebraic Numbers of Small Height: Linear Forms in One Logarithm", J. of Number Theory 47(1), 43-62 (1994);

[V] P. Voutier, "An effective lower bound for the height of algebraic numbers", Acta Arithmetica, Vol. 74 (1), p.81-95 (1996).

Francesco Amoroso,
    Dipartimento di Matematica
    Via Carlo Alberto, 10
    10123 Torino - Italy
    e-mail: amoroso@dm.unipi.it

Transcendence in Positive Characteristic

W. Dale Brownawell

Dedicated to the memory of S. Chowla

ABSTRACT. After a nod in the direction of transcendence in characteristic zero, we give a brief introduction to the Carlitz exponential function, Drinfeld modules, t-modules, and gamma functions in positive characteristic. We survey recent progress in transcendence questions which use analogues of techniques in characteristic zero.

0. Prologue

Transcendence properties in fields of positive characteristic have been investigated off and on for over fifty years using both analogues of tools developed for zero characteristic and also more or less frontal attacks involving little mathematical machinery. Jing Yu revolutionized the field through his investigation of transcendence for Drinfeld modules from the modern point of view in which commutative algebraic groups play a central role. In this article, we aim to give an introduction to research in transcendence for function fields and in particular to give a survey of highlights of the progress not covered in the survey articles [Wl],[Yu9]. Since in these proceedings D.S. Thakur [Tk6] includes a survey of the remarkable progress made with the modern automata techniques, we recommend his article to the interested reader and concentrate here instead on the results obtained by techniques with analogues in characteristic zero. I am indebted to D. Thakur, M. Waldschmidt, and a referee for helpful comments on drafts of this article.

1. Introduction to Transcendence

Transcendence investigations in positive characteristic have a goal similar to that in characteristic 0, namely to test the following working hypothesis:

TRANSCENDENCE EXPECTATION. *Values of ordinary transcendental functions are subject to no unpredictable algebraic relations.*

1991 *Mathematics Subject Classification.* Primary 11J61, 11J81.

This is an updated version of an address at the Tenth Anniversary Meeting of the Ramanujan Mathematical Society, held in Trichy, Tamil Nadu, India. Research supported in part by NSF.

The word "ordinary" indicates that we are not thinking about aberrant transcendental functions, but rather functions arising naturally, say in an analytic or geometric context. Solutions of differential or functional equations or exponential functions of algebraic groups are typical examples. This sweeping expectation has not (yet?) been disappointed.

But we are far from being able to show that it is met in general, even in the most concrete cases, even in characteristic zero. For example, we know the basic algebraic relations for the ordinary exponential function:

$$e^{x+y} = e^x e^y, \quad e^{2\pi i} = 1.$$

In this case, a celebrated conjecture of my esteemed thesis advisor Stephen H. Schanuel (made while he was a graduate student) gives a concrete form to the expectation voiced above.

SCHANUEL'S CONJECTURE. *If $\alpha_1, \ldots, \alpha_n$ are linearly independent (over $\mathbb{Z}$), then at least n of the numbers*

$$\alpha_1, \ldots, \alpha_n, e^{\alpha_1}, \ldots, e^{\alpha_n},$$

are algebraically independent (over $\mathbb{Q}$).

The $n = 1$ case of this conjecture has been known for over 100 years from the work of Hermite and Lindemann. But even today the $n = 2$ case has not yet been established in full generality. For example we have the following:

CANDIDATE FOR THE MOST EMBARRASSING TRANSCENDENCE QUESTION IN CHARACTERISTIC ZERO. *Is $\pi + e$ irrational?*

Since we are stuck at such a rudimentary level in the classical characteristic zero case, it will not be surprising that we are also stuck at a low level in positive characteristic.

Nevertheless, various striking transcendence results have been established in the past dozen decades. Over a period of time most of them have been extended and organized into quite general theorems. One of the main classical theorems harnesses the control imposed on solutions of algebraic differential equations (see [**Bo**] for a deep generalization):

SCHNEIDER-LANG THEOREM. *Let K be a number field. Let $f_1, \ldots, f_n$ be meromorphic functions of order of growth $\leq M$. Assume that at least one of these functions is transcendental while the ring they generate over K is closed under differentiation. Let $S \subset K$ with each $f_i(S) \subset K$. Then*

$$|S| \leq 20[K : \mathbb{Q}]M.$$

An area founded by Siegel concerns the independence of values of so-called *E*-function solutions of *linear* differential equations with coefficients which are rational functions:

$$y_i' = Q_{i,0} + \sum_{j=1}^{n} Q_{ij} y_j, \quad i = 1, \ldots, n.$$

E-functions are entire functions

$$f(z) = \sum_{i=0}^{\infty} c_n \frac{z^n}{n!},$$

where the coefficients c_n are algebraic numbers satisfying sharply curtailed growth with respect to n. All known examples satisfying linear differential equations are based on generalized hypergeometric functions with rational parameters.

Although the area (see [Sh]) cannot be epitomized in a single result, the following theorem is certainly a famous high water mark:

THEOREM. (Shidlovsky) *Suppose that the E-functions $f_1(z), \ldots, f_n(z)$ form a solution of the above system of linear differential equations in which the $f_i(z)$ are algebraically independent over $\mathbb{C}(z)$. Suppose moreover that $\alpha \in \overline{\mathbb{Q}}$ with $\alpha T(\alpha) \neq 0$, where $T(z)$ is a common denominator of the $Q_{ij}(z)$. Then the function values $f_1(\alpha), \ldots, f_n(\alpha)$ are algebraically independent.*

See [NS] for a very recent important advance.

An enormously fruitful modern line of research considers transcendence questions in the setting of commutative algebraic groups. There one has an exponential function as well:

$$\exp_G : Lie_G(\mathbb{C}) \longrightarrow G(\mathbb{C}),$$

where $Lie_G(\mathbb{C}) \approx \mathbb{C}^n$.

This approach was initiated by S. Lang and largely brought to maturity by M. Waldschmidt, D.W. Masser, and G. Wüstholz. One of the towering highlights of this theory is the following result [Wü]:

WÜSTHOLZ'S SUBGROUP THEOREM. *Let G be a commutative algebraic group defined over $\overline{\mathbb{Q}}$. Let $u \in Lie_G(\mathbb{C})$ such that $exp_G(u) \in G(\overline{\mathbb{Q}})$. Define $L(u)$ to be the vector subspace of $Lie_G(\mathbb{C})$ whose elements have coordinates satisfying all the linear relations over $\overline{\mathbb{Q}}$ satisfied by the coordinates of u. Then $L(u)$ is the Lie algebra of an algebraic subgroup of G.*

As an example of its power, we remark that Baker's famous theorem on the linear independence of logarithms of algebraic numbers then follows as an important special case, of which there are many.

BAKER'S THEOREM. *Let $\{u_1, \ldots, u_n\} \subset \mathbb{C}$ be linearly independent (over $\mathbb{Q}$) with each $e(u_i) \in \overline{\mathbb{Q}}$. Then*

$$1, u_1, \ldots, u_n$$

are linearly independent over $\overline{\mathbb{Q}}$.

There is another major approach to transcendence in characteristic zero, due to Mahler, which deals with functions satisfying functional equations. This approach interfaces with automata theory, and the parallels in positive and zero characteristic have not been worked out. See [Be],[De6].

2. Classical Transcendence Results for Carlitz Modules

The story of transcendence in the case of positive characteristic began with Carlitz's definition in the 1930's of an exponential function e_C for rational function fields $k = \mathbb{F}_q(t)$ over finite fields $\mathbb{F}_q$, $|\mathbb{F}_q| = q = p^s$. In this setting, the analogue of $\mathbb{C}$ is played by C_p, the completion of the algebraic closure of $k_\infty = \mathbb{F}_q((1/t))$ with respect to the ultrametric in which $|1/t| = 1/q$. For the convenience of the reader, all the notation of Drinfeld modules is collected below in the Appendix.

Carlitz defined his exponential function in analogy with the ordinary exponential function using the following recursive definition of function analogues D_i of ordinary factorials:

DEFINITION. $D_0 := 1$, $D_i := (t^{q^i} - t)D_{i-1}^q$, $i = 1, 2, \ldots$

THEOREM. (Carlitz) *The Carlitz exponential function $e_C(z)$ defined by*

$$e_C(z) := \sum_{h=0}^{\infty} z^{q^h}/D_h$$

is an entire function with product expansion

$$e_C(z) = z \prod_{\substack{a \in A \\ a \neq 0}} \left(1 - \frac{z}{a\tilde{\pi}_q}\right),$$

where $A := \mathbb{F}_q[t]$ *and*

$$\tilde{\pi}_q := (t - t^q)^{\frac{1}{q-1}} \prod_{i=1}^{\infty} \left(1 - \frac{(t^{q^i} - t)}{(t^{q^{i+1}} - t)}\right) =: (t - t^q)^{\frac{1}{q-1}} \pi_q.$$

Here $(t - t^q)^{\frac{1}{q-1}}$ is the natural analogue of $2i$ and π_q the analogue of the ordinary π.

Using the functional equation

$$e_C(tz) = te_C(z) + (e_C(z))^q,$$

Carlitz's student L.I. Wade established the analogue of the classical theorem of Hermite [**Wd1**], showing in particular that π_q is not algebraic over k. In [**Wd2**] he proved the transcendence of $\sum_{h=0}^{\infty} L_h^{-r}$ for any positive rational number r, where the numbers L_h arise as the coefficients of the Carlitz logarithm function:

$$\log_C(z) := \sum_{h=0}^{\infty} z^{q^h}/L_h,$$

satisfying $e_C(\log_C(z)) = z$. The proofs were not modelled on classical transcendence proofs. Instead they appear as precursors of the modern approach via automata. It was only in his third paper [**Wd3**] that Wade adapted the classical method of Schneider to establish his analogue of the Gelfond-Schneider theorem: If $e_C(u) \in \bar{k}, \beta \in \bar{k} \smallsetminus k$, and $u \neq 0$, then $e_C(\beta u)$ is transcendental (over k).

Although the field was sporadically active after that (see, e.g. [**Gs**]), J. Yu's contributions [**Yu1**] - [**Yu9**] were explosive. He recognized that the considerations of G.V. Drinfeld [**Dr**] in the 1970's furnished a fertile framework in which to develop the analogues of modern transcendence techniques.

To describe this framework, let k be any global function field with subfield of constants $\mathbb{F}_q$, let A be the subring of functions in k regular away from a distinguished place denoted by ∞, and let C_p be the analogue of $\mathbb{C}$ in this setting. The reader is invited to keep the example $A = \mathbb{F}_q[t], k = \mathbb{F}_q(t)$ in mind and to consult [**H1**] or [**H2**] for the details of the general case. Moreover let $C_p\{F\}$ denote the ring of $\mathbb{F}_q$-linear operators, called *twisted polynomials*, i.e. the ring generated over $\mathbb{F}_q$ by F, the q-power Frobenius mapping, whose effect is

$$F : x \mapsto x^q, \quad \forall x.$$

Finally we call a subgroup Λ of C_p an *A-lattice* if
1. Λ is an A-module under the ordinary action of A,
2. Λ is discrete with respect to the topology associated to ∞, and
3. the k-dimension of the k-vector space generated by Λ is finite.

Drinfeld recognized that one could reverse Carlitz's considerations even in this general setting to show that if Λ is an A-lattice, then for every $a \in A$ there is a unique twisted polynomial $\phi_\Lambda(a) \in C_p\{F\}$ such that the entire function

$$e_\Lambda(z) := z \prod_{\substack{\lambda \in \Lambda \\ \lambda \neq 0}} \left(1 - \frac{z}{\lambda}\right)$$

satisfies
$$\phi_\Lambda(a)e(z) = e_\Lambda(az).$$

Moreover $\phi_\Lambda(\zeta) = \zeta F^0$ for all $\zeta \in \mathbb{F}_q$, and so

$$\phi_\Lambda : A \longrightarrow C_p\{F\}$$

is an $\mathbb{F}_q$-homomorphism.

Thus he defined "elliptic modules" to be homomorphisms

$$\phi : A \longrightarrow C_p\{F\},$$

for which $\phi(\zeta) = \zeta F^0$ for all $\zeta \in \mathbb{F}_q$. We will restrict ourselves to the "generic characteristic" case, where $\phi_\Lambda(a) = aF^0 +$ higher degree terms, for each $a \in A$. Maps between elliptic A-modules ϕ_1, ϕ_2 are $\mathbb{F}_q$-linear maps on $\mathbb{G}_a$ commuting with the respective A-actions, i.e.

$$\text{Hom}(\phi_1, \phi_2) = \{T \in C_p\{F\} : T\phi_1 = \phi_2 T\}.$$

In honor of his fundamental work, nowadays these objects are referred to as "Drinfeld A-modules", and the reference to A and Λ may be suppressed if no confusion is likely to arise. In this terminology, Drinfeld showed the following basic "Uniformization Theorem":

THEOREM. (Drinfeld) *There is a category equivalence between the category of generic characteristic Drinfeld A-modules and the category of A-lattices with morphisms*

$$\text{Hom}(\Lambda_1, \Lambda_2) = \begin{cases} \{c \in C_p : c\Lambda_1 \subset \Lambda_2\}, & \text{if } \text{rank}_A \Lambda_1 = \text{rank}_A \Lambda_2 \\ 0, & \text{otherwise} \end{cases}$$

in which
$$\Lambda \leftrightarrow \phi \Leftrightarrow \Lambda = \text{Ker}\,\phi \Leftrightarrow \phi = \phi_\Lambda.$$

Moreover, if $c \in Hom(\Lambda_1, \Lambda_2)$ and $T \in \text{Hom}(\phi_1, \phi_2)$, where $\Lambda_i \leftrightarrow \phi_i$, $i = 1,2$, then

$$c \leftrightarrow T \Leftrightarrow T = cF^0 + \text{higher terms.}$$

This equivalence justifies our thinking of $\mathcal{O} := \text{End}(\Lambda) := \text{Hom}(\Lambda, \Lambda)$ as the *multiplications* of Λ or of the associated Drinfeld module and the non-zero elements of $\text{End}(\Lambda) = \text{Hom}(\Lambda, \Lambda)$ as *isogenies*. We also note that, for $a \in A$

$$\phi(a) = aF^0 + a_1 F + \cdots + a_r F^r, \quad a_r \neq 0,$$

where $r = (\deg a)(\text{rank}_A \Lambda)$. For the remainder of this paper, we *always* assume that the a_i lie in $\bar{k}$, and we sometimes emphasize this assumption by saying that ϕ is *defined over* $\bar{k}$.

In papers [**Yu1**]-[**Yu5**], Yu methodically established all the relevant analogues of the Gelfond-Schneider technique which were obtained before Baker's famous result on linear forms in logarithms of algebraic numbers.

For applications of Wüstholz's Subgroup Theorem in transcendence, it has became crucial to understand the algebraic subgroups of products of interesting commutative algebraic groups. In [**K**], algebraic subgroups of products of simple algebraic groups were shown to be subproducts whose factors have tangent space coordinates satisfying linear relations over the respective endomorphism rings. In particular, it was shown in [**K**] and in [**BK**] (see also [**Co**]) that, for Weierstrass elliptic functions $\wp_i$ corresponding to non-isogeneous elliptic curves $\mathcal{E}_i$, the functions $\wp_i(u_{ij}z)$ are algebraically independent exactly when all the numbers u_{ij} are linearly independent over the multiplication ring of each curve $\mathcal{E}_i$.

In fact, the main independence result for functions in [**BK**] included various $\zeta_i(u_{ij}z)$ (and even $\sigma_i(u_{ij}z)$) as well, where the $\zeta_i(z)$ are Weierstrass quasi-periodic functions. In more modern terms, this deals with the exponential function of products of group extensions E of elliptic curves $\mathcal{E}$ by $\mathbb{G}_a$:

$$0 \longrightarrow \mathbb{G}_a \longrightarrow E \longrightarrow \mathcal{E} \longrightarrow 0,$$

where the exponential function of E is given by

$$(z, t) \mapsto (1, \wp(z), \wp'(z), t - \zeta(z)).$$

Thus the periods of this map are the pairs of the form (ω, η), where $\omega = n_1\omega_1 + n_2\omega_2$ is a period of $\wp$ and $\eta = n_1\eta_1 + n_2\eta_2$ the corresponding quasi-period of ζ, expressed here in terms of a basis ω_1, ω_2 of periods and $\eta_i := 2\zeta(\omega_i/2)$, $i = 1, 2$.

In the Drinfeld setting, work of Deligne, Anderson, and Yu developed a theory of quasi-periodic Drinfeld functions, which was put on a different footing by Gekeler [**Gk**]. If $r = \text{rank}_A \Lambda$, then $r - 1$ is the dimension of the C_p-vector space $D_{sr}(\phi)$ of strictly reduced ϕ-biderivations, i.e. $\mathbb{F}_q$-linear maps $\delta : A \longrightarrow C_p\{F\}F$ satisfying

$$\delta(ab) = a\delta(b) + \delta(a)\phi(b),$$

for all $a, b \in A$, and having $\deg_F \delta(a) < \deg_F \phi(a)$ for all non-constant $a \in A$. Moreover we can choose a basis $\delta_1, \ldots, \delta_{r-1}$ for $D_{sr}(\phi)$ in which $\delta_i(a) \in \bar{k}\{F\}F$ for all $a \in A$. Then for each $i = 1, \ldots, r-1$, there is a unique entire $\mathbb{F}_q$-linear function F_i with no linear term such that

$$F_i(az) - aF_i(z) = \delta_i(a)e_\Lambda(z).$$

These functions exist also for biderivations δ with $\deg \delta(a) \geq \deg_F \phi(a)$, but the corresponding $F_\delta(z)$ are then expressible algebraically in terms of $z, e_\Lambda(z)$, and $F_1(z), \ldots, F_{r-1}(z)$ (cf. [**Br2**]).

The Drinfeld analogues of the results of [**BK**] not involving the σ_i are established in [**Br1**]. This lays the groundwork on which we can start to imagine what the expected relations for values of exponentials of elliptic and quasi-elliptic Drinfeld modules might be. Namely, we expect the analogue of Schanuel's Conjecture to hold here as well. However, we must exclude linear dependence of the functions' arguments $U = \{u_1, \ldots, u_n\}$ not only over A, but also over the ring of multiplications $\mathcal{O}$ of the Drinfeld module. The reason for this necessity is that ϕ extends uniquely to an $\mathbb{F}_q$-linear homomorphism $\phi : \mathcal{O} \longrightarrow \bar{k}\{F\}$ (recall our standing assumption that $\phi : A \to \bar{k}\{F\}$) so that the functional relation $e(\rho z) = \phi(\rho)e(z)$ holds for all $\rho \in \mathcal{O}$. Consequently, any $\mathcal{O}$-linear relation on arguments: $\sum_{j=1}^{m} \rho_j u_j = 0$ leads immediately to a polynomial relation on the values $e_\phi(u_j)$:

$$\sum_{j=1}^{m} \phi(\rho_j) e_\phi(u_j) = 0.$$

Similarly the requirement of $\mathcal{O}$-linear independence is natural from the point of view of the quasi-periodic functions because of their functional equations for $\rho \in \mathcal{O}$ (Equation (7) of [**Br2**]):

$$F_i(\rho z) = S e_\phi(z) - \sigma z + \sum_{j=1}^{r-1} c_{ij} F_j(z),$$

where $c_{ij} \in \bar{k}$ and $S = \sigma F^0 +$ higher degree terms. Having excluded linear dependency of arguments over $\mathcal{O}$, we can conjecturally accomodate several Drinfeld modules at once.

SCHANUEL-LIKE CONJECTURE FOR DRINFELD MODULES. *For $i = 1, \ldots, n$, let Λ_i be an A-lattice of rank r_i over the multiplications $\mathcal{O}_i$ of Λ_i and let the set $U_i \subset C_p$ be $\mathcal{O}_i$-linearly independent. Let $e_i(z)$ denote the corresponding Drinfeld exponential functions and $F_{ij}(z)$, $j = 2, \ldots, r_i$ the quasi-periodic functions defined above. If $\Lambda_1, \ldots, \Lambda_n$ are non-isogenous, then among the following $\sum_{i=1}^{n}(1+r_i)|U_i|$ numbers:*

$$u, e_i(u), F_{ij}(u),$$

$u \in U_i, i = 1, \ldots, n$ at least $\sum_{i=1}^{n} r_i |U_i|$ are algebraically independent.

We are as far from proving this theorem as we are from proving a generalized version of Schanuel's Conjecture in which we also allow values of various elliptic functions parametrizing non-isogenous elliptic curves.

As in the case of zero characteristic, the use of higher dimensional objects allows one to consider transcendence questions involving more than one value. Therefore we turn to the higher dimensional analogue of Drinfeld modules, which has been developed by G. Anderson [**An**].

DEFINITION. A *t-module* of dimension n is a homomorphism

$$\Phi : A \longrightarrow Mat_{n \times n}(C_p\{F\})$$

such that $\Phi(t) = M_0 F^0 + M_1 F^1 + \cdots + M_r F^r$ with $M_i \in Mat_{n \times n}(C_p)$, $d\Phi(t) := M_0$, such that

$$d\Phi(t) = tI + N,$$

where $I, N \in Mat_{n \times n}(C_p)$, I is the $n \times n$ identity matrix, and N is nilpotent.

Anderson shows that, even in this vastly generalized setting, we have an exponential function.

THEOREM. (Anderson) *For every t-module Φ, there is a unique $\mathbb{F}_q$-linear analytic homomorphism*

$$Exp_\Phi : C_p^n \longrightarrow C_p^n$$

such that

$$Exp_\Phi(d\Phi(t)\mathbf{z}) = \Phi(t) Exp_\Phi(\mathbf{z}).$$

We interpret Exp_Φ as the exponential map for G, the algebraic group $\mathbb{G}_a^n$ with the t-action given by Φ:
$$Exp_\Phi =: Exp_G : Lie_G(C_p) \longrightarrow G(C_p)$$
(In the case $n > 1$, the question of surjectivity of Exp_Φ is more subtle. See Theorem 4 of [**An**].) We assume from now on that the $M_i \in Mat_{n \times n}(\bar{k})$, i.e. that Φ is *defined over $\bar{k}$*.

In particular, t-modules provide a natural framework in which quasi-periodic functions appear in an exponential map. A t-module giving rise to these quasi-periodic functions is determined by choosing any non-constant $t \in A$ and setting

$$\Phi_{QP}(t) := \begin{pmatrix} \phi(t) & 0 & 0 & \cdots & 0 \\ \delta_1(t) & tF^0 & 0 & \cdots & 0 \\ \delta_2(t) & 0 & tF^0 & \cdots & 0 \\ \vdots & & & \ddots & \vdots \\ \delta_{r-1}(t) & 0 & \cdots & 0 & tF^0 \end{pmatrix}.$$

The corresponding exponential function is

$$Exp_{QP} : \begin{pmatrix} z_0 \\ z_1 \\ \vdots \\ z_{r-1} \end{pmatrix} \longrightarrow \begin{pmatrix} e(z_0) \\ z_1 + F_1(z_0) \\ \vdots \\ z_{r-1} + F_{r-1}(z_0) \end{pmatrix}.$$

It is in the setting of t-modules that Yu also proved his most powerful results — his zero estimate and the analogue of Wüstholz's Subgroup Theorem:

WÜSTHOLZ-YU THEOREM OF THE INVARIANT VECTOR SPACE. *Let $G = (\mathbb{G}_a, \Phi)$ be a t-module defined over $\bar{k}$. Let $\mathbf{u}$ be a point in $Lie_G G(C_p)$ such that $\exp_G(\mathbf{u}) \in G(\bar{k})$. Then the smallest vector space in $Lie_G G$ defined over $\bar{k}$ which is invariant under $d\Phi(t)$ and which contains $\mathbf{u}$ is the tangent space at the origin of a t-submodule of G.*

We also refer to this result as Yu's Submodule Theorem. The present author and R. Tubbs [**BT**] and L. Denis [**De9**] independently removed the condition that G be regular. Several directions of modern research depend on the astute implementation of this theorem. However still more require that one appeal directly to the main ingredient of the proof of the Submodule Theorem. We state this ingredient to give an indication of its form, but without defining the terms.

THEOREM. (Yu) *Let Φ be an n-parameter analytic submodule of the m-dimensional t-module G. Suppose that Q is a polynomial in m variables which is of total degree at most D and that Q vanishes to order at least $mT + 1$ along Φ at all points of $\Gamma(S)$. Then there exists a proper sub-t-module $H \subset G$ such that*

$$\binom{T + r(\Phi, H)}{r(\Phi, H)} \left| \frac{\Gamma(S - m + 1) + H}{H} \right| \leq C(G) D^{\dim G/H},$$

where $C(G)$ depends only on G.

For example, building on the previous work of Anderson [**An**] and Anderson and Thakur [**AT**], Yu was able to establish [**Yu8**],[**Yu9**] the following very satisfying result involving the Carlitz zeta function:

$$\zeta_C(n) := \sum_{\substack{a \in \mathbb{F}_q[t] \\ a \text{ monic}}} \frac{1}{a^n}, \quad n = 1, 2, \ldots$$

THEOREM. (Yu)

$$\dim_{\bar{k}} \{\zeta_C(1), \ldots, \zeta_C(n), 1, \tilde{\pi}, \ldots, \tilde{\pi}^m\} = 1 + n + m - \left[\frac{\min\{n, m\}}{q-1}\right].$$

Thakur and Anderson had extended Carlitz's definition of the the zeta function to general A, and Thakur [**Tk3**] showed that in the general case $\zeta(n)/\tilde{\pi}^n$ lies in the analogue of the Hilbert class field.

Another of the premier applications of course is Yu's analogue of Baker's theorem mentioned earlier:

BAKER-YU THEOREM. *Let $e(z)$ be the exponential function of a Drinfeld module ϕ defined over $\bar{k}$. If $\{\alpha_1, \ldots, \alpha_n\} \subset C_p$ is $\mathcal{O}$-linearly independent, and each $e(\alpha_i) \in \bar{k}$, then*

$$1, \alpha_1, \ldots, \alpha_n$$

are linearly independent over $\bar{k}$.

L. Denis [**De3**],[**De7**] also removed the restriction of separability from an earlier version of this theorem by other means. Further highlights of the development of transcendence results parallel to the classical ones have been obtained.

3. Analogues of Classical Results

A. Small transcendence degree.

THEOREM. (A. Thiery) *If ϕ is a Drinfeld module defined over $\bar{k}$ with $\mathrm{rank}_A \Lambda = \mathrm{rank}_A \mathrm{End}(\Lambda) = 2$ and ω, η are a non-zero period and quasi-period, respectively, then*

$$\omega, \; \eta$$

are algebraically independent.

The statement [**Tr2**] and the proof parallel an earlier result of G.V. Chudnovsky for a non-zero Weierstrass period and quasi-period in the case of complex multiplication. In this situation, the underlying zero estimate is quite simple. Using the strong zero estimate of [**Yu8**] mentioned above, Denis has generalized Thiery's result in the following manner [**De10**]:

THEOREM. (Denis) *Let $u_1, \ldots, u_r$ be A-linearly independent, where r is the rank of (the lattice Λ of) the Drinfeld module ϕ and where $e(u_1), \ldots, e(u_r) \in \bar{k}$. Let $F(z)$ be any quasi-periodic function defined over $\bar{k}$ which is not algebraically dependent over $\bar{k}(z, e(z))$. Then at least two of the numbers*

$$u_1, \ldots, u_r, F(u_1), \ldots, F(u_r)$$

are algebraically independent.

Thiery's theorem follows from the case $r = 2$. The following result, which is an analogue of a theorem of Gelfond's, is a special case of a more general theorem [**BBT**]:

THEOREM. (P.-G. Becker, Brownawell, R. Tubbs) *When* $\text{rank}_A \text{End}(\Lambda)$ *equals* $\text{rank}_A \Lambda$ *(not necessarily equal to 2)*, $u \neq 0$, $e_\Lambda(u) \in \bar{k}$, *and* $\beta \notin k$ *is cubic over* k, *then* $e_\Lambda(u\beta), e_\Lambda(u\beta^2)$ *are algebraically independent over* k.

This result uses Gelfond's criterion for transcendence (also carried over to this situation in [**Tr2**]). as well the full power of Yu's zero estimate.

B. Large transcendence degree. A. Thiery has shown [**Tr1**] the Drinfeld analogue of the Lindemann-Weierstrass theorem.

THEOREM. (Thiery) *Let ϕ be a Drinfeld module of rank r and let $\alpha_1, \ldots, \alpha_n \in \bar{k}$ be $\mathcal{O}$-linearly independent. Then the transcendence degree of the field generated by the values of the corresponding Drinfeld exponential function over k satisfies*

$$tr.deg._k k(e(\alpha_1), \ldots, e(\alpha_n)) \geq \frac{nr_1}{r},$$

where $r_1 = \text{rank}_A \mathcal{O}$.

When $1 < r_1 < r$, we have "partial" complex multiplication, which has no direct elliptic analogue. This result requires Yu's zero estimate as well as the finite characteristic version [**Ph**] of P. Philippon's criterion for algebraic independence.

C. Linear independence. Using the knowledge [**Br2**] of the possible submodules of products of the quasi-periodic extensions of Drinfeld modules, we can deduce the following generalization of the Baker-Yu Theorem:

THEOREM. *Let ϕ_i, $i = 1, \ldots, n$ denote non-isogenous Drinfeld modules of rank r_i over their respective endomorphism rings $\mathcal{O}_i$. Let F_{ij}, $j = 2, \ldots, r_i$ denote corresponding algebraically independent quasi-periodic functions defined over $\bar{k}$. Let $u_{i\ell}$, $\ell = 1, \ldots, L_i$, be $\mathcal{O}_i$-linearly independent with $e_i(u_{i\ell}) \in \bar{k}$. Then the $1 + \sum r_i L_i$ numbers*

$$1, u_{i\ell}, F_{i2}(u_{i\ell}), \ldots, F_{ir_i}(u_{i\ell})$$

are $\bar{k}$-linearly independent.

D. Diophantine geometry. Very impressive parallels appear in the joint work [**DD**] of S. David and Denis on the analogues of deep results of D.W. Masser and G. Wüstholz [**MW1**], [**MW2**]. The first result uses the notion of height of an element $T = a_0 F^0 + a_1 F + \cdots + a_d F^d \in \bar{k}\{F\}$:

$$h(T) := \frac{1}{m} \sum_w d(w) \max\{0, -w(a_0), \ldots, -w(a_d)\},$$

where the sum runs over all places w of $L = k(a_0, \ldots, a_d)$, $m := [L : k]$, and $d(w)$ denotes the residual degree of the place normalized so that $w(L) = \mathbb{Z} \cup \{\infty\}$.

THEOREM. (David and Denis) *Let ϕ and ϕ' be Drinfeld modules of rank d defined over L where $D := [L : k] < \infty$. If ϕ and ϕ' are isogenous over $\bar{k}$, then there is an isogeny $T \in C_p\{F\}$ from ϕ to ϕ' such that*

$$\deg_F T \leq c(d, q)(Dh(\phi))^{p(d)},$$

where $p(d)$ is polynomial in d. In fact, one can take $p(d) = 10(d + 1)^7$.

This theorem is a corollary of the following result, whose statement uses the straightforward generalization of the notion of degree and height from Drinfeld modules to t-modules:

THEOREM. (David and Denis) *Let the t-module Φ be the product of n Drinfeld modules of rank at most d defined over L with $D := [L : k] < \infty$. If B_ω is the smallest sub-t-module of Φ for which Lie_{B_ω} contains a given period ω of Φ, then*

$$\deg B_\omega \leq c'(q,d,n)(Dh(\Phi)\|\omega\|)^{s(d)},$$

where $\|\omega\|$ denotes the sup norm and $s(d)$ is a polynomial in n and d. In fact, one can take $s(d) = 10d^3n^3$.

The proof itself is a masterful adaptation of the proof of [**MW2**]. It involves a comparison of three different notions of height. As the reader might suspect, the basic Zero Estimate must be invoked.

With this accomplishment, one more main transcendence analogue using the point of view of algebraic groups has been established in positive characteristic. So in a certain sense, Yu's program is nearly accomplished. Not only that, but in various cases, the results have exceeded those know in zero characteristic. An outstanding example is the following result of Denis [**De8**] (see also [**De2**],[**De4**]), which he deduces from the results of [**BBT**] and [**Tr1**].

THEOREM. (Denis) *If $q \neq 2$, the Carlitz analogues of e and π are algebraically independent over k.*

Thus, with exception of $q = 2$, the candidate for the most embarrassing unanswered question has been settled in a very strong manner. However there are also many interesting situations in the case of positive characteristic which have no known analogue in characteristic zero.

4. New Uses for the Submodule Theorem

It was Denis who first realized clearly [**De6**] that the appearance of the variable t in the field k offered new possibilities to (in effect, though he did not originally state it this way) generate new t-modules from a given Drinfeld module and thus obtain transcendence results peculiar to positive characteristic. To describe the first t-module we want to consider, we introduce the notion of divided derivative:

A. Independence of Derivatives.

DEFINITION. For $i = 0, 1, 2, \ldots$, set

$$\Delta_i t^n := \binom{n}{i} t^{n-i}$$

for $n \in \mathbb{Z}_{\geq 0}$.

We retain the notation $\{\Delta_i\}_{i=0}^\infty$ for the family of hyperderivatives on $\bar{k}_\infty^{sep}$, the separable closure of k_∞, extending the action on the powers of t. In other words, this family satisfies

$$\Delta_i(ab) = \sum_{j+\ell=i} \Delta_j(a)\Delta_\ell(b)$$

for all $a, b \in \bar{k}_\infty^{sep}$. We also use the suggestive notation $u^{[i]} := \Delta_i(u)$, for $u \in \bar{k}_\infty^{sep}$. Since for a Drinfeld module ϕ defined over $\bar{k}_\infty^{sep}$, the corresponding exponential function $e(z)$ has coefficients from $\bar{k}_\infty^{sep}$, one can apply the various Δ_i to the coefficients of $e(z)$ to obtain new entire functions $e^{[i]}(z)$ with coefficients from $\bar{k}_\infty^{sep}$.

When we let i range from 0 to $g-1$, where g is the *subdegree* of $\phi(t) = F^0 + a_1 F + \cdots + a_d F^d$, defined by

$$\text{subdeg } \phi = \min_{i>0}\{i : a_i \neq 0\},$$

we obtain the following t-module:

$$\Phi_{[g]}(t) := \begin{pmatrix} tF^0 & 0 & 0 & 0 & \cdots & 0 \\ 0 & \phi(t) & 0 & 0 & \cdots & 0 \\ -F^0 & F^0 & tF^0 & 0 & \cdots & 0 \\ 0 & 0 & F^0 & tF^0 & \cdots & 0 \\ & & \vdots & & \ddots & \vdots \\ 0 & 0 & & \cdots & F^0 & tF^0 \end{pmatrix}.$$

The corresponding exponential function is

$$Exp_{[g]} : \begin{pmatrix} w \\ z_0 \\ z_1 \\ \vdots \\ z_{g-1} \end{pmatrix} \longrightarrow \begin{pmatrix} w \\ e(z_0) \\ z_1 + e^{[1]}(z_0) \\ \vdots \\ z_{g-1} + e^{[g-1]}(z_0) \end{pmatrix},$$

THEOREM. *If $g = \text{subdeg } \phi$ and $u \neq 0$ with $e(u) \in \bar{k}_\infty^{sep}$, then $u \in \bar{k}_\infty^{sep}$ and*

$$1, u, u^{[1]}, \ldots, u^{[g-1]}$$

are $\bar{k}$-linearly independent.

This line of investigation was begun by Denis [**De5**],[**De7**], who worked with ordinary derivatives and thus established the equivalent of the $\bar{k}$-linear independence of $1, u, u^{[1]}, \ldots, u^{[p-1]}$ in the case when ϕ satisfies rather strict technical hypotheses (which hold for the Carlitz modules). In joint work with Denis, we have succeeded in extending the statement to include $u^{[i]}$ for a certain range of larger i. But the structure of the t-module is not the same, and the proof becomes much more tedious. The basic problem lies in the difficulty of understanding the sub-t-modules of the relevant t-module extension of the given Drinfeld module by $\mathbb{G}_a^i$.

However we note here that we can combine this theorem together with the Baker-Yu Theorem to obtain the following:

THEOREM. *Let $\phi_1, \ldots, \phi_n$ be non-isogenous Drinfeld modules of subdegrees s_i and with exponential functions $e_i(z), i = 1, \ldots, n$. For each i, let each of the sets $U_i = \{\ldots, u_{ij}, \ldots\} \subset \mathbb{C}_p$ be linearly independent over $\mathcal{O}_i = \text{End}(\phi_i)$ and let each $e_i(u_{ij}) \in \bar{k}_\infty^{sep}$. Then the $1 + \sum_i s_i |U_i|$ numbers:*

$$1, u_{ij}^{[\ell]}$$

are $\bar{k}$-linearly independent, where $u_{ij} \in U_i, 0 \leq \ell < s_i$.

Denis also establishes various theorems in [**De5**],[**De7**],[**De8**],[**De10**] which show that at least two of certain small sets of numbers are algebraically independent. The following result is particularly pleasing:

THEOREM. (Denis) *If* $\alpha \in \bar{k}_\infty^{sep}$ *is non-zero, then* $\log \alpha, (\log \alpha)^{[1]}$ *are algebraically independent.*

In [**De10**], Denis also makes the intriguing construction of Drinfeld modules having derivatives of periods of Carlitz modules as quasi-periods. We remark here that this construction also extends to higher order divided derivatives.

B. Independence of Values of Carlitz-Bessel Functions. Carlitz used his factorials D_h to define an analogue of Bessel functions [**Ca**]. For every $m \in \mathbb{Z}_{\geq 0}$, he defined the Bessel function of order m to be

$$J_m(z) := \sum_{h=0}^{\infty} \frac{(-1)^h z^{q^{m+h}}}{D_{m+h} D_k^{q^m}}.$$

These functions satisfy a functional equation strongly resembling that of the exponential function of a t-module:

$$\begin{pmatrix} J_m(tz) \\ \Delta J_m(tz) \end{pmatrix} = \begin{pmatrix} t & 1 \\ 0 & t^{q^m} \end{pmatrix} \begin{pmatrix} J_m(z) \\ \Delta J_m(z) \end{pmatrix} + \begin{pmatrix} 0 & 0 \\ -1 & 0 \end{pmatrix} \begin{pmatrix} (J_m(z))^q \\ (\Delta J_m(z))^q \end{pmatrix},$$

where now Δ represents the first forward difference, so that $\Delta J_m(z) = J_m(tz) - J_m(z)$.

Because of the term t^{q^m} in the lower right hand corner of the first matrix, this t-action is not, strictly speaking, a t-module. However after a renormalization, it becomes the second tensor power of the Carlitz module in the sense of Anderson-Thakur [**AT**]. Denis remarks [**De9**] that the proof of the zero estimate above for t-modules actually applies in this situation as well, and he shows that the rest of the transcendence machine carries over to give the following result [**De12**]:

THEOREM. (Denis) *Let* $\alpha \in \bar{k}$ *be non-zero and* $m \in \mathbb{Z}_{\geq 0}$. *Then* $J_m(\alpha), \Delta J_m(\alpha)$ *are algebraically independent over* k.

Since the difference operator can be viewed as a discrete analogue of differentiation, the above functional relation can be viewed as an analogue of the second order differential equation satisfied by the ordinary Bessel functions:

$$\Delta^2 J_m(z) = (t^{q^m} - t) \Delta J_m(z) - (J_m(z))^q.$$

Looked at in this way, Denis' result promises to be the first step in developing counterparts to the theorems of Siegel and Shidlovsky mentioned earlier. Thakur's notion [**Tk4**] of hypergeometric functions and his investigation of its "differential calculus" provides a natural setting for applications.

C. Values of Gamma Functions. It is not quite clear whether this subsection belongs here or in the previous section. The reasons for placing it earlier are that the results here ARE (certain) analogues of objects in $\mathbb{C}$ and the results which have appeared so far are obtained by pre-subgroup methods. The reasons for not placing it here are that the analogues are not unique and there is some basis to hope that consideration of the sub-t-module structure can be made to apply here to give more powerful results.

Thakur defined one analogue of the ordinary Gamma function via the following formula where, once again, $A = \mathbb{F}_q[t]$:

$$\Gamma(z) := \frac{1}{z} \prod_{\substack{a \in A \\ a \text{ monic}}} (1 + \frac{z}{a})^{-1}.$$

For the reasons why this is an appropriate analogue of the classical gamma function, we refer the reader to [**Tk1**],[**Tk2**]. (Thakur's own article in this volume is mainly concerned with the gamma function defined by Goss.) In the very fruitful paper [**Tk1**], he made an important observation (case $q = 2$ of Sinha's theorem below), proved the algebraicity of monomials in gamma values motivated by the classical results, and conjectured which monomials should be transcendental.

In his thesis [**Si**], S.K. Sinha investigates the periods of a t-module originally defined by G. Anderson [**An**] in the equivalent category of $t-motives$, motivated in part by the investigations of [**Tk1**]. Anderson called such motives *soliton t-motives* because of analogies with some aspects of partial differential equations. Since the t-motive is defined over $\bar{k}$, the entries of $\Phi(t)$ lie in $\bar{k}$. In a remarkable calculation, Sinha is able to show that these periods have the form

$$\omega = \Gamma(\frac{a}{f}),$$

where $a, f \in \mathbb{F}_q[t]$ are monic and $\deg a < \deg f$. As a result, he deduces:

THEOREM. (Sinha) *Let $a, b, f \in A = \mathbb{F}_q[t]$ with a, f monic and $\deg a < \deg f$. Then $\Gamma(b + \frac{a}{f})$ is transcendental over $\mathbb{F}_q(t)$.*

Moreover Sinha announces that monomials in Thakur's gamma functions at points of $\mathbb{F}_q(t)$ can also be interpreted as periods of the tensor products of the corresponding t-modules. Therefore Yu's Subgroup Theorem offers an avenue of attack on the question of the algebraic independence of the Thakur's gamma values! This very exciting prospect is tempered only by the difficulty experienced in understanding the sub-t-modules of the other newly appearing t-modules.

In Summary

Some areas of transcendence in positive characteristic, particularly analogues of classical results based on algebraic groups, are already relatively mature. Several further analogues of classical results present intriguing challenges, often related to understanding the structure of sub-t-modules. Meanwhile we find rapid development on several fronts for which no classical analogues are yet apparent. It will be interesting to see whether this vibrant field will yield clues to classical mysteries.

Appendix. Notation

$\mathbb{F}_q$ a finite field of $q = p^s$ elements
$\mathcal{C}$ a smooth projective geometrically irreducible curve over $\mathbb{F}_q$
∞ a closed point on $\mathcal{C}$
k the function field of $\mathcal{C}$ over $\mathbb{F}_q$
A the ring of functions in k regular on $\mathcal{C} \setminus \{\infty\}$
k_∞ the completion of k at ∞
C_p the completion of the algebraic closure of k_∞

References

[An] G. W. Anderson, *t-Motives*, Duke Math. J. **53** (1986), 457-502.

[AT] G.W. Anderson and D.S. Thakur, *Tensor powers of the Carlitz module and zeta values*, Ann. of Math. **132** (1990), 159-191.

[Be] P.-G. Becker, *Transcendence measures for the values of generalized Mahler functions in arbitrary characteristic*, Publ. Math. Debrecen **45** (1994), 269-282.

[BBT] P.-G. Becker, W.D. Brownawell, and R. Tubbs, *Gelfond's theorem for Drinfel'd modules*, Mich. Math. J. **41** (1994), 219-233.

[Br1] W.D. Brownawell, *Drinfeld exponential and quasi-periodic functions*, in Advances in Number Theory (F.Q. Gouvêa and N. Yui, eds.), Oxford University Press, Oxford, 1993.

[Br2] ———, *Submodules of products of quasi-periodic modules*, Schmidt Volume, Rocky Mountain J. Math. **41** (1996), 847-873.

[BK] W.D. Brownawell and K.K. Kubota, *The algebraic independence of Weierstrass functions and some related numbers*, Acta Arith. **33** (1977), 111-149.

[BT] W.D. Brownawell and R. Tubbs, *Zero estimates for t-modules*, Available for anonymous ftp from ftp.math.psu.edu/pub/wdb/.

[Bo] E. Bombieri, *Algebraic values of meromorphic maps*, Invent. Math. **10** (1970), 267-287.

[Co] R.F. Coleman, *On a stronger version of the Schanuel-Ax theorem*, Am. J. Math. **102** (1980), no. 4, 595-624.

[DD] S. David and L. Denis, *Isogénie minimale entre modules de Drinfel'd*, manuscript.

[De1] L. Denis, *Géometrie diophantienne sur les modules de Drinfel'd*, The Arithmetic of Function Fields (D. Goss, D.R. Hayes, and M.I. Rosen, eds.), De Gruyter, Berlin, New York, 1992, pp. 285-303.

[De2] ———, *Remarques sur la transcendance en caractéristique finie*, C.R.A.S. du Canada **14** (1992), 157-162.

[De3] ———, *Théorème de Baker et modules de Drinfel'd*, J. Number Th. **43** (1993), 203-215; C.R. Acad. Sci. Paris Sér. I Math. **311** (1990), 473-475.

[De4] ———, *Indépendance algébrique sur le module de Carlitz*, C.R. Acad. Sci. Paris Sér. I Math. **317** (1993), 913-915.

[De5] ———, *Transcendance et dérivées de l'exponentielle de Carlitz*, Séminaire Théorie des Nombres, Paris, 1991-1992 (S. David, ed.), Prog. Math. 116, Birkhäuser, Basel, New York, 1993, pp. 1-21.

[De6] ———, *Méthodes fonctionnelles pour la transcendance en caractéristique finie*, Bull. Austral. Math. Soc. **50** (1994), 273-286.

[De7] ———, *Dérivées d'un module de Drinfeld et transcendance*, Duke Math. J. **80** (1995), 1-13.

[De8] ———, *Indépendance algébrique et exponentielle de Carlitz*, Acta Arith. **69** (1995), 75-89.

[De9] ———, *Lemmes de multiplicité et T-modules*, Mich. Math. J. **43** (1996), 67-79.

[De10] ———, *Indépendance algébrique en caractéristique 2*, manuscript.

[De11] ———, *Problèmes diophantiens sur les t-modules*, J. Théor. Nombres de Bordeaux.

[De12] ———, *Valeurs transcendantes des fonctions de Bessel-Carlitz*, manuscript.

[Dr] V.G. Drinfeld, *Elliptic modules*, Mat. Sbornik **94** (1974), 594-627, Engl. transl. Math. USSR Sbornik **23** (1974), 561-592.

[Gs] J.M. Geijsel, *Transcendence in fields of finite characteristic*, Mathematical Centre Tracts, vol. 91, Mathematisch Centrum, Amsterdam, 1979.

[Gk] ———, *De Rham isomorphism for Drinfeld modules*, J. für die reine und angew. Math. **401** (1989), 188-208.

[H1] D.R. Hayes, *Explicit class field theory in global function fields*, in Studies in Algebra and Number Theory (G.-C. Rota, ed.), Academic Press, New York, 1979, pp. 1-32.

[H2] ———, *A brief introduction to Drinfeld modules*, in The Arithmetic of Function Fields (D. Goss, D.R. Hayes, M. Rosen, eds.), De Gruyter, Berlin, New York, 1992, pp. 173-217.

[K] E. Kolchin, *Algebraic groups and algebraic independence*, Amer. J. Math. **90** (1968), 1151-1164.

[MW1] D.W. Masser and G. Wüstholz, *Periods and minimal abelian subvarieties*, Ann. of Math. **137** (1993), 407-458.

[MW2] ———, *Isogeny estimates for abelian varieties and finiteness theorems*, Ann. of Math. **137** (1993), 459-472.

[NS] Yu.V. Nesterenko and A.B. Shidlovsky, *Linear independence of values of E-functions*, Sbornik: Math. **187** (1996), no. 8, 1197-1211; Russian orig. in Mat. Sbornik **187** (1996), no. 8, 93-108.

[Ph] P. Philippon, *Critères pour l'indépendance algébrique dans les anneaux diophantiens*, C. R. Acad. Sci. Paris Sér. Math. I **315** (1992), 511-515.

[Sh] A.B. Shidlovsky, *Transcendental Numbers*, De Gruyter, Berlin, New York, 1989, transl. of *Trantsendentnye Chisla*, Nauka, Moscow, 1987.

[Si] S.K. Sinha, *Periods of t-motives and special functions in characteristic p*, Thesis, Univ. Minnesota (1995).

[Tk1] D.S. Thakur, *Gamma function for function fields and Drinfeld modules*, Ann. of Math. **134** (1991), 25-64.

[Tk2] _____, *On Gamma functions for function fields*, The Arithmetic of function Fields (D. Goss, D.R. Hayes, and M.I. Rosen, eds.), De Gruyter, Berlin, New York, 1992, pp. 75-86.

[Tk3] _____, *Drinfeld modules and arithmetic in function fields*, International Math. Res. Notices (1992), 185-197; Duke Math. J. **68** (1992), 185-197.

[Tk4] _____, *Hypergeometric functions for function fields*, Finite Fields and their Appl. **1** (1995), 219-231.

[Tk5] _____, *Automata and transcendence*, in this issue.

[Tr1] A. Thiery, *Théorème de Lindemann-Weierstrass pour les modules de Drinfeld*, Compositio Math. **95** (1995), 1-42.

[Tr2] _____, *Indépendance algébrique des périodes et quasi-périodes d'un module de Drinfeld*, The Arithmetic of Function Fields (D. Goss, D.R. Hayes, and M.I. Rosen, eds.), De Gruyter, Berlin, New York, 1992, pp. 265-284.

[Tr3] _____, *Transcendance de quelques valeurs de la fonction gamma pour les corps de fonctions*, C.R. Acad. Sci. Paris Sér. I Math. **314** (1992), 973-976.

[Wd1] L.I. Wade, *Certain quantities transcendental over $GF(p^n, x)$*, Duke Math. J. **8** (1941), 701-729.

[Wd2] _____, *Certain quantities transcendental over $GF(p^n, x)$, II*, Duke Math. J. **10** (1943), 587-594.

[Wd3] _____, *Transcendence properties of the Carlitz ψ-functions*, Duke Math. J. **13** (1946), 79-85.

[Wl] M. Waldschmidt, *Transcendence problems connected with Drinfeld modules*, Istanbul Üniv. Fen Fak. Mat. Derg. **49** (1993), 57-75.

[Wü] G. Wüstholz, *Some remarks on a conjecture of Waldschmidt*, Diophantine Approximations and Transcendental Numbers (Luminy 1982) (D. Bertrand and M. Waldschmidt, eds.), Progr. Math. 31, Birkhäuser, Basel, New York, 1983, pp. 329-336.

[Yu1] J. Yu, *Transcendental numbers arising from Drinfeld modules*, Mathematika **30** (1983), 61-66.

[Yu2] _____, *Transcendence theory over function fields*, Duke Math. J. **52** (1985), 517-527.

[Yu3] _____, *A six exponentials theorem in finite characteristic*, Math. Ann. **272** (1985), 91-98.

[Yu4] _____, *Transcendence and Drinfeld modules*, Inv. Math. **83** (1986), 91-98.

[Yu5] _____, *Transcendence and Drinfeld modules: several variables*, Duke Math. J. **58** (1989), 559-575.

[Yu6] _____, *On periods and quasi-periods of Drinfeld modules*, Compositio Math. **74** (1990), 235-245.

[Yu7] _____, *Transcendence and special zeta values in characteristic p*, Ann. of Math. **134** (1991), 1-23.

[Yu8] _____, *Analytic homomorphisms into Drinfeld modules*, preprint (1991).

[Yu9] _____, *Transcendence in finite characteristic*, in The Arithmetic of Function Fields (D. Goss, D.R. Hayes, M. Rosen, eds.), Walter De Gruyter, Berlin, New York, 1992, pp. 253-264.

PENN STATE UNIVERSITY, UNIVERSITY PARK, PA 16802
E-mail address: wdb@math.psu.edu

Minorations des hauteurs normalisées des sous-variétés de variétés abéliennes[1]

Sinnou DAVID & *Patrice* PHILIPPON

Résumé : nous décrivons les progrès récents obtenus dans la direction de la conjecture de BOGOMOLOV généralisée et ses liens avec l'annulation des hauteurs normalisées ; nous détaillons tout d'abord comment on peut obtenir directement dans le cas de multiplications complexes une minoration effective de la hauteur normalisée d'une sous-variété d'une variété abélienne (admettant des multiplications complexes) ; ensuite, nous décrivons l'équivalence entre le problème de BOGOMOLOV et celui de la minoration de la hauteur normalisée. Enfin, dans le dernier paragraphe, nous donnons une minoration effective de la hauteur normalisée d'une sous-variété d'une variété abélienne dans le cas général, rendant ainsi effectifs les travaux récents d'ULLMO et ZHANG. Cette minoration est monomiale inverse en le degré de la variété, et de nature géométrique, au sens où elle ne dépend pas du degré du corps de définition de la variété considérée, ni de sa hauteur.

Abstract : we describe the recent results obtained towards the generalized BOGOMOLOV conjecture and its links with the vanishing of the normalized height ; we first show how one can obtain directly a lower bound for the normalized height of a subvariety of an abelian variety of CM type ; we then describe the equivalence between the normalized height and the BOGOMOLOV problems. Finally, in the last paragraph we give an effective lower bound for the normalized height of a subvariety of an abelian variety, thus making effective the recent results obtained by ULLMO and ZHANG. Our bound is inverse monomial in the degree of the variety and geometric (we mean here that the lower bound does not depend on the degree of a field of definition of the variety studied, nor its height).

---

[1] Math. subject classification 11 G 10, 11 J 81, 14 G 40

© 1998 American Mathematical Society

1 Introduction et résultats

Soit A une variété abélienne de dimension g définie sur un corps de nombres K et plongée comme sous-variété *projectivement normale* dans un espace projectif $\mathbb{P}_n$. On suppose que la classe d'équivalence linéaire du diviseur donnant le plongement de A dans $\mathbb{P}_n$ est symétrique, on a alors défini dans [Ph1]–I, §3 une *hauteur normalisée* $\hat{h}$ associée à cette classe de diviseur, étendant la hauteur de NÉRON–TATE classique aux variétés de dimensions supérieures. Si $S = \sum_{i=1}^{s} \ell(V_i).[V_i]$ est un cycle défini sur $\overline{\mathbb{Q}}$ dont le support est contenu dans A on posera $\hat{h}(S) = \sum_{i=1}^{s} \ell(V_i).\hat{h}(V_i)$. On désignera ici par X une sous-variété algébrique de A, définie sur $\overline{\mathbb{Q}}$. En particulier, il existe un réel $C_A > 0$ ne dépendant que de A comme variété plongée, tel que pour toute sous-variété X de A, on ait $\left|\hat{h}(X) - h(X)\right| \leqslant C_A d(X)$, où $h(X)$ désigne la hauteur projective et $d(X)$ le degré de X dans $\mathbb{P}_n$ (*cf.* [Ph1]–I).

La question laissée en suspens dans [Ph1]–I demande de déterminer les sous-variétés X de A de hauteur normalisée $\hat{h}(X)$ nulle. Plus précisément, considérons la propriété suivante :

Propriété du $\hat{h}$. *On dira que la variété abélienne A a la propriété du $\hat{h}$ lorsque pour toute sous-variété algébrique X de A, définie sur $\overline{\mathbb{Q}}$, on a $\hat{h}(X) = 0$ si et seulement si X est translatée d'une sous-variété abélienne par un point de torsion de A.*

La conjecture naturelle est alors :

Conjecture 1.1 *Toutes les variétés abéliennes définies sur $\overline{\mathbb{Q}}$ ont la propriété du $\hat{h}$.*

Une question très liée a été posée par BOGOMOLOV. Dans sa version originale, le problème est le suivant. Soit X une courbe algébrique définie sur $\overline{\mathbb{Q}}$, de genre > 1, plongée dans sa jacobienne $J(X)$, alors les points de $X(\overline{\mathbb{Q}})$ sont discrets[2] pour la topologie induite par la métrique de NÉRON–TATE.

---
[2] La topologie n'étant pas séparée, on entendra par ce mot le fait que tout point admet un voisinage de cardinal fini.

Cette question a été généralisée au cas des sous-variétés quelconques d'une variété abélienne (*voir* par exemple [Zh3]). Dans ce contexte introduisons :

Propriété de Bogomolov. *On dira que la variété abélienne A a la propriété de Bogomolov si pour toute sous-variété algébrique X de A, définie sur $\overline{\mathbb{Q}}$, qui n'est pas translatée d'une sous-variété abélienne par un point de torsion de A il existe un réel $\varepsilon > 0$ tel que l'ensemble des points $x \in X(\overline{\mathbb{Q}})$ de hauteur $\hat{h}(x) \leqslant \varepsilon$ ne soit pas Zariski dense dans X.*

La question formulée par S. ZHANG peut alors s'énoncer ainsi :

Conjecture 1.2 *Toutes les variétés abéliennes définies sur $\overline{\mathbb{Q}}$ ont la propriété de Bogomolov.*

De la formulation ci-dessus, on déduit facilement (à l'aide de translations *ad hoc*) les variantes que l'on trouve souvent dans la littérature, par exemple :

Conjecture 1.3 *Soient A une variété abélienne définie sur $\overline{\mathbb{Q}}$ et X une sous-variété algébrique de A également définie sur $\overline{\mathbb{Q}}$. Notons X° l'ouvert de Zariski de X formé du complémentaire dans X de la réunion des translatées de sous-variétés abéliennes propres de A contenues dans X. Alors $X^\circ(\overline{\mathbb{Q}})$ est discret pour la topologie induite par la métrique de Néron–Tate.*

Si l'on désigne par $X^\star$ le complémentaire dans X de la réunion des translatés de sous-variétés abéliennes par des points de torsion de A contenus dans X, on remarquera que $X^\circ \subset X^\star \cup \{X^\circ \cap A_{\text{Torsion}}\}$, mais $X^\star(\overline{\mathbb{Q}})$ n'est pas nécessairement discret pour la métrique de NÉRON–TATE. D'un point de vue quantitatif, on peut demander une minoration effective de :

$$\delta^\circ(X) = \inf\left\{\hat{h}(x-y); x-y \notin A_{\text{Torsion}}, x, y \in X^\circ(\overline{\mathbb{Q}})\right\}$$

(resp. $\mu^\star(X) = \inf\left\{\hat{h}(x); x \in X^\star(\overline{\mathbb{Q}})\right\}$), des degrés et hauteurs des composantes de $X \setminus X^\circ$ (resp. $X \setminus X^\star$). Egalement, on peut chercher à minorer le minimum essentiel, $\mu^{\text{ess}}(X) = \sup_Z \inf\{\hat{h}(x); x \in X(\overline{\mathbb{Q}}) \setminus Z(\overline{\mathbb{Q}})\}$ où le supremum est pris sur tous les diviseurs Z de X définis sur $\overline{\mathbb{Q}}$. On a $\mu^{\text{ess}}(X) \geqslant \mu^\star(X)$ et $\delta^\circ(X) \geqslant \inf_{y \in X^\circ(\overline{\mathbb{Q}})} \mu^\star(X - y)$.

Ces questions peuvent se transposer dans le cas des sous-variétés algébriques d'un tore (*i. e.* $\mathbb{G}_m^n$). Dans ce cadre, l'analogue du problème de BOGOMOLOV a été résolu par S. ZHANG (*voir* [Zh1] et [Zh2]). Par la suite, W. M. SCHMIDT (*voir* [Schm2]) ainsi que E. BOMBIERI et U. ZANNIER (*voir* [Bo–Za1]) ont donné des preuves « élémentaires » de ce résultat, obtenant de surcroît des minorations de l'écart entre deux points de X° en fonction de la variété X. Fait remarquable, ces minorations ne dépendent que du degré de X et non de sa hauteur, ni de son corps de définition. Ce dernier point est obtenu par l'intermédiaire des « variétés déterminantes », une technique introduite dans ce type de contexte par H. P. SCHLICKEWEI (*voir* [Schl]).

Revenons au cas des variétés abéliennes. Des progrès décisifs ont étés accomplis sur ce problème ces dernières années. Dans [Zh3], théorème 1.10, S. ZHANG montre en particulier :

Théorème 1.4 *Les propriétés du $\hat{h}$ et de Bogomolov sont équivalentes. De plus, si X est une sous-variété algébrique de A, définie sur $\overline{\mathbb{Q}}$, on a :*

$$\frac{\hat{h}(X)}{(\dim(X)+1)d(X)} \leqslant \mu^{\text{ess}}(X) \leqslant \frac{\hat{h}(X)}{d(X)}.$$

Dans ce même texte, S. ZHANG prouve la conjecture 1.2 dans le cas où les conditions suivantes sont satisfaites :

- $X - X$ engendre A ;

- la flèche : $\mathrm{NS}(A) \otimes \mathbb{R} \longrightarrow \mathrm{NS}(X) \otimes \mathbb{R}$ n'est pas injective.

Auparavant, L. SZPIRO (*voir* [Sz1], [Sz2]) avait partiellement établi la conjecture 1.2 dans le cas des courbes lisses et a été le premier à remarquer le lien entre les deux propriétés (mis en évidence pour les courbes lisses) ; J.-F. BURNOL (*voir* [Bu]) a démontré la conjecture 1.2 pour les courbes plongées dans leurs jacobiennes dans le cas où cette dernière admet des multiplications complexes. E. BOMBIERI et U. ZANNIER ont prouvé la conjecture 1.3 dans le cas des sous-variétés des variétés abéliennes admettant des multiplications complexes (et utilisant la technique des variétés déterminantes, obtiennent une borne pour l'écart entre deux points de $X^\circ(\overline{\mathbb{Q}})$ ne faisant intervenir que A et le degré de X). La conjecture 1.1 est par ailleurs montrée par le deuxième auteur dans [Ph1]–III, §3 pour les variétés abéliennes A isogènes à un produit de courbes elliptiques.

Très récemment, E. ULLMO (*voir* [Ul]) a étendu le résultat de J.-F. BURNOL à toutes les courbes lisses, démontrant ainsi la conjecture de BOGOMOLOV originale. Ce résultat a été immédiatement généralisé par S. ZHANG (*voir* [Zh4]) démontrant les conjectures 1.1 et 1.2, ainsi que ses variantes.

Dans ce texte, nous précisons les divers aspects quantitatifs de ces résultats en donnant des preuves indépendantes de la conjectures 1.1. Dans le cas des variétés abéliennes à multiplications complexes, qui est particulièrement simple à traiter (une application du théorème de BÉZOUT arithmétique suffit), nous obtenons le résultat suivant :

Théorème 1.5 *Soit A une variété abélienne à multiplications complexes, définie et plongée sur un corps de nombres K comme sous-variété projectivement normale d'un espace projectif $\mathbb{P}_n$, par un diviseur symétrique. Il existe alors une constante $c(A, K) > 0$ telle que toute sous-variété algébrique irréductible X de A, définie sur une extension K' de K et qui n'est pas translatée d'une sous-variété abélienne par un point de torsion de A satisfait*

$$\frac{\hat{h}(X)}{d(X)} \geqslant \exp(-c(A,K)d(X)[K':\mathbb{Q}]^2).$$

En particulier, A a la propriété du $\hat{h}$ et celle de Bogomolov.

Remarque – Le théorème 1 de [Bo–Za2] montre l'existence d'un réel $\gamma(A, d(X))$ tel que les points $x \in X(\overline{\mathbb{Q}})$ de hauteurs $\hat{h}(x) \leqslant \gamma(A, d(X))$ ne soient pas ZARISKI dense dans X lorsque X n'est pas translatée d'une sous-variété abélienne(par un point quelconque) de A. Ce résultat, combiné au corollaire 3.2 (ii), entraîne la formule $\frac{\hat{h}(X)}{d(X)} \geqslant \gamma(A, d(X))$. Cette minoration est meilleure que celle du théorème 1.5 lorsque $[K':K]$ est grand, mais moins précise pour ce qui est de la dépendance en A et $d(X)$ lorsque $[K':K]$ est borné. En fait, le théorème 1.5 ci-dessus précise la minoration qui se déduit du théorème 2 de [Bo–Za2] via le corollaire 3.2 (ii). La dépendance en A pour sa part provient de deux points. Tout d'abord, elle fait intervenir la constante de comparaison C_A entre les hauteurs normalisées et les hauteurs projectives ; ensuite (et c'est là que l'arithmétique de K intervient) elle tient compte des places v de K pour lesquelles le FROBENIUS se relève en un endomorphisme ; on trouvera des précisions au paragraphe 2. Enfin, comme il est remarqué dans [Bo–Za2], l'énoncé reste vrai pour toutes les variétés abéliennes sur lesquelles le FROBENIUS se relève en un endomorphisme de A pour une infinité de places.

Passons maintenant au cas général. Nous obtenons :

Théorème 1.6 *Soit A une variété abélienne définie et plongée sur un corps de nombres K comme sous-variété projectivement normale d'un espace projectif $\mathbb{P}_n$, par un diviseur symétrique. Alors, il existe un réel $c(A, K) > 0$ ne dépendant que de A ainsi plongée tel que pour toute sous-variété algébrique X de A, définie sur une extension finie de K, qui n'est pas translatée d'une sous-variété abélienne de A, on ait :*

$$\frac{\hat{h}(X)}{d(X)} \geqslant c(A,K).d(X)^{-\max(1,2(k_X-1)).(\dim B_X+1)}.(\log d(X))^{-\dim B_X} ,$$

où B_X désigne la sous-variété abélienne de A engendrée par $X-X$ et k_X le nombre minimal de copies de $X-X$ dont la somme vaut B_X. En particulier, A a la propriété du $\hat{h}$ et celle de Bogomolov.

Remarques – La variété $Y = X - X$ satisfait $Y = -Y$ et $0 \in Y$ ce qui assure l'existence de l'entier k_X tel que la somme de k_X copies de Y soit la sous-variété abélienne B_X de A, on vérifie aisément $1 \leqslant k_X \leqslant \dim B_X - \dim(X - X) + 1$. Lorsque $k_X = 1$ on a un résultat plus précis (*cf.* § 4.3), le cas général s'en déduira par récurrence sur k_X. On notera qu'outre une minoration bien plus précise que celle du theorème 1.5 en ce qui concerne la dépendance en $d(X)$ (monomiale en $d(X)^{-1}$ au lieu d'exponentielle en $-d(X)$), cet énoncé est de nature géométrique, au sens où la minoration ne dépend pas du degré d'un corps de définition de X. Comme déjà remarqué, lorsque A est à multiplications complexes le théorème 1 de [Bo–Za2] entraîne une minoration de $\hat{h}(X)$ de même nature géométrique, mais contrairement à cette référence notre minoration est obtenue sans avoir recours aux variétés déterminantes. Cela permet d'avoir une minoration monomiale en $d(X)^{-1}$ au lieu d'une borne double exponentielle (la première exponentielle étant perdue du fait que la minoration dans le cas de multiplications complexes est plus faible (la méthode est plus élémentaire) et la perte de la deuxième exponentielle semble inhérente à la méthode des variétés déterminantes).

Notre méthode utilise un ingrédient commun aux preuves qualitatives d'E. ULLMO et S. ZHANG : il s'agit du recours à un morphisme d'une puissance de X vers une puissance de A pour lequel une fibre au moins a une dimension supérieure à celle de la fibre générique. Si E. ULLMO considère la flèche naturelle de C^g vers $J(C)$ (où $J(C)$ est la jacobienne de la courbe C), et se sert des points de WEIERSTRASS, S. ZHANG considère la flèche $X^m \longrightarrow A^{m-1}$ donnée par $(x_1,\ldots,x_m) \longmapsto (x_1-x_2, x_2-x_3,\ldots,x_{m-1}-x_m)$, apparaissant déjà dans la preuve de la conjecture de S. LANG par G. FALTINGS. Nous nousramenons au cas où $X - X$ est une sous-variété abélienne de A, condition qui intervenait dans un travail antérieur de S. ZHANG (voir [Zh3]), et nous considérons le morphisme naturel $X^2 \longrightarrow X - X$. Par contre les méthodes d'équidistribution de petits points n'interviennent pas dans notre construction, qui est plus proche de l'approximation diophantienne classique. Une dernière remarque enfin. Contrairement aux hypothèses des conjectures 1.1 ou 1.2, nous supposons que X n'est pas translatée d'une sous-variété abélienne sans préciser s'il s'agit d'un point de torsion. Il faut penser que dans le cas où X est translatée par un point d'une sous-variété abélienne de A, sa hauteur normalisée est controlée par celle du point en question. On ne peut donc espérer (dans le cas où le point est d'ordre infini) de minoration ne faisant pas intervenir le degré d'un corps de définition de X. Mais on peut utiliser dans ce cas la minoration de la hauteur du point qui se déduit de l'approche de D.W. MASSER [Ma] via [Da2].

On trouvera au paragraphe 2 une preuve du théorème 1.5 ; au paragraphe 3, nous donnerons une preuve plus élémentaire du théorème 1.4. Avec les théorèmes de MINKOWSKI et BÉZOUT arithmétique, le seul outil fin utilisé est en effet une minoration asymptotique de la fonction de HILBERT (arithmétique), qui n'est pas la partie de [B-G-S] nécéssitant le plus de «bagages théoriques» (voir [Ab-Bo] pour une preuve dans le cas général). Enfin, au paragraphe 4, nous donnerons une preuve du théorème 1.6.

Terminons cette introduction par un problème, qui devrait ouvrir des directions de recherche intéressantes pour la compréhension de l'approximation diophantienne en dimension supérieure. Soulignons que, si pour les points les conjectures 1.1 et 1.2 sont triviales, leurs versions quantitatives sont encore largement conjecturales (conjecture de LANG, problème de LEHMER), même si l'on se restreint au cas elliptique, voire au cas multiplicatif (pour le problème de LEHMER). De plus, comme le montre le théorème 1.6, l'arithmétique des variétés de dimension au moins 1 est radicalement différente de celle des points.

Problème 1.7 *Soient g, d et δ, des entiers $\geqslant 1$ et h un nombre réel > 0. Soient encore Δ un entier $\geqslant 1$, r un entier $\geqslant 1$ et D un entier $\geqslant 1$. Quelles sont les fonctions «optimales» $f(g, d, \delta, h, r, \Delta)$ et $\varphi(g, d, \delta, h, r, \Delta, D)$ telles que, pour toute variété abélienne A de dimension g définie sur un corps de nombres K de degré d sur $\mathbb{Q}$, de hauteur de FALTINGS $\leqslant h$ et munie d'une polarisation de degré δ et toute sous-variété algébrique X de A, irréductible, de dimension $r - 1$, de degré Δ et définie sur une extension de K de degré relatif au plus D, si $\hat{h}(X) < f(g, d, \delta, h, r, \Delta)$ alors X est la translatée d'une sous-variété abélienne de A, et si de plus $\hat{h}(X) < \varphi(g, d, \delta, h, r, \Delta, D)$ alors ce point est de torsion ?*

On conviendra pour cette question que si $\mathcal{L}$ est un fibré en droite ample induisant la polarisation, on voit A comme une sous-variété d'un projectif, le plongement étant défini par $\mathcal{L}^{\otimes 4}$.

On notera que pour un point P (*i. e.* $r = 1$, $\Delta = 1$), une version optimiste combinant les problèmes de LEHMER et LANG donnerait pour φ la fonction $\varphi(g, d, \delta, h, 1, 1, D) = c(g, d, \delta) h D^{-\frac{1}{g'}}$, où g' est la dimension du plus petit sous-groupe algébrique (non néccessairement connexe) contenant P, et $c(g, d, \delta)$ n'est pas précisé (le problème est déjà bien assez difficile comme ça !). En dimension > 0 (*i. e.* $r \geqslant 2$), le théorème 1.6 montre que la fonction $f(g, d, \delta, h, r, \Delta)$ peut être prise monomiale en Δ^{-1}. On peut voir par ailleurs que l'on ne peut espérer de dépendance en Δ meilleure que $\Delta^{\frac{g'-r}{g'-r+1}}$ où g' est maintenant la dimension du plus petit sous-groupe algébrique contenant X.

On notera enfin qu'il est vraisemblable que les constantes $c(A,K)$ introduites dans les théorèmes 1.5 et 1.6 puissent être explicitées en fonction des données h, δ, etc... Cela nécéssite en particulier de préciser la nature arithmétique des formes représentant la multiplication par 2, et devrait pouvoir se faire (sauf pour la dépendance en g qui est plus délicate) en utilisant les techniques utilisées dans [Da2].

Au cours de nombreuses discussions, nous avons bénéficié des éclaircissements d'A. CHAMBERT-LOIR et des remarques de D. BERTRAND et E. ULLMO. C'est un plaisir pour nous de les remercier chaleureusement.

2 Les variétés abéliennes à multiplications complexes

Si w est une place finie de K on note $\mathcal{O}_w$ l'anneau des entiers du complété K_w de K en w, π_w une uniformisante de $\mathcal{O}_w$, $\mathrm{N}w$ la norme de K sur $\mathbb{Q}$ de l'idéal w, $\mathbb{C}_w$ le complété d'une clôture algébrique de K_w et $\|.\|_w$ la norme du maximum pour w. On posera également $g = \dim A$, $\mathfrak{A}_w$ l'idéal de définition de A dans $\mathcal{O}_w[X_0,\ldots,X_n]$, $\Pi_w = \pi_w.\mathcal{O}_w[X_0,\ldots,X_n]$, et $\mathcal{A}_w = \mathcal{O}_w[X_0,\ldots,X_n]/\mathfrak{A}_w$. On posera également pour tout polynôme homogène $P \in \mathcal{O}_w[X_0,\ldots,X_n]$, $\mathrm{M}_w(P) = \sup_i\{|a_i|_w\}$, où les a_i sont les coefficients de P.

Commençons par un lemme. Quitte à se placer sur une extension de K de degré borné on peut supposer que A est semi-stable sur K. On supposera également dans tout ce paragraphe que les conditions suivantes sont réalisées :

- K contient le corps des multiplications complexes ;

- l'anneau des endomorphismes de A contient l'anneau des entiers de K ;

- la place w n'est pas ramifiée dans K.

Pour réaliser la seconde condition, il suffit de remplacer A par une variété isogène. Quitte à modifier la constante $c(A, K)$ du théorème 1.5 pour tenir compte du degré de formes représentant une telle isogénie, une telle réduction est licite. Il suffit de faire une extension de degré relatif borné pour réaliser la première condition, et de prendre Nw assez grand (plus grand que le discriminant $D_{K/\mathbb{Q}}$ de K sur $\mathbb{Q}$) pour la troisième. Cela suffit à nos besoins.

Lemme 2.1 *Soit w une place finie de K comme ci-dessus, il existe une famille $\boldsymbol{F}$ de formes dans $\mathcal{O}_w[X_0, \ldots, X_n]$ de degré Nw représentant l'endomorphisme de Frobenius associé à w sur $A(\mathbb{C}_w)$ et telle que $\|\boldsymbol{F}(z')\|_w = \|z'\|_w^{Nw}$ pour tout $z' \in \mathbb{C}_w^{n+1} \setminus \{0\}$.*

Démonstration : on sait, vu les réductions effectuées (*voir* [Sh–Ta]), que le FROBENIUS sur la variété réduite modulo w se relève en un endomorphisme α_w sur A. De plus, il existe un système de formes $\boldsymbol{F} = (F_0, \ldots, F_n) \in \mathcal{O}_w[X_0, \ldots, X_n]^{n+1}$ de degré Nw, représentant α_w sur tout A (*voir* [Hi1], lemmes 1 et 2, pages 21 à 23, *voir* aussi [Da-Hi], § 2). En particulier, $F_0, \ldots, F_n$ n'ont pas de zéros communs dans A.

Quitte à diviser les formes F_i par une puissance convenable de π_w on peut supposer $\max_{0 \leqslant j \leqslant n} M_w(F_j) = 1$. De cette façon, on a $\|\boldsymbol{F}(z')\|_w \leqslant \|z'\|_w^{Nw}$ pour tout $z' \in \mathbb{C}_w^{n+1} \setminus \{0\}$, montrons l'inégalité inverse. Pour ce faire, nous allons normaliser par un changement linéaire l'extension $\mathcal{A}_w$ de $\mathcal{O}_w[X_0, \ldots, X_n]$, et écrire le théorème des zéros de HILBERT sous une forme préparée par une division euclidienne (relations (1) et (2)). Un calcul de congruences nous permettra de conclure. Soit $U = (u_{i,j}) \in \mathrm{GL}_{n+1}(\mathcal{O}_w)$, posons
$$Y_i = u_{i,0} X_0 + \ldots + u_{i,n} X_n \ , \quad i = 0, \ldots, n \ ,$$
d'après le lemme de normalisation, pour U parcourant un ouvert de ZARISKI de $\mathrm{GL}_{n+1}(\mathcal{O}_w)$ les formes $Y_0, \ldots, Y_g$ forment une base de transcendance de $\mathcal{A}$ qui est une extension entière de $\mathcal{O}_w[Y_0, \ldots, Y_g]$; fixons un tel U. Considérons le système de formes :
$$\boldsymbol{G} = U \boldsymbol{F}(U^{-1} \boldsymbol{Y}) \ ,$$

où $Y = (Y_0, \ldots, Y_n)$. On a encore :
$$\max_{0 \leqslant i \leqslant n}\{M_w(G_i)\} = \max_{0 \leqslant i \leqslant n}\{M_w(F_i)\} = 1 \ .$$

Ces formes représentent le FROBENIUS en w dans le plongement de A dans $\mathbb{P}_n$ déduit du plongement initial par la transformation linéaire U. D'après le choix de U, et comme $F_0, \ldots, F_n$ n'ont pas de zéros communs dans $A(\overline{K_w})$, les formes $G_0, \ldots, G_g$ n'ont pas de zéro commun dans $A(\overline{K_w})$. On en déduit l'existence d'un élément $\lambda \in \mathcal{O}_w \setminus \{0\}$, d'un entier N et de formes $A_{i,j} \in \mathcal{O}_w[Y_0, \ldots, Y_n]$ de degré N tels que pour $i = 0, \ldots, n$ on ait :

$$\sum_{j=0}^{g} A_{i,j} G_j - \lambda . Y_i^{N+Nw} \in \mathfrak{A}_w + \Pi_w \ . \tag{1}$$

Mais, pour $l = g+1, \ldots n$, Y_l est entier sur $\mathcal{O}_w[Y_0, \ldots, Y_g]$ de degré disons d_l ; on peut donc supposer $d^\circ_{Y_l} A_{i,j} < d_l$ pour $l = g+1, \ldots, n$. Quitte à diviser toutes les formes $A_{i,j}$ par une puissance convenable de π_w on peut également supposer $\max_{0 \leqslant j \leqslant n}\{M_w(A_{i,j}), |\lambda|_w\} = 1$. Enfin, comme $(G_0, \ldots, G_n)$ représentent le FROBENIUS en w on a :

$$G_k Y_j^{Nw} - G_j Y_k^{Nw} \in \mathfrak{A}_w + \Pi_w \ . \tag{2}$$

On peut supposer $d^\circ_{Y_l} A_{i,j} < Nw$ pour $l > j$: sinon, on écrit $A_{i,j} = A'_{i,j} + Y_l^{Nw} B_{i,j}$ et l'on remplace dans (1) $Y_l^{Nw} G_j$ par $Y_j^{Nw} G_l$ pour $j < l$. Finalement, on remplace $A_{i,j}$ par $A'_{i,j}$ pour $j < l$ et $A_{i,l}$ par :

$$A_{i,l} + \sum_{j<l} Y_j^{Nw} B_{i,j} \ . \tag{3}$$

Montrons que $\lambda \notin \pi_w \mathcal{O}_w$; si ce n'était pas le cas, on aurait d'après la relation (1), et pour $k = 0, \ldots, n$,

$$Y_k^{Nw} . \sum_{j=0}^{g} A_{i,j} G_j \in \mathfrak{A}_w + \Pi_w$$

et donc, d'après la relation (2),

$$G_k . \sum_{j=0}^{n} A_{i,j} X_j^{Nw} \in \mathfrak{A}_w + \Pi_w$$

pour $k = 0, \ldots, n$ et avec $d^\circ_{Y_l} A_{i,j} < \min(d_l, Nw)$ pour $l > j$. Comme les variables $Y_0, \ldots, Y_g$ sont indépendantes modulo l'idéal premier $\mathfrak{A}_w + \Pi_w$ et Y_l est de degré d_l sur $\mathcal{O}_w[Y_0, \ldots, Y_g]$ on en déduit que :

- ou bien $G_0, \ldots, G_n \in \mathfrak{A}_w + \Pi_w$,

- ou bien $\sum_{j=0}^n A_{i,j} X_j^{Nw} \in \mathfrak{A}_w + \Pi_w$.

La première possibilité est exclue car on a choisi les F_i de sorte que l'on ait $\max_{0 \leqslant i \leqslant n} M_w(F_i) = \max_{0 \leqslant i \leqslant n}\{M_w(G_i)\} = 1$ et la seconde est exclue car on a choisi les $A_{i,j}$ de sorte que $\max_{0 \leqslant j \leqslant n} M_w(A_{i,j}) = 1 > |\lambda|_w$. Cette impossibilité montre que λ est une unité de $\mathcal{O}_w$, et donc pour $z' \in \mathbb{C}_w^{n+1} \setminus \{0\}$ on a, par la relation (1) :

$$1 \leqslant \frac{\|\boldsymbol{G}(Uz')\|_w}{\|Uz'\|_w} \cdot \max_{i,j}\left(\frac{|A_{i,j}(Uz')|_w}{\|Uz'\|_w^N}\right)$$

$$\leqslant \frac{\|U\boldsymbol{F}(z')\|_w}{\|Uz'\|_w} = \frac{\|\boldsymbol{F}(z')\|_w}{\|z'\|_w} \cdot \quad \square$$

Lemme 2.2 *Soit A une variété abélienne à multiplications complexes définie sur un corps de nombres K, V une sous-variété algébrique irréductible et F un endomorphisme de Frobenius en une place w de K. Si $F(V) = V$, alors V est un translaté d'une sous-variété abélienne de A par un point de torsion.*

Démonstration : comme $F(V) = V$, on a $F^{(l)}(V) = F \circ \cdots \circ F(V) = V$ pour tout $l \in \mathbb{N}^\star$, et $\deg(V) = \deg(F^{(l)}(V))$. Notons $\mathrm{St}(V)$ le stabilisateur de V, et $\mathrm{St}(V)^0$ sa composante neutre ; on a :

$$\deg(F^{(l)}(V)) = \frac{Nw^{l\dim(V)}}{\ker(F^{(l)}) \cap \mathrm{St}(V)} \deg(V)$$

(*voir* par exemple [Hi1] lemme 3, page 25 ou [Da-Hi], §2). Par ailleurs,

$$\mathrm{card}(\ker(F^{(l)}) \cap \mathrm{St}(V)) \leqslant \mathrm{card}(\mathrm{St}(V)/\mathrm{St}(V)^0) Nw^{l\dim(\mathrm{St}(V))} \ .$$

On tire de ces deux inégalités lorsque $l \longrightarrow \infty$:

$$\dim(V) = \dim(\mathrm{St}(V)) \ .$$

On en déduit donc que V est un translaté d'une sous-variété abélienne B de A, disons $V = P + B$. Comme $F(B) = B$, on tire de $F(V) = V$ que

$F(P)-P \in B$. En notant Q la projection de P dans A/B et F le morphisme de FROBENIUS induit par F sur A/B, on a $F(Q) = Q$, c'est-à-dire que Q est de torsion. □

Démonstration du théorème 1.5 – Soient K un corps de nombres, A une variété abélienne à multiplications complexes définie sur K et X une sous-variété algébrique de A définie sur une extension K' de K, $f \in K'[\boldsymbol{u}]$ une forme de Chow de X et $\mathfrak{d} : K'[\boldsymbol{u}] \to K'[\boldsymbol{s}][X_0, \ldots, X_n]$ l'opérateur défini par $\mathfrak{d}(u_j^{(i)}) = \sum_{k=0}^{n} s_{j,k}^{(i)} X_k$, alors les coefficients dans $K'[X_0, \ldots, X_n]$ de $\mathfrak{d}f$ définissent chacun une hypersurface de $\mathbb{P}^n$ de degré $\leqslant rd(X)$ et hauteur $\leqslant h(X) + 10rd(X)\log(n+1)$ où $r = \dim X + 1$ (*voir* [Ph1]–III). De plus, d'après le théorème de l'élimination (*voir* par exemple [Ph2]), l'intersection (ensembliste) de toutes ces hypersurfaces avec A est précisément X. Soit F un endomorphisme de FROBENIUS associé à une place w de K' dont la restriction à K n'est pas ramifiée ; si l'intersection de ces hypersurfaces contient $F(X)$, on en déduit que $F(X) \subset X$ et comme X est une variété de même dimension que $F(X)$ on a $F(X) = X$. Il en résulte en tenant compte du lemme 2.2 que X est translatée d'une sous-variété abélienne par un point de torsion de A (et donc qu'en particulier, $\hat{h}(X) = 0$).

Sinon, soit q l'équation d'une de ces hypersurfaces H ne contenant pas $F(X)$, on applique le théorème de BÉZOUT arithmétique (proposition 4 de [Ph1]–III) qui donne :

$$0 \leqslant \frac{h(F(X) \cdot H)}{d(F(X))} \leqslant \frac{h(F(X))}{d(F(X))} . d^\circ q + h_{F(X)}(q) \qquad (4)$$

où :

$$h_{F(X)}(q) = \sum_{v \in \mathcal{P}(K')} \frac{[K'_v : \mathbb{Q}_v]}{[K' : \mathbb{Q}]} . \int_{\sigma_v(F(X))(\mathbb{C}_v)} \log\left(\frac{|\sigma_v(q)(z)|_v}{\|z\|_v^{d^\circ q}}\right) . \Omega_{F(X),v}(z)$$

$$\leqslant h(q) + \frac{d^\circ q . \log(n+1)}{2} +$$

$$+ \frac{[K'_w : \mathbb{Q}_w]}{[K' : \mathbb{Q}]} . \int_{\sigma_w(F(X))(\mathbb{C}_w)} \log\left(\frac{|\sigma_w(q)(z)|_w}{\mathrm{M}_w(q)\|z\|_w^{d^\circ q}}\right) . \Omega_{F(X),w}(z) \ ,$$

où pour toute place v de K', σ_v est le plongement de K' dans $\mathbb{C}_v$ associé à v. En effet, $h(q) = \sum_v \frac{[K'_v : \mathbb{Q}_v]}{[K' : \mathbb{Q}]} . \log \mathrm{M}_v(q)$, pour toute place v de K' on a

$\dfrac{|\sigma_v(q)(z)|_v}{\mathrm{M}_v(q)\|z\|_v^{d^\circ q}} \leqslant 1$ si v est ultramétrique et $\leqslant (n+1)^{\frac{d^\upsilon q}{2}}$ si v est archimédienne et la mesure $\Omega_{F(X),v}(z)$ est positive.

Soit $q_w = \sigma_w(q)/\pi_w^{\log \mathrm{M}_w(q)} \in \mathcal{O}_w[X_0,\ldots,X_n]$. Le lemme 2.1 fournit une famille de formes $\boldsymbol{F}$ représentant le FROBENIUS telles que pour tout système de coordonnées projectives z d'un point de $\sigma_w(F(X))(\mathbb{C}_w)$ on ait :

$$\frac{|\sigma_w(q)(z)|}{\mathrm{M}_w(q)\|z\|_w^{d^\circ q}} = \frac{|q_w(z)|_w}{\|z\|_w^{d^\circ q}} = \frac{|q_w(\boldsymbol{F}(z'))|_w}{\|\boldsymbol{F}(z')\|_w^{d^\circ q}} = \frac{|q_w(\boldsymbol{F}(z'))|_w}{\|z'\|_w^{Nw d^\circ q}},$$

où z' est un système de coordonnées projectives d'un point de $\sigma_w(X)(\mathbb{C}_w)$ satisfaisant $z = \boldsymbol{F}(z')$. Du fait que $\sigma_w(q)$ s'annule sur $\sigma_w(X)(\mathbb{C}_w)$ et que, par définition du FROBENIUS,

$$q_w(\boldsymbol{F}) \equiv q_w^{Nw} \bmod (\Pi_w) \; ,$$

on a :

$$\frac{|\sigma_w(q)(z)|_w}{\mathrm{M}_w(q)\|z\|_w^{d^\circ q}} = \frac{|q_w(\boldsymbol{F}(z')) - q_w(z')^{Nw}|_w}{\|z'\|_w^{Nw d^\circ q}} \leqslant |\pi_w|_w$$

pour tout $z \in \sigma_w(F(X))(\mathbb{C}_w)$. Reportant dans la relation (4) on obtient

$$0 \leqslant r\frac{h(F(X))}{d(F(X))}.d(X) + 11 r d(X)\log(n+1) + h(X) + \frac{[K'_w:\mathbb{Q}_w]}{[K':\mathbb{Q}]}.\log|\pi_w|_w \; ,$$

car $\int_{\sigma_w(F(X))(\mathbb{C}_w)} \Omega_{F(X),w}(z) = 1$ (voir [Ph1]–II, § 2b).

Posons $E := \dfrac{\hat{h}(X)}{d(X)}$, d'après la proposition 9 (iv) de [Ph1]–I, on a donc $h(X) \leqslant (C_A + E)d(X)$. Mais F étant une isogénie (très particulière), il suit de la (démonstration de la) proposition 14 de [Ph1]–III que $\dfrac{\hat{h}(F(X))}{d(F(X))} = Nw\dfrac{\hat{h}(X)}{d(X)}$ (pour une preuve, on pourra également se reporter à [Da-Hi], § 2) et donc, on a aussi $h(F(X)) \leqslant (Nw.E + C_A)d(F(X))$. Finalement, l'inégalité ci-dessus et la relation $-[K'_w:\mathbb{Q}_w]\log|\pi_w|_w = \log(Nw)$ entraînent :

$$\log(Nw) \leqslant (r+1)(C_A + 11\log(n+1) + NwE)d(X)[K':\mathbb{Q}] \; . \qquad (5)$$

Soit

$$U = \max\left(D_{K/\mathbb{Q}}, \exp((r+1)(C_A + 11\log(n+1)+1)[K':\mathbb{Q}]d(X))\right)$$

il existe un premier $U \leqslant p \leqslant 2U$ de $\mathbb{Z}$, fixons une place w de K' au-dessus de p, sa restriction à K n'est pas ramifiée et on a $U \leqslant Nw \leqslant (2U)^{[K':\mathbb{Q}]}$. En reportant dans (5) on obtient, après simplifications :

$$NwE \geqslant 1 ,$$

on en déduit $E \leqslant Nw^{-1} \leqslant (2U)^{-[K':\mathbb{Q}]}$ d'où la minoration du théorème 1.5. □

3 Lien avec la conjecture de Bogomolov généralisée

Soit A une variété abélienne définie sur un corps de nombres K et plongée comme sous-variété projectivement normale dans un espace projectif $\mathbb{P}_n$ par un diviseur symétrique. Nous montrons dans ce paragraphe que la propriété du $\hat{h}$ pour une variété abélienne A est équivalente à la propriété de BOGO-MOLOV. Nos arguments s'inpirent de ceux de [Zh2], thm. 5.2. Nous commençons par un résultat sur les variétés projectives.

Théorème 3.1 *Soient $X \subset \mathbb{P}_n$ une variété algébrique définie sur un corps de nombres K, de dimension d, posons pour η réel $\geqslant 0$*

$$X(\eta) = \{x \in X(\overline{\mathbb{Q}});\ h(x) \leqslant \eta\} .$$

Soit $\varepsilon > 0$, alors :

(i) *l'ensemble $X\left(\frac{h(X)}{(d+1)d(X)} - \varepsilon\right)$ n'est pas Zariski dense dans X.*

(ii) *l'ensemble $X\left(\frac{h(X)}{d(X)} + \varepsilon\right)$ est Zariski dense dans X.*

On en déduira le corollaire suivant.

Corollaire 3.2 *Soient* $X \subset A \subset \mathbb{P}_n$ *une sous-variété algébrique d'une variété abélienne définie sur un corps de nombres* K, *et* d *la dimension de* X ; *posons pour tout* η *réel* $\geqslant 0$

$$\widehat{X}(\eta) = \{x \in X(\overline{\mathbb{Q}}); \hat{h}(x) \leqslant \eta\} .$$

Soit $\varepsilon > 0$, *alors* :

(i) *l'ensemble* $\widehat{X}\left(\frac{\hat{h}(X)}{(d+1)d(X)} - \varepsilon\right)$ *n'est pas Zariski dense dans* X.

(ii) *l'ensemble* $\widehat{X}\left(\frac{\hat{h}(X)}{d(X)} + \varepsilon\right)$ *est Zariski dense dans* X.

Remarques – On peut de façon similaire définir pour $l \geqslant 0$, $X_l(\eta) = \{V \subset X(\overline{\mathbb{Q}}), \dim(V) = l, h(V) \leqslant \eta\}$. Il est facile de voir que les énoncés ci-dessus restent vrais pour $l \geqslant 1$ (se reporter aux preuves ci-dessous). D'un autre côté, on peut se demander ce qu'il en est pour l'ensemble $\widehat{X}(\eta)$ lorsque $\frac{\hat{h}(X)}{(d+1)d(X)} \leqslant \eta \leqslant \frac{\hat{h}(X)}{d(X)}$. Dans quels cas est-il ZARISKI dense dans X ? On notera que A a la propriété du $\hat{h}$ si et seulement si pour toute sous-variété algébrique X telle que $\hat{h}(X) = 0$, $\widehat{X}(0)$ est ZARISKI dense dans X, puisque, d'après le théorème de RAYNAUD (*voir* [Ra]), si $\widehat{X}(0)$ (*i. e.* l'ensemble des points de torsion de A sur X) est ZARISKI dense dans X alors X est translatée d'une sous-variété abélienne par un point de torsion de A.

Démonstration du théorème 3.1 –(i). Soit I l'idéal premier des formes dans $\mathbb{Z}[Y_0, \ldots, Y_n]$ s'annulant sur X, pour $\delta \in \mathbb{N}^*$ on note

$$E_\delta = (\mathbb{Z}[Y_0, \ldots, Y_n]/I)_\delta \otimes_\mathbb{Z} \mathbb{R}$$

le $\mathbb{R}$-espace vectoriel engendré par les éléments homogènes de degré δ de $\mathbb{Z}[Y_0, \ldots, Y_n]/I$. De cette façon, $\dim_\mathbb{R} E_\delta$ est la valeur en δ de la fonction de HILBERT de I et on a donc $\dim_\mathbb{R} E_\delta = \frac{d(X) \cdot \delta^d}{d!} + O(\delta^{d-1})$ où $d = \dim X$. On munit l'espace E_δ de la norme suivante :

$$\|q\|_X := \sup\left\{\frac{|q(x)|}{\|x\|^\delta}, x \in \bigcup_{\sigma: K \hookrightarrow \mathbb{C}} \sigma(X)(\mathbb{C})\right\} .$$

D'après le théorème 3.2.5 de [B–G–S] (*voir* aussi [Ab–Bo]), on sait que le co-volume $\mathcal{V}$ de $(\mathbb{Z}[Y_0,\ldots,Y_n]/I)_\delta$ pour cette norme satisfait $\log \mathcal{V} = -\frac{h(X).\delta^{d+1}}{(d+1)!} + O(\delta^d \log \delta)$. Lorsque δ est assez grand, on déduit du théorème de MINKOWSKI (*voir* par exemple [Schm1], chap 4) l'existence d'une forme $q \in \mathbb{Z}[Y_0,\ldots,Y_n]_\delta \setminus I_\delta$ (i. e. ne s'annulant pas sur X) telle que $\log \|q\|_X \leqslant -\frac{h(X)}{(d+1)d(X)}.\delta + O(\log \delta)$. Pour $x \in X(\overline{\mathbb{Q}}) \setminus \mathcal{Z}(q)$ on écrit (on note $\mathcal{P}(K(x))$ l'ensemble des places du corps $K(x)$) :

$$-\delta.h(x) = \sum_{v \in \mathcal{P}(K(x))} \frac{[K(x)_v : \mathbb{Q}_v]}{[K(x) : \mathbb{Q}]} \cdot \log \frac{|q(x)|_v}{\|x\|_v^\delta}$$

$$\leqslant \sum_{v \mid \infty} \frac{[K(x)_v : \mathbb{Q}_v]}{[K(x) : \mathbb{Q}]} \cdot \log \|q\|_X$$

$$\leqslant \log \|q\|_X \ ,$$

car q est à coefficients dans $\mathbb{Z}$. On a donc $-\delta.h(x) \leqslant -\frac{h(X)}{(d+1)d(X)}.\delta + O(\log \delta)$. Soit $\varepsilon > 0$, en prenant δ assez grand de sorte que $\frac{O(\log \delta)}{\delta} < \varepsilon$ on obtient $h(x) > \frac{h(X)}{(d+1)d(X)} - \varepsilon$ lorsque $x \in X(\overline{\mathbb{Q}}) \setminus \mathcal{Z}(q)$. Ainsi, $X\left(\frac{h(X)}{(d+1)d(X)} - \varepsilon\right)$ est contenu dans la sous-variété propre $X \cap \mathcal{Z}(q)$ et n'est donc pas ZARISKI dense dans X.

Passons maintenant au point (ii). Soient $\delta \in \mathbb{N}^*$ et f une forme éliminante d'indice $(\delta,\ldots,\delta) \in \mathbb{N}^{d+1}$ de X, cette forme étant de degré $d(X)\delta^d$ par rapport à chaque variable, il existe des formes $p_1,\ldots,p_{d+1} \in \mathbb{Z}[Y_0,\ldots,Y_n]$ de degré δ telles que $p_i = \sum_{|\alpha|=\delta} p_{i,\alpha} \boldsymbol{Y}^\alpha$, $\max_{i,|\alpha|=\delta} |p_{i,\alpha}| \leqslant d(X).(\delta+1)^d$ et $f(p_1,\ldots,p_{d+1}) \neq 0$. Autrement-dit, les formes $p_1,\ldots,p_{d+1}$ n'ont pas de zéro commun sur X. Raisonnons par l'absurde, soit $\varepsilon > 0$ et Z un diviseur de X contenant $X\left(\frac{h(X)}{d(X)} + \varepsilon\right)$. Soient $q_1,\ldots,q_d$ des combinaisons linéaires des p_i ci-dessus telles que $\dim(X \cap \mathcal{Z}(q_1) \cap \ldots \cap \mathcal{Z}(q_d)) = 0$ et $Z \cap \mathcal{Z}(q_1) \cap \ldots \cap \mathcal{Z}(q_d) = \emptyset$. Pour construire ces q_j on peut choisir des entiers $\lambda_{1,1},\ldots,\lambda_{d,d+1}$ n'annulant pas une spécialisation convenable d'une forme éliminante d'indice $(\delta,\ldots,\delta)$ du cycle $[X] + [Z]$ et on pose $q_j = \sum_{i=1}^{d+1} \lambda_{j,i} p_i$. Cette forme éliminante étant de degré $\leqslant (d(X) + d(Z))\delta^d$ par rapport à chaque variable il existe de tels $\lambda_{j,i}$ majorés en valeurs absolues par $(d(X)+d(Z))(\delta+1)^d$ et on peut donc assurer $|q_j| := \left(\sum_{|\alpha|=\delta} |q_{j,\alpha}|^2 / \binom{\delta}{\alpha}\right)^{\frac{1}{2}} \leqslant (d+1)(d(X)+d(Z))^2.(\delta+1)^{n+2d}$. Considérons maintenant le cycle intersection $Z' = X \cdot \mathcal{Z}(q_1) \cdots \mathcal{Z}(q_d)$ de dimension 0 supporté par $X \cap \mathcal{Z}(q_1) \cap \ldots \cap$

$\mathcal{Z}(q_d)$, d'après le théorème de BÉZOUT arithmétique (*voir* [Ph1]–III, prop. 4), on a :

$$\frac{h(Z')}{d(Z')} \leqslant \frac{h(X)}{d(X)} + \sum_{j=1}^{d} \frac{\log|q_j|}{\delta} \leqslant \frac{h(X)}{d(X)} + O\left(\frac{\log \delta}{\delta}\right) .$$

En effet, pour $j = 1, \ldots, d$ et pour toute composante V de :

$$X \cdot \mathcal{Z}(q_1) \cdots \mathcal{Z}(q_{j-1}),$$

on a $h_V(q_j) \leqslant \log|q_j|$ étant donné que q_j est à coefficients dans $\mathbb{Z}$ et sachant que $\frac{|q_j(x)|_v}{\|x\|_v^\delta} \leqslant |q_j|$ (en tenant compte de l'inégalité de CAUCHY–SCHWARZ) pour toutes les places v divisant l'infini.

Choisissons maintenant δ assez grand de sorte que $O(\frac{\log \delta}{\delta}) \leqslant \varepsilon$, il existe $y \in \mathrm{Supp}(Z')$ tel que $h(y) \leqslant \frac{h(Z')}{d(Z')} \leqslant \frac{h(X)}{d(X)} + \varepsilon$. Ainsi $y \in X(\frac{h(X)}{d(X)} + \varepsilon)$, mais comme $\mathrm{Supp}(Z') \subset \mathcal{Z}(q_1) \cap \ldots \cap \mathcal{Z}(q_d)$ est d'intersection vide avec Z (par construction des q_j), on obtient une contradiction qui montre que $X\left(\frac{h(X)}{d(X)} + \varepsilon\right)$ est bien ZARISKI dense dans X. □

Démonstration du corollaire 3.2 – Considérons le plongement projectif φ_L de la variété abélienne A :

$$\varphi_L : A \hookrightarrow A^L \hookrightarrow \mathbb{P}_n^L \underset{\text{Segre}}{\hookrightarrow} \mathbb{P}_N$$
$$x \longmapsto (x, [p]x, \ldots, [p^{L-1}]x)$$

avec p un nombre premier, L un entier assez grand et $N + 1 = (n+1)^L$. On a, d'après [Ph1]–III, prop. 7 et 9,

$$\frac{\hat{h}(\varphi_L(X))}{d(\varphi_L(X))} = \frac{p^{2L}-1}{p^2-1} \cdot \frac{\hat{h}(X)}{d(X)} , \qquad \left|\frac{h(\varphi_L(X))}{d(\varphi_L(X))} - \frac{\hat{h}(\varphi_L(X))}{d(\varphi_L(X))}\right| \leqslant C_A L ,$$

et d'après les propriétés de la hauteur de NÉRON–TATE, $\widehat{\varphi_L(X)}\left(\eta\frac{p^{2L}-1}{p^2-1}\right) = \varphi_L\left(\widehat{X}(\eta)\right)$ et :

$$\varphi_L(X)(\eta - C_A L) \subset \widehat{\varphi_L(X)}(\eta) \subset \varphi_L(X)(\eta + C_A L)$$

pour tout réel $\eta > 0$. On en déduit :

$$\varphi_L\left(\widehat{X}\left(\frac{\hat{h}(X)}{(d+1)d(X)} - \varepsilon\right)\right) \subset \varphi_L(X)\left(\frac{h(\varphi_L(X))}{(d+1)d(\varphi_L(X))} + 2C_A L - \varepsilon \cdot \frac{p^{2L}-1}{p^2-1}\right)$$

et :

$$\varphi_L\left(\widehat{X}\left(\frac{\hat{h}(X)}{d(X)} + \varepsilon\right)\right) \supset \varphi_L(X)\left(\frac{h(\varphi_L(X))}{d(\varphi_L(X))} - 2C_A L + \varepsilon \cdot \frac{p^{2L}-1}{p^2-1}\right).$$

On prend L suffisamment grand de sorte que $\varepsilon \cdot \frac{p^{2L}-1}{p^2-1} > 2C_A L$ et le corollaire est alors conséquence du théorème 3.1 appliqué à $\varphi_L(X) \subset \mathbb{P}_N$, car $\widehat{X}(\eta)$ est ZARISKI dense dans X si et seulement si $\varphi_L(\widehat{X}(\eta))$ l'est dans $\varphi_L(X)$. □

Pour conclure, passons à :

Démonstration du théorème 1.4 – D'après le corollaire 3.2 (*ii*), pour tout $\varepsilon > 0$ et tout diviseur Z de X, on a : $\widehat{X}\left(\frac{\hat{h}(X)}{d(X)} + \varepsilon\right) \cap (X \setminus Z)(\overline{\mathbb{Q}}) \neq \emptyset$, et donc $\mu^{\text{ess}}(X) \leqslant \frac{\hat{h}(X)}{d(X)}$. Réciproquement, le corollaire 3.2 (*i*) montre que pour tout $\varepsilon > 0$, il existe un diviseur Z de X contenant $\widehat{X}\left(\frac{\hat{h}(X)}{(d+1)d(X)} - \varepsilon\right)$, et donc $\mu^{\text{ess}}(X) \geqslant \frac{\hat{h}(X)}{(d+1)d(X)}$. Enfin, la propriété de BOGOMOLOV est clairement équivalente à l'inégalité $\mu^{\text{ess}}(X) > 0$. □

4 Le cas général

Afin d'établir le théorème 1.6, nous rassemblons dans les deux sous-paragraphes qui suivent, les préliminaires nécessaires.

4.1 Hauteurs et fibres spéciales

Considérons l'isomorphisme :

$$s : \begin{array}{ccc} A^2 & \longrightarrow & A^2 \\ (x_1, x_2) & \longmapsto & (x_1, x_1 - x_2) \end{array}.$$

On munit A^2 de la polarisation carrée de celle de A, ce qui revient à plonger A^2 dans $\mathbb{P}_{(n+1)^2-1}$ par le plongement de SEGRE $\mathbb{P}_n^2 \hookrightarrow \mathbb{P}_{(n+1)^2-1}$. Comme s est un isomorphisme de A^2, on a (*voir* par exemple [Ph1]–III, proposition 1) :

$$d(s(X^2)) = d(X^2) = \frac{(2\dim X)!}{(\dim X)!^2}.d(X)^2 ,$$

et comme le plongment de A dans $\mathbb{P}_n$ est projectivement normal cet isomorphisme s (ainsi que son inverse) est représenté par des formules biquadratiques, d'où :

$$4^{-\dim X}.\hat{h}(X)d(X) \leqslant \hat{h}(s(X^2)) \leqslant 2.4^{\dim X + 1}.\hat{h}(X)d(X) .$$

Soit π la projection linéaire de $\mathbb{P}_n^2$ sur le second facteur $\mathbb{P}_n$, on a alors $\pi \circ s(X^2) = X - X$ et :

$$d(X - X) = d(\pi \circ s(X^2)) \leqslant d(s(X^2)) \leqslant 4^{\dim X}.d(X)^2 ,$$

$$\hat{h}(X - X) = \hat{h}(\pi \circ s(X^2)) \leqslant \hat{h}(s(X^2)) \leqslant 2.6^{2\dim X + 1}.\hat{h}(X)d(X) .$$

Lorsque $k_X = 1$ on a $B_X = X - X$ et π induit un morphisme surjectif $\pi : s(X^2) \longrightarrow B_X$ dont la fibre générique est de dimension $2\dim X - \dim B_X$ et la fibre spéciale $s(X^2) \cap \pi^{-1}(0)$ est de dimension $\dim X > 2\dim X - \dim B_X$. On notera que cette fibre spéciale s'identifie à la projection X de $s(X^2)$ sur le premier facteur et est donc de degré $d(X)$.

Plus généralement (*voir* la proposition 4.1), pour établir une minoration non triviale de la hauteur, il suffit de disposer des données suivantes :

- une variété abélienne A définie et plongée dans $\mathbb{P}_n$ sur un corps de nombres K ;

- une sous-variété abélienne B de A, de dimension $d' > 0$. La variété produit $A \times B$ est alors naturellement plongée dans $\mathbb{P}_n^2$;

- une sous-variété algébrique W de $A \times B$ de dimension d, définie sur une extension K' de K, telle que (c'est le point crucial) la projection

$$\pi \ : \ A \times B \longrightarrow B \ ,$$

induise un morphisme surjectif $\pi \ : \ W \longrightarrow B$ de fibre générique de dimension $d - d'$, et de fibre spéciale $W \cap \pi^{-1}(0)$ de dimension $d_0 > d - d'$, coïncidant avec la projection de W sur A.

4.2 Hauteurs et intersections

On suppose fixé dans tout ce sous-paragraphe un quadruplet (A, B, W, π) comme ci-dessus, et l'on reprend les plongements φ_L utilisés aux paragraphes précédents :

$$\varphi_L \ : \ A \times B \ \hookrightarrow \ A^L \times B^L \ \hookrightarrow \ \mathbb{P}_n^L \times \mathbb{P}_n^L \underset{\text{Segre}}{\hookrightarrow} \mathbb{P}_N \times \mathbb{P}_N$$

$$x \ \longmapsto \ (x, [p]x, \ldots, [p^{L-1}]x)$$

où $N + 1 = (n+1)^L$, p est un nombre premier et L est un entier $\geqslant 1$. On notera dans la suite $c_1, c_2, \ldots, c'_1, c'_2, \ldots$ des réels > 0 ne dépendant pas de B, ni de L.

Dans la situation ci-dessus, on dispose de la :

Proposition 4.1 *Il existe des réels $c_1, c_2 > 0$ ne dépendant que de A plongée dans $\mathbb{P}_n$ tels que pour tout $0 < \varepsilon < c_1$ et $L \geqslant c_1^{-1}$, on ait :*

$$\frac{h(\varphi_L(W))}{d(\varphi_L(W))} \ \geqslant \ p^{(2-(d'+1)\varepsilon)L} \left(c_1 \varepsilon^{2(d-d_0)} p^{(d_0+d'-d)\varepsilon L} \cdot \frac{d(W \cap \pi^{-1}(0))}{d(W)} \right.$$
$$\left. - c_2 (L + \log d(B)) \right) \ .$$

Avant de démontrer la proposition 4.1 nous établissons deux lemmes. Nous notons $W_L = \varphi_L(W) \subset \mathbb{P}_N^2$, $B_L = \varphi_L(B) \subset \mathbb{P}_N$, $d = \dim W$, $d' = \dim B$. On fixe un plongement de K dans $\mathbb{C}$ et à toute place à l'infini v de K'/K on associe un plongement σ_v de K' dans $\mathbb{C}$ au-dessus de celui de K. Nous notons $\Theta = (\theta_0, \ldots, \theta_n)$ les séries thêta associées au plongement de $A(\mathbb{C})$ dans $\mathbb{P}_n(\mathbb{C})$. Quitte à faire une transformation linéaire projective définie sur K on peut supposer, sans perte de généralité, que l'origine 0 de A a pour coordonnées projectives $(1 : 0 : \ldots : 0)$, que l'espace tangent T_A à A en 0 a pour équations $\{X_{g+1} = \ldots = X_n = 0\}$ et que $f_i(z) := \frac{\theta_i(z)}{\theta_0(z)} = z_i + O(\|z\|^2)$ pour $i = 1, \ldots, g$ et $= O(\|z\|^2)$ pour $i = g+1, \ldots n$.

Le premier lemme auxiliaire est typique des méthodes d'approximation diophantienne ; il s'agit de construire une hypersurface auxiliaire ayant un ordre de contact adéquat en l'origine avec B_L. Dans le contexte des hauteurs normalisées, un lemme de ce type a déjà été utilisé par le deuxième auteur dans [Ph1]-III.

Lemme 4.2 *Soit $0 < \varepsilon < 2/(d'+1)$ et $\delta \in \mathbb{N}$, $\delta \geqslant d(B)p^{2Ld'}$; avec les notations ci-dessus, il existe une forme $q \in K'[Y_0, \ldots, Y_N]$ de degré δ, de hauteur :*

$$h(q) \leqslant c_3(L + \log \delta + \log d(B)).\delta p^{(2-(d'+1)\varepsilon)L} ,$$

qui a un contact d'ordre $> T := c_3^{-1}\delta p^{(2-\varepsilon)L}$ en l'origine avec B_L, mais qui ne s'annule pas identiquement sur cette dernière.

Démonstration : comme $\delta \geqslant d(B)p^{2Ld'} \geqslant d(B_L)$ l'espace des formes de degré δ en $N+1$ variables ne s'annulant pas identiquement sur $\varphi_L(B)$ est de dimension :

$$\geqslant c_4 d(B)(\delta.p^{2L})^{d'} .$$

La condition d'avoir avec B_L un contact d'ordre $> T$ en 0 s'écrit via un système linéaire de $\binom{T+d'}{d'}$ équations en les coefficients de q. Les coefficients de ce système linéaire sont dans K', de hauteurs $\leqslant c_5(T+\delta)\log(T\delta d(B))$. En effet, le système s'écrit

$$\partial(q \circ \varphi_L \circ \Theta)(0) = \sum_\lambda a_\lambda.\partial\left(\prod_{\ell=0}^{L-1} \Theta(p^\ell z)^{\lambda_\ell}\right)\bigg|_{z=0} = 0 ,$$

où ∂ décrit l'espace des opérateurs différentiels le long de B d'ordres $\leqslant T$ et $|\lambda_\ell| = \delta$, $\ell = 0, \ldots, L-1$. Le lemme de «l'espace tangent» (voir [Ma–Wü], page 442) permet d'assurer que l'on peut choisir une base de l'espace tangent $T_{B(\mathbb{C})}$ de B à l'origine formée d'opérateurs différentiels de hauteur au plus $c_6 \log(d(B))$. On vérifie, par induction qu'un système complet d'opérateurs différentiels d'ordre au plus T peuvent ainsi être choisis de hauteurs $\leqslant c_7 T \log(T d(B))$. Par ailleurs, la valeur de T permet d'assurer que les termes parasites comme par exemple $T\log(\delta \log p^L)$ sont négligeables devant $c_7 T \log(T d(B))$. Des calculs classiques permettent d'arriver à la majoration annoncée ci-dessus.

Enfin, l'exposant de DIRICHLET du système linéaire (défini comme le quotient entre le nombre de contraintes par celui des degrés de liberté) est majoré par $c_8 \frac{(T+1)^{d'}}{(\delta p^{2L})^{d'}} \leqslant c_9 p^{-\varepsilon d' L}$, le lemme de SIEGEL (voir par exemple [Schm1], lemme 5B, page 127 pour une version convenant à ce cas de figure) permet de conclure. $\square$

L'énoncé suivant permet de tirer parti de la dimension «exceptionnelle» de la fibre spéciale. On obtient ainsi qu'un voisinage tubulaire autour de cette dernière a un volume plus important que ce qui se passerait génériquement (on notera que le rayon est affecté d'un exposant $d - d_0$ en lieu et place de $d - d'$). Plus précisément :

Lemme 4.3 *Fixons une place v de K'/K archimédienne, notons $W_L = \sigma_v(W_L)$, $\Omega_{\mathbb{P}^2_N}$ la forme de Fubini-Study sur $\mathbb{P}^2_N$ et $B(r)$ l'image dans B_L de la boule de rayon r, centrée à l'origine, dans l'espace tangent à l'origine de B_L. Alors, il existe un réel $c_{10} > 0$ tel que si $r < c_{10}$ on a*

$$\mathcal{V} := \int_{W_L \cap \pi^{-1}(B(r))} \Omega_{\mathbb{P}^2_N}^{\wedge d} \geqslant c_{10} d(W \cap \pi^{-1}(0)) p^{2Ld} r^{2(d-d_0)} .$$

Démonstration : on a $\Omega_{\mathbb{P}^2_N}(x,y) = \Omega_{\mathbb{P}_N}(x) + \Omega_{\mathbb{P}_N}(y)$ et, comme les formes

$\Omega_{\mathbb{P}_N}^{\wedge \alpha}(x) \wedge \Omega_{\mathbb{P}_N}^{\wedge \beta}(y)$, $\alpha + \beta = d$, sont positives sur W_L, on en déduit :

$$\begin{aligned}
\mathcal{V} &\geq \int_{W_L \cap \pi^{-1}(B(r))} \Omega_{\mathbb{P}_N}^{\wedge d_0}(x) \wedge \Omega_{\mathbb{P}_N}^{\wedge (d-d_0)}(y) \\
&\geq \int_{W_L \cap \pi^{-1}(0)} \left(\int_{\pi(W_L \cap (\{x\} \times \mathbb{P}_N)) \cap B(r)} \Omega_{\mathbb{P}_N}^{\wedge (d-d_0)}(y) \right) \Omega_{\mathbb{P}_N}^{\wedge d_0}(x) ,
\end{aligned}$$

car rappelons que $W_L \cap \pi^{-1}(0)$ coïncide avec la projection de W_L sur A_L. Mais $\pi(W_L \cap (\{x\} \times \mathbb{P}_N))$ est une sous-variété algébrique de B_L de dimension $d - d_0$ passant par 0, nous allons vérifier :

$$\int_{\pi(W_L \cap (\{x\} \times \mathbb{P}_N)) \cap B(r)} \Omega_{\mathbb{P}_N}^{\wedge (d-d_0)}(y) \geq c_{11}(p^L r)^{2(d-d_0)} , \qquad (6)$$

tandis que d'après la formule de WIRTINGER (*voir* [Gr-Ha], page 31) :

$$\int_{W_L \cap \pi^{-1}(0)} \Omega_{\mathbb{P}_N}^{\wedge d_0}(x) = d(W_L \cap \pi^{-1}(0)) = \left(\frac{p^{2L} - 1}{p^2 - 1} \right)^{d_0} . d(W \cap \pi^{-1}(0)) .$$

Mises ensemble, ces inégalités démontrent le lemme.

Plus généralement, soit V une sous-variété algébrique de dimension d_1 de B passant par l'origine et $V_L = \varphi_L(V)$. On remarque que pour $x \in B$ on a $\Omega_{\mathbb{P}_N}(\varphi_L(x)) = \sum_{\ell=0}^{L-1} \Omega_{\mathbb{P}_n}([p^\ell]x)$ et, pour les mêmes raisons de positivité que ci-dessus on a pour tout ρ assez petit :

$$\begin{aligned}
\int_{V_L \cap B(\rho)} \Omega_{\mathbb{P}_N}^{\wedge d_1}(x) &\geq \int_{V \cap B_0(\rho)} ([p^{L-1}]^* \Omega_{\mathbb{P}_n}^{\wedge d_1})(y) \\
&\geq \int_{\exp_A^{-1}(V \cap B_0(\rho))} ([p^{L-1}] \circ \exp_A)^* (\Omega_{\mathbb{P}_n}^{\wedge d_1})(z) ,
\end{aligned} \qquad (7)$$

où $B_0(\rho)$ désigne l'image dans B de la boule de rayon ρ de l'espace tangent à B en 0. Rappelons qu'on a supposé le plongement projectif de A et la base de $T_{A(\mathbb{C})}$ choisis de sorte que $f_i(z) = \frac{\theta_i(z)}{\theta_0(z)} = z_i + O(\|z\|^2)$ pour $i = 1, \ldots, g$ et $= O(\|z\|^2)$ pour $i = g+1, \ldots, n$. Ecrivons :

$$\exp_A^*(\Omega_{\mathbb{P}_n})(z) = dd^c \log(1 + \|f(z)\|^2) = \frac{1}{-2i\pi} \sum_{1 \leq j,k \leq g} g_{j,k}(z) . dz_j \wedge \overline{dz_k} ,$$

où :

$$g_{j,k}(z) = \frac{\sum_{i=1}^n \frac{\partial f_i}{\partial z_j}(z) . \overline{\frac{\partial f_i}{\partial z_k}(z)}}{1 + \|f(z)\|^2} - \frac{\left(\sum_{i=1}^n f_i(z) . \overline{\frac{\partial f_i}{\partial z_k}(z)} \right) . \left(\sum_{i=1}^n \overline{f_i(z)} . \frac{\partial f_i}{\partial z_j}(z) \right)}{(1 + \|f(z)\|^2)^2} .$$

Soit Λ le réseau des périodes de $\exp_A$ et notons $\|\cdot\|_\Lambda$ la distance au plus proche élément de Λ, alors $g_{j,k}(z) = \delta_{j,k} + O(\|z\|_\Lambda)$ où $\delta_{j,k}$ est le symbole de KRONECKER. Mais $([p^{L-1}] \circ \exp_A)^*(\Omega_{\mathbb{P}_n})(z) = \exp_A^*(\Omega_{\mathbb{P}_n})(p^{L-1}z)$ et $\|p^{L-1}z\|_\Lambda \leqslant p^{L-1}\|z\|_\Lambda$; on a donc, en posant $\mu = \sum_{j=1}^{g} \frac{dz_j \wedge \overline{dz_j}}{-2i\pi}$,

$$|([p^{L-1}] \circ \exp_A)^*(\Omega_{\mathbb{P}_n})(z) - p^{2(L-1)}.\mu| \leqslant c_{12}.p^{3(L-1)}\|z\|_\Lambda.\mu \ .$$

Reportant dans l'inégalité (7) on obtient, pour ρ assez petit devant p^{-L},

$$\int_{V_L \cap B(\rho)} \Omega_{\mathbb{P}_N}^{\wedge d_1}(x) \geqslant p^{2(L-1)d_1}(1 - c_{12}.\rho p^L).\int_{\exp_A^{-1}(V \cap B_0(\rho))} \mu^{\wedge d_1}$$

$$\geqslant c_{13}.p^{2Ld_1}.\rho^{2d_1}$$

en effet, $\exp_A^{-1}(V)$ étant une variété analytique passant par le centre de $\exp_A^{-1}(B_0(\rho))$, on a $\int_{\exp_A^{-1}(V \cap B_0(\rho))} \mu^{\wedge d_1} \geqslant \rho^{2d_1}$ (voir [Gr-Ha], pages 390-391 ou [Ph1]-I, fait 1).

Soit $r > 0$, prenant ρ arbitrairement petit on recouvre $V \cap B_0(r)$ par des boules de rayons ρ centrées en des points de V de sorte que les boules de mêmes centres et rayons $\rho/2$ soient disjointes. Le nombre I de telles boules est minoré par $I \geqslant c_{14}(r/\rho)^{2d_1}$. Appliquant l'inégalité précédente aux variétés $V'_1, \ldots, V'_I$ déduites de V par des transformations unitaires envoyant chaque centre des boules fixées sur 0 on obtient la relation (6) :

$$\int_{V_L \cap B(r)} \Omega_{\mathbb{P}_N}^{\wedge d_1}(x) \geqslant \sum_{i=1}^{I} \int_{(V'_i)_L \cap B(\rho/2)} \Omega_{\mathbb{P}_N}^{\wedge d_1}(x)$$

$$\geqslant c_{15}.\left(\frac{r}{\rho}\right)^{2d_1}.p^{2Ld_1}.\rho^{2d_1} = c_{15}(p^L r)^{2d_1} \ . \ \square$$

Démonstration de la proposition 4.1 – Soit q la forme obtenue au lemme 4.2 pour $\delta = d(B)p^{2Ld'}$, pour chaque place à l'infini v de K'/K on applique le lemme de SCHWARZ à la fonction $F := \sigma_v(q) \circ \varphi_L \circ \Theta(z)$ restreinte à $\sigma_v(T_B)(\mathbb{C})$, qui s'annule à un ordre $> T = c_3^{-1}\delta p^{(2-\varepsilon)L}$ en 0, avec des rayons $0 < \eta < e\eta < 1$, on obtient :

$$\log \|F\|_\eta \leqslant -T + \log \|F\|_{e\eta}$$

où $\|F\|_r = \sup(|F(z)|; z \in \sigma_v(T_B)(\mathbb{C}), \|z\| < r)$. Mais, pour tout $z \in \sigma_v(T_B)(\mathbb{C})$ on a (voir [Da1], théorème 3.1) :

$$\exp(-c_{16}) \leqslant \|\Theta(z)\|_v \leqslant \exp(c_{17}(1 + \|z\|^2)) ,$$

on en déduit

$$\|F\|_{e\eta} \leqslant (N+1)^\delta . M_v(q) . \exp\left(c_{17}\delta \sum_{i=0}^{L-1}(1 + (ep^\ell\eta)^2)\right)$$

$$\leqslant M_v(q) . \exp\left(c_{18}\delta(L + (p^L\eta)^2)\right) ,$$

et donc finalement :

$$\log\left(\frac{|\sigma_v(q) \circ \varphi_L \circ \Theta(z)|_v}{M_v(q).\|\varphi_L \circ \Theta(z)\|_v^\delta}\right) \leqslant \log\left(\frac{\|F\|_\eta}{M_v(q).\prod_{\ell=0}^{L-1}\|\Theta(p^\ell z)\|_v^\delta}\right)$$

$$\leqslant -T + c_{19}\delta(p^{2L}\eta^2 + L) ,$$

pour tout $z \in \sigma_v(T_B)(\mathbb{C}), \|z\| < \eta$. De plus, pour toute place finie v de K' et pout tout $x \in \sigma_v(B_L)(\mathbb{C}_v)$, on a $\log\left(\frac{|\sigma_v(q)(x)|_v}{M_v(q).\|x\|_v^\delta}\right) \leqslant 0$ si v est ultramétrique et $\leqslant \frac{\delta}{2}\log(N+1)$ sinon.

D'après le lemme 4.3, le volume de l'intersection de W_L et de l'image par $\pi^{-1} \circ \varphi_L \circ \Theta$ de la boule $B(\eta)$ de $\sigma_v(T_B)(\mathbb{C})$ pour la métrique de FUBINI-STUDY de $\mathbb{P}_N^2$ est minoré par $c_{20}.d(W \cap \pi^{-1}(0)).p^{2Ld}.\eta^{2(d-d_0)}$. On choisit $\eta = \varepsilon p^{-L\varepsilon/2}$, $\varepsilon^{-2} > 3c_{19}c_3$ et $p^{(2-\varepsilon)L} > 3Lc_{19}c_3$ de sorte que pour toute place à l'infini v de K'/K on ait :

$$\log\left(\frac{|\sigma_v(q)(x)|_v}{M_v(q).\|x\|_v^\delta}\right) \leqslant -c_{21}\delta p^{(2-\varepsilon)L}$$

pour $x \in \sigma_v(B_L) \cap B(\eta)$. On déduit en sommant sur v :

$$h_{W_L}(\pi^*(q)) = \frac{1}{d(W_L)} \cdot \sum_v \frac{[K'_v : \mathbb{Q}_v]}{[K' : \mathbb{Q}]} \cdot \int_{\sigma_v(W_L)} \log\left(\frac{|\sigma_v(q)(x)|_v}{\|x\|_v^\delta}\right) . \Omega_{W_L,v}(x)$$

$$= h(q) + \frac{1}{d(W_L)} \cdot \sum_v \frac{[K'_v : \mathbb{Q}_v]}{[K' : \mathbb{Q}]} \cdot \int_{\sigma_v(W_L)} \log\left(\frac{|\sigma_v(q)(x)|_v}{M_v(q). \|x\|_v^\delta}\right) . \Omega_{W_L,v}(x)$$

$$\leqslant h(q) + \frac{L\delta \log(n+1)}{2} - c_{22} \frac{p^{(2-\varepsilon)L}}{d(W_L)} \cdot \sum_{v|\infty} \frac{[K'_v : \mathbb{Q}_v]}{[K' : \mathbb{Q}]} \cdot \int_{\sigma_v(W_L) \cap \pi^{-1}(B(\eta))} \Omega_{\mathbb{P}_N^2}^d$$

$$\leqslant h(q) + \frac{L\delta \log(n+1)}{2} - c_{23} \varepsilon^{2(d-d_0)} \delta p^{(2-(d-d_0+1)\varepsilon)L} \cdot \frac{d(W \cap \pi^{-1}(0))}{d(W)}.$$

On a donc, d'après le théorème de BÉZOUT arithmétique et en tenant compte de $h(W_L \cap \mathcal{Z}(\pi^*(q))) \geqslant 0$ (car $\pi^*(q)$ n'est pas identiquement nulle sur W_L), de $d^\circ \pi^*(q) = d^\circ q = \delta$ et de la majoration de $h(q)$ dans le lemme 4.2,

$$\frac{h(W_L)}{d(W_L)} \geqslant -\frac{h_{W_L}(\pi^*(q))}{d^\circ \pi^*(q)}$$

$$\geqslant c_1 \varepsilon^{2(d-d_0)} p^{(2-(d-d_0+1)\varepsilon)L} \cdot \frac{d(W \cap \pi^{-1}(0))}{d(W)}$$

$$- c_2 (L + \log d(B)) . p^{(2-(d'+1)\varepsilon)L},$$

ce qui achève de montrer la proposition 4.1. $\square$

4.3 Démonstration du théorème 4.1

Commençons par établir un résultat plus précis que le théorème 1.6 lorsque $k_X = 1$.

Théorème 4.4 *On reprend les notations et hypothèses du théorème 1.6, si $k_X = 1$ (i.e. si $X - X$ est une sous-variété abélienne B_X de A de dimension $> \dim X$), on a :*

$$\frac{\hat{h}(X)}{d(X)} \geqslant c(A, K) . d(X)^{-\frac{\dim B_X + 1}{\dim B_X - \dim X}} . (\log d(X))^{-\frac{\dim X + 1}{\dim B_X - \dim X}}.$$

Démonstration : on applique la proposition 4.1 avec $\varepsilon = c_1/2$, $W = s(X^2)$, $A = A$, $B = B_X$, on a $d = 2\dim X$, $d_0 = \dim X$, et $d' = \dim B_X$, ainsi que $d(W), d(B_X) \leqslant 4^{\dim X} . d(X)^2$, et $d(W \cap \pi^{-1}(0)) = d(X)$. On en déduit, avec les propositions 7 et 9 de [Ph1]-III et les inégalités du sous-paragraphe 4.1,

$$4^{2(\dim X+1)} . \frac{\hat{h}(X)}{d(X)} \geqslant \frac{\hat{h}(W)}{d(W)} = \frac{p^2-1}{p^{2L}-1} . \frac{\hat{h}(W_L)}{d(W_L)} \geqslant p^{-2L} . \left(\frac{h(W_L)}{d(W_L)} - C_A L \right)$$

$$\geqslant p^{-(d'+1)\varepsilon L} . \left(2^{1-2\dim X} \varepsilon^{2(d-d_0)+1} p^{(d_0+d'-d)\varepsilon L} . d(X)^{-1} \right.$$

$$\left. - (c_2 + C_A)(L + \log d(B_X)) \right) .$$

En choisissant L minimal de sorte que :

$$p^{(d_0+d'-d)\varepsilon L} > 4^{\dim X} \frac{c_2 + C_A}{\varepsilon^{2(d-d_0)+1}} . (L + \log d(B_X)) . d(X) ,$$

ce qui est possible car $d_0 > d - d'$, on a $L \geqslant c_{24} . \log d(X)$ et on obtient la minoration :

$$\frac{\hat{h}(X)}{d(X)} \geqslant c_{25} . L . p^{-(d'+1)\varepsilon L}$$

$$\geqslant c_{26} . d(X)^{-\frac{d'+1}{d_0+d'-d}} . (\log d(X))^{\frac{d_0-d-1}{d_0+d'-d}}$$

$$\geqslant c(A,K) . d(X)^{-\frac{\dim B_X + 1}{\dim B_X - \dim X}} . (\log d(X))^{-\frac{\dim X + 1}{\dim B_X - \dim X}} ,$$

qui établit le théorème 4.4. □

Démonstration du théorème 1.6 – On procède par réccurence sur k_X, le cas $k_X = 1$ se déduisant du théorème 4.4 car $\dim X < \dim B_X$. Si $k_X > 1$ la variété $Y = X - X$ n'est pas une sous-variété abélienne de A (et comme $Y = -Y$ ce n'est pas non plus une translatée de sous-variété abélienne). On a $k_Y \leqslant [(k_X+1)/2]$, $d(Y) \leqslant 4^{\dim X} . d(X)^2$, $\hat{h}(Y) \leqslant 2.4^{2\dim X+1} . \hat{h}(X) d(X)$ et enfin $B_Y = B_X$. On applique l'hypothèse de récurrence à Y ce qui donne :

$$\hat{h}(X) \geqslant 4^{-2(\dim X+1)} . \frac{\hat{h}(Y)}{d(X)}$$

$$\geqslant 4^{-2(\dim X+1)} . c(A,K)^{k_Y} . \frac{d(Y)^{-\max(1, 2(k_Y-1)) . (\dim B_Y + 1)+1}}{d(X) . (\log d(Y))^{\dim B_Y}}$$

$$\geqslant c(A,K)^{k_X} . d(X)^{-2(k_X-1) . (\dim B_X + 1)+1} . (\log d(X))^{-\dim B_X} ,$$

car $\max(1, 2(k_Y - 1)) \leqslant k_X - 1$. Ceci achève d'établir le théorème 1.6. □

Bibliographie

[Ab–Bo]   A. ABBÈS et T. BOUCHE. *Théorème de Hilbert–Samuel « arithmétique ».* C. R. Acad. Sci. Paris Sér. I, t. **317**, pages 589–591, 1993 ; Ann. Inst. Fourier (Grenoble), t. **45** (2), pages 375-401, 1995.

[B–G–S]   J.-B. BOST, H. GILLET, et C. SOULÉ. *Heights of projective varieties and positive Green forms.* J. Amer. Math. Soc., t. **7** (4), pages 903–1022, 1994.

[Bo–Za1] E. BOMBIERI et U. ZANNIER. *Algebraic points on subvarieties of $\mathbb{G}_m^n$.* Internat. Math. Res. Notices, t. **7**, pages 333–347, 1995.

[Bo–Za2] E. BOMBIERI et U. ZANNIER. *Heights of algebraic points on subvarieties of abelian varieties.* typographié, 1996.

[Bu]   J.-F. BURNOL. *Weierstrass points on arithmetic surfaces.* Invent. Math., t. **107**, pages 421–432, 1993.

[Da1]   S. DAVID. *Fonctions thêta et points de torsion des variétés abéliennes.* Compositio Math., t. **78**, pages 121–160, 1991.

[Da2]   S. DAVID. *Minorations de hauteurs sur les variétés abéliennes.* Bull. Soc. Math. France, t. **121** (4), pages 509–544, 1993.

[Da-Hi]   S. DAVID et M. HINDRY. *Sur le problème de Lehmer pour les variétés abéliennes à multiplications complexes.* manuscrit, 1997.

[Gr-Ha]   P. GRIFFITHS et J. HARRIS. *Principles of algebraic geometry.* Wiley-Interscience, 1978.

[Hi1]   M. HINDRY. *Géométrie et hauteurs dans les groupes algébriques.* Thèse de Doctorat, Université de Paris VI, mai 1987.

[Hi2]   M. HINDRY. *Autour d'une conjecture de S. Lang.* Invent. Math., t. **94**, pages 575–603, 1988.

[Ma]   D.W. MASSER *Counting points of small height on elliptic curves.* Bull. Soc. Math. France, t. **117**, pages 247–265, 1989.

[Ma–Wü]   D. MASSER et G. WÜSTHOLZ. *Periods and minimal abelian subvarieties.* Ann. of Math., t. **137**, pages 407–458, 1993.

[Ph1]   P. PHILIPPON. *Sur des hauteurs alternatives I; II; III.* Math. Ann., t. **289**, pages 255–283, 1991; Ann. Inst. Fourier (Grenoble), t. **44** (4), pages 1043–1065, 1994; J. Math. Pures Appl., t. **74** (4), pages 345–365, 1995.

[Ph2]   P. PHILIPPON. *Quatre exposés sur la théorie de l'élimination. Publications du Laboratoire de Mathématiques Discrètes*, CIRM, Luminy, 1994.

[Ra] M. Raynaud. Sous-variétés d'une variété abélienne et points de torsion. *Arithmetic and Geometry, papers dedicated to I.R. Shafarevich on the occasion of his sixtieth birthday, Volume 1*, M. Artin et J. Tate, éditeurs, Progr. Math. n° **35**, pages 327–352, Birkhäuser, Boston–Basel–Stuttgart, 1983.

[Schl] H. P. Schlickewei. *Multiplicities of algebraic linear recurrences.* Acta Math., t. **170**, pages 151–180, 1993.

[Schm1] W. M. Schmidt. *Diophantine approximation.* Lecture Notes in Math. n° **785**, Springer-Verlag, Berlin–Heidelberg–New-York, 1980.

[Schm2] W. M. Schmidt. Heights of points on subvarieties of $\mathbb{G}_m^n$, *Number Theory 93–94*, S. David éditeur, London Math. Soc. Lecture Notes Ser. t. **235**, pages 157–187, Cambridge University Press, 1996.

[Sh–Ta] G. Shimura et Y. Taniyama. *Complex multiplication of abelian varieties and its applications to number theory.* Publ. Math. Soc. Japan, t. **6**, 1961.

[Sz1] L. Szpiro. Small points and torsion points. *The Lefschetz Centennial Conference, Part I*, Contemp. Math., t. **58**, pages 251–260, Amer. Math. Soc., 1986.

[Sz2] L. Szpiro. Sur les propriétés numériques du dualisant-relatif d'une surface arithmétique. *The Grothendieck Festschrift, Volume III, a collection of articles written in honour of A. Grothendieck*, P. Cartier et al. éditeurs, Progr. Math. n° **88**, pages 229–246, Birkhäuser, Boston–Basel–Stuttgart, 1990.

[Ul]   E. ULLMO. *A propos de la conjecture de Bogomolov.* Ann. of Math., à paraître, 1996.

[Zh1]   S. ZHANG. *Positive line bundles on arithmetic surfaces.* Ann. of Math., t. **136**, pages 569–587, 1992.

[Zh2]   S. ZHANG. *Positive line bundles on arithmetic varieties.* J. Amer. Math. Soc., t. **8** (1), pages 187–221, 1995.

[Zh3]   S. ZHANG. *Small points and adelic metrics.* J. Algebraic Geom., t. **4**, pages 281–300, 1995.

[Zh4]   S. ZHANG. *Equidistribution of small points on abelian varieties.* Ann. of Math., à paraître, 1996.

Sinnou DAVID et Patrice PHILIPPON

UMR 9994 du C. N. R. S. - UFR 920,

Problèmes Diophantiens,

Tour 46-56, 5-ième étage, case 247,

Université Pierre et Marie CURIE,

4, Place JUSSIEU,

75252 PARIS cedex 05,

tel. : (1) 44277520 & 44277521

adresses électroniques :

david@mathp6.jussieu.fr

pph@math.jussieu.fr

Klein polygons and geometric diagrams

Gilles LACHAUD

ABSTRACT. We show that continued fractions with negative signs give the minimal as well as the extremal points of the *Klein polygon* of an irrational line. The *geometric diagram* of such a line is the picture of the distribution of extremal points among the minimal ones in the two Klein polygons defined by that line ; we show that the geometric diagram determines the line. In particular, we can "see" on that diagram if an irrational number is a quadratic surd.

1. Introduction

Let α be an irrational positive number. Define the vector $a = (1, \alpha)$ in the plane and the half-line $D(a)$ generated by a in the first quadrant. According to the classical geometrical representation of usual continued fractions (with positive signs), we get the sequence of vectors with integer coordinates approximating $D(a)$ from the convergents to α (cf. Klein [5], and Poincaré [12]). But this description can be made more precise if one uses the algorithm of *continued fractions with negative signs* which has been introduced in the past century by Mœbius [11] and later by Hurwitz [4], as well as the sequence of vectors (x_n) defined by the associated convergents. This algorithm generates in a selective way the upper approximating usual convergents, and the *intermediate convergents* between them (cf. prop. 3).

The convergent vectors x_n occur in the following way in the geometrical representation. Consider the angle C defined by the half-line $D(a)$ and the vertical axis. The *Klein polygon* P of C is the convex hull of the integer vectors contained in that angle, and the *Klein line* V is the piecewise linear curve which is the boundary of P. The vectors which are extremal points of P are the upper approximating usual convergents. Moreover, the following sets are equal (cf. prop. 1) :

- the set of minimal points in P,
- the set of integer points in V,
- the set of convergents x_n,

and the intermediate convergents correspond to minimal points which are not extremal. Hence the convexity properties of the Klein line can be completely described by the continued fractions with negative signs of α.

1991 *Mathematics Subject Classification*. Primary 11J70; Secondary 11A55, 11E16, 11H06, 11R11, 11Y65.

© 1998 American Mathematical Society

The Klein line of the angle defined by the half-line $D(a)$ and the horizontal axis can be defined by the continued fractions with negative signs of $1/\alpha$. The *geometric diagram* of the line $D(a)$ is the picture of the distribution of extremal points among the minimal ones in the two Klein lines (cf. prop. 2). We show that the geometric diagram determines the line $D(a)$ and the number α (theorem 3). Hence we can recover an irrational number from the convexity properties of the Klein lines on their own, without any consideration of the properties related to the euclidean distance.

According to Lagrange theorem, a number is a real quadratic surd if and only if the Klein lines are invariant under some transformation of the special modular group. We show that we can in a more elementary way "read" or "see" on the geometric diagram if an irrational number is a quadratic surd or not, by looking at the distribution of extremal points among minimal ones (theorem 4).

The stated here are proved in [7], where one can also find the generalisations of these results (stated in [6] and [8]) for irrationals of higher degree.

2. Continued fractions

If $\alpha \in \mathbf{R}$, we denote by $\lceil \alpha \rceil$ the least integer which exceeds α, in such a way that $\lceil \alpha \rceil - 1 < \alpha \leq \lceil \alpha \rceil$. Let $\alpha = \alpha_0$ and if $n \geq 1$ define a sequence α_n of real numbers by the identities

$$\alpha_{n-1} = b_{n-1} - \frac{1}{\alpha_n}, \quad b_{n-1} = \lceil \alpha_{n-1} \rceil,$$

in such a way that $\alpha_n > 1$ if $n \geq 1$. In that way, α is written as a *continued fraction with negative signs* :

$$\alpha = b_0 - \cfrac{1}{b_1 - \cfrac{1}{\cdots - \cfrac{1}{b_{n-1} - \cfrac{1}{\alpha_n}}}}$$

We write such a continued fraction in the form

$$\alpha = [[b_0, b_1, \ldots, b_{n-1}, \alpha_n]].$$

We call $b_0, b_1, \ldots, b_{n-1}, \ldots$ the *partial quotients* of that continued fraction ; then $b_n \geq 2$ if $n \geq 1$. (If (r_n) and (s_n) are defined by

$$r_{-2} = 0, \quad r_{-1} = 1, \quad r_0 = b_0, \quad r_n = b_n r_{n-1} - r_{n-2},$$

$$s_{-2} = -1, \quad s_{-1} = 0, \quad s_0 = 1, \quad s_n = b_n s_{n-1} - s_{n-2},$$

then

$$\xi_n = \frac{r_n}{s_n} = [[b_0, b_1, \ldots, b_n]].$$

We call ξ_n the nth *convergent* (with negative signs) to α.

Let **I** be the set of irrational numbers ≥ 1.

PROPOSITION 1. *Assume $\alpha \in \mathbf{I}$. The sequences (r_n) and (s_n) are strictly increasing. Moreover $r_n > s_n$ and*

$$r_{n-1}s_n - r_n s_{n-1} = 1.$$

The sequence of convergents is strictly decreasing. The sequence $(r_n - s_n\alpha)$ is strictly decreasing to 0, and satisfies $0 < r_n - s_n\alpha < 1$.

We need also the expansion in *continued fraction with positive signs* of a number, which is the usual one (cf. e.g. [2], [9]). In this case we define if $k \geq 1$ a sequence α'_k as

$$\alpha'_{k-1} = c_{k-1} + \frac{1}{\alpha'_k}, \qquad c_{k-1} = \lfloor \alpha'_{k-1} \rfloor.$$

Here $\lfloor \alpha \rfloor$ is the largest integer which does not exceed α, and the expansion of α as a continued fraction with positive signs is

$$\alpha = c_0 + \cfrac{1}{c_1 + \cfrac{1}{\cdots + \cfrac{1}{c_{k-1} - \cfrac{1}{\alpha'_k}}}}$$

We write

$$\alpha = [c_0, c_1, \ldots, c_{k-1}, \alpha'_k].$$

If (p_k) and (q_k) are defined by the usual relations

$$p_{-2} = 0, \quad p_{-1} = 1, \quad p_0 = c_0, \quad p_k = c_k p_{k-1} + p_{k-2},$$
$$q_{-2} = 1, \quad q_{-1} = 0, \quad q_0 = 1, \quad q_k = c_k q_{k-1} + q_{k-2},$$

then

$$\xi'_k = \frac{p_k}{q_k} = [c_0, c_1, \ldots, c_k].$$

We call ξ'_k the nth convergent (with positive signs) to α. The relation between the two expansions is as follows:

PROPOSITION 2. *Let $k \geq 2$ be an even integer and $c_0, \ldots, c_{k+1}$ positive integers. If $\beta > 0$, then*

$$[c_0, c_1, c_2, \ldots, c_{k-1}, \beta] = [[c_0 + 1, \overbrace{2, \ldots, 2}^{c_1 - 1}, c_2 + 2, \ldots, c_{k-2} + 2, \overbrace{2, \ldots, 2}^{c_{k-1} - 1}, \beta + 1]].$$

3. Intermediate convergents

Let α be an irrational positive number. Define the vector $a = (1, \alpha)$ in the plane and the half-line $D(a)$ generated by a lying in the first quadrant. If $n \geq -1$ and $k \geq -1$, define the two sequences of vectors

$$x_n = (s_n, r_n), \qquad x'_k = (q_k, p_k).$$

The approximation of α by its convergents means that these vectors are more and more close to the line $D(a)$. In fact, if $w = (-\alpha, 1)$, then

$$\mathrm{dist}(x_n, D(a)) = \frac{(w \mid x_n)}{\| w \|} = \frac{r_n - s_n \alpha}{\sqrt{1 + \alpha^2}} \to 0.$$

We call x_n (resp. x'_k) the *convergent vectors with negative signs (resp. with positive signs)* to $D(a)$. The vectors x_n are located above the line $D(a)$. If $0 \leq c \leq c_{k+1}$, the *intermediate convergents* (with positive signs) to α are the numbers

$$\frac{p_{k-1} + cp_k}{q_{k-1} + cq_k}.$$

This is a classical notion, cf. [12], [5], [1]. These approximations to α apparently appear for the first time in the *Triparty* by Nicolas Chuquet from XVe century (cf. [1]). The associated vectors $x'_{k-1} + cx'_k \in \mathbf{Z}^2$ are located on the segment $[x'_{k-1}, x'_{k+1}]$ since

$$x'_{k+1} = x'_{k-1} + c_{k+1} x'_k.$$

We call these vectors the *intermediate convergent vectors* of order k to $D(a)$.

PROPOSITION 3. *The convergent vectors with negative signs to $D(a)$ are the intermediate convergent vectors of even order to that line. More precisely, for every even integer $k \geq 2$, let*

$$\nu(k) = c_1 + c_3 + \cdots + c_{k-1}.$$

and let $\nu(0) = 0$. If k is even and if $0 \leq c \leq c_{k+1}$, then

$$x_{\nu(k)+c-1} = x'_{k-1} + cx'_k,$$

$$\xi_{\nu(k)+c-1} = \frac{p_{k-1} + cp_k}{q_{k-1} + cq_k}.$$

The position of the various convergents is shown on figure 1.

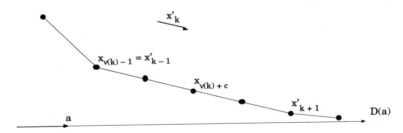

FIGURE 1. Intermediate convergents

4. Klein polygon

The *Klein polygon* of an angle C is the convex hull P of $C \cap \mathbf{Z}^2$. The *Klein line* V of C is the boundary of P. Let (w_1, w_2) be a basis of E and let C be the angle defined by

$$(w_1 \mid x) > 0, \qquad (w_2 \mid x) > 0.$$

A vector $x \in C \cap \mathbf{Z}^2$ is a *minimal vector* of P if $x = y$ whenever

$$y \in C \cap \mathbf{Z}^2, \quad (w_1 \mid y) \leq (w_1 \mid x), \quad (w_2 \mid y) \leq (w_2 \mid x).$$

It is easy to see that the minimal vectors of P are exactly the points of $V \cap \mathbf{Z}^2$.

Recall that $w = (-\alpha, 1)$. Let $C(a)$ be the angle between the vertical axis and the line $D(a)$, defined by

$$(w \mid x) > 0, \qquad (e_1 \mid x) > 0,$$

and denote by $P(a)$ and $V(a)$ the Klein polygon and the Klein line of $C(a)$.

THEOREM 1. *Under the preceding notations :*

(i) *The extremal vectors of $P(a)$ are e_2 and the convergent vectors with positive signs of odd order of $D(a)$.*

(ii) *The minimal vectors of $P(a)$ are the convergent vectors with negative signs of $D(a)$.*

(iii) *The Klein line $V(a)$ is made up of an half-line on the vertical axis and of the union of the segments $[x'_{k-1}, x'_{k+1}]$ with k even.*

(iv) *There are $c_{k+1}-1$ minimal and non extremal vectors of $P(a)$ on the preceding segment.*

Let $\alpha^- = \alpha^{-1}$ and $a^\pm = (1, \alpha^\pm)$. We mark by the exponent $+$ or $-$ the various objects defined above. Since

$$\alpha^+ = [c_0, c_1, \ldots], \qquad \alpha^- = [0, c_0, c_1, \ldots],$$

the partial quotients with positive signs of α^- are those of α^+, with a shift of one place on the right. We define the angles

$$C^+(a^\pm) : (w^\pm \mid x) > 0 \text{ and } (e_1 \mid x) > 0,$$

$$C^-(a^\pm) : (w^\pm \mid x) < 0 \text{ and } (e_2 \mid x) > 0,$$

in such a way thay $C^+(a)$ is the angle denoted above by $C(a)$, and we denote by $P^\pm(a^\pm)$ the Klein polygon of $C^\pm(a^\pm)$. If $(a, b) \in E$, let

$$W(a, b) = (b, a).$$

PROPOSITION 4. *If k is an even integer, let*

$$\nu^-(k) = c_0 + \cdots + c_{k-2}.$$

If $0 \leq c \leq c_k$, then

$$W x^-_{\nu^-(k)+c-1} = x'_{k-2} + c x'_{k-1};$$

in other words, the transforms under W of the convergent vectors with negative signs of $D(a^-)$ are the intermediate convergent vectors with positive signs of odd order of $D(a^+)$.

THEOREM 2. *Under the preceding notations :*

(i) *The extremal vectors of $P^-(a^+)$ are e_1 and the convergent vectors with positive signs of even order of $D(a^+)$.*

(ii) *The minimal vectors of $P^-(a^+)$ are the vectors $W x^-_n$ $(n \geq -1)$.*

(iii) *The boundary of $P^-(a^+)$ is made up of an half-line on the horizontal axis and of the union of the segments $[x'_{k-2}, x'_k]$, where k is even.*

(iv) *There are $c_k - 1$ minimal and non extremal vectors of $P^-(a^+)$ on the preceding segment.*

Figure 2 is a picture representing the line $D(a)$ and the two Klein polygons $P^\pm(a)$. The extremal points are labelled by a black mark, and the minimal points by a white one. The distances are not correct, because it is not possible to distinguish between the Klein line and the line $D(a)$ on an exact picture.

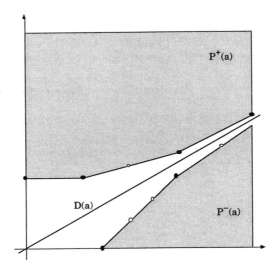

FIGURE 2. Approximation of an irrational line

5. Geometric diagrams

We are now going to show that an irrational number $\alpha \in \mathbf{I}$ or equivalently the line $D(a)$ generated by $(1, \alpha)$ is not only determined by the metric properties of the vectors which are approximations of this line, but more simply by the convexity properties alone of the associated Klein polygons. More precisely, we will see that the knowledge of the distibution of extremal points among minimal ones allows to define completely the line $D(a)$ and consequently the number α. We call *geometric diagram* a couple of half-lines, where the integer points are labelled by a black or white mark, the first point being black (cf. the figure 3). If we set black = 1 and white = 0 for instance, we can also define a geometric diagram as a couple of two functions
$$s^{\pm} : \mathbf{N} \to \{0, 1\}.$$
We denote by $\mathbf{S}$ the set of geometric diagrams. We associate to the line $D(a)$ the following geometric diagram : $s^{\pm}(n) = 1$ if the minimal vector of $P^{\pm}(a)$ of order n is extremal ; else $s^{\pm}(n) = 0$.

THEOREM 3. *The map* $\mathbf{I} \mapsto \mathbf{S}$ *who sends a number α to the geometric diagram of the line $D(a)$ generated by $(1, \alpha)$ is injective.*

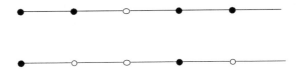

FIGURE 3. Geometric diagram

6. Consequences of Lagrange theorem

We give now a variant of Lagrange theorem which fits in the present framework. Recall that a *real quadratic surd* is a real number which is an irrational root of a second degree trinomial equation with integer coefficients.

PROPOSITION 5. *Assume $\alpha \in \mathbf{I}$. Then the following are equivalent :*

(i) *the number α is a real quadratic surd ;*
(ii) *There is a matrix $T \in SL(2, \mathbf{Z})$ and two natural numbers m and N such that, for every $q \geq 0$:*

$$x_{N+qm} = T^q x_N.$$

Recall what are the matrices T involved here. Let D be a nonsquare natural integer, and let $K = \mathbf{Q}(\sqrt{D})$ be the corresponding real quadratic field. Every number $\lambda \in K$ is written

$$\lambda = \frac{r + s\sqrt{D}}{2}$$

with r and s in $\mathbf{Q}$. If $\mathcal{O}_D$ is the order of discriminant D, then

$$\mathcal{O}_D = \{\, \lambda \in K \mid r, s \in \mathbf{Z},\, r \equiv sD \,(\mathrm{mod}\, 2)\,\},$$

and the group $\mathcal{U}_D$ of units in $\mathcal{O}_D$ is made up of numbers $\lambda \in \mathcal{O}_D$ such that

$$r^2 - Ds^2 = 4.$$

Let

$$\alpha = \frac{B + \sqrt{D}}{2C}$$

be a real quadratic surd in K, and

$$a_1 = (1, \alpha), \quad a_2 = (1, \alpha'),$$

where α' is the conjugate of α in K. We define a representation $T : K \to \mathbf{M}(2, \mathbf{Q})$ in the following way :

$$T\left(\frac{r + s\sqrt{D}}{2}\right) = \begin{bmatrix} \dfrac{r - Bs}{2} & Cs \\ -As & \dfrac{r + Bs}{2} \end{bmatrix}.$$

Let $\mathcal{U}_D^+$ be the infinite cyclic group of totally positive units of $\mathcal{O}_D$. The group

$$G = T(\mathcal{U}_D^+) \subset SL(2, \mathbf{Z})$$

is cyclic as well. The matrices lying in G leave invariant the lines $D(a_1)$ and $D(a_2)$, the angle between these lines, and the corresponding Klein polygons and Klein lines. Hence they satisfy condition (ii) of proposition 5. We deduce from the same proposition :

THEOREM 4. *Let α be an irrational number. Then α is a real quadratic surd if and only if the geometric diagram of α is periodic.*

References

[1] *Encyclopédie des Sciences Mathématiques, vol. I : Arithmétique*, Gauthier-Villars, Paris & Teubner, Leipzig, 1904-1909.

[2] Hardy, G.H., Wright E.M., *An Introduction to the Theory of Numbers,* 4th ed., Clarendon Press, Oxford, 1960.

[3] Hirzebruch, F.E.P., *Hilbert modular surfaces*, L'Enseignement Mathé-matique, **19** (1973), p. 183-281.

[4] Hurwitz, A., *Über eine besondere Art der Kettenbruch-Entwicklung reeller Grössen*, Acta Mathematica **12** (1889), p. 367-405 ; = Mathematische Werke, n° 50, t. 2, p. 84-115.

[5] Klein, F., *Ausgewählte Kapitel der Zahlentheorie*, autographié, Teubner Verlag, Leipzig 1907.

[6] Lachaud, G., *Polyèdre d'Arnol'd et voile d'un cône simplicial*, C. R. Acad. Sci. Paris **317** (1993), 711-716.

[7] Lachaud, G., *Voiles et Polyèdres de Klein*, Hermann, Paris, to appear.

[8] Lachaud, G., *Sails and Klein polyhedra*, these Proceedings.

[9] Lang, S., *Introduction to diophantine approximation*, Addison-Wesley, Reading, 1966.

[10] Minkowski, H., *Über periodische Approximationen algebraischer Zahlen*, Acta Math. **26** (1902), 333-351 ; = Ges. Abh. **I**, p. 357-371.

[11] Möbius, A.F., *Beiträge zu der Lehre von den Kettenbrüchen, nebst einem Anhange dioptrischen Inhalts*, J. Reine ang. Math. **6** (1830), p. 215-243 ; = Werke 4, p. 503-539.

[12] Poincaré, H., *Sur un mode nouveau de représentation géométrique des formes quadratiques définies ou indéfinies*, Journal de l'École Polytechnique **47** (1880), p. 177-245 ; = Œuvres, tome 5, p. 117-183.

[13] Zagier, D., *Zetafunktionen und quadratische Körper*, Springer, Berlin 1981.

ÉQUIPE "ARITHMÉTIQUE ET THÉORIE DE L'INFORMATION"
I.M.L., C.N.R.S.
LUMINY CASE 930, 13288 MARSEILLE CEDEX 9 - FRANCE
E-mail address: lachaud@iml.univ-mrs.fr

Sails and Klein Polyhedra

Gilles LACHAUD

ABSTRACT. We present here a generalization of continued fractions to higher dimensions including a generalization of Lagrange's theorem. The *sail* V of a simplicial cone C in the euclidean space of dimension d with irrational faces is the boundary of the convex hull of the integer points included in C. We show that the walls (or the generators) of C are algebraic if and only if they admit a periodic approximation by a path of chambers (or of vertices) in V. If C is defined over a totally real field of degree d, we define a group stabilizing V with a fundamental set which is the union of a finite number of simplexes, and we give an algorithm for the construction of that fundamental set. The calculations are written down in the case of simplest cubic fields.

1. Introduction

The geometric representation of approximation of irrational numbers by rational ones is a classical question. The continued fraction algorithm gives a representation of approximation of an irrational line by rational vectors, as developed in the works of Klein and Poincaré ; this representation is revisited in [15]. The most remarkable phenomenon appearing in this algorithm is Lagrange's theorem, who states that the process of approximation is periodic if and only if the approximated number is a real quadratic surd.

On the other hand, we have at our disposal, since a long time, several multidimensional approximation algorithms : the Jacobi-Perron algorithm, developed and enhanced by Minkowski (cf. [16], [17]) and Voronoï [26], and also some other algorithms introduced by more recent authors (cf. e.g. Bergmann [3], Brentjes [4], Appelgate and Onishi [1], Hellegouarch and Paysant-Le Roux [8], Buchmann [6]).

An higher-dimensional geometric generalization of the continued fraction algorithm has been suggested by Klein and Poincaré themselves (cf. [9] and [19], [20]), but these ideas apparently did not received much consideration. Our aim is to write down some results of that program.

The starting point is not a line but merely a simplicial cone, and we consider the convex hull of the integer points included in this cone. This set is supposed to contain all the information about the approximation of the boundary hyperplanes.

1991 *Mathematics Subject Classification.* Primary 11H06; Secondary 05C25, 11E76, 11H46, 11J70, 11R80, 52B20.

© 1998 American Mathematical Society

The most striking results are obtained when we consider a *Klein form*, that is a form N in d variables of degree d with rational coefficients which splits in d real linear factors. Such a form generalize the binary Pell-Fermat form $x^2 - Dy^2$. If N is a Klein form, a connected component of the open set $N > 0$ we call a *split simplicial cone*.

This work begun in early 1992 from a talk with Vladimir Arnol'd, who was led to the study of split cones by the problem of classification of graded algebras [2]. Then I became aware of other works on split cones, namely those of Tsuchihashi [25], Korkina [10], [11] and Bruno and Parushnikov [5]. Some of the results presented here were announced in [13], and a detailed exposition of the theory can be found in [14].

2. The Klein polyhedron

Let C be an open simplicial cone generated by a basis $(a_1, \ldots, a_d)$ of the euclidean space $E = \mathbf{R}^d$, and let $(w_1, \ldots, w_d)$ be the dual basis. We have

$$\begin{aligned} C &= \{x = \lambda_1 a_1 + \ldots + \lambda_d a_d \in E \mid \lambda_1, \ldots, \lambda_d > 0\} \\ &= \{x \in E \mid (w_i \mid x) > 0, 1 \leq i \leq d\}. \end{aligned}$$

Let $C^* = \bar{C} - \{0\}$. The *walls* of C are the intersections of C^* with some hyperplane $(w_i \mid x) = 0$, and the *generators* of C are the rays consisting of positive multiples of some a_i.

DEFINITION. The *Klein polyhedron* of C is the convex hull P of $C^* \cap \mathbf{Z}^d$. The *sail* V of C is the boundary of P.

If $a \in E$, we denote by $H(a)$ the hyperplane orthogonal to a, in such a way that for $1 \leq i \leq d$, the faces of C lie on the hyperplanes $H(w_i)$, and the generating half-lines of C lie on the lines $D(a_i)$. The vector $a \in E$ or the hyperplane $H(a)$ is called *irrational* if $H(a) \cap \mathbf{Q}^d = 0$; the cone C is called *irrational* if the hyperplanes $H(w_i)$ and $H(a_i)$ are such for $1 \leq i \leq d$.

In order to state the forthcoming results, it is worth introducing some terminology. Recall that a *cellular decomposition* of a subset $X \subset E$ is a set Σ of convex compact polyhedra of E satisfying the following conditions :

(i) every face of a polyhedron belonging to Σ is also an element of Σ ;
(ii) the intersection of two polyhedra belonging to Σ is either empty or a face of each of them ;
(iii) the set Σ is locally finite, that is every compact subset of E intersects only a finite number of polyhedra belonging to Σ ;
(iv) the union of all the polyhedra belonging to Σ is equal to X.

An *euclidean complex* of E is a set $X \subset E$ together with a cellular decomposition of X. The dimension of an euclidean complex is the maximum dimension of the polyhedra belonging to its cellular decomposition. A *generalized polyhedron* is a convex set X such that the intersection of X with any polyhedron is itself a polyhedron. A generalized polyhedron is an euclidean complex, with the set of its faces of any dimension (including X itself) as cellular decomposition.

THEOREM 1. *If C is an irrational simplicial cone, then P is the convex hull of its extremal points. This is a generalized polyhedron of dimension d with compact faces.*

Convergents in the theory of continued fractions are geometrically represented by extremal points of the Klein line (cf. [15], sec. 4). Moreover two consecutive convergents form a segment of this line, which is made up of these segments. In higher dimensions, common sense leads to expect that by the very construction of the sail, the generators of C are approximated by paths of vertices of V and the walls of C are approximated by paths of simplexes of V. In order to put this expectation on a firm basis, we describe in section 3 the geometric structure of the sail, in section 4 the algebraic structure of paths, and in section 5 the appropriate definitions of approximation.

3. Geometry of the sail

We associate two (combinatorial) graphs to the sail (cf. the book of Serre [21] for definitions on graphs). The first one is the graph $\Gamma_e(V)$ whose set of vertices is the set of vertices in V, i.e. the set Ext P of extremal points in the Klein polyhedron P, and whose edges are the edges of P. A point $x \in$ Ext P is a *minimal point* of the sail, that is

$$y \in C^* \cap \mathbf{Z}^d \text{ and } (w_i \mid y) \leq (w_i \mid x) \text{ if } 1 \leq i \leq d \Rightarrow y = x.$$

Paths in this graph hold the role of sequences of convergents in dimension 2.

For the second one, we use the terminology of *Tits buildings*, which is classical for abstract simplicial complexes. If X is an euclidean complex, a polyhedron in X with maximum dimension is called a *chamber*. The faces of a chamber σ are called the *panels* of σ. Two different chambers σ_1 and σ_2 are *adjacent* if they have a common panel τ; in such a case $\sigma_1 \cap \sigma_2 = \tau$. We denote by $\Gamma_f(X)$ the graph whose set of vertices is the set of chambers of X, and whose edges are the couples of adjacent chambers. A *gallery* of X is a path γ of $\Gamma_f(X)$, given by a map $C_m \to \Gamma_f(X)$ where C_m is the standard path of length m ($1 \leq m \leq \infty$). In other words, a *gallery* of length m of X is a sequence $\gamma = (\sigma_0, \ldots, \sigma_n, \ldots)$ of $m+1$ chambers such that σ_i and σ_{i+1} are adjacent if $0 \leq i \leq m-1$.

PROPOSITION 1. *The sail V of an irrational simplicial cone C satisfies the following conditions:*

(i) *every polyhedron of V is a face of some chamber of V;*
(ii) *there is a gallery joining any two chambers of V;*
(iii) *any panel belongs to exactly two chambers.*

An euclidean complex satisfying conditions (i), (ii) and (iii) is called a *pseudovariety* (without boundary). The galleries of the sail hold the role of sequences of couples of consecutive convergents vectors in dimension 2.

4. Paths in the sail

We need to replace the original cellular decomposition of the sail by a simplicial triangulation. This simplicial complex, which we still denote by V, is a pseudovariety as well, and it enjoys some more properties.

PROPOSITION 2. *The sail V of an irrational simplicial cone satisfies the following conditions:*

(i) *The vertices in the simplexes in V are in $\mathbf{Z}^d$;*

(ii) *If σ_1 and σ_2 are two simplexes in V, then*
$$C(\sigma_1) \cap C(\sigma_2) = C(\sigma_1 \cap \sigma_2),$$
where $C(\sigma)$ is the cone generated by the simplex σ.

A simplicial complex satisfying conditions (i) and (ii) is called a *fan*.

Let $k = \mathbf{Z}$ or $k = \mathbf{Q}$. We define two subgroups in $GL(d,k)$, namely the group W of permutation matrices (the Weyl group) and the group $\mathsf{U}(k)$ of matrices

$$U(c) = \begin{bmatrix} 1 & \cdots & 0 & c_1 \\ \cdots & \cdots & \cdots & \cdots \\ 0 & \cdots & 1 & c_{d-1} \\ 0 & \cdots & 0 & c_d \end{bmatrix},$$

where $c = (c_1, \ldots, c_d) \in k^d$ and $c_d \in k^\times$. Let Υ be the graph whose set of vertices is $GL(d,k)$ and whose set of edges is made of the couples

$$(M, MUW), \quad U \in \mathsf{U}(k), \quad W \in \mathsf{W}, \quad UW \neq 1.$$

A path (finite or not) in Υ is a sequence $\mu = (M_0, \ldots, M_n, \ldots)$ of matrices in $GL(d,k)$ such that

(4.1) $\qquad M_n = M_0 U_1 W_1 \cdots U_n W_n, U_n \in \mathsf{U}(\mathbf{Q}), \quad W_n \in \mathsf{W}.$

hence, a path in Υ is given by a matrix M_0 and by a word (finite or not)

$$U_1 W_1 \ldots U_n W_n$$

on the alphabet made up of elements from W and $\mathsf{U}(k)$. It is well-known that $GL(d,k)$ is generated by W and $\mathsf{U}(k)$: every $M \in GL(d,k)$ can be written (in many ways) as a word like above, hence the graph Υ is connected.

A *generating matrix* of a chamber $\sigma \subset V$ is a matrix

$$M(\sigma) \in GL(d, \mathbf{Q}) \cap M(d, \mathbf{Z})$$

such that $\sigma = M(\sigma)\delta$, where $\delta \subset E$ is the standard simplex of dimension $d-1$. If $(\sigma_0, \ldots, \sigma_n, \ldots)$ is a gallery of V, the *transition matrix* $R(\sigma_{n-1}, \sigma_n)$ between the chambers σ_{n-1} and σ_n is given by

$$M(\sigma_n) = M(\sigma_{n-1}) R(\sigma_{n-1}, \sigma_n).$$

We say that a (finite or infinite) gallery of the sail

$$\gamma = (\sigma_0, \sigma_1, \sigma_2, \ldots)$$

is *normalized* if the transition matrices are such that

$$R(\sigma_{n-1}, \sigma_n) = U_n W_n \qquad U_n \in \mathsf{U}(\mathbf{Q}), \qquad W_n \in \mathsf{W}.$$

One proves that any gallery in the sail can be normalized if the panels of each chamber are properly ordered. If γ is a normalized gallery in the sail, we associate it the sequence of generating matrices $M_n = M(\sigma_n)$. This defines a path μ in Υ, and to a normalized gallery of length m is associated a diagram

$$\begin{array}{ccc} C_m & \longrightarrow & \mu \subset \Upsilon \\ & \searrow & \downarrow \\ & & \gamma \subset \Gamma_f(V) \end{array}$$

The matrices $U_n W_n$ hold the role played by partial quotients in dimension 2.

In the same way, we say that a path $\gamma' = (x_0, x_1, x_2, \ldots)$ in the graph $\Gamma_e(V)$ of vertices of the sail is *normalized* if
$$x_n = M_n e_0, \quad e_0 = (1, \ldots, 1)$$
with a sequence of generating matrices M_n like in (4.1), where we allow loops for these paths. One proves that any path in $\Gamma_e(V)$ can be normalized. If γ' is a normalized path of length m we have a diagram

$$\begin{array}{ccc} C_m & \longrightarrow & \mu \subset \Upsilon \\ & \searrow & \downarrow \\ & & \gamma' \subset \Gamma_e(V) \end{array}$$

5. Approximation

Let $a \in E$ and D the line generated by a. The sequence $(x_n)_{n \geq 0}$ where $x_n \in \mathbf{Q}^d$ for every n is called a *strong* (resp. *bounded*, *weak*) *approximation* of $a \in E$ if it is bounded from below and if
$$d(x_n, D) \to 0, \quad \text{resp.} \quad d(x_n, D) = 0(1), \quad d(x_n, D) = o(d(x_n, H))$$
when $n \to \infty$, where d is the euclidean distance. Here is an example. If $x_0 \in \operatorname{Ext} P$, the *principal neighbor* of x_0 in the direction a_i is the point $x_1 = v_i(x_0)$ of the non-empty set
$$\{x \in \operatorname{Ext} P \mid (w_j \mid x) < (w_j \mid x_0) \text{ if } j \neq i\}$$
where the minimum of $(w_i \mid x)$ is attained. The sequence defined by the recurrence relation $x_{n+1} = v_i(x_n)$ is called the *Voronoï sequence* starting from x_0 in the direction following the generator a_i : this is a bounded approximation of a_i.

If C is a simplicial cone as before, the *dual cone* of C is
$$C^\circ = \{x \in E \mid (a_i \mid x) > 0, 1 \leq i \leq d\}.$$
If P is the Klein polyhedron of C, we set
$$P' = \{y \in E \mid (y \mid x) \geq 1 \text{ if } x \in P\} \quad ;$$
this is a generalized polyhedron included in C°. Let $\gamma = (\sigma_0, \ldots, \sigma_n, \ldots)$ be a gallery of V. An equation of the support hyperplane of σ_n is $(y_n \mid x) = 1$, where y_n belongs to the set $\operatorname{Ext} P'$ of extremal points of P'. Such a gallery is called a *strong* (resp. *bounded*, *weak*) approximation of $H(w_i)$ if the sequence $(y_0, \ldots, y_n, \ldots)$ is a strong (resp. bounded, weak) approximation of w_i in P'.

6. Algebraic hyperplanes and vectors

Let H be an irrational hyperplane of E with orthogonal vector a. If k is any subring of the field of complex numbers, define
$$\mathcal{A}_H(k) = \{T \in \mathbf{M}(d, k) \mid {}^t T H \subset H\}.$$
There is a unique real field K with degree dividing d and a unique bijective representation
$$T_H : K \longrightarrow \mathcal{A}_H(\mathbf{Q})$$
such that $T_H(\lambda)a = \lambda a$ if $\lambda \in K$. We call K the *characteristic field* of H (or of a); the hyperplane H (or the vector a) is called *algebraic* if $[K : \mathbf{Q}] > 1$, and *totally algebraic* if $[K : \mathbf{Q}] = d$. The group $\mathcal{A}_H^\times(k)$ is the group of k-rational points of a Cartan subgroup, i.e. a maximal torus $\tilde{G}$ in $GL(d)$ defined over $\mathbf{Q}$.

By a *module* in K we mean a $\mathbf{Z}$-module in K of rank $[K:\mathbf{Q}]$. If H is a totally algebraic hyperplane, we can assume that the coordinates of a form a basis of K, hence generate a module M in K. If $\mathcal{U}$ is the group of units of the order $\mathcal{O}$ of M, the representation T_H defines an isomorphism

$$T_H : \mathcal{U} \longrightarrow \mathsf{G}(\mathbf{Z}).$$

Define a *linearly periodic approximation* of a as a sequence $T^n x_0$ $(n \geq 0)$, where $T \in GL(d, \mathbf{Z})$ and $x_0 \in E$, which is an approximation of a. If r is the rank of $\mathcal{U}$, there are r PV-numbers $\eta_1, \ldots, \eta_r$ in $\mathcal{U}$ such that $T^n x_0$ is a strong linearly periodic approximation of a for any $x_0 \in E$ if T is anyone of the matrices $T_H(\eta_1), \ldots, T_H(\eta_r)$. In fact :

THEOREM 2. *An irrational vector a is totally algebraic if and only if there is a weak (resp. strong) linearly periodic approximation of a. The field K of a is totally real if and only if there are $d-1$ independent hyperbolic matrices in $SL(d, \mathbf{Z})$, each of them giving a weak (resp. strong) linearly periodic approximation of a.*

7. Split simplicial cones

Let $K \subset \mathbf{R}$ be a totally real algebraic number field of degree d, and denote by $\lambda^{(1)}, \ldots, \lambda^{(d)}$ the d real imbeddings of K. We say that the basis $(w_1, \ldots, w_d)$ of E is *split* over K if $(w_i \mid e_1) = 1$ for $1 \leq i \leq d$ and if there is a representation

$$T : K \to GL(E)$$

such that ${}^t T(\lambda) w_i = \lambda^{(i)} w_i$ for $1 \leq i \leq d$. The simplicial cone C is *split* over K if

$$C = \{x \in E \mid (w_i \mid x) > 0, 1 \leq i \leq d\},$$

where $(w_1, \ldots, w_d)$ is a split basis of E over K. If $(\omega_1, \ldots, \omega_d)$ is a basis of K over $\mathbf{Q}$ with $\omega_1 = 1$ and if

$$w^{(i)} = (\omega_1^{(i)}, \ldots, \omega_d^{(i)}),$$

then the basis $(w^{(1)}, \ldots, w^{(d)})$ is a split basis of E over K, and conversely every split basis is of this form. The map $\Psi : x \mapsto (w_1 \mid x)$ induces an isomorphism of $\mathbf{Z}^d$ on the module M generated by the basis $(\omega_1, \ldots, \omega_d)$ of K, and the norm of $\Psi(x)$ is equal to

$$N(x) = (w_1 \mid x) \ldots (w_d \mid x).$$

The *Klein form* $N \in \mathbf{Q}[X_1, \ldots, X_d]$ of degree d is decomposable and irreducible over $\mathbf{Q}$, and every such form can be constructed as a Klein form. The simplicial cone C is a connected component of the set $N(x) > 0$.

Let $\mathcal{U}^+$ be the group of totally positive units of the order of M. Let $\Phi : M \to \mathbf{Z}^d$ be the inverse of the map Ψ. Since

(7.1) $$\Phi(\lambda \alpha) = T(\lambda) \Phi(\alpha) \quad (\lambda \in \mathcal{U}^+, \alpha \in M),$$

the group $G = T(\mathcal{U}^+) \subset SL(d, \mathbf{Z})$ leaves invariant the cone C and its sail. We call G the *group of arithmetic automorphisms* or the *Dirichlet group* of C.

PROPOSITION 3. *The simplicial cone C is split if and only if there is a free abelian group of rank $d-1$ in $SL(d, \mathbf{Z})$ leaving invariant the sail of C.*

Let $\mu = (M_0, \ldots, M_n, \ldots)$ be an infinite path of the graph Υ defined in section 4, with
$$M_n = M_0 U_1 W_1 \cdots U_n W_n, \quad (U_n \in \mathsf{U}(\mathbf{Q}), W_n \in \mathsf{W})$$
and let $R_n = U_n W_n$. We say that μ is *ultimately periodic* if there are two integers m and N such that
$$R_{n+qm} = R_n \quad (q \geq 0, \ n \geq N).$$
If $N = 0$, then μ is called *periodic*. In this case, we say that
$$T = M_0 R_1 \cdots R_m M_0^{-1}$$
is the *period matrix* of μ, since
$$M_{n+qm} = T^q M_n \quad (q \geq 0, \ n \geq 0).$$
In the same way, a normalized gallery of the sail (resp. a normalized path of extremal points) is a *periodic gallery* (resp. a *periodic path of extremal points* if the associated path of Υ is periodic. The following result generalizes Lagrange's theorem.

THEOREM 3. *The following conditions are equivalent :*
(i) *The simplicial cone C is split.*
(ii) *the following conditions are fulfilled :*
  (a) *for $1 \leq i \leq d$, there is a periodic gallery of the sail which is a strong approximation of $H(w_i)$.*
  (b) *the period matrices of these galleries commute.*
(iii) *the following conditions are fulfilled :*
  (a) *for $1 \leq i \leq d$, there is a periodic path of vertices of the sail which is a strong approximation of a_i.*
  (b) *the period matrices of these paths commute.*

When C is a split simplicial cone, here is an algorithm giving a fundamental set for the Dirichlet group G in the sail V by a method coming from Shintani [**23**], [**24**] (and also Zagier [**28**] in the two-dimensional case). If $x \in C$, let
$$f(x) = (w_1 \mid x) + \ldots + (w_d \mid x),$$
and define the *Dirichlet domain* of G in C as
$$\mathcal{D} = \{x \in C \mid f(x) \leq f(Tx) \text{ if } T \in G\}.$$

THEOREM 4. *Let C be a split simplicial cone. The Dirichlet domain $\mathcal{D}$ of G in C is a polyhedral cone. If $T \in G$ and $T \neq 1$, then*
$$T\overset{\circ}{\mathcal{D}} \cap \overset{\circ}{\mathcal{D}} = \emptyset.$$
If $\mathcal{F}$ is the finite set of faces of V meeting $\mathcal{D}$, every face of V is the image of a face of $\mathcal{F}$ under some element of G.

Theorem 4 has the following consequences. The quotient set $X = V/G$ is a compact and connected pseudovariety. The projection map $\pi : V \to X$ is a covering with structural group G, inducing a covering of graphs
$$\Gamma_f(V) \to \Gamma_f(X),$$
where $\Gamma_f(X)$ is the finite graph of faces of X. For $1 \leq i \leq d$, there is a closed path in $\Gamma_f(X)$ whose preimage is a strong approximation of the hyperplane $H(w_i)$.

COROLLARY 1. *A simplicial cone is split if and only if the Voronoï sequences for every generator of C are linearly periodic, and if the period matrices commute. In that case for every generator of C there is a circuit in the graph $\Gamma_e(X)$ of vertices of X whose preimage is a path of $\Gamma_e(V)$ containing the Voronoï sequence.*

We denote by p the projection of E on the projective space $\mathbf{P}(E)$. If
$$\mu = (M_0, \ldots, M_n, \ldots)$$
is a path of Υ the sequence
$$(p(M_0 M_n^{-1} a_1), \ldots, p(M_0 M_n^{-1} a_d)) \in \mathbf{P}(E)^d$$
hold the role played by complete quotient in dimension 2. Thus we call
$$(p(M(\sigma)^{-1} a_1), \ldots, p(M(\sigma)^{-1} a_d))$$
The *complete quotient* of a chamber σ of V.

COROLLARY 2. *A simplicial cone is split if and only if the set of complete quotients of the chambers of the sail of that cone is finite.*

8. Algorithms for the simplest cubic fields

Let $m \in \mathbf{Z}$ and $m \geq 1$. The *simplest cubic field* of index m is the totally real field K of roots of the irreducible polynomial
$$f_m(x) = x^3 - (m-2)x^2 - (m+1)x - 1.$$
The square root of the discriminant of f_m is equal to $D = m^2 - m + 7$. If ω is a root of $f_m(x)$, the two other roots are
$$\omega' = \frac{-1}{\omega + 1}, \quad \omega'' = -1 - \frac{1}{\omega}.$$
The roots $\omega, \omega', \omega''$ are units of K with $N(\omega) = 1$. Choose ω as the root of $f_m(x)$ which is > 1. The number $\omega + 1$ is a PV-number. If $m = 1$, then $D = 7$ and
$$f_1(x) = x^3 + x^2 - 2x - 1, \quad \omega = 2\cos\frac{2\pi}{7}.$$
If $m = 2$, then $D = 9$ and
$$f_2(x) = x^3 - 3x - 1, \quad \omega = 2\cos\frac{\pi}{9}.$$
The corresponding fields are respectively the field of resolution of the heptagon and of the nonagon. It is well-known (cf. e.g Shanks [22], Washington [27]) that if $m \not\equiv 5 \pmod 9$ and if D is squarefree when $m \geq 3$, then $1, \omega, \omega^2$ is a basis of the ring $\mathcal{O}_K$ of integers of K and $-1, \omega, \omega''$ generate the group $\mathcal{U}_K$ of units of K. This implies that $\eta = \omega^2$ and $\varepsilon = \omega''^2$ generate the group $\mathcal{U}_K^+$ of totally positive units of K. The number $\theta = \varepsilon\eta = (\omega + 1)^2$ is both a positive unit and a PV-number > 1. Let $w_1 = (1, \omega, \omega^2), w_2 = w_1', w_3 = w_1''$, and
$$C = \{x \in E \mid (w_i \mid x) > 0, 1 \leq i \leq 3\}.$$
The representation T associated to C is calculated by means of relation (7.1) : if $\lambda = a + b\omega + c\omega^2 \in K$, then
$$T(\lambda) = \begin{bmatrix} a & c & b + (m-2)c \\ b & a + (m+1)c & (m+1)b + (m^2 - m - 1)c \\ c & b + (m-2)c & a + (m-2)b + (m^2 - 3m + 5)c \end{bmatrix}.$$

In particular, if $m = 1$:

$$T(\theta) = \begin{bmatrix} 1 & 1 & 1 \\ 2 & 3 & 3 \\ 1 & 1 & 2 \end{bmatrix}, \quad T(\varepsilon) = \begin{bmatrix} 2 & 0 & 1 \\ 1 & 2 & 2 \\ 0 & 1 & 1 \end{bmatrix}.$$

The Dirichlet group $G = T(\mathcal{U}_K^+) \subset SL(3, \mathbf{Z})$ is a free abelian group of rank 2, and every basis of $\mathcal{U}_K^+$, for instance the basis (θ, ε), gives a basis of G. Let

$$\begin{aligned}
\alpha &= \Phi(1) &&= (1, 0, 0), \\
\beta &= \Phi(\eta) &&= (0, 0, 1), \\
\gamma_1 &= \Phi(\eta\eta') &&= (1 - m, -m^2 + m - 1, m), \\
\gamma_2 &= \Phi(\eta\eta'') &&= (1, 2, 1), \\
\zeta &= \Phi(\theta\eta) &&= (m, m^2 + m + 1, m^2 - m + 2),
\end{aligned}$$

These are vertices of the following chamber of the sail

$$\sigma_1 = (\alpha, \gamma_1, \beta), \quad \sigma_2 = (\alpha, \gamma_2, \beta), \quad \sigma_3 = (\gamma_2, \beta, \zeta).$$

The sequence $T(\theta)^n \alpha$ is a strong approximation of a_1 in V and one can construct a periodic path in the sail of period 4 whose vertices include this sequence. On the other hand, the periodic gallery

$$(\sigma_3, \sigma_2, \sigma_1 = T(\theta)^{-1}\sigma_3, T(\theta)^{-1}\sigma_2, \ldots)$$

is a strong approximation of $H(w_1)$. The associated path in Υ is of period 2:

$$M(\sigma_1) U_1 W_1 U_2 W_2 U_1 W_1 U_2 W_2 \ldots.$$

If $m = 1$, then

$$U_1 = \begin{bmatrix} 1 & 0 & 3 \\ 0 & 1 & 1 \\ 0 & 0 & -2 \end{bmatrix}, \quad U_2 = \begin{bmatrix} 1 & 0 & 1/2 \\ 0 & 1 & 3/2 \\ 0 & 0 & -1/2 \end{bmatrix},$$

$$W_1 = \begin{bmatrix} 0 & 0 & 1 \\ 0 & 1 & 0 \\ 1 & 0 & 0 \end{bmatrix}, \quad W_2 = \begin{bmatrix} 1 & 0 & 0 \\ 0 & 0 & 1 \\ 0 & 1 & 0 \end{bmatrix}.$$

THEOREM 5. *Let $\mathcal{F} = \sigma_1 \cup \sigma_2$. The sail of C is the orbit of $\mathcal{F}$ under the group G. The group G operates transitively on the set of vertices of the sail of C.*

REMARK. Introduce the matrix

$$M = \begin{bmatrix} 3 & 2 & 1 \\ 2 & 2 & 1 \\ 1 & 1 & 1 \end{bmatrix}.$$

The eigenvalues of M are ε and its conjugates. If

$$P = \begin{bmatrix} 0 & 1 & 0 \\ 1 & 0 & 1 \\ 0 & 0 & 1 \end{bmatrix},$$

and if $\lambda = a + b\omega + c\omega^2 \in K$, define $\widetilde{T}(\lambda) = PT(\lambda)P^{-1}$; then

$$\widetilde{T}(\lambda) = \begin{bmatrix} a + 2c & b & b - c \\ b & a + c & c \\ b - c & c & a - b + 2c \end{bmatrix},$$

in such a way that $M = \widetilde{T}(\theta)$. Let
$$\widetilde{a}_1 = {}^t P^{-1} w_1 = -\omega' P a_1 = (\omega, 1, \omega^2 - 1).$$
The basis $(\widetilde{a}_1, \widetilde{a}_2, \widetilde{a}_3)$ defined by $\widetilde{a}_1$ and its conjugates is self-dual, hence the split simplicial cone $\widetilde{C}$ defined by this basis is equal to its dual $\widetilde{C}^\circ$. The corresponding Dirichlet group $\widetilde{G} = \widetilde{T}(\mathcal{U}_K^+)$ is included in the space of symmetric matrices. It is worth noticing that in even dimensions, the Dirichlet group of a split simplicial cone is never included in the space of symmetric matrices. A basis of $\widetilde{G}$ is for instance given by M and by
$$\widetilde{T}(\varepsilon) = \begin{bmatrix} 2 & 1 & 1 \\ 1 & 2 & 0 \\ 1 & 0 & 1 \end{bmatrix}.$$

9. Caliber of modules

If K is a totally real field, denote by $\Theta : K \to \mathbf{R}^d$ the map
$$\Theta(\lambda) = (\lambda^{(1)}, \ldots, \lambda^{(d)}),$$
where $\lambda^{(1)}, \ldots, \lambda^{(d)}$ are the d distincts embeddings of K into $\mathbf{R}$. If M is a module of K, let M^+ be the set of totally positive elements of M and P_M the convex hull of $\Theta(M^+)$. We say that $\lambda \in M^+$ is *extremal* if $\Theta(\lambda)$ is an extremal point of P_M.

A module M is *integral* if it is contained in $\mathcal{O}_K$. A natural integer a *divides* M if $M \subset d\mathcal{O}_K$. A module M is *primitive* if M is not divisible by any integer > 1. A module M is *reduced (of extremal type)* if it is integral, reduced, and if the least natural integer in K is an extremal element of M^+. Every module is equivalent to a reduced module. The *caliber* of M is the number of reduced modules equivalent to M. This definition generalizes the one given for quadratic fields (cf. [12]) and has to be compared to those given by Buchmann [7] and Paysant-Le Roux [18].

THEOREM 6. *Let M be a module in a totally real number field K with basis $(w_1, \ldots, w_d)$, let C be the split simplicial cone defined by that basis, V the sail and G the Dirichlet group of C. Then the caliber of M is equal to the number of vertices of $X = V/G$.*

From theorem 5 and 6 we deduce :

COROLLARY 3. *The rings of integers of the simplest cubic fields have caliber one.*

REMARK. The ring generated by the *golden number* $\dfrac{1 + \sqrt{5}}{2} = 2\cos\dfrac{\pi}{5}$ has caliber one ; the preceding corollary shows that the rings generated by the numbers $2\cos\dfrac{2\pi}{7}$ and $2\cos\dfrac{\pi}{9}$ share this property with the golden number.

10. Figures

Figure 1 shows the simplicial cone C of section 8 when $m = 1$ and six simplexes of the sail V. The fundamental set $\mathcal{F}$ is the vertical diamond-shaped couple of two simplexes on the left. Figure 2 shows the graph of vertices which is the 1-skeleton of the sail V (24 simplexes are shown). Figure 3 is a three-dimensional representation of the part of the sail shown in figure 2.

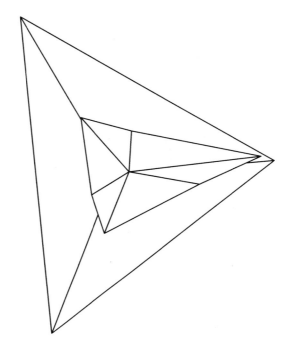

FIGURE 1

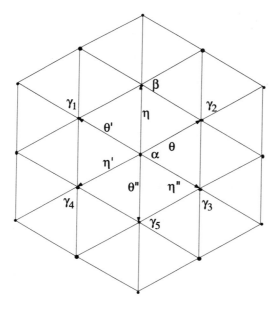

FIGURE 2

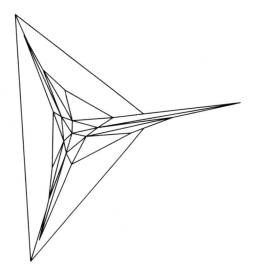

FIGURE 3

References

[1] Appelgate, H., Onishi, H. *Periodic expansion of modules and its relation to units*, J. of Number Theory **15** (1982), p. 283-294.
[2] Arnol'd, V.I., *A-graded algebras and continued fractions*, Comm. Pure and Appl. Math. **42** (1989), 993-1000.
[3] Bergmann, G., *Theorie der Netze*, Math. Ann. **149** (1963), p. 361-418.
[4] Brentjes, A.J., *Multi-dimensional continued fraction algorithms*, Math. Centre Tracts **145**, Math. Centrum, Amsterdam, 1981.
[5] Bruno, A.D., Parushnikov, V.I., *Klein polyhedrals for two cubic Davenport forms*, Mat. Zametki, **56**, No. 4 (1994), 9-27 ; = Math. Notes **56**, No 3-4 (1994), 994-1007.
[6] Buchmann, J., *On the Computation of Units and Class Numbers by a Generalization of Lagrange's Algorithm*, J. Number Theory **26** (1987), p. 8-30.
[7] Buchmann, J., *On the Period Length of the Generalized Lagrange's Algorithm*, J. Number Theory **26** (1987), p. 31-37.
[8] Hellegouarch, Y., Paysant-Le Roux, R., *Invariants arithmétiques des corps possédant une formule du produit*, Journées Arithmétiques de Besançon, Astérisque **147-148** (1987), p. 291-300.
[9] Klein, F., *Über eine geometrische Auffassung der gewönlichen Kettenbruchentwicklung*, Göttingen Nachr. (1895) ; = Ges. Abh., t.II, p. 209-211 ; = *Sur une représentation géométrique du développement en fraction continue ordinaire*, Nouv. Ann. Math. (3) **15** (1896), p. 327-331.
[10] Korkina, E., *The simplest 2-dimensional continued fractions*, Proceedings of the International Geometrical Colloquium, Moscow, 1993 (to appear).
[11] Korkina, E., *La périodicité des fractions continues multidimensionnelles*, C. R. Acad. Sci. Paris **319** (1994), 777-780.
[12] Lachaud, G., *On real quadratic fields*, Bull. Amer Math. Soc. **17** (1987), p. 307-311.
[13] Lachaud, G., *Polyèdre d'Arnol'd et voile d'un cône simplicial : analogues du théorème de Lagrange*, C.R. Acad. Sci. Paris **317** (1993), p. 711-716.
[14] Lachaud, G., *Voiles et Polyèdres de Klein*, Hermann, Paris, to appear.
[15] Lachaud, G., *Klein polygons and geometric diagrams*, these Proceedings.

[16] Minkowski, H., *Über periodische Approximationen algebraischer Zahlen*, Acta Math. **26** (1902), 333-351 ; = Ges. Abh. I, p. 357-371.
[17] Minkowski, H., *Généralisation de la théorie des fractions continues*, Ann. Ec. Sup., 3e série **13** (1896), p. 41-60 ; = *Zur Theorie der Kettenbrüche*, Ges. Abh. I, p. 278-292.
[18] Paysant Le Roux, R., *Calibre d'un corps global*, J. of Number Theory **30** (1988), p. 267-287.
[19] Poincaré, H. *Analyse de ses travaux sur l'algèbre et l'arithmétique*, Acta Math. **38** (1921), p. 89-100 ; = Œuvres, tome 5, p. 1-16.
[20] Poincaré, H., *Sur une généralisation des fractions continues*, C. R. Acad. Sci. Paris **99** (1884), p. 1014-1016 ; = Œuvres, tome 5, p. 185-188.
[21] Serre, J.-P., *Arbres, amalgames, SL_2*, Astérisque **46** (1977) ; = *Trees*, Springer, Berlin, 1980.
[22] Shanks, D., *The Simplest Cubic Fields*, Math. Comp. **28** (1974), p. 1137-1152.
[23] Shintani, T., *On evaluation of zeta functions of totally real algebraic number fields at nonpositive integers*, J. Fac. Sci. Univ. Tokyo, Sect. IA, **24** (1976), 393-417.
[24] Shintani, T., *A remark on zeta functions of algebraic number fields*, Proc. Coll. Automorphic Forms, Representation Theory and Arithmetic, Bombay, 1979, Springer, Berlin, 1979, p. 255-260.
[25] Tsuchihashi, H., *Higher dimensional analogues of periodic continued fractions and cusp singularities*, Tôhoku Math. J. **35** (1983), p. 607-639.
[26] Voronoï, G., *Über eine Verallgemeinerung des Kettenbruchalgorithmus*, dissertation, University of Warschau, 1896.
[27] Washington, L.C., *Class numbers of the simplest cubic fields*, Math. Comp. **48** (1987), p. 371-384.
[28] Zagier, D., *Valeurs des fonctions zêta des corps quadratiques réels aux entiers négatifs*, Journées Arithmétiques de Caen, Astérisque **41-42** (1977), p. 135-151.

Équipe "Arithmétique et Théorie de l'Information",
I.M.L., C.N.R.S.,
Luminy Case 930, 13288 Marseille Cedex 9 - FRANCE
E-mail address: lachaud@iml.univ-mrs.fr

Automata and Transcendence

DINESH S. THAKUR

Interesting quantities of number theory or geometry are often defined by analytic processes such as integration or infinite series or product expansions. A central question which transcendence theory addresses is whether they can also be described by simpler ('finite') algebraic methods; in particular, whether they turn out to be algebraic. For example, the most common analytic description of a number is by its infinite decimal or p-adic expansion and that of a function is by its power series expansion. It is well-known that any expansion that is eventually periodic represents a rational number (or a function). On the other hand, rational numbers and rational functions over a finite field give expansions that are eventually periodic.

More generally, we can ask for a similar, simple characterization of digit patterns for algebraic numbers or functions. We begin by describing an automata-theoretic criterion of discrete mathematics for this, and will prove three transcendence results in number theory as applications.

The first major advance in answering this general question was made by Furstenberg [F67]:

For $r = \sum r_{n_1, \cdots, n_k} x_1^{n_1} \cdots x_k^{n_k}$, define the 'diagonal' $Dr := \sum r_{n, \cdots, n} x^n$.

THEOREM 1. *For $k = \mathbb{C}$ or $\mathbb{F}_q$, the set of algebraic power series $f(x)$ over $k(x)$ is the same as the set of diagonals Dr of two-variable rational functions $r(x_1, x_2)$. The diagonal of a rational function of many variables over $\mathbb{F}_q$ (but not over $\mathbb{C}$) is algebraic.*

PROOF. : Over $\mathbb{C}$, for small ϵ and $|x|$, we have

$$Dr(x) = \frac{1}{2\pi i} \int_{|z|=\epsilon} r(z, \frac{x}{z}) \frac{dz}{z}.$$

1991 *Mathematics Subject Classification.* 11G07, 11R58; Secondary 11J89, 11B85, 11J91.
Key words and phrases. elliptic curve, gamma, function field.
This paper is in final form and no version of it will be submitted for publication elsewhere

Evaluating by residues gives an algebraic function. On the other hand, suppose f is algebraic, satisfying a polynomial equation $P(x, f(x)) = 0$. Also assume $f(0) = 0$ and that 0 is an isolated root of $P(0, w) = 0$, then expressing

$$f(x) = \int_\gamma w \frac{\partial P}{\partial w}(x, w)/P(x, w) dw$$

as an integral above, we see that f is a diagonal of a rational function. Also, if we have more than two variables, the resulting contour integration of an algebraic function of two variables is in general transcendental. Though this proof does not work over $\mathbb{F}_q$, the resulting formulae can be directly checked. □

Deligne [D84] generalized the last statement of the Theorem to algebraic functions of many variables.

From now on, we restrict to a coefficient field $\mathbb{F}_q$ of characteristic p.

COROLLARY 1. *(1) Algebraic power series are closed under Hadamard (term-by-term) product.*
(2) $\sum a_n x^n$ is algebraic if and only if $\sum_{a_n=a} x^n$ is algebraic for each $a \in \mathbb{F}_q$.

PROOF. : If $\sum a_n x^n = Dr_1(x_1, x_2)$, $\sum b_n x^n = Dr_2(x_1, x_2)$, then $\sum a_n b_n x^n = D(r_1(x_1, x_2) r_2(x_3, x_4))$. This implies (1). Then (2) follows from (1) by expanding out $\sum_{a_n=a} x^n = \sum (1 - (a_n - a)^{q-1}) x^n$. □

By the Corollary, we can focus on characteristic sequences on subsets of natural numbers, i.e., on the series $\sum x^{n_i}$, if we wish. The main theorem giving the automata-theoretic criterion for algebraicity is the following theorem due to Christol [C79], [CKMR80]. In fact, the equivalence of the last two conditions (as well as another description in terms of substitutions) of the Theorem is due to Cobham [CO72]. We sketch the proof later.

THEOREM 2. *(i) $\sum f_n x^n$ is algebraic over $\mathbb{F}_q(x)$ if and only if (ii) $f_n \in \mathbb{F}_q$ is produced by a q-automaton if and only if (iii) there are only finitely many subsequences of the form $f_{q^k n + r}$ with $0 \leq r < q^k$.*

Here, an m-automaton (we shall usually use $m = q$) consists of a finite set S of states, a table of how the digits base m operate on S, and a map τ from S to $\mathbb{F}_q$ (or some alphabet in general). For a given input n, fed in digit by digit from the left, each digit changing the state by the rule provided by the table, the output is $\tau(n\alpha)$ where α is some chosen initial state. For more details, see the expository article [A87].

Example: The following table, where $2 \leq i < p$, together with $\tau(\alpha) = \tau(\gamma) = 0$ and $\tau(\beta) = 1$ defines a p-automaton whose output f_n is the characteristic sequence of $\{p^m\}$, i.e., the numbers of the form $1000\cdots$ base p.

	α	β	γ
0	α	β	γ
1	β	γ	γ
i	γ	γ	γ

It is easy to see directly that $f := f(x) := \sum x^{p^n}$ satisfies $f^p - f + x = 0$, in accordance with Theorem 2. In fact, this f is Mahler's famous counterexample to the analogue of Roth's theorem in characteristic p. The partial sums to f approximate f with Liouville bounds.

From the proof of the first theorem, we find that $f = D(x_1/(1-(x_1^{p-1}+x_2)))$. We leave to the reader the interesting exercise of verifying this equation directly from the definition by expanding the right hand side via a geometric series, and proving the necessary divisibility properties of the binomial coefficients thus arising.

As a warm-up to the proof in the second application, we now prove that f is transcendental in finite characteristic $\ell \neq p$. Clearly, there are infinitely many k's such that $0 < p^m - \ell^k \mu < \ell^k$, for some m and some $0 < \mu < \ell$. So the subsequence $f_{\ell^k n + (p^m - \ell^k \mu)}$ assumes the value 1, for $n = \mu < \ell$. The next n for which it is 1 corresponds to $\ell^k n + (p^m - \ell^k \mu) = p^{m+w}$, with $w > 0$ the least such. So ℓ^k divides $p^w - 1$ and hence $n = \mu + p^m(p^w - 1)/\ell^k > p^m \to \infty$ as k tends to ∞. Hence there are infinitely many such subsequences, and (iii) of Theorem 2 finishes the proof.

In fact, this is a special case of a very general result of Cobham [Co69]:

THEOREM 3. *Non-periodic sequences produced by m-automata cannot be produced by n-automata, if m and n are multiplicatively independent.*

We do not sketch the proof. To quote Eilenberg [E74], "The proof is correct, long and hard. It is a challenge to find a more reasonable proof of this fine theorem". See [BHMV94] for survey of other proofs based on logic.

Together with Theorem 2, this implies

COROLLARY 2. *If $\sum x^{n_i}$ is irrational and algebraic in one finite characteristic, then it is transcendental in all other finite characteristics.*

The natural question is whether corresponding real numbers, e.g. the decimal $\sum 10^{-n_i}$, are transcendental. For our example $f(x)$ above, instead of using the algebraicity equation $f(x)^p - f(x) + x = 0$, which is special to characteristic p, Mahler [M29] used the characteristic-free functional equation $f(x^p) - f(x) + x = 0$ and established (for $p = 2$) the transcendence of various values of $f(x)$. It implies the transcendence of real numbers $f(1/k) = \sum k^{-2^n}$, for any integral base $k > 1$. See [LP77] for a nice exposition of the proof of a more general result.

Loxton and van der Poorten generalized Mahler's method, but it should be noted here that the proof of the often quoted result of Loxton and van der Poorten that under the hypothesis of the Corollary, the real number $\sum 10^{-n_i}$

is transcendental, has a gap, as van der Poorten has mentioned to the author. Thus, this statement and similar p-adic statement remain as challenging open questions.

We now present an example due to Richie [Ri63], which we shall use, and which does not look as special as the first example.

THEOREM 4. *The characteristic sequence of the set of squares is not produced by any 2-automaton.*

PROOF. : Consider the 2-automaton given by the following table and $\tau(f) = 1$, $\tau(\text{rest}) = 0$.

	α	s_1	s_2	s_3	s_4	f	n
0	n	s_3	s_4	s_4	s_3	n	n
1	s_1	s_2	s_1	n	f	n	n

A straight entry chase in the table shows that this produces the characteristic function χ_A of the set $A = \{1^n 0^m 1 : n, m > 0, n+m \text{ odd}\}$. It is easy to see that the intersection of A with the set of squares is $B = \{1^n 0^{n+1} 1 : n > 0\}$ and also that in general, the intersection corresponds to the direct product of automata or the Hadamard product of series.

But χ_B can not be produced by a 2-automaton: As there are only a finite number of states, $1^\ell \alpha = 1^{\ell+m} \alpha$, for some ℓ, m. But $1^{\ell+m} 0^{\ell+1} 1$ is not in B whereas $1^\ell 0^{\ell+1} 1 \in B$. □

We finish this introduction by giving a very brief sketch of the ideas involved in the proof of Theorem 2. For more details, see [CKMR80] or [A87].

(ii) implies (iii): There are only finitely many possible maps $\beta : S \to S$ and any $f_{q^k n + r}$ is of the form $\tau(\beta(n\alpha))$.

(iii) implies (i): Let V be the vector space over $\mathbb{F}_q(x)$ generated by monomials in $\sum f_{q^k n + r} x^n$. Then V is finite dimensional with $fV \subset V$, so f satisfies its characteristic polynomial.

(i) implies (iii): For $0 \leq r < q$, define C_r (twisted Cartier operators) by $C_r(\sum f_n x^n) = \sum f_{qn+r} x^n$. Considering the vector space over $\mathbb{F}_q$ generated by the roots of the polynomial satisfied by f, we can assume that $\sum_{i=0}^{k} a_i f^{q^i} = 0$, with $a_0 \neq 0$. Using $g = \sum_{r=0}^{q-1} x^r (C_r(g))^q$ and $C_r(g^q h) = g C_r(h)$, we see that

$$\{h \in \mathbb{F}_q((x)) : h = \sum_{i=0}^{k} h_i (f/a_0)^{q^i}, h_i \in \mathbb{F}_q[x], \deg h_i \leq \max(\deg a_0, \deg a_i a_0^{q^{i-2}})\}$$

is a finite set containing f and stable under C_r's.

(iii) implies (ii): If there are m subsequences $f_n^{(i)}$ with $f_n^{(1)} = f_n$ say, put $S := \{\alpha := \alpha_1, \cdots, \alpha_m\}$. Define a digit action by, $r\alpha_i := \alpha_k$ if $f_{qn+r}^{(i)} = f_n^{(k)}$. Define $\tau(\alpha_i) := f_n$, if $n^- \alpha_1 = \alpha_i$ with n^- being the base q expansion of n written in the reverse order. □

1. Application I

Let k be an algebraic closure of $\mathbb{F}_p$, q be a variable and let

$$a_4 := a_4(q) := \sum_{n \geq 1} \frac{-5n^3 q^n}{1-q^n}, \quad a_6 := a_6(q) := \sum_{n \geq 1} \frac{-(7n^5 + 5n^3)q^n}{12(1-q^n)}$$

THEOREM 5. *The period q of the Tate elliptic curve $y^2 + xy = x^3 + a_4 x + a_6$ over $K := k(a_4, a_6)$ is transcendental over K.*

This Theorem was proved by Voloch [V96] using Igusa theory and we provided another proof [T96a]. The Theorem can be considered as an analogue of the result of Siegel and Schneider on the transcendence of the period of elliptic curve (over a number field) over its field of definition. If, more appropriately, we consider q as the multiplicative version of the period, then it can be considered as an analogue of a conjecture of Mahler (in p-adic setting) and a conjecture of Manin (in the complex setting), as pointed out to us by Waldschmidt. These conjectures themselves were settled since then by Barre-Sirieix, Diaz, Gramain, Philibert [BDGP96]. In particular, they show that the '$\log_p q$' appearing in the p-adic Birch-Swinnerton-Dyer conjectures of Mazur, Tate and Teitelbaum (Theorem of Stevens/Greenberg) does not vanish, so that the order of vanishing is exactly as predicted in the conjectures.

PROOF. : First, let $p = 2$. Then

$$a_4 = a_6 = \sum_{n \text{ odd} \geq 1} q^n/(1-q^n) = \sum_{n \text{ odd} \geq 1} \sum_{k=0}^{\infty} q^{kn} = \sum_{m=1}^{\infty} d_o(m) q^m,$$

where $d_o(m)$ = number of odd positive divisors of m. Hence, if $m = 2^k \prod p_i^{m_i}$, then $d_o(m) = \prod (m_i + 1)$. So $d_o(m)$ is odd if and only if $m = n^2$ or $2n^2$. Hence, with $f := \sum q^{n^2}$ (essentially theta), we have

$$a_4 = \sum_{n=1}^{\infty} (q^{n^2} + q^{2n^2}) = f + f^2,$$

Now Theorems 2 and 4 imply that f and so $a_4 = a_6 = f + f^2$ is transcendental over $k(q)$, i.e., q is transcendental over $K = k(a_4)$, finishing the proof when $p = 2$.

Now consider the case of general p. We will show:

(1) f is transcendental over $k(q)$: This follows from the generalization [E74] of the Theorem 4 to any base.

(2) a_4 and a_6 are algebraically dependent over k.

(3) f is algebraic over $k(\bar{a}_4, \bar{a}_6)$, where $\bar{a}_4 := a_4(q^2)$, $\bar{a}_6 := a_6(q^2)$.

It is easy to see that (1), (2) and (3) imply the Theorem.

Proof of (2): Elementary congruences show that $a_4 = a_6$ if $p = 2$, $a_4 = 0$ if $p = 5$ and $a_4 = 5a_6$ if $p = 7$.

Swinnerton-Dyer [S-D73] noticed that expressing the fact that the Hasse invariant of the Tate elliptic curve is one in terms of a_4 and a_6 gives (2) for $p > 3$.

For $p = 3$, the Hasse invariant is identically one and we do not get a relation this way. Nonetheless, we claim that $a_6 + a_4 + 2a_4^2 = 0$. In fact, $a_4 + a_6$ is

$$\sum_{n \equiv 2,4(9)} \frac{q^n}{1-q^n} + 2 \sum_{n \equiv 5,7(9)} \frac{q^n}{1-q^n} = \sum_{n \not\equiv 0(3)} \frac{n + (-1)^{n-3\lfloor n/3 \rfloor}}{3} \frac{q^n}{1-q^n} = a_4^2$$

The first two equalities follow by analyzing n modulo 9 and the last equality is a rearrangement of Ramanujan's identity (19) of [R16].

Proof of (3): Note that $\overline{a_4}$, $\overline{a_6}$, f are connected to the well-known Eisenstein series and theta function: $\theta = 1 + 2f$, $e_4 = 1 - 48\overline{a}_4$ and $e_6 = 1 - 72\overline{a}_4 + 864\overline{a}_6$.

We have the following explicit algebraic dependency relation between the three modular forms:

$$4e_6^2 - 4e_4^3 + 27e_4^2\theta^8 - 54e_4\theta^{16} + 27\theta^{24} = 0$$

A straight translation to the original variables implies (3). $\square$

2. Application II

Here we give an application to the transcendence of the gamma monomials (in the function field case) evaluated at fractions.

The most well-known classical result is : $\Gamma(1/2) = (-1/2)! = \sqrt{\pi}$. Chudnovsky [Chu84] showed that Γ values at proper fractions with denominator 4 or 6 are transcendental. This is basically all that is known about the transcendence of the individual gamma values at proper fractions.

We shall consider $A = \mathbb{F}_q[T]$, $K = \mathbb{F}_q(T)$, $K_\infty = \mathbb{F}_q((t))$ (with $t = 1/T$) and $\Omega = $ the completion of an algebraic closure of K_∞. These are analogues of $\mathbb{Z}$, $\mathbb{Q}$, $\mathbb{R}$, $\mathbb{C}$ respectively, in the theory of function fields.

The Carlitz factorial $\Pi(n) \in A$ for $n \in \mathbb{Z}_{\geq 0}$ is defined by

$$[n] := T^{q^n} - T, \quad D_n := \prod_{k=1}^{n} [k]^{q^{n-k}}$$

$$\Pi(n) := \prod D_i^{n_i}, \text{ for } n = \sum n_i q^i, \ 0 \leq n_i < q.$$

David Goss noticed that

$$\overline{D_i} := \frac{D_i}{T^{\deg D_i}} = 1 - t^{q^i - q^{i-1}} + \text{higher degree terms} \to 1 \text{ as } i \to \infty$$

gives an interpolation: $\overline{\Pi}(n) \in K_\infty$ for $n \in \mathbb{Z}_p$ by

$$n! := \overline{\Pi}(n) := \prod \overline{D_i}^{n_i} \text{ for } n = \sum n_i q^i, \ 0 \leq n_i < q.$$

Why is this gamma function good? We mention some reasons as catchphrases: analogous prime factorization, divisibility properties, analogous occurrence in the Taylor coefficients in the relevant exponential (of Carlitz-Drinfeld module), right functional equations, interpolations at finite primes and connection with Gauss sums (Gross-Koblitz type formula), and connection with periods (Chowla-Selberg type formula). See [T92] for details and references.

For our factorial $(-1/2)! = \sqrt{\tilde{\pi}}$, when $p \neq 2$, where $\tilde{\pi} \in \Omega$ is the period of the Carlitz-Drinfeld exponential (the analogue of $2\pi i \in \mathbb{C}$). This is known [W41] to be transcendental. See also [A90] for an automata-theoretic proof.

In fact, the correct analogue for classical $(-1/2)!$ is $(1/(1-q))!$, because 2 and $q-1$ are the number of roots of unity in $\mathbb{Z}$ and $\mathbb{F}_q[T]$, respectively. And the analogue of Chudnovsky's $4, 6$ (which are numbers of roots of unity in quadratic imaginary fields) is $q^2 - 1$ (the number of roots of unity in $\mathbb{F}_{q^2}[T]$).

We have an analogue of the Chowla-Selberg formula for constant field extensions, expressing periods of Drinfeld modules with complex multiplication in terms of gamma values at particular fractions. Combining with their results on transcendence of periods, Jing Yu [Y92] and Thiery [Thi92] (by different techniques) proved that $(1/(1-q^2))!$ is transcendental. So far, the results are parallel in the classical and the function field case.

Using Christol's criterion, we proved [T96b] the transcendence of gamma values at rationals with any denominator, but with some restrictions on numerators. Using logarithmic derivatives on the product formula, which makes exponents relevant only modulo p, instead (a trick also used by L. Denis), Allouche [A96] then proved the transcendence for all values at fractions. Finally, we have shown how Allouche's technique, in fact, settles completely the question of which monomials in gamma values at fractions are algebraic and which are transcendental. Let us describe the result in detail and sketch the simplest case. For the full details and the history, we refer to [A96].

For a proper fraction f, let $\langle f \rangle$ denote its fractional part. For a finite formal sum $\underline{f} = \sum m_i [f_i]$, with $m_i \in \mathbb{Z}$ and $f_i \in \mathbb{Q}$, put $m(f) = \sum m_i \langle -f_i \rangle$ and $\Gamma(\underline{f}) = \prod \Gamma(f_i)^{m_i}$. Also, for $\sigma \in \mathbb{Z}$, put $\underline{f}^{(\sigma)} = \sum m_i [f_i \sigma]$

Let $\underline{f}$ be given, with all f_i's having a common denominator, say N.

THEOREM 6. *(Usual Gamma)* If $m(\underline{f}^{(\sigma)}) = 0$ for all σ relatively prime to N, then $\Gamma(\underline{f})$ is algebraic.

The way we have presented it, this was conjectured (together with Galois action) by Deligne [D79] (proved in [D82]). But using the ideas of Lang and Kubert on distributions, it was shown in the appendix by Koblitz and Ogus to [D79] that the algebraicity also follows by taking the correct combinations of multiplication and reflection formulae. The converse is not known, but is conjectured, because it follows from the general belief that functional equations force all the relations and also from conjectures [D82] in algebraic geometry.

THEOREM 7. *(Our case)* $m(\underline{f}^{(\sigma)}) = 0$ *for all* $\sigma = q^j$ *if and only if* $\Gamma(\underline{f})$ *is algebraic.*

The conditions of the two theorems are analogous, since the Galois group in the relevant cyclotomic theory is $(\mathbb{Z}/N\mathbb{Z})^* = \text{Gal}(\mathbb{Q}(\zeta_N)/\mathbb{Q})$ in the classical case and $\text{Gal}(\mathbb{F}_q(T)(\zeta_k)/\mathbb{F}_q(T))$ (generated by q-power Frobenius) in our case. In our case, the monomials are not obtained by combinations of naive analogues of multiplication and reflection formulae, but the 'only if' part was proved directly [T91] by showing that in the multiplicative basis of factorials of $1/(1-q^i)$ our monomial turns out to be a trivial monomial. Automata theory takes care of the converse, as mentioned above. We explain the simplest case below:

Claim: $(1/(1-q^\mu))!$ is transcendental.

PROOF. : After some simple manipulations, we get

$$P := (\frac{1}{1-q^\mu})!^{1-q^\mu} = \prod_{i=1}^{\infty}\prod_{j=0}^{\mu-1}(1-t^{q^{\mu i}-q^j})$$

Write $P = \sum a_n t^n$ and use Christol's criterion (Theorem 2).
Consider the representations

$$n = \sum(q^{\mu i} - q^j), \text{ all terms distinct, } 0 \leq j < \mu.$$

If such a representation is impossible, then $a_n = 0$, whereas if such a representation is unique (not always the case), then $a_n = \pm 1$. (In a special case, when the representations are always unique, a different type of proof for the next claim is given in [A90]).

Claim: There are infinitely many subsequences of the form $b_n := a_{q^k n + (q^k - k)} = a_{q^k(n+1)-k}$. (Hence P is transcendental).

An imbalance between what $q^{\mu i}$'s can add up to and what q^j's can subtract makes it impossible to have representations of $q^k(n+1) - k$ of the required form, for at least the first $k/q^{\mu-1} - 1$ values of n, if k is sufficiently large. This implies that for sufficiently large k, b_n is 0 for at least the first $k/q^{\mu-1} - 1$ values of n. Since $k/q^{\mu-1} - 1 \to \infty$ as $k \to \infty$, to show that there are infinitely many distinct subsequences (b_n), it is enough to show that infinitely many of these subsequences are not identically zero.

Let $m := k/q^{\mu-1} - 1$ and $n := q^\mu + q^{2\mu} + \cdots + q^{m\mu}$. Then

$$q^k(n+1) - k = (q^k - q^{\mu-1}) + (q^{k+\mu} - q^{\mu-1}) + \cdots + (q^{k+m\mu} - q^{\mu-1})$$

is the unique representation of the form required, and so $b_n = \pm 1 \neq 0$. □

Since then another proof of this case has been given by Hellegouarch [H95], using de Mathan's criterion [Ma95] instead. Note also that Koskas [K95] has given an automata-theoretic proof of de Mathan's criterion.

In fact, there is another gamma function in function field theory, and a breakthrough in establishing the transcendence of its values at proper fractions has been recently achieved by S. Sinha [S95], using Anderson's theory of t- motives and solitons, in particular. The proof expresses the gamma values as periods on analogues of Fermat Jacobians in the setting of t- motives and uses Jing Yu's transcendence results for this. In a sense the reason why we can do better than the classical case by similar (philosophically!) methods is that in the setting of Drinfeld modules and t- motives, we can have arbitrary fractions (and not just half integers) as 'weights'.

3. Application III

Recently, using the automata-theoretic techniques, as well as the logarithmic differentiation trick used in Allouche's proof, Mendès France and Yao gave a very nice proof [MY] proving that $n! = \overline{\Pi}(n)$ is transcendental for all $n \in \mathbb{Z}_p - \mathbb{Z}_{\geq 0}$. This settles the question of the transcendence of the values of $\overline{\Pi}(n)$ completely, as it is easy to see that $\overline{\Pi}(n) \in K$ if $n \in \mathbb{Z}_{\geq 0}$.

We cannot expect to have analogous result for the classical gamma function, because its domain and range are archimedean, and continuity is a quite strong condition in the classical case. In other words, a non-constant continuous real valued function on an interval cannot fail to take on algebraic values. Since, as we have seen above, the values at proper fractions are expected to be transcendental, many values at irrationals should be algebraic.

Morita's p-adic gamma function has domain and range $\mathbb{Z}_p$, which being non-archimedean is closer to our situation. In fact, let us also look at interpolation of $\Pi(n)$ at a finite prime v of $A = \mathbb{F}_q[T]$:

Carlitz showed that D_i is the product of all monic polynomials in A of degree i. Following Morita's idea of throwing out terms divisible by p, David Goss defined $D_{i,v}$ to be the product of all monic polynomials of degree i, which are not divisible by v. He showed (an analogue of Wilson-type theorem) that $-D_{i,v}$ approaches 1 as i tends to infinity and defined $\Pi_v(n) \in A_v$ for $n \in \mathbb{Z}_p$ by

$$\Pi_v(n) = \prod(-D_{i,v})^{n_i}, \text{ for } n = \sum n_i q^i, \ 0 \leq n_i < q.$$

We refer to [T92] and references there for several properties of this interpolation, as well as a discussion of the transcendence question at fractions.

In particular, we have proved that if d is the degree of v, then $\Pi_v(1/(1-q^d))$ is algebraic (connected to function field Gauss sums).

As our third application of the automaton method, we will prove

THEOREM 8. *If v is a prime of degree 1, then $\alpha_k := \Pi_v(1/(1-q^k)) \in A_v$ is transcendental for all $k > 1$.*

Before we begin the proof, let us give direct proofs of some results quoted above, in the case where v is of degree one. Using the automorphism sending T to $T + \theta$, $\theta \in \mathbb{F}_q$, we can assume without loss of generality that $v = T$. Now, for a general monic prime v of degree d, we have $D_{i,v} = D_i/v^w D_{i-d}$, where w is such that $D_{i,v}$ is a unit at v. So in our case,

$$-D_{n,T} = (\prod_{i=0}^{n-1}(1 - T^{q^n - q^i}))/(\prod_{i=0}^{n-2}(1 - T^{q^{n-1} - q^i})) \to 1, \text{ as } n \to \infty$$

and

$$\Pi_T(\frac{1}{1-q}) = \lim \prod_{i=0}^{N} -D_{i,T} = -\lim \prod_{j=0}^{N-1}(1 - T^{q^N - q^j}) = -1$$

because the product telescopes after the first term $-D_{0,T} = -1$.

PROOF OF THE THEOREM. We have

$$-\alpha_k = \lim \prod_{i=1}^{N} -D_{ki,T} = \prod_{i=1}^{N} \prod_{0 \le j < ki}(1 - T^{q^{ki} - q^j})/\prod_{0 \le j < ki-1}(1 - T^{q^{ki-1} - q^j})$$

Notice that the logarithmic derivative with respect to T of $1 - T^{q^i - q^j}$, when $i > j \ge 0$, is $-(q^i - q^j)T^{q^i - q^j - 1}/(1 - T^{q^i - q^j}) = \delta_{j0} T^{q^i - 2}/(1 - T^{q^i - 1})$. Hence we have

$$T\frac{\alpha'_k}{\alpha_k} = \sum \frac{T^{q^{ki} - 1}}{1 - T^{q^{ki} - 1}} - \sum \frac{T^{q^{ki-1} - 1}}{1 - T^{q^{ki-1} - 1}} = \sum c(n) T^n$$

where $c(n)$ is the number of divisors of n of the form $q^{ki} - 1$ minus the number of divisors of the form $q^{ki-1} - 1$, as we can see by expanding the sums above in geometric series. We shall show using Theorem 2 that $T\alpha'_k/\alpha_k$ is transcendental, which is enough to prove the Theorem.

Let us first do the simplest case: $p = k = 2$. Then $c(n)$ is just the number of divisors of n of the form $q^i - 1$. Note that $q^i - 1$ divides $q^j - 1$ if and only if i divides j. Let $\mathcal{P}$ be the set of primes. Since this is an infinite set, by the Theorem 2, it is enough to show that if $p_1, p_2 \in \mathcal{P}$, $p_1 > p_2$, then the subsequences c_{p_1} and c_{p_2} are distinct, where we define $c_a(n) := c(q^a n + q^a - 1)$. Let n be so large that $n^2 - (p_1 - p_2)$ is not a square and put $u = n^2 - p_1$. Then we claim that $c_{p_1}(q^u - 1) \ne c_{p_2}(q^u - 1)$. The left hand side is $c(q^{n^2} - 1)$, which is the number of divisors of n^2 and hence odd (equals one, since $p = 2$), whereas the right hand side is $c(q^{n^2 - (p_1 - p_2)} - 1)$, which is the number of divisors of the non-square $n^2 - (p_1 - p_2)$ and hence even. This finishes the proof of the simplest case.

To do the general case, we use lemma 1 of [MY], which states that for positive integers a, b, c; $q^c - 1$ divides $q^a(q^b - 2) + 1$ if and only if c divides (a, b), the greatest common divisor of a and b. We do not repeat the proof, which is an elementary exercise in divisibility.

First we assume that $k > 2$. Let S be the set of primes which are not congruent to -1 modulo k and which are greater than k. By Dirichlet's theorem, S is infinite, so it is enough to show that if $p_1 > p_2$ are members of S, then the subsequences C_{kp_1} and C_{kp_2} are distinct, where we define $C_a(n) := c(q^a n + 1)$. In fact, we claim that $C_{kp_1}(q^{kp_2} - 2) \ne C_{kp_2}(q^{kp_2} - 2)$. The left hand side is

$c(q^{kp_1}(q^{kp_2}-2)+1)$, which equals, by the lemma quoted above, the number of divisors of k of the form ki minus the number of divisors of k that are congruent to -1 modulo k, and so is $1-0=1$. On the other hand, the right hand side is $c(q^{kp_2}(q^{kp_2}-2)+1)$, which equals, again by the lemma quoted above, the number of divisors of kp_2 of the form ki minus the number of divisors of kp_2 that are congruent to -1 modulo k, and so is $2-0=2$. This establishes the claim and finishes the case, when $k > 2$.

Finally, we settle the remaining case $k=2$. As before, it is enough to show that if $p_1, p_2 \in \mathcal{P}$ and $p_1 > p_2 > 2$, then $C_{p_1}(q^{p_1}-2) \neq C_{p_2}(q^{p_1}-2)$. The left hand side is the number of even divisors of p_1 minus the number of odd divisors of p_1 and so is $0-2=-2$, whereas the right hand side is the number of even divisors of 1 minus the number of odd divisors of 1 and so is $0-1=-1$. □

The referee has pointed out that the proof could have been made shorter (but less self-contained) by appealing to Theorem 3 of [MY], which says that for a sequence $n_j \in \mathbb{F}_q$ which is not ultimately zero, $\sum_{j=1}^{\infty} n_j/(x^{q^j}-x) \in \mathbb{F}_q((1/x))$ is transcendental over $\mathbb{F}_q(x)$. With $T = 1/x$ and with n_j being 1, -1 or 0 depending on whether j is congruent to 0, 1 modulo k or otherwise, we immediately get the transcendence of $T\alpha'_k/\alpha_k$.

In fact, by a similar argument, Yao has recently proved (private communication) that with $0 \leq n_j < q$, and for v of degree one, $\Pi_v(\sum n_j q^j)$ is transcendental if (and only if, by the results quoted above) n_j is not ultimately constant.

The author would be grateful to learn about any progress by the reader on these questions or related questions of transcendence of gamma or zeta values for rational function fields or higher genus function fields, by automata-theoretic or other methods. For the results on the zeta values, the reader should look at [Y92], [B94], [B95] and the references there.

What should be the implications for the Morita's p-adic gamma function? As explained in [T92], the close connection to cyclotomy leads us to think that the situation for values at proper fractions should be parallel. But then this implies that the algebraic values in the image not taken at fractions (conjecturally (see [T92]) the only algebraic values at fractions arise at fractions with denominators dividing $p-1$, and we know these values by the Gross-Koblitz theorem and functional equations) should be taken at irrational p-adic integers. Thus we do not expect a Mendès France-Yao type result for Morita's p-adic gamma function, but it may be possible to have such a result for Π_v's. This breakdown of analogies seems to be due to an important difference: in the function field situation, the range is a 'huge' finite characteristic field of Laurent series over a finite field, and the resulting big difference in the function theory prevents analogies being as strong for non-fractions.

Acknowledgements: The portion of the paper before the third application is based on the talk given at the International Conference on Discrete Mathematics and Number Theory, January 96. I thank the organizers from the Ramanujan

Mathematical Society, as well as the Number Theory session organizers Kumar Murty and Michel Waldschmidt. I am obliged to Jean-Paul Allouche for pointing out several related references, and to the referee and Robert Maier for correcting several mistakes related to grammar and style. I am also grateful to the Vaughn Foundation and the University of Arizona for the travel support.

REFERENCES

[A87] Allouche J. -P. - Automates finis en théorie des nombres, Expo. Math. 5 (1987), 239-266.

[A90] Allouche J. -P. - Sur la transcendance de la série formelle Π, Séminaire de Théorie des Nombres, Bordeaux 2 (1990), 103-117.

[A96] Allouche J. -P. - Transcendence of the Carlitz-Goss gamma function at rational arguments, J. Number Theory 60 (1996), 318-328.

[BDGP96] Barre-Sirieix, Diaz, Gramain, Philibert - Une Preuve de la conjecture de Mahler-Manin, Invent. Math. 124 (1996), 1-9.

[B94] Berthé V. - Automates et valeurs de transcendance du logarithme de Carlitz, Acta Arith. 66 (1994), 369-390.

[B95] Berthé V. - Combinaisons linéaires de $\zeta(s)/\Pi^s$ sur $\mathbb{F}_q(x)$, pour $1 \leq s \leq q - 2$, J. Number Theory 53 (1995), 272-299.

[BHMV94] Bruyère V., Hansel G., Michaux C., Villemaire R. - Logic and p-recognizable sets of integers, Bull. Belg. Math. Soc. 1 (1994), 191-238. (Corrigendum pp. 577)

[C79] Christol G. - Ensembles presque periodiques k-reconnaissables, Theoretical Computer Science, t. 9 (1979), 141-145.

[CKMR80] Christol G., Kamae T., Mendes-France M., Rauzy G. - Suites algebriques, automates et substitutions, Bull. Soc. Math. France. 108 (1980) 401-419.

[Chu84] Chudnovsky G. - Contributions to the Theory of transcendental Numbers, Mathematical surveys and monographs, no. 19, (1984), American Math. Soc. , Providence.

[Co69] Cobham A. - On the base-dependence of sets of numbers recognizable by finite automata, Math. Systems theory, 3 (1969), 186-192.

[Co72] Cobham A. - Uniform tag sequences, Math. Systems theory, 6 (1972), 164-192.

[D79] Deligne P. - Valeurs de fonctions L, in Proc. Symp. Pure Math. 33 (1979), 313-346.

[D82] Deligne P. - Hodge cycles on Abelian varieties, in Lecture Notes in Math. 900 (1982).

[D84] Deligne P. - Intégration sur un cycle évanescent, Invent. Math., 76 (1984), 129-143.

[E74] Eilenberg S. - Automata, Languages and machines, vol. A, Academic Press, New York, 1974.

[F67] Furstenberg H. - Algebraic functions over finite fields, J. of Algebra 7 (1967), 271-277.

[GHR92] Goss D., Hayes D., Rosen M. (Ed.) - The Arithmetic of Function Fields, Walter de Gruyter, Berlin, NY 1992.

[H95] Hellegouarch Y. - Une généralisation d'un critère de DeMathan, C. R. Acad. Sci. Paris, t. 321, I (1995), 677-680.

[K95] Koskas M. - Une démonstration par automates d'un critère de transcendance, Thèse, Université de Bordeaux, Chapter III (1995).

[LP77] Loxton J.H. and van der Poorten - Transcendence and algebraic independence by a method of Mahler, pp. 211-226 in Transcendence theory: Advances and applications, Edited by Baker A. and Masser D. W. , Academic Press (1977), London.

[M29] Mahler K. - Arithmetische Eigenschaften der Lösungen einer Klasse von Funktionalgleichungen, Math. Ann. 101 (1929), 342-366.

[Ma95] de Mathan B. - Irrationality measures and transcendence in positive characteristic, J. Number Theory 54 (1995), 93-112.

[MY] Mendès France M. and Yao J. - Transcendence of the Carlitz-Goss Gamma function, To appear in J. Number Theory.

[R16] Ramanujan S. - On certain arithmetical functions, Trans. Cambridge Phil. Soc.

XXII no. 9, 1916, 159-184. Also, paper 18 in Collected Papers of Srinivasa Ramanujan, Ed. by Hardy G. H. et al, Chelsea Pub. Co. , NY 1962.

[Ri63]   Ritchie R. - Finite automata and the set of squares, J. Assoc. Comput. Mach. 10 (1963), 528-531.

[S95]   Sinha S. - Periods of t- motives and special functions in characteristic p, Ph.D. thesis, U. of Minnesota, (1995).

[S-D73]   Swinnerton-Dyer H. P. F. - On l-adic representations and congruences of modular forms, Springer LNM 350 (1973), 1-55.

[T91]   Thakur D. - Gamma function for function fields and Drinfeld modules, Ann. Math. 134 (1991), 25-64.

[T92]   Thakur D. - On Gamma functions for function fields, in [GHR] (1992), 75-86.

[T96a]   Thakur D. - Automata-style proof of Voloch's result on Transcendence, J. Number Theory, 58 (1996), 60-63.

[T96b]   Thakur D. - Transcendence of Gamma values for $\mathbb{F}_q[T]$, Ann. Math. 144 (1996), 181-188.

[Thi92]   Thiery A. - Indépendance algébrique des périodes et quasi-périodes d'un module de Drinfeld, in [GHR] (1992), 265-284.

[V96]   Voloch J. F. - Transcendence of elliptic modular functions in characteristic p, J. Number Theory, 58 (1996), 55-59.

[W41]   Wade L. - Certain quantities transcendental over $GF(p^n, x)$, Duke Math. J. 8 (1941), 701-720.

[Y92]   Yu J. - Transcendence in finite characteristic, in [GHR] (1992), 253-264.

DEPARTMENT OF MATHEMATICS, UNIVERSITY OF ARIZONA, TUCSON, AZ 85721
E-mail address: thakur@math.arizona.edu

Selected Titles in This Series

(Continued from the front of this publication)

183 **William C. Connett, Marc-Olivier Gebuhrer, and Alan L. Schwartz, Editors,** Applications of hypergroups and related measure algebras, 1995

182 **Selman Akbulut, Editor,** Real algebraic geometry and topology, 1995

181 **Mila Cenkl and Haynes Miller, Editors,** The Čech Centennial, 1995

180 **David E. Keyes and Jinchao Xu, Editors,** Domain decomposition methods in scientific and engineering computing, 1994

179 **Yoshiaki Maeda, Hideki Omoro, and Alan Weinstein, Editors,** Symplectic geometry and quantization, 1994

178 **Hélène Barcelo and Gil Kalai, Editors,** Jerusalem Combinatorics '93, 1994

177 **Simon Gindikin, Roe Goodman, Frederick P. Greenleaf, and Paul J. Sally, Jr., Editors,** Representation theory and analysis on homogeneous spaces, 1994

176 **David Ballard,** Foundational aspects of "non"standard mathematics, 1994

175 **Paul J. Sally, Jr., Moshe Flato, James Lepowsky, Nicolai Reshetikhin, and Gregg J. Zuckerman, Editors,** Mathematical aspects of conformal and topological field theories and quantum groups, 1994

174 **Nancy Childress and John W. Jones, Editors,** Arithmetic geometry, 1994

173 **Robert Brooks, Carolyn Gordon, and Peter Perry, Editors,** Geometry of the spectrum, 1994

172 **Peter E. Kloeden and Kenneth J. Palmer, Editors,** Chaotic numerics, 1994

171 **Rüdiger Göbel, Paul Hill, and Wolfgang Liebert, Editors,** Abelian group theory and related topics, 1994

170 **John K. Beem and Krishan L. Duggal, Editors,** Differential geometry and mathematical physics, 1994

169 **William Abikoff, Joan S. Birman, and Kathryn Kuiken, Editors,** The mathematical legacy of Wilhelm Magnus, 1994

168 **Gary L. Mullen and Peter Jau-Shyong Shiue, Editors,** Finite fields: Theory, applications, and algorithms, 1994

167 **Robert S. Doran, Editor,** C^*-algebras: 1943–1993, 1994

166 **George E. Andrews, David M. Bressoud, and L. Alayne Parson, Editors,** The Rademacher legacy to mathematics, 1994

165 **Barry Mazur and Glenn Stevens, Editors,** p-adic monodromy and the Birch and Swinnerton-Dyer conjecture, 1994

164 **Cameron Gordon, Yoav Moriah, and Bronislaw Wajnryb, Editors,** Geometric topology, 1994

163 **Zhong-Ci Shi and Chung-Chun Yang, Editors,** Computational mathematics in China, 1994

162 **Ciro Ciliberto, E. Laura Livorni, and Andrew J. Sommese, Editors,** Classification of algebraic varieties, 1994

161 **Paul A. Schweitzer, S. J., Steven Hurder, Nathan Moreira dos Santos, and José Luis Arraut, Editors,** Differential topology, foliations, and group actions, 1994

160 **Niky Kamran and Peter J. Olver, Editors,** Lie algebras, cohomology, and new applications to quantum mechanics, 1994

159 **William J. Heinzer, Craig L. Huneke, and Judith D. Sally, Editors,** Commutative algebra: Syzygies, multiplicities, and birational algebra, 1994

158 **Eric M. Friedlander and Mark E. Mahowald, Editors,** Topology and representation theory, 1994

157 **Alfio Quarteroni, Jacques Periaux, Yuri A. Kuznetsov, and Olof B. Widlund, Editors,** Domain decomposition methods in science and engineering, 1994

156 **Steven R. Givant,** The structure of relation algebras generated by relativizations, 1994

155 **William B. Jacob, Tsit-Yuen Lam, and Robert O. Robson, Editors,** Recent advances in real algebraic geometry and quadratic forms, 1994

(See the AMS catalog for earlier titles)